Contestable Markets and the Theory of Industry Structure

Revised Edition

William J. Baumol
Princeton and New York Universities

John C. Panzar
Northwestern University

Robert D. Willig
Princeton University

with contributions by
Elizabeth E. Bailey
Dietrich Fischer
Herman C. Quirmbach

HBJ

HARCOURT BRACE JOVANOVICH, PUBLISHERS
and its subsidiary, Academic Press
San Diego   New York   Chicago   Austin   Washington, D.C.
London   Sydney   Tokyo   Toronto

To the newest generation *(in order of appearance):*

Act One   John Roland Panzar
Jared Mason Willig
Luke William Panzar
Naomi Leah Lee Baumol

Act Two   Scott Mason Willig
Alexander Solomon Lee Baumol
Brent Mason Willig

338.6
B34 ra

ISBN: 0-15-513911-8

Library of Congress Catalog Card Number: 87-81152

Printed in the United States of America

Copyrights and Acknowledgments and Illustration Credits appear on pages 523–525, which constitute a continuation of the copyright page.

Preface to the Revised Edition

With the publication of this paperback edition of *Contestable Markets*, we decided to revise the book slightly. The revision consists of the addition of Chapter 17, describing the small flood of writings on the subject of contestability that has appeared since this book was first published in 1982. The chapter is a slightly modified version of an article written by two of us for *Oxford Economic Papers*. In that chapter, we also try to correct some misunderstandings of our position and provide a few remarks about our general views on contestability.

Only one afterthought occurs to us as we write this preface. Frank Ramsey once wrote a charming essay entitled "Is There Anything to Discuss?" and concluded that there really wasn't. In somewhat the same spirit, we have concluded, though there is a widely held view that contestability theory is a subject marked by great controversy, that it really entails no controversy. Of course, it still does and perhaps always will provide some unsettled issues. For example, we do not know how large a proportion of U.S. industry should be considered highly contestable. That is an empirical question, and no one, at least for the moment, has convincing evidence on the subject. How rapidly does an imperfectly contestable industry lose the attributes of the theoretical state of perfect contestability? There are some applicable criteria, but again, no one is sure of answers that hold universally. Moreover, as far as we can see, on neither of these conclusions is there any disagreement about what is known and what is not.

Similarly, we think there are no *theoretical* controversies now surrounding the subject. Contestability theory was never intended, nor is it now proposed, to serve as a substitute for the fruitful game theoretic analyses of strategic behavior in markets that are not contestable. Nor is the theory in any way a rival to the illuminating transactions costs analysis contributed by Professor O.E. Williamson. These are all complementary lines of investigation, each with its appropriate domain.

Perhaps it is a pity to dispel the aura of controversy that seemed to surround the subject. After all, it does remove some of the fun. But sufficient fun remains in seeking answers to the questions that persist.

WILLIAM J. BAUMOL
JOHN C. PANZAR
ROBERT D. WILLIG

Foreword

This book provides the building blocks of a new theory of industrial organization, one which, in my view, will transform the field and render it far more applicable to the real world. By bringing the large enterprise firmly into the body of microeconomic theory, this volume makes a major contribution to generalization of microanalysis. The world of single-product firms with U-shaped average cost curves simply is not the world of reality. Industrial organization has long awaited a theory dealing explicitly with the variety of outputs and prices and production processes that comprise our economy. This book begins the difficult transition to a theory of industrial organization that can encompass the richness and breadth of actuality while retaining a strong underpinning in theory. Moreover, the fundamentally new theoretical concepts it explores permit, for the first time, an endogenous determination of industry structure.

By describing the research process behind this book I hope to facilitate the work of future scholars. I hope to make clear, for example, that puzzled states and fuzzy thoughts are not manifestations of mental deficiency, but are the very stuff of which research is made.

The first step in the creative process behind the book was, characteristically, recognition of a puzzle or an apparent inconsistency in our analysis. In following the problem stubbornly to its source, new ideas and insights were brought to light. These ideas often appeared obvious and even trivial at first. Sometimes their power and startling implications became clear only a year or more later, and then, often, in the hands of some member of the group other than the originator of the idea. Through continuous interaction, new ideas were converted into systematic theory, the development of whose major concepts is recounted here.

Subadditivity of Costs

One of the fundamental insights on which this book is based is the finding that it is subadditivity of costs, and not scale economies, that determines when society can be served more economically by a monopoly firm.

The insight was attained by a roundabout process. The first contributor was William Baumol, who in July 1970 devised a formal "burden test" for the prices charged by a multiproduct firm. This test proved to be a catalyst not only for the notion of subadditivity but also for that of sustainability. Baumol's aim was to determine whether or not, at a given set of prices, some one of a firm's products received cross subsidies from the consumers of the firm's other products. He devised what appeared to be a straightforward formula, whose use in practice was relatively simple. Basically, the rule rests on the conclusion that when a firm's profits are constrained by a regulatory ceiling, if at the pertinent prices a product contributes net incremental revenues that exceed its incremental cost (taking into account cross-elasticity effects), then the supply of that product must reduce the net (incremental) profits required from other products. The good in question will then, by definition, constitute no "burden" upon the customers of those goods. Thus, provision of this product can in no sense be considered the source of a cross-subsidy payment.

Two years later—in the summer of 1972—while examining this test, it occurred to Edward Zajac and Gerald Faulhaber, both of Bell Laboratories, that Baumol's insight on the nature of cross subsidy might be deceptively simple. Faulhaber, in particular, tried to persuade himself that if two products of a firm passed the burden test individually, they would pass it jointly. He began using game-theoretic tools to explore this hypothesis.

By considering the price-setting issues examined in the burden test as a cooperative game in which subsidy-free prices were the core, Faulhaber was able to deduce three important conclusions. First, he demonstrated that the burden test must be carried out not just for each product in isolation but also for all products in every combination. Second, the game-theoretic framework made it clear that subadditivity of costs is the right criterion for determining when efficiency is served by having only a single producer in an industry. Third, he reached the surprising and disturbing conclusion that even if a firm's cost function is subadditive, there might exist *no* subsidy-free prices; that is, the core might be empty. Thus, subsets of the group of consumers might well find it more economical to buy from a second supplier, even though a single (exclusive) supplier might minimize total social costs.

In the fall of 1972, Faulhaber reported these results to Baumol, who was stimulated by it, but somewhat disturbed: stimulated by the role of subadditivity, but disturbed by the conclusion that a subadditive cost function might preclude all subsidy-free prices. Baumol felt instinctively that this result must be attributable to some perversity of the cross elasticities, which were the sole source of complexity in the burden test. But Faulhaber was able to construct a devastating counterexample. Although it seemed obvious that a multiproduct firm with a cost advantage can always find prices that preclude entry, the counterexample showed that this is not so. (This example was eventually published; see Faulhaber, 1975b.)

Faulhaber posited a situation in which three towns desire water supplies in given quantities. The cost of supplying any one of the towns by itself is $300, of supplying any two of them via one facility is $400, and of all three towns jointly is $660. He pointed out that, from the point of view of society, it was cheaper for only one company to supply all three towns because $660 is less than the cost of supply by three separate plants (3 × $300 = $900), or of supply of any two towns by one plant with the other town supplied by a second plant ($400 + $300 = $700). However, there are then no fixed prices at which a single firm can keep the market entirely for itself. For example, at a price per town of $660/3 = $220, it is cheaper for some two of the towns to split off and supply themselves (at a cost of $400/2 = $200 apiece) than for them to join the third town as customers of the single supplier. Thus, Faulhaber proved that there can be a situation in which there is a natural monopoly in the sense of subadditivity of production costs, but in which monopoly supply may be difficult to preserve in the free marketplace.

Economies of Scale and Scope

It was not until the late fall of 1974 and the early winter of 1975 that the attempt was made to build a bridge between the subadditivity concept and the cost concepts traditional in the study of industrial organization. This bridge-building effort resulted in a wealth of new ideas about cost concepts in a multiproduct environment.

Baumol had received a grant from the Office of Science Information Services of the National Science Foundation to study the economics of information dissemination. This required him to reexamine the nature of any possible economies of joint publication of technical journals. He had naturally supposed that the idea of scale economies could be generalized directly to the multiproduct firm, and that it would serve his purpose. Instead, he found himself forced to devise what were apparently totally new cost concepts, such as decreasing ray average costs and trans-ray convexity of cost functions.

At about the same time, at Bell Laboratories, John Panzar produced an important insight while working on a peak-load pricing model. He showed that many traditional results, including some of my own, followed only from the special cost structure assumed in the literature. He showed that these results must be modified if a neoclassical production function describes the supply of peak and off-peak services, so that marginal cost increases gradually as quantity demanded approaches capacity (Panzar, 1976). Panzar noted that the cost structure in the traditional peak-load pricing model involves constant returns because both capacity costs and operating costs per unit are taken to be constant. Yet because capacity, once built, can be used in both periods, it is normally better to have the same firm produce in both periods rather than to use two separate firms—one producing in the peak and the other in the off-peak period. Thus, Panzar realized that a cost function involving constant returns can lead to production by multiproduct firms.

Panzar and Robert Willig began a collaborative process. Panzar was puzzled because he had always believed that scale economies imply that marginal cost prices do not cover costs. Yet, here was a case in which there were clear economies to single-firm production relative to multifirm production; yet average costs were not decreasing and marginal cost pricing yielded all the revenues the firm needed to cover its costs. A series of mathematical explorations followed, leading to clarifications in the definition of scale economies for the multiproduct firm, and, equally important, to the identification of other fundamental properties of cost functions that had not been recognized before. The most significant of these, in my view, is that of "economies of scope" (Panzar and Willig, 1975). This concept refers to situations in which the cost of producing two products in combination, $C(A, B)$, is less than the total cost of producing each product separately, $C(A, 0) + C(0, B)$.

Working independently, Baumol at New York University and Panzar and Willig at Bell Laboratories began to explore the interplay between subadditivity and other cost concepts, both old and new. In the course of this research, summarized in Chapters 3, 4, and 7 of this book, they discovered that scale economies were neither necessary nor sufficient for monopoly to be the least costly form of productive organization. Subadditivity of costs was the requisite concept; yet neither ray concavity nor ray average costs that decline everywhere were even necessary for strict subadditivity. But Baumol did demonstrate that declining ray average costs, together with a form of complementarity which he called "trans-ray convexity," were sufficient for the purpose (Baumol, 1975, 1977). Panzar and Willig (1977b, 1979) showed that the ideas of economies of scope and overall and product-specific returns to scale are inextricably related. Indeed, economies of scope and decreasing average incremental costs in each product line together imply both overall scale economies and strict subadditivity.

Over the next few years, the exploration of cost conditions continued. Dietrich Fischer and Baumol at New York University devised a series of graphical and numerical examples exploring the conditions under which multiproduct firms exhibit economies of scale and trans-ray convexity. Baumol and Yale Braunstein conducted the first empirical study of scale economies and production complementarity, using the new theoretical tools (Baumol and Braunstein, 1977). Panzar and Willig (1981), as well as David Teece (1981), conducted both theoretical and empirical studies of sources of economies of scope, and others began to join this line of work, conducting increasingly sophisticated empirical analyses of multi-output production (see Chapter 15 and references therein for a discussion of some of this work).

Sustainability

Meanwhile, as an adjunct professor at New York University as well as a supervisor at Bell Laboratories, I provided liaison, communicating many of these discoveries back and forth. I became hooked myself, the form of my addiction requiring delving in more depth into the significance of the Faulhaber counterexample. I was struck by the fact that in Faulhaber's example, market failure occurred when there were economies in single-firm production, yet the demand curve intersected a U-shaped average cost curve in its upward sloping portion, a region not ordinarily associated with scale economies.

The more I thought about it, the more I was persuaded that there must be ways in which natural monopolies can hold together. Faulhaber's pathology did not arise in the downward sloping portion of the U-shaped average cost curve. By seeking to discover the counterpart of this region in the world of the multiproduct firm, I stumbled upon the notion that there must be a set of conditions under which a multiproduct natural monopoly can be sustained.

I could not, however, figure out how to provide a workable definition of this concept. I had long discussions at Bell Laboratories with Panzar and Willig and at New York University with Baumol. As it turned out, the formal definition of sustainability was invented independently—on the same day—by Baumol and me at New York University and by Panzar and Willig at Bell Laboratories. Both groups arrived at the definition that takes the announced prices of a monopolist to be sustainable if the monopoly is viable financially at these prices and if there exists no output-price vector for any potential entrant that can be expected to yield economic profits covering the costs of entry. Thus was born what was to become an important new way of looking at potential entry. Potential entry was judged in terms of enterprises facing the same market demands and having access to the same productive techniques as those available to incumbent firms. The attractiveness of entry was to be evaluated at the incumbent firms' pre-entry prices. Taken literally, the definition implies that the entrant supposes the monopolist to be constrained from adjusting prices in response to entry. The entering firms can thus be interpreted to be price takers of a sort, price takers who are aware of the negative slopes of the industry demand curves and who therefore know that their contribution to industry output can lead to a new equilibrium with reduced prices.

Panzar and Willig soon began to use this concept in rather different ways than did Baumol and I (see Panzar and Willig, 1977a). They explored its relationship to Faulhaber's concept of subsidy-free prices and derived a set of economically significant conditions necessary for sustainability of a monopoly. By focusing on the problems of the regulator, they showed that, contrary to conventional wisdom, a regulated natural monopoly may be vulnerable to entry by noninnovative competitors even if it is producing and pricing efficiently and earning zero economic profits. Panzar and Willig were also able to show that strong demand substitutability and product-specific scale economies work against sustainability and that oligopoly cannot be sustained where there is a natural monopoly that is unsustainable.

In contrast, Baumol and I began to seek sets of conditions, both on the cost and the demand sides, that yield sustainable prices. We were more optimistic about finding relevant conditions for the cost side than those applicable to the demand side, for we knew of some disturbing results that had been contributed by Edward Zajac. In raising his questions about the burden test, Zajac had begun by noting that the burden test implicitly used as its standard of comparison the case in which the price of the product being tested is so high that all demand for it is choked off. But if that service were instead offered by a different supplier, its price might be expected to be somewhat higher than it would be if offered as part of the product line of a natural monopolist, but far lower than the price necessary to reduce its demand to zero. This lower price might very well have a more moderate effect upon the demands for other (substitute or complementary) products of the natural monopoly and could thus require a revision of the burden test's calculations.

Zajac had also been studying Frank Ramsey's rule for optimal pricing under a budget constraint and was led to wonder whether such Ramsey prices would automatically pass the burden test. In an illuminating graph with the prices of two products on the axes, Zajac was able to depict the locus of prices subsidy-free under the burden test and to construct examples in which the point representing the Ramsey prices can lie at a point not on this locus.

Since Baumol was well aware of these results, he did not expect Ramsey-optimal prices to play much of a role in the new sustainability theory. He was as surprised as I was when (while standing in line at a theater) he suddenly envisioned the outlines of a proof of a startling new result involving Ramsey pricing. The result was to play a critical role in the development of this book. The theorem asserted that there is a set of circumstances—apparently not implausible—that may lead even a monopolist to adopt Ramsey-optimal prices, for such prices can then prevent the entry of competing firms.

Thus, a fundamental new idea was conceived. However, a rigorous mathematical formulation turned out to be difficult. Our initial assumptions included both strict global ray concavity and strict global trans-ray convexity, which proved to be mutually inconsistent. Willig pointed this out, but was so intrigued by the result that he set out to devise a set of mathematical conditions that were capable of yielding the result. Thus, at its (eventual) publication, the authors of our paper were Baumol, Bailey, and Willig (1977).

Panzar has gone on to apply the notion of sustainability to problems that span the borderline between theory and application, in particular exploring its relevance for the issue of vertical integration (Panzar, 1980). He has also examined its relevance for nonlinear pricing.

Contestability

After some period of work on the theory of subadditivity and sustainability, Baumol began to conjecture that it constituted the basis for a systematic theory of industry structure. The notion was pursued in an all-night discussion with Willig on the long flight to a conference in Leningrad. The first fruit of this approach was a paper by Baumol and Fischer (1978), now summarized in Chapter 5, in which they examined the number of firms required to produce an industry's vector of outputs at minimum cost, given the characteristics of its cost function. This paper is, in a sense, normative in analyzing the character of the industry structure dictated by productive efficiency. But its relevance stemmed from the conjecture—not then proved, except for the case of monopoly—that any inefficient industry structure *must* be unsustainable. If so, the analysis must be considered behavioral and not merely normative.

The new theory of industry structure is the chief contribution of this book. The notion of contestable markets offers a generalization of the notion of purely competitive markets, a generalization in which fewer assumptions need to be made to obtain the usual efficiency results. Using contestability theory, economists no longer need to assume that efficient outcomes occur only when there are large numbers of actively producing firms, each of whom bases its decisions on the belief that it is so small as not to affect price. What drives contestability theory is the possibility of costlessly reversible entry. Where such entry is possible, efficient outcomes are shown to be consistent with the relatively large scales of operation that characterize many industrial technologies.

The notion of contestability was contributed by Willig. In 1978, Willig, who had by then moved to Princeton University, agreed to present an analysis of a relatively simple policy approach to postal issues, resting on free markets (see Willig, 1980). In essence, he set out to explore how well economic markets can deal with such traditional problems as inefficiencies in the provision of products and cross subsidization of services. He sought conditions under which markets can generally be relied upon to exercise effective control over these problems and looked to sustainability theory to provide him an entrée. But this time, he did not want to prejudge whether the industry was a natural monopoly.

The result of his thinking was the invention of a type of idealized economic market, later termed a "contestable" market. A perfectly contestable market is defined as one in which entry and exit are easy and costless, which may or may not be characterized by economies of scale or scope, but which has no entry barriers, as discussed by Baumol and Willig (1981b). Potential entrants are assumed to face the same set of productive techniques and market demands as those available to incumbent firms. There are no legal restrictions on market entry or exit and no special costs that must be borne by an entrant that do not fall on incumbents as well. An entrepreneur will enter the market if he expects to obtain a positive profit by undercutting the incumbent's price and serving the entire market demand at the new lower price. If the incumbent readjusts his price, reducing it beneath that of the entrant, then the new competitor can readily exit from the market without loss of investment. Thus, potential entrants are undeterred by prospects of retaliatory price cuts by incumbents and instead are deterred only when the existing market prices leave them no room for profitable entry.

Willig distinguished between his concept of idealized markets and the textbook notion of a perfectly competitive market. Both concepts involve markets in which there is frictionless free entry. But, the purely competitive model assumes that there exist in the market a large number of firms each of whom considers its production decisions to have no effect on market prices. In contrast, both incumbents and potential entrants in a contestable market recognize their power over prices and realize that they cannot sell more than consumers demand at given prices, without bidding market prices down. Consequently, a contestable market need not be populated by a great many firms; indeed, contestable markets may contain only a single monopoly enterprise or they may be comprised of duopolistic or oligopolistic firms.

Thus, it was only during the last phases of the writing of this book that the theorems of Chapter 2 about sustainable industry configurations were derived. Only then was it discovered that the absence of entry barriers can lead to just the socially "right" amount of entry. Willig was able to prove that for a feasible industry configuration to be sustainable, it must minimize total industry cost for the production of the aggregate industry output (Willig, 1979a). Moreover, the authors discovered the applicability to contestable markets of the traditional results of the theory of perfect competition. All firms producing any given product must select output levels at which marginal cost of all the firms are equal. Moreover, these marginal costs must equal the market price of that product, so that profits must be zero when there are constant returns to scale, locally. But these results now are shown to hold not only for an industry with a large number of firms, but for any contestable market in which each good is supplied by two or more firms. It is noteworthy that the results are derived not from models based on assumptions about how incumbent firms behave *vis-à-vis* one another, but from models in which such assumptions are largely irrelevant. The powerful, yet simple, theorems followed only after years of piecing together the diverse elements of the new theory.

It would have been satisfying to find that equilibrium solutions generally exist in the case of two or a few firms. Unfortunately, the authors show that sustainable equilibria may exist only rarely if average cost curves are distinctly U-shaped, as they are depicted to be in nearly every economics textbook. However, Thijs ten Raa at NYU first noticed that the assumption of "flat-bottomed" cost curves, in which there are substantial regions over which average costs are roughly constant, provides one solution to the existence problem. Since this assumption is consistent with the findings of much of the empirical and statistical literature, it adds validity to the contestable markets model.

During the summer of 1977, I became a Commissioner at the Civil Aeronautics Board. I continued to be interested in the research for this book—especially in how it might apply to industries such as the airlines that are structurally competitive. I was fascinated by the notion of idealized economic markets that are open to entry by entrepreneurs who face no disadvantages *vis-à-vis* incumbent firms. Baumol and Willig, working together at Princeton, began to distinguish more systematically between the concept of entry barriers, sunk costs, and fixed costs as a source of entry barriers (Baumol and Willig, 1981b; now Chapter 10). Upon reading this work, it struck me that the theory of perfectly contestable markets had direct relevance for policy.

In my view, one unfortunate feature of the research was that the theory was likely to encounter substantial problems in application and observability. If one needed to know global properties of cost and demand functions, as subadditivity requires, how can the theory ever be used without great difficulty? In contrast, contestability theory appeared to lend itself more readily to empirical observation.

Panzar and I began to collaborate on a paper arguing that contestability theory is particularly applicable to aviation. We reasoned that city-pair airline markets are characterized both by easy entry and exit and significant economies of scale. We suggested that potential competition may be an adequate policeman in such markets. Even if a route is flown by a single carrier, other carriers who have stations at both end-point cities can readily enter if monopoly profits become evident. Airline markets can be contestable because their capital costs, while substantial, are not sunk costs. That is, the major portion of the capital—the aircraft—can be recovered from any particular market at little or no cost. Panzar and I used data generated late in 1979 and early in 1980 to show that potential (rather than actual) competition by trunk carriers had provided an effective competitive check on the pricing behavior of local service carriers in long- and medium-haul routes during this period (Bailey and Panzar, 1981).

I then went on in the summer of 1980 to offer the proposition that because of the observability of its properties, the theory of contestable markets can be extraordinarily helpful in the design of public policy (Bailey, 1981). The key element in the discussion is the conclusion of contestability theory that sunk costs, not economies of scale, constitute the entry barrier that confers monopoly power. The trick is for regulators to adopt policies that enhance the contestability of markets. Increased freedom of entry and exit are clearly important. But other policies are equally important. For there may well be a need to regulate access rules, for example, by requiring lease or shared use of sunk cost facilities. Moreover, the theory has a great deal to offer in the antitrust arena, where it can be valuable in turning the courts away from exclusive reliance upon traditional market share measures to evaluate mergers and toward reliance on the degree of structural contestability in the industry.

Each of us involved in the research underlying this book was open to interaction with the others. We each brought strengths—whether it was a question asked by Zajac, an intuitive integration achieved by Baumol, a counterexample constructed by Faulhaber, a complex example worked out by Fischer, a deep "triviality" noted by Panzar, a stubborn belief held by myself, or a "how do I do this right?" by Willig. Together we produced results that exceeded the sum of those that could have been derived by our individual abilities. It was an exciting process. The result, I believe, is marvelous theory.

Elizabeth E. Bailey

Contents

3
Ray Behavior and
Multiproduct Returns to Scale   47

4
Cost Concepts Applicable to
Multiproduct Cases   65

5
The Cost-Minimizing Industry Structure   97

6

Input-Price Changes, Cost Functions, and Efficient Industry Structure   151

7

Natural Monopoly: Sufficient Conditions for Subadditivity   169

8

Monopoly Equilibrium   191

9
Equilibrium in the
Multiproduct Competitive Industry   243

10
Fixed Costs, Sunk Costs,
Entry Barriers, Public Goods,
and Sustainability of Monopoly   279

11
Sustainable Industry Configurations:
General Industry Structures
in Contestable Markets   311

12
Powers of the Market Mechanism   347

13
Intertemporal Sustainability   371

14

Intertemporal Unsustainability   405

Part I: The Role of Scale Economies in Construction   407

Part II: Other Sources of Intertemporal Unsustainability   429

15

Toward Empirical Analysis   445

16

Toward Application of the Theory   465

17

Developments since the First Edition   485

Bibliography   511

Index   527

Contestable Markets
and the Theory of Industry Structure

Revised Edition

1
Objectives and Orientation

This book is an ambitious undertaking in which we seek to provide a substantial contribution both to value theory (along with its implications for social welfare) and to the understanding of industrial organization.

It is only natural that, in describing our objectives and results, we stress their novel elements. But this does not mean that we pretend to have sprung from nowhere. Our debt to our predecessors is enormous, and much of it is obvious. In value theory, anyone from Cournot and Bertrand to Sonnenschein and Stiglitz who has contributed to the literature of market forms clearly has helped us directly or indirectly. In industrial organization, our most obvious debts are to Joe S. Bain, George J. Stigler, and all of those who have written on the subjects of limit pricing and the role of entry. But there are many more on whose ideas we also rely.

It is also important to emphasize that, by stressing the portions of our analysis that are new, we are not implying that our work will show substantial portions of the literature to be wrong or obsolete. Although, of course, we have some disagreements with others on details, in our view this volume, by and large, only underscores the significance of what others have done and serves to supplement their work.

1A. The Intended Contribution to Value Theory

With respect to value theory, our hope is to help in filling three substantial gaps. First, much standard analysis of the determination of output and prices takes the structure of particular industries to be determined outside the domain of the analysis, in effect to be imposed by the fates in a manner that requires no explanation. Once it is presumed that a particular industry is, say, monopolistic or perfectly competitive, one can proceed to investigate the implication of this datum, along with other pertinent information, for the prices and output vectors that will emerge.[1] Since we believe that the structure of an industry is in reality determined primarily by economic forces, we have sought to proceed differently. We take that structure to be determined endogenously and simultaneously with the vectors of industry outputs and prices. Thus a central task of our work is the integration of the process of structure determination into our model and the extraction of theoretical and policy implications from the resulting expanded construct.

Second, we provide a formal analytic structure that, we hope, fully encompasses the insight offered long ago, notably by Bain: that potential competition, that is, the *mere threat* of entry, can have enormous consequences for the general welfare and that it can affect the behavior of firms significantly and beneficially. Upon this foundation we build some new welfare analysis that may modify substantially the standard views on the limits of the domain within which the "invisible hand" wields its scepter. In this area, we follow tradition in its emphasis on freedom of entry. But we also show that freedom of exit is of comparable importance, though its crucial role seems to have received inadequate attention.

Third, we identify a segment of oligopoly analysis within which one is not troubled by the usual problems of indeterminacy and conjectural variations and surmises about competitors' reactions play little or no overt role. Here, it is the behavior of potential entrants that provides the determinacy, and their behavior can be a natural response to the freedom of both entry and exit.

Fourth, we explore the intertemporal allocation of resources brought about by time sequences of prices and depreciation charges established by firms. We find that, in some specified circumstances, the invisible hand can be relied upon to solve intertemporal allocation problems of great subtlety. However, we are left with a wide range of circumstances in which adequate solutions to such problems require elements of what are usually regarded as market failures.

---

[1] For example, Seade (1980) provides an elegant analysis of the effects of (presumably) *exogenous* changes in the number of firms on industry price and output which is valid for a wide class of behavioral assumptions.

1B. The Intended Contribution to Industrial Organization

We devote a considerable part of this work to the construction of basic tools for the analysis of industries comprised of multiproduct firms. In this we are motivated by the fact that, although much of received theory focuses on single-product firms, virtually all firms in reality produce and sell more than one good or service. This multiplicity of outputs can take the form of a variety of physically dissimilar offerings, a wide variety of offerings of similar outputs (such as shoes of different sizes) adapted to the demands of individual customers, or just physically similar outputs sold at various places or times. For all of these cases the received theory of demand is well developed, but the theory of production, as it relates to industry structure, is not.

Several characteristics of multiproduct cost functions will be identified and shown to be crucial for the analysis of the behavior of multiproduct industries. We will discuss the underlying features of productive techniques associated with these characteristics, show the nature of their dependence on relative factor prices, and seek to contribute to the design of tools for their empirical assessment.

We describe some of the connections between the nature of the set of available productive techniques and the character of the industry structure that is efficient for the production of the output vectors consistent with market demands. These connections provide the beginnings of a theory of the determination of industry structure. Then, we define a special form of behavior by potential entrants that may be a rational response to idealized, reversible, and frictionless entry and exit. This degree of freedom of entry forces the industry in equilibrium to adopt the structure that is efficient (in the sense of minimization of the cost of producing the industry's output vector), and it imposes a number of other surprising and desirable properties on any industry equilibrium.

Leaving specifics until later, we can say only that we hope to provide a unifying framework for a pure theory of industrial organization where none was available before. In saying this, we intend not the slightest denigration of the many brilliant pieces that this literature contains. But while brilliant, they are generally disconnected, each casting a clear light on a particular corner (though often an important corner) of the subject. It seems to us that this book, whatever its shortcomings, does not suffer from disconnectedness or concentration on a few special portions of the subject matter. Rather, it seeks to proceed in a systematic manner from its premises about the nature of costs and demands to their implications about the different market forms, the areas of the economy they can be expected to occupy, and their welfare consequences. In doing this, we deal, of course, with familiar concepts such as pure competition, oligopoly, and the like; but we may perhaps succeed in throwing some new light on each of these subjects.

1C. Contestable Markets and Industry Structure

Having described the ambitious area of our endeavor, we turn next to the nature of our approach and its relation to work that is being done by others.

Theoretical industrial organization today can be said to focus upon two types of strategic interaction—that among the firms already incumbent in a market and that between incumbent firms and potential entrants. Increasingly sophisticated analytic methods, recent advances in game theory, and rational expectations equilibrium concepts have all provided fuel for the booming study of industry structure.[2]

The treatment of entry in the models that focus on incumbent interactions follows the classical tradition of monopolistic competition. Some pattern of behavior is postulated for incumbents; the resulting equilibrium is determined on the premise that the population of firms is fixed. Equilibrium in the number of firms is then taken to require all enterprises to earn nonnegative profits, while the introduction of one additional firm is taken to cause each of the incumbent firms to earn profits that are strictly negative. It must be emphasized that here entrants and potential entrants are assumed to believe that they, as well as the others, will abide by the rules of behavior that are postulated.

The second class of models examines entrant behavior in a manner that is far deeper and more intriguing. The assumed patterns of post-entry behavior are not taken to be necessarily the same as those that ruled before entry. Entrants are expected to calculate the profits that entry can bring them, and incumbents are assumed to precommit themselves to investments and business tactics that maximize profit for them, taking into account the effects of those actions on the likelihood of entry. In all of these models, entry is assumed to be free in the sense that the act exacts no explicit costs and that entrants suffer from no disadvantages in the techniques available to them. Despite their assumption of the absence of such impediments to entry, writers using this model confirm the earlier intuitive conclusion of Bain and Stigler: that a variety of first-arrival advantages serve as barriers to the entry of those who seek to come in later. These findings call into question the meaning of the term "free entry" and raise afresh the need to determine what constitutes an entry barrier. The two leading candidate definitions of an

---

[2] Important recent work on strategic interactions among incumbent firms has been provided by Bresnahan (1981), Dixit (1979), Dixit and Stiglitz (1977), Flaherty (1980), Friedman (1977), Grossman (1981), Koenker and Perry (1981), Loury (1979), Salop (1979a), Spence (1976), Varian (1980), and others. Those who have recently made important contributions on strategic interactions between incumbent firms and potential entrants include Dixit (1980), Eaton and Lipsey (1980), Gilbert and Stiglitz (1979), Hay (1976), Kamien and Schwartz (1971, 1972), Kreps and Wilson (1980), Milgrom and Roberts (1979), Prescott and Visscher (1977), Reynolds (1980), Salop (1979b), Salop and Shapiro (1980), Schmalensee (1978), Spence (1977, 1979) Stiglitz (1981), and von Weizsäcker (1980a,b).

entry barrier are those most clearly espoused by George J. Stigler (1968) and Christian von Weizsäcker (1980a,b). Stigler defines an entry barrier to be present when the potential entrants face costs greater than those incurred by a firm now incumbent in the industry. We shall refer to this, for brevity, as the "entry costs" definition of an entry barrier. In contrast, von Weizsäcker defines an entry barrier as an impediment to the flow of resources into the industry, arising as a result of socially excessive protection of incumbent firms. In effect, a barrier is an undefined object whose presence is to be judged only in terms of its undesirable consequences for social welfare.

As will be shown next, our analysis employs for its free entry concept an extremely strong criterion and explores the implications for industry equilibrium of the potential entry associated with that criterion. Markets in which entry is free under this strong definition we refer to as "perfectly contestable." We will see that in such markets the Stigler and von Weizsäcker definitions of entry barriers can become, in effect, one and the same. That is, we will find conditions under which free entry involving the absence of barriers in Stigler's sense forces socially optimal behavior upon the incumbent firms in an industry.

We define a perfectly contestable market as one that is accessible to potential entrants and has the following two properties: First, the potential entrants can, without restriction, serve the same market demands and use the same productive techniques as those available to the incumbent firms. Thus, there are no entry barriers in the sense of the term used by Stigler. Second, the potential entrants evaluate the profitability of entry at the incumbent firms' pre-entry prices. That is, although the potential entrants recognize that an expansion of industry outputs leads to lower prices—in accord with the market demand curves—the entrants nevertheless assume that if they undercut incumbents' prices they can sell as much of the corresponding good as the quantity demanded by the market at their own prices. This is an extension of the axioms on entry behavior in the classical model of perfect competition that makes it possible to deal with the small-numbers case.

We define a sustainable industry configuration to be a price vector and a set of output vectors, one for each of the firms in the configuration, with the following properties: First, the quantities demanded by the market at the prices in question must equal the sum of the outputs of all the firms in the configuration. Second, the prices must yield to each active firm revenues that are no less than the cost of producing its outputs. And, last, there must be no opportunities for entry that appear profitable to potential entrants who regard the prices of the incumbent firms as fixed.

It is evident from the definitions of perfectly contestable markets and sustainable industry configurations that in such a market only sustainable configurations are consistent with equilibrium. In a perfectly contestable market, a configuration that is not sustainable will invite entry that perturbs or alters the configuration. Thus, the study of sustainable industry configurations is the study of the control that free entry in contestable markets exercises over incumbent firms in equilibrium. Just what the requirement of sustainability implies about market structure and about pricing depends on the nature of the available set of production techniques and market demands. The requirements of contestability in a market also depend on its available production techniques and its demand relationships.

In particular, the classical case of a perfectly competitive market also satisfies the requirements of perfect contestability. In a perfectly competitive market, potential entrants evaluate the profitability of entry on the basis of parametric market prices because they are themselves so small relative to the size of the market that they expect to leave prices unperturbed by their entry and, therefore, to leave the behavior of the incumbent firms completely unaffected. It can be demonstrated formally that competitive equilibria are the only configurations sustainable in the limit as the minimum efficient scale of production shrinks relative to the quantities demanded by the market.

At the other extreme of the spectrum of market forms is the case in which the productive techniques yield a natural monopoly. In this case, only a configuration comprised of a single seller can be sustainable, as we will show. However, here, the traditional welfare problems of monopoly behavior are solved by the pressure exerted by the presence of potential entrants. To achieve sustainability, even a natural monopolist must operate in an efficient manner and must earn no more than a normal rate of return on its capital investments. That is, in contestable markets a monopoly firm can only earn zero economic profit and must operate efficiently. It is also shown in our weak invisible hand theorem that, if certain conditions on the available set of production techniques and the nature of market demands are satisfied, then the Ramsey-optimal prices for the monopoly firm (that is, the prices which maximize consumer welfare, subject to the financial viability of the enterprise) are guaranteed to be sustainable and are therefore guaranteed to effect an equilibrium in a contestable market.

Because these results sound counterintuitive and thoroughly at odds with conventional wisdom on the subject of monopoly, it is useful to consider the plausibility of, and the preconditions for, the contestability of a market which is a natural monopoly. The key requirement of contestability in markets in which the set of techniques dictates that the size of incumbent firms be large relative to market demand is that the entry process be entirely, or almost entirely, reversible without cost. With reversible entry—that is, with costless exit—unsustainable prices will afford incentives for rational entrepreneurs to enter in fact. Such entrants need not fear changes in prices by the incumbent firms for, if and when such reactions do occur, even if they preclude all further profit to the entrant, that firm need only exit. With reversibility, this process involving entry, the earning of possibly temporary profits at the initial prices of incumbents, and then exit will be profitable overall to the entrant. Consequently, we may conclude that where the productive techniques and market demands available to incumbents are also freely available to potential entrants, markets must be perfectly contestable if entry is costlessly reversible. Moreover, these markets will be (almost perfectly) contestable if entry is inexpensively reversible.

Viewed in this way, it becomes clear that there exist, at least in theory, important and intriguing cases in which markets may be natural monopolies and remain contestable. But there are also real cases for which both of these conditions are plausible approximations to the facts. A clear example is provided by small, and therefore naturally monopolistic, airline markets. Consider two towns between which the demand for travel is only sufficient to support one flight a day. This is a natural monopoly market. And yet, because airline equipment (virtually "capital on wings") is so very freely mobile, entry into the market can be fully reversible. In principle, faced with a profitable opportunity in such a market, an entrant need merely fly his airplane into the airport, undercut the incumbent's price, and fly the route profitably. Then, should the incumbent respond with a sufficient price reduction, the entrepreneur need only fly his airplane away to take advantage of some other lucrative option—even if he only returns his rented aircraft or resells it in the well-functioning secondary aircraft market. Thus, it is highly plausible that air travel provides real examples of contestable markets. In fact, recent empirical work by Elizabeth E. Bailey (1981) and one of the present authors indicates that the facts actually do lend support to this hypothesis.

This example suggests the role of sunk costs in determining whether or not a market is contestable. Clearly, when entry requires the sinking of substantial costs, it will not be reversible because, by definition, the sunk costs are not recoverable. However, if efficient operation requires no sunk outlays, then entry can, by and large, be presumed to be reversible, and the market can be presumed to be contestable.

Having discussed the meaning and implications of contestability for perfect competition and natural monopoly, the two polar cases of market structure, we turn briefly to some industry structures that lie in between. Here, too, we will show that sustainability requires an industry configuration to be efficient. That is, the division of the total industry output among its firms must minimize the industry's total production cost. Moreover, for a configuration to be sustainable, if at least two firms produce the same good, then its price must equal its marginal costs.

While these results have many interesting implications for multiproduct firms, their implications are simplest and most striking for industries offering a single, homogeneous product. In such cases, the only sustainable configurations with more than one active firm must have perfectly competitive properties: each firm must produce an output at which marginal cost is equal to average cost, which is, in turn, equal to the industry price. But this must be so not because the incumbents are small relative to the size of market. Rather, it follows because, in the presence of two or more incumbent firms, if the price were above the competitive level, we will see that profitable entry would become possible. Even reactions by incumbents that make continued production by the entrant unprofitable are of no concern in contestable markets where there are no irreversible entry costs. For, at worst, that firm can then leave the industry, happy to have earned profits for some period of time. It is the pressure exerted by such freedom of entry and exit that makes the competitive solution the only sustainable outcome in a perfectly contestable industry of this kind. And this is true even where the number of actively producing firms is as small as two.

Thus, in the single-product oligopoly, in perfectly contestable markets, there are no entry barriers in the normative sense of von Weizsäcker. Moreover, there are no indeterminacies of the kind that have plagued oligopoly theory since its inception in the controversies between the Cournot and Bertrand positions. The pressures of reversible free entry enforce socially optimal behavior by incumbent firms, whatever their behavioral proclivities would be in more protected circumstances. Furthermore, in an industry that is a natural monopoly, under the hypotheses of the weak invisible hand theorem, the Ramsey (1927) optimum is sustainable. Thus, even here, albeit in a somewhat weaker sense, the reversible potential entry of contestable markets is consistent with maximal welfare.

1D. Normative and Behavioral (Positive) Elements in the Analysis

All of our analysis of industry structure rests upon two pillars, one normative and one behavioral. Their character is easily indicated. Suppose one wishes to investigate whether or not a particular industry is "inherently oligopolistic." This ambiguous question can mean (at least) one of two things. First, it may ask the normative question: whether or not that industry's vector of outputs is produced most cheaply by a small number of firms, say, something between two and ten enterprises. Second, we can ask the behavioral question: whether market forces lead to the establishment of such a small number of firms. First, following the normative avenue, if it would be (significantly) more expensive for that output vector to be produced either by one or 500 firms, we will say that the industry is a natural oligopoly. More generally, we will use the following definitions throughout the book.

Definition 1D1   An industry is a *natural monopoly* for industry output vector y^I if single-firm production is the least costly way of supplying y^I. If the least costly number of firms is two or greater, but still "quite small," the industry is *naturally oligopolistic*. If a larger number of firms is required to minimize the cost of producing y^I, we say the industry is *naturally competitive*.

These concepts are basically normative in character since it is by no means obvious in advance that actual market behavior will (tend to) force any particular industry to adopt the market structure that is least costly. For example, an industry that is naturally competitive might conceivably be taken over by an oligopoly. Normatively speaking, then, the industry should be naturally competitive; but in its actual behavior it would be oligopolistic. Thus, the second issue to be investigated as part of the theory is the structure to which market behavior can actually be expected to lead.

Throughout this book, the only restrictions placed on market behavior of firms arise from their financial viability and from their participation in an industry configuration that affords no opportunities for profitable entry. The concept of sustainability is the analytic device that summarizes these stipulations for market behavior. It is designed to describe an equilibrium *vis-à-vis* potential entrants who take as (temporarily) fixed the prices of incumbent firms. As we have noted, a price-output vector is said to be sustainable against entry if it satisfies two conditions: (1) It permits incumbents to operate without loss; and (2) so long as incumbents remain in the market, no entrant can hope to produce without loss at or below those prices. The crucial point is that, in the absence of exogenous changes, if incumbents adopt prices sustainable against entry, then, in principle, they need never resort to strategic price responses and countermoves in order to prevent profitable entry opportunities. In this sense, sustainable prices, and sustainable prices alone, can be kept from changing when entry threatens and can keep entry from perturbing the industry situation. This may appear to make sustainable prices equivalent to equilibrium prices. But, for reasons that will now be discussed, they are generally not equivalent. Rather, only in the benchmark case of contestable markets can it be said that sustainability is a necessary condition for equilibrium.

1E. Sustainability versus Equilibrium

Before discussing further the concept of sustainability and its interpretation, it is appropriate to offer a formal definition:

Definition 1E1 A vector of prices p^* is *sustainable* for a set of incumbent firms in an industry if the incumbent firms are financially viable at those prices and if no potential entrant can find a marketing plan for which the anticipated economic profits, $p^e y^e - C(y^e)$, exceed the costs of entry $E(y^e)$. Here, we define the marketing plan of a potential entrant to be the choice of a subset A of the industry's product set N and vectors of prices p_A^e and outputs y_A^e for the goods in A. The entrant cannot sell anything at prices higher than current prices p^*, so that $p_A^e \leq p_A^*$. The entrant's sales cannot exceed total quantities demanded by the market at the relevant prices, so that $y_A^e \leq Q_A(p_A^e, p^*_{N-A})$, where $N - A$ is the set of industry products not offered by the entrant and where $Q(\cdot)$ is the vector market demand function.

By selecting sustainable prices, incumbent firms can protect themselves from successful entry by rivals who regard their prices as fixed. Any such firm that is sufficiently foolish to enter cannot expect to earn a positive rent. But this simple view of the matter conceals vast complications. While an entrant is in operation, some incumbents *must* also lose money at the sustainable prices if there are no entry barriers or other differences in the entrant's and the incumbents' cost functions. For some incumbent would presumably lose some sales to the entrant, and by the definition of sustainability, any output vector smaller than the incumbent's initial vector must be unprofitable at the sustainable prices. Thus, even sustainable prices may not protect incumbents against a financially stronger rival determined to enter the field by outlasting them during a period of mutual losses.

The meaning of *unsustainable* prices is, however, still more elusive. It is tempting to conclude that an unsustainable price vector can only be temporary, for when potential rivals discover a profitable entry opportunity, they will take advantage of it and thereby alter the market prices or force incumbents to change their prices.

But this leaves out of consideration the state of knowledge and the expectations of potential entrants that may inhibit their willingness to take advantage of the opportunity presented by unsustainable prices. Even if prices are unsustainable, the identity of the entry strategies that can be successful under those prices may be far from obvious. It is easy to construct cases in which profitable entry is possible only in a very small region of output space, one that may be quite distant from any output vector in recent experience and is, therefore, unlikely to be noticed by potential entrants.

Second, and perhaps more important, an entrant who is aware of the entry opportunities provided by incumbents' current price vector p^1 can also be expected to recognize that entry is likely to lead to a change in those prices, a change which may well destroy the profitability of the output vectors that would have been attractive to entry had the incumbents' prices remained at p^1. True, if no sustainable prices exist, the second set of prices p^2 must, by definition, provide other profitable entry opportunities. But an output vector y^2 that is profitable with prices p^2 may be very different from the output vectors that were profitable with prices p^1 and may require totally different plant and equipment. Moreover, an attempt by the entrant to produce y^2 may elicit still a third price vector p^3 from the incumbents, and so forth. Potential entrants that recognize these possibilities may therefore elect to stay out of the industry, even though they are aware that the original price vector is not sustainable.

Nevertheless, unsustainability can confidently be expected to imply absence of equilibrium, and sustainability is necessary for equilibrium under any one of the following significant sets of circumstances:

1. Antitrust or regulatory policy which actually inhibits price changes by incumbents in response to entry.

2. Bertrand-Nash expectations on the part of potential entrants that mean that they will *assume* that incumbents will *not* change prices in response to entry. Then any unsustainable price vector will indeed induce entry that alters the industry situation. Here, paradoxically, it is the entrant's belief that incumbents' prices will not change which may make such price changes inevitable.

It should be emphasized that these are strong and often implausible assumptions. As has just been emphasized, Bertrand-Nash expectations are not always fulfilled, and in some cases they are unlikely to be. For, as we have seen, competitive entry can impose significant losses upon incumbents and thereby force a change in their prices. However, if an entrant's output is "small" relative to that of the industry, the magnitude of these required adjustments may also be "small," and hence it may be justifiable for the entrant to ignore them.

There is a third and very important set of circumstances under which unsustainability of prices implies disequilibrium, and under which sustainability is a necessary condition for equilibrium:

3. Circumstances in which entry and exit are costless, or are virtually so, so that the market is contestable. If an entrant can quickly take advantage of a profit opportunity offered by current prices and can withdraw quickly without exit cost if prices are adjusted to eliminate the profit opportunity, incumbents will not be able to protect themselves from the potential-entry pressures by threatening strategic price responses.

No doubt because of lack of clarity in our earlier writings on this subject, the intended role of the Bertrand-Nash assumption in sustainability analysis has been widely misunderstood. Let us try, in summary, to clear up the issue.

As already noted, oligopoly models rest upon some critical assumptions about behavior and expected behavior, for example, an assumption that a rival's output or a rival's reaction function is absolutely fixed.

It is not our intention to substitute for these an unqualified statement that incumbents' prices will never vary in response to entry, or that entrants believe this to be so. Reality simply is not generally like that, and it would be nonsense to maintain the contrary.

Rather, we introduce the *possibility* of nonresponsive price behavior by incumbents only in order to examine three questions:

1. Under what circumstances are stationary equilibrium prices even possible (when exogenous changes are assumed away) in an industry in which entry is not precluded?

2. What are the social benefits and costs of a public policy that rules out responsive price behavior, explicitly or indirectly?
3. What are the properties of a market in which entry and exit can occur costlessly and more rapidly than somewhat sticky prices can change in response?

These are all important questions, and their investigation obviously requires the assumption that prices will not change, at least for the pertinent period. But that is quite different from a naive assumption that, say, oligopolists will in fact never offer a price response to entry. On the contrary, where the answer to the first of our questions is negative, that is, where no sustainable prices exist, one must expect responsive prices to be the rule, and certainly not to be an exception. Indeed, even where sustainable prices do exist, incumbents may disdain them and may instead choose to follow a responsive price policy in the hope of earning profits higher than those permitted by the sustainable prices. But, as we will see, in a contestable market like that just described, such an attempt will be futile.

1F. Responsive Pricing and Equilibrium in Market Structure

Where markets are contestable and sustainable prices exist, the direction of the analysis is clear. In these circumstances, the threat posed by potential competition will determine the behavior of firms and will tend to push them toward the adoption of sustainable prices. The remaining issue in this benchmark case is to determine the details of the resulting behavioral patterns, and much of our analysis is devoted to that objective.

However, one of the problems that will recur at various points is the existence of cost and demand combinations that preclude sustainability, that is, under which no sustainable prices exist. In such circumstances, a variety of scenarios becomes possible. Strategic moves and countermoves can yield many different courses of development for the industry. But this does not mean that disequilibrium is the inevitable outcome. It is true that nonexistence of sustainable prices rules out an equilibrium that is maintained by a set of prices known to be fixed and unresponsive to entry attempts. But an industry structure may nevertheless be preserved under these conditions. If so, it will, however, be preserved not by stationarity of prices but by the *threat* that prices will change when entry is attempted. If potential entrants know that any move into the industry will elicit price changes that are more than likely to cause losses to them, then entry may never even be attempted and so no responsive price change may ever be necessary. Here, paradoxically, it is the threat of price *changes* which permits stationarity of prices as well as stationarity of industry structure.

This observation should induce us to reexamine the standard evaluation of responsive pricing. Economists are inclined to deplore this practice, regarding it as an instrument that helps to preserve monopoly power, with all of its undesirable welfare consequences.[3] Despite the preceding observations, for the most part this judgment probably remains valid. But where no sustainable prices exist, responsive pricing may be essential for the achievement of any sort of stability in industry structure. In its absence, chaotic restructuring and re-restructuring of industry becomes a likely scenario, a process that can prove extremely wasteful and costly. In Chapter 14 we will examine such a case in detail, showing that where there are economies of scale in the construction of durable plant and equipment there may exist no intertemporal price vectors that are sustainable. We will conclude that the absence of responsive price behavior in such circumstances is likely to lead to costly market failure. We will also conclude that such problems need not be rare and pathological curiosa, but may, on the contrary, arise with disturbing frequency.

1G. Concluding Comment

Joe S. Bain began his classic volume, *Barriers to New Competition* (1956), with the profound observation "that most analyses of how business competition works and what makes it work have given little emphasis to the force of the potential or threatened competition of possible new competitors, placing a disproportionate emphasis on competition among firms already established in any industry; [and] that so far as economists have recognized the *possible* importance of [the former] they have no very good idea of how important it actually is." (page 1; Bain's italics.) Although Bain cites this weakness as the reason for his study, our analysis will suggest that even he did not realize fully how valid and significant this statement was.

The power of potential competition to extend the beneficent sway of the invisible hand is the central theme of our book which undertakes to provide the theoretical underpinnings of this conclusion and to draw out some of its implications for public policy. To explore the limits of this power, we have provided the concept of contestable markets in which potential competition is unencumbered by frictions, entry costs, or exit costs. We offer the concept of perfectly contestable markets as a new widely applicable benchmark that both encompasses and transcends the concept of perfectly competitive markets. In particular, unlike perfect competition, perfect contestability can provide a standard for the performance of markets in which concentration

---

[3] Recent discussions on predatory pricing have been collected in Vol. X of *The Journal of Reprints for Antitrust Law and Economics* (1980). See also Salop (1981).

is inevitable because of the nature of the production technology. Thus, we are motivated to understand the details of value determination and of industry structure in contestable markets, not because we believe that most markets are perfectly contestable (although many may be approximately so), but because we believe that prices and industry structure in most markets can usefully be compared to what they would be if those markets were perfectly contestable.[4]

---

[4] In his classic work, *The Economics of Regulation* (1970, Vol. 1, p. 17), Alfred Kahn states that the proper goal of regulation is the achievement of as-if competitive behavior. From our point of view, such a goal is a *non sequitur* because, generally, as-if competitive behavior is infeasible under the technological conditions of natural monopoly that justify regulation in the first place. In contrast, as-if contestable behavior seems always to be apt as a standard of comparison.

2

Industry Structure and Performance in Perfectly Contestable Markets

The Single-Product Case

The purpose of this book, as was indicated in Chapter 1, is to design a methodological approach capable of providing a unifying pure theory of industrial organization. This pure theory is intended to serve as a benchmark against which other theories and empirical industry analyses can be compared. The purpose of this chapter is to describe the outlines of our methodological approach by examining its workings in the limiting case of perfect contestability. We confine ourselves, for simplicity, to the literature's most familiar form, the single-product industry. We do this in spite of the fact that much of the original impetus for writing this book and some of our most original results flow from the multiproduct analyses that follow in later chapters.

We hope to impart the following message in the course of this chapter: The theory of contestable markets is a substantive generalization of the classical theory of perfect competition because, (i) when cost and demand conditions are such that the theory of perfect competition is applicable (the "large-numbers" case), the theory of perfect contestability yields the same results with fewer behavioral assumptions; and (ii) unlike competitive theory, the methodological approach of *perfect contestability theory* is, *without* modification, also applicable to monopolistic and oligopolistic markets.

The chapter begins by defining and exploring the cost concepts that play a crucial role in the analysis of single-product industries. We then use these concepts to analyze the behavior of contestable markets under competitive, oligopolistic, and monopolistic conditions, indicating at the same time how the market form is determined endogenously and simultaneously with the outputs and prices that characterize an equilibrium. Finally, we end the chapter by contrasting our analysis with those of two oligopoly models typical of the current literature and with some of the writings that are predecessors of our work.

The examination of perfectly contestable markets and its surprising welfare implications undoubtedly constitutes the most novel element in this chapter. But even the formal discussion of costs with which the chapter begins offers some clarification of the relationships among economies of scale, natural monopoly, and the viability of marginal cost pricing. Conventional wisdom sometimes suggests that scale economies and natural monopoly are synonymous and that a natural monopoly cannot cover its costs via marginal cost pricing. Along with this, it is frequently implied that, if the monopoly's average cost curve is horizontal near its current output, scale economies will have been exhausted so that total industry costs need not be increased by division of the firm into several smaller enterprises.[1]

We will see in this chapter that none of these views is quite correct. For example, while the existence of global economies of scale is a sufficient condition for a single-product firm to be a natural monopoly, the former is not necessary for the latter. (In the multiproduct case we will see that scale economies are neither necessary nor sufficient for an industry to be a natural monopoly.) As a corollary, it will follow that an industry may be a natural monopoly even though economies of scale are exhausted and the average cost curve is horizontal or rising near the industry output. We will also see that natural monopoly is not, in general, a sufficient condition for marginal cost pricing to be unprofitable. However, we are able to describe the pertinent conditions by means of a precise definition of the degree of scale economies.

2A. Basic Cost Concepts

We begin our formal analysis by defining three cost concepts, one of which is not completely familiar but which is easily understood intuitively. We will show their interrelationships and the ways in which they can be used to answer questions about both *efficient* and *equilibrium* market structure.

Our first and most basic cost concept is defined for the single-output, scalar case in precisely the same way as in the general multiproduct case. (Thus, for future reference, the reader should note that the output levels in the following definition may in fact be interpreted as vectors.)

---

[1] See, for example, Kahn (1970, p. 123). The literature on natural monopoly has been plagued with imprecision since its inception. See Lowry (1973) and Sharkey (1982) for discussions.

Definition 2A1: Strict Subadditivity A cost function[2] $C(y)$ is strictly subadditive at y if for any and all quantities of outputs $y^1, \ldots, y^k, y^j \neq y, j = 1, \ldots, k,$ such that

$$\sum_{j=1}^{k} y^j = y,$$

we have

(2A1) $$C(y) < \sum_{j=1}^{k} C(y^j).$$

Intuitively, if we interpret y^j as the quantity of output produced by firm j, a cost function is subadditive at output y if it is more expensive for two or more firms to produce y than it is for y to be produced by a single firm (the cost of producing the whole is less than the sum of the costs of producing the parts).[3]

Subadditivity can then be taken as the obvious criterion of natural monopoly. That is,

Definition 2A2: Natural Monopoly An industry is said to be a natural monopoly if, over the entire relevant range of outputs, the firms' cost function is subadditive.

It is important to note that subadditivity is a local concept in one sense but not in another. It is a local concept (meaning that it refers to a particular point on the cost surface rather than to the entire surface), because costs may be subadditive at one output level and not at another. However, to determine whether costs are subadditive at a particular output y, we require *global* information about the cost function; that is, it is necessary to know the behavior of costs at levels of operation well below those currently observed. There is a simple intuitive reason for this demanding data requirement. To know whether single-firm production of y is (or is not) cheaper than its production by *any* combination of smaller firms, as subadditivity requires, we must know the magnitudes of the costs that would be incurred by any smaller firms; that is, we must know $C(y^*)$ for every $y^* \leq y$.

---

[2] To avoid excessive complications, in this chapter we ignore the functional dependence of costs on the vector of input prices. These important effects are discussed in Chapter 6.

[3] It may seem at first that a case of superadditivity cannot occur in reality since the firm that finds it more expensive to produce two lots together will turn them out separately (in different plants). But administrative and communication costs can still make it more expensive to produce this way than in two totally independent firms. That is why, in reality, some industries are characterized by many small firms, each with their different specializations. The cost functions are such that larger firms just cannot compete.

Definition 2A3: Declining Average Costs   Average costs are strictly declining at y if there exists a $\delta > 0$ such that

(2A2) $C(y')/y' < C(y'')/y''$ for all y' and y'' with $y - \delta < y'' < y' < y + \delta$.

Average costs are said to be *globally declining* if relation (2A2) holds for all $y > 0$. A somewhat weaker, but more useful condition corresponds to the stipulation that average costs decline through the output in question; that is,

Definition 2A4   Average costs decline through output y if

(2A3) $C(y')/y' < C(y'')/y''$ for all y' and y'' with $0 < y'' < y' \leq y$.

We will often be interested in the level of output beyond which average costs no longer decline, that is, the largest y for which (2A3) holds. Our next cost concept is based on the behavior of the marginal cost curve $MC(y)$.

We assume that the cost function is differentiable, so that marginal cost is well-defined, at all positive levels of output. However, we leave open the possibility that fixed costs render the cost function discontinuous and nondifferentiable at $y = 0$.

Definition 2A5: Declining Marginal Costs   Marginal costs decline through output y if

(2A4) $MC(y') < MC(y'')$ for all y' and y'' with $0 < y'' < y' \leq y$.

This becomes a global concept if it holds for all y.

In nontechnical discussions in the past, Definitions 2A1, 2A4, and 2A5 have all been used to described the case of natural monopoly, perhaps creating the impression that the three were equivalent. In fact, the concepts are distinct from one another, and in the scalar output case they constitute a simple hierarchy. Since subadditivity, by definition, means that one firm can produce a given output more cheaply than any combination of many firms, it is clearly equivalent to what is commonly meant by the term "natural monopoly." The following proposition (Faulhaber, 1975a) clarifies the connection of this basic concept with the other two cost concepts.

Proposition 2A1   (i) *Decreasing marginal cost through output y implies that average costs decrease through y. (ii) This, in turn, implies that the cost function is subadditive at y. (iii) However, the converses of both these conclusions are false—subadditivity does not imply declining average cost, and declining average cost does not imply declining marginal cost.*

Proof (i) *Declining marginal costs through y imply declining average costs through y.* The cost function can be written $C(y') = F + \int_0^{y'} MC(t)\,dt$, where F represents any fixed cost that may be present, and where $0 \le F = \lim_{y \to 0} C(y)$. If marginal costs are declining through y, $\int_0^{y'} MC(t)\,dt > y'MC(y')$ for $0 < y' \le y$. Then, for $0 < y' \le y, d[C(y')/y']/dy' = (y')^{-2}(y'MC(y') - C(y')) = (y')^{-2}[y'MC(y') - F - \int_0^{y'} MC(t)\,dt] < (y')^{-2}(-F) \le 0$. Thus, average costs are strictly decreasing for output levels between 0 and y.

(ii) *Declining average cost implies subadditivity.* Let $y^1, \ldots, y^k$ be any nontrivial way of subdividing output y, so that $\sum y^i = y$ and $y > y^i > 0$. Because average cost is declining and because $y^i < y$, $C(y)/y < C(y^i)/y^i$, so that $(y^i/y)C(y) < C(y^i)$. Then, summing over i, $\sum C(y^i) > \sum(y^i/y)C(y) \equiv C(y)$, which is the definition of subadditivity.

(iii) As is well known, marginal costs may be rising even when average costs are falling, as is clearly true in the standard diagram, Figure 2A1, over the range of output between v and w. More specifically, consider the cost function

$$C(y) = F + ay^2 \qquad \text{for } y > 0, \, C(0) = 0, \text{ where } a > 0.$$

Average cost is $AC = F/y + ay$. Its derivative is $-F/y^2 + a = AC'$, which is negative through the point where $AC' = 0$, that is, where $y = \sqrt{F/a}$, at which

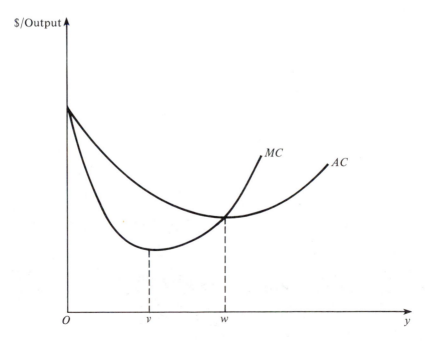

FIGURE 2A1

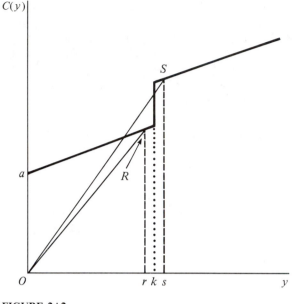

FIGURE 2A2

AC attains a minimum. However, marginal cost is obviously rising for all $y > 0$.

This same example can also be used to demonstrate that the converse of (ii) is false.[4]

However, this is more easily seen intuitively with the aid of the following example: Consider the cost function

$$C(y) = \begin{cases} a + cy > 0 & \text{for } 0 < y < k \\ a + b + cy & \text{for } y \geq k, \text{ where } 0 < b < a. \end{cases}$$

(The graph of this cost function is shown in Figure 2A2.) Now, this cost function is clearly subadditive, because any output can be produced by a single firm at a total cost no greater than $a + b + cy$. However, no two or more firms can each produce positive outputs at a total cost less than $2a + cy > a + b + cy$. Thus, subadditivity follows.

---

[4] *Proof* Average costs are rising for outputs greater than $y_m = \sqrt{F/a}$. However, it can be shown that C is subadditive through $y_s = \sqrt{2F/a} > y_m$. To see this, we first note that for cost functions that have rising marginal costs, like this one, an industry output y^I is divided in positive portions most cheaply among k different firms if each firm produces the same amount, y^I/k. Next, note that for two firms $2C(y/2) = 2F + ay^2/2 > F + ay^2 = C(y)$ for all $y < y_s$. A similar argument holds for $k > 2$. Q.E.D.

Yet, there is an evident range of outputs over which average cost is increasing. For example, at $y = s$, average cost, as given by the slope of ray OS is greater than average cost at output $y = r < s$. Hence, it follows that subadditivity throughout any range of outputs does not imply that average cost must decline throughout that range.

In sum, of the three basic cost concepts, declining marginal cost is the strongest and subadditivity is the weakest.

Of the three single-output cost concepts, the behavior of average costs, $AC(y)$, is the most familiar *and*, as we shall see, the most important for the analysis of market structure. Let us begin our study of its role by noting how it is related to the concept of returns to scale.

Scale economies are often defined[5] to be present when a k-fold *proportionate* increase in every input quantity yields a k'-fold increase in output, where $k' > k > 1$. This definition is certainly stronger than that of declining average cost. That is, so defined, economies of scale through output y imply that average cost will decline through output y, but not vice versa. The reason is straightforward. If one wishes to increase any output y^* by the factor k', the cheapest way to do so need *not* be a proportionate increase in all inputs. Thus, even if average cost does not fall when output is increased by expanding all inputs proportionately, it may nevertheless fall when output is expanded in the most efficient manner, changing input proportions if appropriate.

For our purposes, it is desirable to define scale economies in a way that differs somewhat from the standard textbook concept.[6]

Definition 2A6 The degree of scale economies at y is $S = C(y)/yC'(y) = AC(y)/C'(y) = $ (average cost)/(marginal cost). Returns to scale are increasing, constant, or decreasing as S is greater than, equal to, or less than unity.

Since $dAC/dy = (yC' - C)/y^2$, a statement equivalent to the last sentence in Definition 2A6 is that returns to scale are increasing, decreasing, or constant at output y as dAC/dy, the derivative of the average cost curve, is less than, greater than, or equal to zero, respectively.

The degree of scale economies at y, S, is the elasticity of output at y with respect to the cost incurred to produce it. S is also the elasticity of output with respect to the magnitude of a proportionate expansion in all inputs, from any combination of input levels that is efficient for the production of y. With this definition, S reflects the standard intuitive view of scale economies mentioned above. However, S avoids the ambiguities mentioned there by its definition in terms of local changes in input quantities about their efficient levels. It is well known that any local change in any combination of inputs from any

---

[5] See, for example, Debreu (1959), Lancaster (1968), Mansfield (1975), or Menger (1954).
[6] It does correspond to that of Ferguson (1969).

set of efficient quantities yields the same ratio of output change to cost change. Thus, S is invariant to the direction of change in input quantities from their efficient levels.

Definition 2A6 would seem, at first glance, to make increasing returns to scale at y equivalent to decreasing average cost at y. But a closer look shows that under this definition, too, scale economies remain a stronger condition than decreasing average cost. This follows from the fact that we have defined scale economies in terms of the derivative of AC, and that a function may be strictly decreasing at a point and, yet, have a zero derivative there. For example, consider the cost function $C(y) = y[2 - (y - 1)^3]$, depicted in Figure 2A3. At $y = 1.0$, $C' = AC = 2$, and $S = 1$; yet the function is strictly decreasing since, clearly, $AC(1 + \varepsilon) = 2 - \varepsilon^3 < AC(1) = 2 < 2 + \varepsilon^3 = AC(1 - \varepsilon)$ for any small, positive ε. All of this adds up to

Proposition 2A2 *Locally, economies of scale are sufficient but not necessary for declining average cost.*

We also obtain from Propositions 2A1 and 2A2

Proposition 2A3 *Global economies of scale are sufficient but not necessary for subadditivity of costs (natural monopoly).*

At this point it is natural to inquire whether or not marginal cost pricing must always be financially ruinous for a natural monopoly. From Definition 2A6, it is clear that the revenues collected by pricing at marginal cost $yC'(y)$ cover the firm's costs $C(y)$ if and only if $S \leq 1$, that is, if there are nonincreasing returns to scale. However, we just provided examples in which average costs are not decreasing throughout, even though the firm is a natural monopoly, because costs are subadditive. Thus, contrary to widespread belief, a natural monopoly can sometimes set optimal prices in accord with the Hotelling (1938) rule, $p = MC$, and yet be profitable. Figure 2A3 also shows that marginal cost pricing can be financially viable even when average cost is strictly decreasing.

Now let us assume for simplicity that all average cost curves are *roughly* U-shaped, first falling, then reaching a minimum at some point, and finally rising, although the last two stages may occur only at output levels far beyond current experience. Let y_m be the smallest output at which AC attains its minimum value, AC_m. This premise permits us to illustrate most effectively the power of AC as a tool for explanation of market structure. We shall do this in part by describing informally some results that emerge as special, single-product cases of propositions proved rigorously for the multiproduct world in later chapters.

In this book, a central issue is what determines the optimal number of firms for the production of any given industry output y^I. As we shall see, this

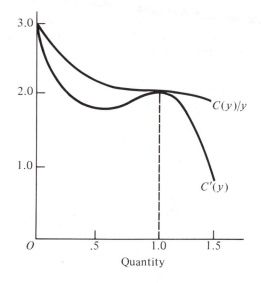

FIGURE 2A3

is a very complicated problem in the multiproduct case. But, where only a single product is involved, it can be shown that

Proposition 2A4 *m^*, the cost-minimizing number of firms when industry output is y^I and y_m is the unique point of minimum average cost, must equal y^I/y_m if the latter is an integer. Otherwise m^* will either be the integer just below y^I/y_m or it will be the integer just above it.*

A more precise formulation of this result (and its proof) is provided in Chapter 5, Proposition 5D1. This result tells us that the optimal number of firms is, as one would expect intuitively, essentially determined by the ratio of industry output to the most efficient firm size, that is, by y^I/y_m. That is, if industry output is $y^I = 103$ and the point of minimum average cost is $y_m = 20$, then it is cheapest for either five or six firms to produce industry output, with the firms tending to produce outputs close to $y_m = 20$. Once again, the requisite information, the magnitude of y_m, is provided by the AC curve.

At this point, the traditional analyst of industry structure would conclude the discussion by classifying, as we did in Chapter 1, the industry as structurally competitive or oligopolistic, depending upon whether or not the market demand at $p = AC_m$ was large or small relative to y_m. The analyst would reserve the category natural monopoly for the case in which AC falls throughout the relevant range. Only in the competitive case would the analyst be confident in predicting the properties of the actual industry equilibrium.[7] As we shall see in Section 2B, we can carry the analysis much further in the case of perfectly contestable markets.

2B. Industry Equilibrium

In the preceding section we introduced and discussed the cost concepts that are essential to understand the technological attributes pertinent to the determination of the structure, conduct, and performance of the industry. While such considerations have always played an important role in the traditional analysis of these three classical issues in the literature of industrial organization, in our analysis of contestable markets they move to center stage. Indeed, it is the combination of the pertinent cost information with the assumption of unfettered, truly free entry that permits us to achieve our central objective: the *endogenous* and *simultaneous* determination of the size and number of firms (the structure of the industry), of their behavior, and, hence, of the performance of the industry in a contestable market. Thus, our analysis seeks to answer the following question: Given the pertinent information about the cost function and the industry demand function $Q(\cdot)$, what market structure will actually tend to emerge as the consequence of the relevant economic forces? The remainder of this chapter attempts to answer this question for the limiting, but important case in which entry and exit are absolutely free—the case of perfect contestability.

The power of the perfect-contestability hypothesis will prove to be enormous. It will be helpful to construct the argument slowly, letting its implications unfold gradually, step by step. We begin by defining a basic property that any equilibrium industry structure in the private sector must surely satisfy:

Definition 2B1: Feasible Industry Configuration A feasible industry configuration is composed of m firms respectively producing the output quantities $y^1, \ldots, y^m$ for sale at a price p such that

$$\sum_{i=1}^{m} y^i = Q(p) \quad \text{and} \quad py^i - C(y^i) \geq 0 \qquad \text{for } i = 1, \ldots, m.$$

Thus, the vector of firms' outputs, $y = (y^1, \ldots, y^m)$, is feasible at price p if its components add up to the quantity demanded at price p and permit each firm to cover its costs. Feasibility is a rather weak requirement, which we would expect any reasonable concept of industry equilibrium to include. All theoretical industry analyses implicitly incorporate the feasibility condition.

---

[7] See, for example, Samuelson (1973, p. 484).

For a feasible industry configuration to be consistent with equilibrium in a contestable market, it must not offer any opportunities for profitable entry. That is, it must be *sustainable* by Definition 1E1, even when entry costs are zero. Thus, for the case of a single-product industry with a perfectly contestable market, we employ the concept of a *sustainable configuration*.

Definition 2B2: Sustainable Industry Configuration   A feasible industry configuration with price p and firm outputs $y^1, \ldots, y^m$ is *sustainable* if $p^e y^e \le C(y^e)$ for all $p^e \le p$ and $y^e \le Q(p^e)$.

In other words, no *feasible entry plan*, $p^e \le p$, $y^e \le Q(p^e)$, can be expected to yield positive profit under the assumption that the prevailing prices charged by incumbents will not change as a result of entry. Recall from our discussion in Section 1C that in a contestable market a potential entrant faces no disadvantages *vis à vis* incumbents with respect to either the available production techniques or the perceptions of consumers as to the desirability of his product. Thus, if he quotes a price below[8] that formerly prevailing in the market, he can sell any quantity he wishes, constrained only by the limitations inherent in the market demand function.

We begin our discussion of output determination, pricing, and efficiency in a contestable market with the following fundamental result:

Proposition 2B1: Equilibrium and Sustainability   *In a perfectly contestable market with profit-seeking firms, only a sustainable configuration can constitute an equilibrium.*

The reason is straightforward. Suppose the industry's configuration were unsustainable. Then, because entry and exit are costless, it would pay entrepreneurs to enter and take advantage of the profitable entry opportunities presented by the unsustainability. Thus, the configuration would be changed either by the presence of additional firms or by price responses or other measures designed to drive them out. Even if such countermeasures were taken, however, the entrants would have reaped at least a temporary profit from a hit-and-run incursion into the industry which, by the definition of perfect contestability, requires no entry or exit cost.

The power of our free entry and exit (contestability) assumption is also evident in the following strong efficiency result.

---

[8] Our formal mathematical analysis requires only that the entrant *meet* the prevailing price. However, since we are dealing with the case of a single homogeneous product, the reader may take the entrant to shade price by an arbitrarily small amount. The effect is the same and it avoids the notation problem arising when one deals with what is, formally, an open set of price options available to the entrant.

Proposition 2B2   *A sustainable configuration must minimize the total cost to the industry of producing the total industry output. That is, no different number, size distribution, output quantities, or productive techniques for the industry's firms can provide the industry's output at a lower total cost than that incurred by the firms in a sustainable configuration.*[9]

The intuitive argument behind this important result is quite clear. Assume that no incumbent firm is losing money, as feasibility requires. Then, if there existed an alternative industry configuration that could produce the same total output more cheaply, that *group* of alternative producers would *in toto* earn a positive profit at the prevailing industry price. This implies that *at least one* firm, *A*, in the alternative configuration would earn a positive profit at the going price and would therefore embody a profitable entry plan relative to the initial (supposedly) sustainable configuration. That is, an entrant is thereby shown to have the opportunity to begin operations in the industry and make a profit by doing exactly what is done by firm *A* in the hypothetical configuration. Thus, the original configuration cannot be invulnerable to entry, and that establishes the result.

There is an important and revealing corollary to this proposition.

Proposition 2B3   *If two or more firms each produce positive amounts of the same commodity in a sustainable industry configuration, their outputs must be such that their marginal costs are equal.*

To show this, suppose, on the contrary, that firm 1's marginal costs were greater than firm 2's. Then the configuration would not be sustainable because total industry costs could be reduced by transferring a small amount of output from firm 1 to firm 2. We turn next to our pricing results, beginning with

Proposition 2B4   *In any sustainable industry configuration, $p \geq MC(y^i)$, $i = 1, \ldots, m$.*

That is, sustainability requires the market price to equal or exceed each firm's marginal cost.

For an intuitive proof of the proposition,[10] assume that there is a sustainable configuration containing a firm *B* which (as required) earns nonnegative profits, $\pi_B \geq 0$, but whose output quantity is such that its marginal cost exceeds the market price. Then an entrant could earn a positive profit by operating exactly as firm *B* does, but reducing its output below *B*'s by the

---

[9] The statement and proof of this important result are formally identical with those of Proposition 11B2, which is expressed in more general multiproduct terms.

[10] This result is proved formally in Chapter 8 in a slightly different context.

small quantity Δy. For the entrant's profit would then be $\pi_e = \pi_B + (MC - p)\Delta y > 0$. Thus, the configuration cannot be sustainable, as the proposition asserts.

This result is significant for economic policy because of its relation to the issue of "predatory pricing," which has long played an important role in antitrust policy. Roughly speaking, a price charged by an incumbent firm threatened by entry is said to be "predatory" if it is, in some sense, unfairly low—it is not justifiable in terms of the incumbent's normal business interests, but makes sense only as a punitive *response* to an entrant's activity. Areeda and Turner (1975) of the Harvard Law School have proposed a test of predatory pricing that has been accepted by many courts (though not by all of them). They argue that no price can be predatory if it exceeds the corresponding (short-run) marginal cost. Proposition 2B4 shows that in a perfectly contestable market, market forces will guarantee that prices will not remain predatory in the Areeda-Turner sense. That is, any such prices will invite reversible entry that will eliminate them.

We come next to a proposition that is still more remarkable.

Proposition 2B5   *In a sustainable industry configuration involving two or more producing firms, all firms must produce outputs at which $p = MC(y^i)$ and $py^i = C(y^i)$, $i = 1, \ldots, m \geq 2$.*

The logic behind this result is straightforward.[11] We have already seen in the previous proposition that price cannot be less than marginal cost. To prove that it also cannot be greater than marginal cost when two or more firms sell the same product, suppose that firm 1 in an industry configuration of two or more firms produces an output y^1 at which price p is greater than marginal cost. Then this automatically guarantees the existence of a profitable entry plan because a potential entrant can offer to sell a slightly larger quantity $y^1 + \Delta y$ at the going price (or a price a tiny bit lower) and earn an amount *strictly* greater than the nonnegative profits initially earned by firm 1. The presence of at least one producer, firm 2, other than firm 1 is necessary for the proof since it guarantees the entrant the opportunity to sell more than firm 1 did without a substantial reduction in market price; that is, it guarantees the existence of a $\Delta y > 0$ such that, if y is the amount demanded by the market at the initial price p, then the entrant's proposed output can be sold at that price since $y^1 + \Delta y < y^1 + y^2 \leq y$. In sum, since in a contestable market in which two or more firms produce any price above (or below) marginal cost creates an opportunity for profitable entry, sustainability requires $p = MC$, as the proposition asserts.

---

[11] This result is proved formally in Chapter 11 for the multiproduct case, as Proposition 11B6.

Furthermore, if py^i were less than $C(y^i)$, the firm could not remain solvent and the configuration would not be feasible. If $py^i > C(y^i)$, an entrant could undercut the incumbent's price and still earn a positive profit. Thus, for sustainability the incumbents must earn zero profit with $py^i = C(y^i)$.

Thus, we have established the surprising result that in perfectly contestable markets, even a duopoly must satisfy the fundamental pricing rule familiar from the theory of perfect competition: For each firm, price must equal both average cost and marginal cost—this is a *necessary* condition for equilibrium. This result contrasts sharply with the standard view that the smaller the number of firms in a market, the greater the difference between price and marginal cost is likely to be.[12] It is usually held that as the number of firms increases, price tends gradually to approach marginal cost. We have shown, on the contrary, that in this respect there is a sharp break between monopoly and duopoly in contestable markets. As soon as the number of firms equals or exceeds two, the equilibrium price must equal marginal cost.

It is particularly noteworthy that this unexpected result, with its strong welfare implications, does not require any special assumptions about strategic interactions among incumbent firms. It is attributable entirely to the discipline enforced by the pool of potential entrants, poised to exploit any opportunity to reap a profit. It does not matter whether the decision strategies of the firms initially in the market are very simple, extraordinarily complex, consciously parallel, coordinated, or random. The constant threat of entry, at least on a hit-and-run basis, ensures that only firms which practice marginal cost pricing, for whatever reasons, can be present in any lasting multifirm equilibrium. Finally, it should be noted that, in contrast with the underpinnings of the classical model of perfect competition, contestability theory does not need the assumption that each firm produces a very small output relative to the aggregate quantity demanded by the market. It is potential entry which, even in the absence of large numbers, can serve to discipline the market.

Thus the $p = MC$ condition of (first-best) Pareto optimality must always be satisfied in equilibrium in multifirm perfectly contestable markets. Not quite the same thing can be said about performance in natural monopoly markets, but they offer results that are similar in spirit.

Proposition 2B6   *For production by a single firm to be sustainable, that firm must (i) produce an output quantity for which it is a natural monopoly (i.e., at which cost is subadditive), and (ii) it must set the lowest price which is equal to the average cost of producing the output demanded by the market.*

---

[12] See Seade (1980) for a recent and rigorous analysis of this issue. This conventional wisdom, of course, dates back at least to Cournot (1838).

Proof Condition (i) is a direct corollary of the industry cost minimization required by Proposition 2B2; condition (ii) follows because price in excess of average cost means that the incumbent earns a positive profit. Then, an entrant can slightly undercut his price, take away his market, and earn a slightly lower profit that is still positive, thus undermining sustainability. Specifically, if $p > C(y)/y$, then by continuity, a profitable, feasible entry plan exists. Moreover, even with a price equal to average cost, the incumbent would be vulnerable to entry if there existed a lower price equal to (or greater than) the average cost of producing the output demanded at that price.

Thus, in the case of natural monopoly, we find that the rule $MC \leq p = AC$ emerges as a necessary condition for equilibrium in perfectly contestable markets. This behavior satisfies the Ramsey principle for (second-best) Pareto optimal pricing, where marginal cost pricing might impose a loss upon the seller. In particular, since we are dealing here with a case of natural monopoly, we should not be surprised to find the firm's output to be such that average cost is declining, with $MC < AC$. Obviously, then, if $p = MC$, the firm must lose money. The price nearest MC that yields a nonnegative profit to the supplier is $p = AC$, and for a single-product firm this is the Ramsey rule for second-best pricing under a zero-profit constraint.[13] We conclude, then, that in a perfectly contestable market even a (single-product) monopoly is required to behave in a manner consistent with Pareto optimality to this degree. We will find that this result is weakened somewhat, but still remains powerful, in the multiproduct case.

2C. Existence of Sustainable Industry Configurations: U-Shaped *AC* Curves

During our enumeration of the desirable properties that must be satisfied at equilibrium in contestable markets, the skeptical reader may have been led to wonder whether all this is not too good to be true. Is it, in general, possible for a feasible industry configuration to satisfy *simultaneously* the conditions of zero firm profits, industry cost minimization, and marginal (or average) cost pricing? In short, do sustainable equilibria exist? As we shall see, the answer is by no means trivial. Indeed, a considerable portion of the remainder of this volume is devoted to this issue. However in the single-product case we can arrive, without complications, at the heart of one of the crucial existence problems. We find, paradoxically, that in a model that employs the standard analytical construct of classical (partial) industry

---

[13] That is, it is the necessary condition for welfare optimality subject to the constraints that profits be nonnegative and that neither discriminatory nor multipart (nonlinear) prices be used.

equilibrium analysis, the U-shaped average cost curve, such equilibria may rarely exist when the number of firms is small. But if we modify the model only slightly in accord with accumulated empirical wisdom in this area, we find that, in the single-product case, the existence problem recedes and may well disappear altogether. The form of the problem we will discuss here (as well as its solution) results from the inescapable fact that the number of firms in any feasible industry configuration must be an integer.

We begin with the monopoly case, where the nature of the existence problem emerges immediately. We know from Proposition 2B6 that sustainability requires $p = AC(y)$, while Proposition 2B4 requires that $p \geq MC(y)$. Thus, it is necessary for its sustainability that a monopoly produce an output at which $AC(y) \geq MC(y)$. This is, of course, precisely equivalent to the requirement that returns to scale be nondecreasing. However we know from Proposition 2A1 and the subsequent example that the industry will, in general, remain a natural monopoly for some output quantities at which returns to scale are decreasing. Hence, if the industry demand curve intersects the average cost curve beyond its minimum point, but within the natural monopoly region, no sustainable equilibrium exists.

This is easily shown diagrammatically. In Figure 2C1, the demand curve DD intersects average cost at point R, which entails an output greater than y_m, where average cost is at its minimum, but less than y_s, the largest output at which costs are subadditive. If the incumbent is a monopolist and cannot ration output, it cannot charge a price lower than p_r at which the demand and average cost curves meet, since at any lower price it can only satisfy market demand at a loss. Nevertheless, if the incumbent charges a price $p' \geq p_r$, an entrant can offer to sell the quantity y_m at a price lower than p' but above p_m, take away a portion of the incumbent's market, and earn a positive profit. There is, therefore, no stationary price that can cover the incumbent's cost and yet prevent profitable entry. That is, there is no sustainable price.

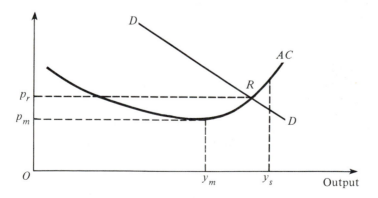

FIGURE 2C1

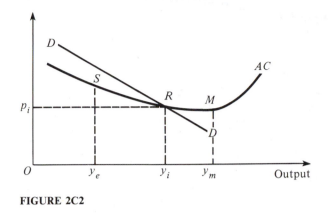

FIGURE 2C2

However, if the average cost curve decreases through the output at which it intersects the market demand curve, a natural monopoly will be able to find a stationary price that covers its own costs and yet protects it from entry. This is shown in Figure 2C2. There, we see an industry demand curve DD which cuts the average cost curve to the left of the point, M, of minimum average cost. In that case, if the monopoly selects the price p_i at which the average cost curve and the demand curve intersect, any entrant attempting to capture a segment of the market by supplying some smaller output y_e must incur an average cost S that is above the incumbent's price p_i. Since the entrant cannot hope to sell anything at a price higher than the incumbent's any such attempt at entry must result in a loss for the entrant. In addition, no entrant can cover its costs if it supplies more than the incumbent, because the prices at which such quantities can be sold lie below the associated average costs. We say then that, given the market conditions depicted in Figure 2C2, the price p_i is sustainable for the incumbent monopoly.

With U-shaped average costs, the existence of sustainable configurations is even more problematic. The nature of the difficulty can readily be seen with the aid of Figure 2C3, which depicts (minimized) average industry costs $AC^I(y^I)$ as a function of industry output y^I. In general, this curve must assume the scalloped shape that is indicated because of the character of the optimal allocation of the industry's output among firms. For outputs no greater than y_s, cost minimization requires that only a single firm be producing. Then, typically, it is efficient to bring in a second firm, with the two firms dividing the industry output between them. There is then an industry output level beyond which it becomes less costly to bring in a third firm, and so forth. The usual shape of the resulting AC^I curve is that depicted, with both the height of the scallops above $AC(y_m)$ and the distance between peaks shrinking as industry output expands. We are now in a position to see the nature of the existence problem. If the industry demand curve intersects

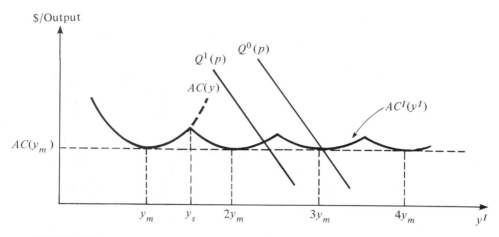

FIGURE 2C3

AC^I at an output level that is an integer multiple of y_m [as does $Q^0(p)$ in Figure 2C3], that intersection point corresponds to a sustainable industry configuration characterized by efficiency, marginal cost pricing, and zero profits for each firm. However, if, as in the case of the demand curve $Q^1(p)$, the intersection lies elsewhere, no sustainable multifirm configuration exists. To see this, note that feasibility and free entry combine to ensure that, if a sustainable configuration exists, it must satisfy $p = AC^I(y^I)$, $y^I = Q(p)$; that is, it must lie at an intersection point of the demand curve and the average cost curve. However if this intersection does not occur at an integer multiple of y_m, then the multifirm sustainability requirement $p = AC = MC$ established in Proposition 2B5 cannot be satisfied since where AC^I is not minimized, $AC \neq MC$ for some firm. Consequently, in such a case a profitable entry plan must exist; for example, $p^e = AC(y_m) < p$ and $y_e = y_m < Q(p^e)$. Thus, no sustainable equilibrium exists. Since a fortuitous intersection between the demand curve and a point of minimum industry AC would seem unlikely, *a priori*, it would appear that the prospects for the existence of equilibrium in perfectly contestable markets are rather poor.[14]

2D. Existence and Optimality of Sustainable Industry Configurations with Flat-Bottomed AC Curves

However, a rather minor and empirically justifiable modification in the underlying cost assumptions permits us to surmount this difficulty. Instead of accepting the classic Vinerian U-shape for the firm's average cost curve,

---

[14] However, several possible remedies for this kind of existence problem, especially in multiproduct industries, are discussed in Chapter 11.

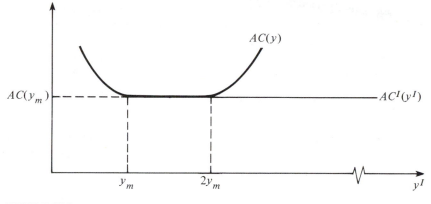

FIGURE 2D1

we can assume that the curve has a flat bottom—that average cost declines through some output y_m and then remains horizontal for an interval. Let us first adopt the strong assumption that it remains horizontal, at least up to output $2y_m$, leaving until later the case in which the flat-bottomed region is shorter than the interval $[y_m, 2y_m]$. The assumption that the *AC* curve is flat-bottomed is consistent, at least qualitatively, with the mass of empirical evidence accumulated over the last 25 or 30 years, beginning with the pioneering work of Joe S. Bain.[15]

This change alters the situation dramatically. Figure 2D1 depicts this type of flat-bottomed average cost curve, $AC(y)$, for a firm. It also indicates that the industry average cost curve, $AC^I(y)$, which coincides with $AC(y)$ at all outputs not exceeding $2y_m$, becomes *perfectly flat thereafter*.[16] To see that this must be so, suppose that $Q(p)$ intersects the horizontal line given by $AC(y_m)$ at some noninteger multiple of y_m, for example, at y^I, where $ny_m < y^I < (n + 1)y_m$, for some integer n. To show that this point is indeed on the average cost curve for the industry, we consider the following industry configuration: $(n - 1)$ firms, each producing y_m, and one firm that produces the remaining industry output, $y^I - (n - 1)y_m$. This output lies between y_m and $2y_m$. Thus, all the firms incur average costs equal to $AC(y_m)$, and the

---

[15] See Bain (1954); see also the discussion in Scherer (1980), and note, especially, his Figure 4-1.

[16] Our assertions here and later in the book that in reality the industry average and marginal cost curves are approximately flat should not be taken as a denial of phenomena such as Ricardian rents and pecuniary diseconomies for the industry which lead to rising supply curves for the industry, both in theory and in practice. We generally ignore such phenomena for expository purposes only. (The reason for this will become obvious in our discussions of the multiproduct case.) With very little modification, our basic analytic structure can deal with them directly.

associated value of $AC^I(y^I)$ must also be $AC(y_m)$. In this way every output greater than y_m can be produced at this same average industry cost. Then, with the industry average cost curve horizontal beyond y_m, every intersection between it and the market's demand curve becomes sustainable. Since we have already seen that every output to the left of y_m is consistent with the sustainability of a natural monopolist, it follows that the existence problem disappears if the firms' AC curves are flat-bottomed over the interval indicated.

One consequence of the assumption that firms' AC curves are flat-bottomed is that we must modify the previously presented bounds upon the number of firms required for industry cost minimization, as is discussed more fully in an appendix to Chapter 5.

Proposition 2D1[17]   *When $AC(y)$ is flat between y_m and $2y_m$, industry cost minimization for output y^I requires production by only one firm for $0 < y^I < 2y_m$, and for larger industry outputs it requires at least $\underline{m}$ firms and at most $\bar{m}$ firms. Here, $\underline{m}$ is the smallest integer $\geq y^I/2y_m$ and $\bar{m}$ is the largest integer $\leq y^I/y_m$.*

With an AC^I like that in Figure 2D1, we see that equilibrium will be one of three basic types in a contestable market, depending upon the greatest output at which the market's demand curve intersects AC^I. If that intersection point lies in $(0, y_m)$, equilibrium must involve a natural monopolist earning zero profits with $p = AC[Q(p)] > MC$, a result that is "second-best." That is, it represents a deviation from the first-best optimal result, which, as is well known, requires $p = MC$. This deviation is required here since with a negatively sloping AC curve marginal cost pricing is not financially feasible for the firm. However, given the character of production costs, market demands, and the standard price system, this solution is socially optimal (Ramsey-optimal) under the constraint that firms must be financially viable. If the intersection lies in $[y_m, 2y_m)$, equilibrium again requires the operation of only a single natural monopoly firm, but here first-best optimality prevails, since $p = AC(Q(p)) = MC$. If it occurs at $2y_m$, equilibrium requires either *one* or *two* firms using marginal cost pricing. Finally, if the intersection lies in $(2y_m, \infty)$, equilibrium requires a multiplicity of firms with the output of each satisfying $p = AC^I(Q(p)) = MC(y^i) = AC(y^i)$. The bounds upon the number of firms are those indicated in Proposition 2D1. However, regardless of the actual number of firms, the marginal cost pricing result holds whenever $Q[AC(y_m)] \geq y_m$.

---

[17] This result is due to Herman Quirmbach (unpublished).

Since this exhausts all possibilities, we have established the following fundamental result:

Proposition 2D2   *In industries that produce a single output, a configuration is sustainable if and only if it is socially (Ramsey) optimal, subject to the constraint that firms be financially viable, provided that firms' average cost curves have flat bottoms from minimum efficient scale to at least twice that output. Thus, if this proviso is satisfied, equilibria in single-product markets that are perfectly contestable must be Ramsey-optimal.*

This proposition yields the strongest conclusion that is possible here. For it asserts that market equilibria and social optima must coincide. In this arena it supports completely the preconception that the workings of decentralized market forces will, by themselves, bring about the best results possible for society. And here, it is essential to recognize, the result does not require the stereotyped textbook market populated by a host of relatively small firms. Rather, the conclusion applies with equal force to markets that are monopolized, or that hold only two sellers, or that include any larger number of firms.

However, the conclusion of Proposition 2D2 does, of course, depend on some very strong assumptions. Perfect contestability is not likely to be satisfied exactly by any real market. Yet, it does provide a standard against which actual markets can be compared, no matter if the relevant production techniques and market demands dictate production by a single giant firm or by a multitude of independent enterprises. As is discussed in Chapters 12 and 16, perfect contestability is therefore an appropriate target for public policy.

The proviso in Proposition 2D2, that firms' average cost curves be flat-bottomed from minimum efficient scale to at least twice that output level, can be weakened considerably.[18] Suppose the flat bottom extends from minimum efficient scale, y_m, to $(1 + k)y_m$, where k may be less than unity. If the Ramsey-optimal level of industry output y^{I*} is sufficiently small that $y^{I*} \leq y_m$, or if it can be supplied in total by an integer number of firms, each producing at minimal average cost, then it follows from our earlier reasoning that this optimum is the unique sustainable solution. As explained in the next section, the latter case will hold if $(y^{I*}/y_m)/\lfloor y^{I*}/y_m \rfloor \leq 1 + k$ (where $\lfloor y^{I*}/y_m \rfloor$ denotes the largest integer not greater than y^{I*}/y_m), that is, if the optimal output is not too far from an integer multiple of y_m. Where this inequality holds, y^{I*} can be supplied by $\lfloor y^{I*}/y_m \rfloor$ firms, each producing an

---

[18] The material that follows is an extension of an analysis by Herman Quirmbach and an independent suggestion by Dietrich Fischer. Much of it is implicit in Thijs ten Raa's dissertation, "A Theory of Value and Industry Structure" (1980).

equal output between y_m and $(1 + k)y_m$, at minimum average cost. This establishes the following generalization of Proposition 2D2:

Proposition 2D3   *In a single-product industry, if the Ramsey-optimal industry output y^{I*} satisfies either $(y^{I*}/y_m)/\lfloor y^{I*}/y_m \rfloor \leq 1 + k$, or $y^{I*} \leq y_m$, then the associated industry configuration is sustainable, and any sustainable configuration is Ramsey-optimal.*

The last condition that underlies Propositions 2D2 and 2D3 is that the industry in question have only a single output. Matters turn out to be very much more complex in the more realistic case of multiproduct industries, and much of the remainder of this book is devoted to their analysis. We will find no normative results as strong and unqualified as those derived in this section. In fact, we will describe special circumstances in which even perfectly contestable markets cannot be relied upon blindly to yield socially optimal results. However, our general normative conclusion will be that perfectly contestable markets do permit the invisible hand to control matters in accord with the public interest in circumstances far wider than those recognized by conventional economic wisdom.

2E. Flat-Bottomed AC Curves and Industry Structure

In this section, we discuss just how the industry's cost function and cost-minimizing industry structure are affected if the flat-bottomed portion of the firm's cost function is shorter than the interval $[y_m, 2y_m]$ assumed in Proposition 2D1. For this purpose we proceed with an example that we also formulate simultaneously in more general terms.

We will see that even if the flat bottom of the firm's AC curve extends over only a small interval, it flattens the industry's AC curve over regions that are surprisingly large, and there is an industry output beyond which its AC curve remains horizontal thereafter. Moreover, this transition value of industry output tends to be surprisingly small.

To show this, suppose that the firm's AC curve is flat over the interval $[y_m, (1 + k)y_m]$, that it is declining in the interval $[0, y_m)$, and that it is rising for all outputs greater than $(1 + k)y_m$. However, aside from this, we presume nothing about the actual shapes of the average (or total) cost curves. For convenience, we select our unit of measurement of output so that $y_m = 1$.

In Figure 2E1 we take $k = 0.25$ and let A be the value of AC for the firm in the output interval $[1, (1 + k)]$. Then, while we cannot infer the shape of the entire AC curve for the industry, we can deduce its shape for the intervals shown in the diagram. That is, we know that the industry's average cost, $AC^I = A$ for the intervals $[1, (1 + k)] = [1, 1.25]$; $[2, (2 + 2k)] = [2, 2.5]$; $[3, (3 + 3k)] = [3, 3.75]$; and $[4, \infty)$ as shown, respectively, by line segments

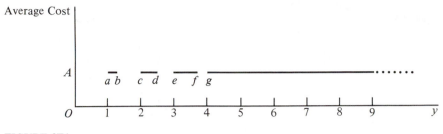

FIGURE 2E1

ab, cd, ef, and g . . . in the diagram. It may be noted that the gaps *Aa, bc, de,* and *fg* decline in length, their lengths constituting the sequence 1, $1 - k =$ 0.75, $1 - 2k = 0.5$, $1 - 3k = 0.25$, and $1 - 4k = 0$. In these in-between intervals, $AC^I > A$.

To see why this is so, consider the industry output $y^I = 2.0$. This output is obviously produced most cheaply by two firms, each producing $y = 1$, at which AC^I attains its minimum value, A. Similarly, industry output $y^I = 2.5 = 2y_m(1 + k)$ is produced most cheaply by two firms, each turning out $y_m(1 + k) = 1.25$ units of output. Any industry output $y^I, 2 < y^I < 2.5$ can be produced at the same average cost if, for example, each of the two firms produces $y^I/2$. Thus, the industry *AC* curve must be flat between $y^I = 2$ and $y^I = 2.5$. Moreover, it cannot be flat anywhere just outside this interval. Thus, for ε sufficiently small, if $y^I = 2 - \varepsilon$, then industry *AC* must exceed *A* because either it will be produced by two or more firms with at least one of them necessarily producing $y < 1$, or it will be produced by one firm whose output will be $2 - \varepsilon > 1.25$. In either case, at least one firm's average cost must be larger than *A* and, by hypothesis, for no firm is $AC < A$ possible. A similar relationship holds for any industry output $y^I = 1 + k + \varepsilon$. This proves our result for the interval $[0, 1 + k + \varepsilon]$, and the same argument obviously holds for the remainder of the diagram.

One implication of the analysis so far that is particularly noteworthy is

Proposition 2E1 *If the firm's AC curve is flat-bottomed in the interval* $[1, 1 + k]$, *then the industry's AC curve must be horizontal* (*its total cost must be linearly homogeneous*) *for all industry outputs* $y^I \geq h$ *where h is the smallest integer for which* $hk \geq 1$. *In particular, if* $k \geq 1$ *then the industry's AC curve is horizontal for all* $y^I \geq 1$. *The industry's AC curve must also be horizontal for all intervals* $[v, v(1 + k)]$ *for any integer v with* $vk \leq 1$.

Figure 2E2 shows in more detail the nature of the relationships involved. Again, selecting output units so that $y_m = 1$, the 45° ray *OF* is the locus relating industry output y^I to the number of firms necessary to produce y^I if each firm produces $y_m = 1$. Thus, the line *OF* satisfies $m = y^I$.

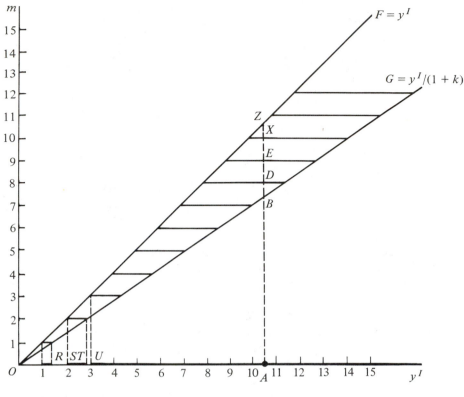

FIGURE 2E2

Similarly, the line OG gives the number of firms necessary to produce y^I if each firm produces output $1 + k$, the maximal output in the flat-bottomed range of the AC curve. Thus, the equation of OG is $m = y^I/(1 + k)$, whose slope is $1/(1 + k)$.

Now, for any industry output y^I these lines give the minimum and the maximum number of firms capable of producing y^I at minimum total cost. For example, at $y^I = A$ this requires $B \leq m \leq Z$. However, these upper and lower values may not be integers and may once again lead us into the fractional firm problem. Thus, in the figure, in which we have this time set $k = 0.4$, and $A = y^I = 10.5$, the height of point Z is also 10.5, while that of B is $10.5/1.4 = 7.5$.

However, there do exist integer solutions. To find them, between rays OF and OG draw in all the horizontal line segments corresponding to integer values of m. Then any point on such a line segment above y^I represents an integer number of firms capable of producing y^I at minimum cost. For example, at $y^I = 10.5$ (point A) this can be done by 8 firms (point D) or by 9 firms (point E) or by 10 firms (point X). This is obviously so since the procedure just described yields all integer values of m satisfying $B \leq m \leq Z$. Thus, we see that with a flat-bottomed *AC* curve the cost-minimizing number of firms need not be unique.

Indeed, following the logic of our construction we see why at some industry outputs there may exist *no* integer number of firms which can produce it at minimum average cost $AC(y_m)$. For example, at any output between $y^I = 1.4$ (point *R*) and $y^I = 2$ (point *S*) the vertical line does not intersect any integer line segment between *OF* and *OG*. The same is clearly true for $2.8 = T < y^I < U = 3.0$. However, for any $y^I \geq 3.0$ there always exists an integer number of firms which can produce y^I at average cost $AC(y_m)$. That is, for any larger industry output, there will always exist a horizontal line segment representing an integer number of firms which lies between *OF* and *OG* for the corresponding y^I.

We can generalize this result for any value of k. Within the range of y^I for which an integer solution always exists, one possible solution is provided by an equal division of y^I among the largest possible number of firms, each producing at least y_m. This number of firms is equal to $\lfloor y^I/y_m \rfloor$, the largest integer that does not exceed $y^I/\hat{y}_m$, because if there were more firms than that at least one of them would have to produce an output less than y_m, at an average cost above the minimal level. Thus, the test determining whether such a solution is possible is whether an equal division of y^I among $\lfloor y^I/y_m \rfloor$ firms enables each to produce an output $\bar{y}$ with $\bar{y} \geq y_m$ and $\bar{y} \leq (1 + k)y_m$, at which average cost is minimized.

In this proposed solution each of the $\lfloor y^I/y_m \rfloor$ firms produces an output equal to

$$\bar{y} = y^I/\lfloor y^I/y_m \rfloor \geq y^I/(y^I/y_m) = y_m$$

so that $\bar{y}$ satisfies the first of the required inequalities. It satisfies the second required inequality $\bar{y} \leq (1 + k)y_m$ if and only if

$$\bar{y} = y^I/\lfloor y^I/y_m \rfloor \leq (1 + k)y_m$$

or

(2E1) $$(y^I/y_m)/\lfloor y^I/y_m \rfloor \leq 1 + k.$$

The inequality in (2E1) must certainly hold if

(2E2) $$(\lfloor y^I/y_m \rfloor + 1)/\lfloor y^I/y_m \rfloor \leq 1 + k$$

because

$$y^I/y_m \leq \lfloor y^I/y_m \rfloor + 1.$$

Multiplying (2E2) through by $\lfloor y^I/y_m \rfloor$ and subtracting this from both sides of the inequality, we obtain

(2E3) $$1 \le \lfloor y^I/y_m \rfloor k.$$

Thus, for any industry output y^I that satisfies this inequality, the existence of an integer solution is assured for the given value of k.

Obviously, the greater the value of k, that is, the greater the length of the flat bottom of the AC curve relative to the value of y_m, the smaller the value of y^I beyond which the existence problem for sustainable solutions disappears.

It is to be noted that the lower-bound ray OG in Figure 2E2 will be lower the greater the value of k, approaching the horizontal axis in the limit as k approaches infinity. Moreover, the distance between OG and the 45° line increases linearly with y^I. This means that if k, the length of the flat bottom of the firm's AC curve, is fairly big or if y^I is fairly large, the range of numbers of firms that minimize industry cost will be fairly wide. Society will be in a position to choose between a large and a small number of firms without any cost penalty. The implication for policy is that a lengthy flat bottom gives the antitrust authorities scope for measures that increase the number of firms. For in such a case not only can this be done without social loss, but this is the one situation in which deliberate proliferation in the number of firms need not lead to unsustainability.

2F. Contrast with Traditional Analysis: An Illustrative Example

Let us now illustrate briefly some of the differences between our analytic approach and those of the received writings. We will see that the standard analyses are highly dependent upon special assumptions about the behavior of incumbents and entrants. By limiting the options open to entrants in this way, freedom of entry and exit is implicitly inhibited, and this accounts for much of the difference between those discussions and contestability theory. To illustrate this, we describe the equilibria of two variants of the well-known Cournot-Nash oligopoly model. We select this example because it provides a convenient vehicle for an analysis of market equilibrium in the most familiar versions of two regimes which appear (superficially) to involve complete freedom of entry. These were alluded to in Chapter 1.

In the first regime, incumbents enjoy a type of first-mover advantage, while in the second regime profits are driven to zero by the entry of new firms whose behavior is similar to that of the incumbents. When the number of firms is fixed at n, with no possibility of entry, market equilibrium is characterized by the usual first order condition for profit maximization,

(2F1) $$\frac{\partial \pi^i}{\partial y^i} = y^i P'(\bar{y}) + P(\bar{y}) - MC(y^i) = 0,$$

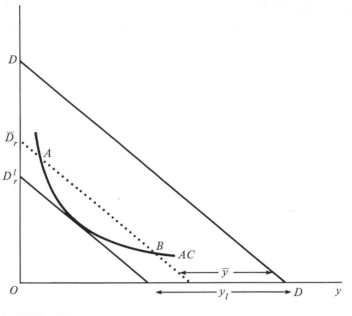

FIGURE 2F1

where π^i is the profit of firm i, y^i its rate of output, $P(\cdot)$ the inverse market demand function, and $\bar{y} = \sum_{i=1}^{n} y^i$, which, under symmetry, reduces to $\bar{y} = ny^i$.

Now suppose that this industry is not protected from entry, but must instead consider the possibility that entrants with access to the same productive techniques would choose to enter if it were to appear profitable *to them* to do so. However, unlike the potential entrants in perfectly contestable markets, they believe that the incumbents will keep their total output fixed at $\bar{y}$ even after entry. This is a classic example of a first-mover model of entry behavior. It is attributable in varying degrees to Bain (1956), Modigliani (1958), and Sylos-Labini (1962).[19]

We begin a diagrammatic description of this model by positing that the incumbents find it profitable to prevent entry and that the Cournot equilibrium output will not do so. Such a situation is depicted in Figure 2F1. Here the upper line DD is the market demand curve. Under the assumption that $\bar{y}$ will remain fixed after entry, the potential entrant can make decisions with the aid of the residual demand curve $\bar{D}_r$ (the dotted lower line parallel to DD), which lies $\bar{y}$ units to the left of DD. Since this curve contains points lying above the AC curve (segment AB), entry will occur with incumbents'

---

[19] See Scherer (1980, Chapter 8) for an extensive discussion of this and related models.

output fixed at $\bar{y}$. To deter such entry, the incumbents must therefore initially select a larger total output, one that is at least equal to y_l, the smallest output (yielding the highest price) that produces a residual demand curve (like D_r^l) that does not intersect AC. Notice that although entrants suffer no cost disadvantage relative to incumbents, this model permits positive profits to the incumbents without causing entry to occur.

This model, which has been employed extensively in the literature, clearly depends in a fundamental way on the assumption that entrants will expect incumbents' outputs to be fixed and unresponsive to entry. In general, however, such an assumption (the "quantity-taking" premise) is not very plausible. We believe that any discussion that succeeds in imparting plausibility to the quantity-taking assumption must ultimately involve some special costs that are incurred upon entry, costs of adjustment or capital costs, which cannot be completely salvaged upon exit. For if there were no such costs, the availability of profits would, as we have argued, invite hit-and-run incursions by entrants. These entrants could undercut the prices of the incumbents and expect, as a result, to divert to themselves some or all of the incumbents' market, in violation of the model's assumption that potential entrants expect incumbents' sales to be fixed and impossible to reduce. Thus, barriers to entry or to exit are vital for the logic of this model, which breaks down in a market that is perfectly contestable.

Now let us turn to the second of the standard models examined in this section. Here it is assumed that the entrants who take advantage of freedom of entry behave in exactly the same manner as do the incumbents and expect to share the market equally with them.[20] Under these conditions, and ignoring the fact that the number of firms must be an integer, entry or exit continues until profits are zero. Thus, in addition to the necessary condition for profit maximization (2F1), the equilibrium is now characterized by the zero profit condition

(2F2) $$\pi^i = P(ny^i)y^i - C(y^i) = 0.$$

Again ignoring the requirement that n be an integer, equations (2F1) and (2F2) together determine the number of firms and their outputs in equilibrium. It is easily seen from (2F2) that this equilibrium requires $P = AC(y^i)$, while from (2F1) we obtain $p > MC(y^i)$ because the firms set their marginal revenues (rather than price) equal to marginal cost. Thus, it is clear that while the profits of the firms are zero, the industry's total costs for the output selected are not minimized. For we know that each firm will then produce an

---

[20] This assumption is most often employed in models of monopolistic competition. In addition to Chamberlin's classic treatment (1962), see Koenker and Perry (1981) and the references cited therein. We discuss that issue in Chapter 11. Here we examine its implications in the single-product case. See Novshek (1980) for a more theoretical treatment.

output involving excess capacity in Chamberlin's sense, since $AC(y^i) > MC(y^i)$. Hence, there must be too many firms, each operating at an inefficiently small scale. This configuration is clearly unsustainable since it does not minimize industry costs. One obvious profitable entry plan permitted by it is

$$AC(y_m) < p^e < p; \qquad y^e = y_m.$$

That is, the equilibrium configuration of this model permits profitable entry by a firm that produces the output at which average cost is minimized and sells it at a price lower than that prevailing in the market.

This last example illustrates the pitfalls of extrapolation to the behavior of potential entrants, without clear justification, of even rather reasonable assumptions about the strategic interactions of incumbents. It also reveals that zero profits and "freedom" of entry (constrained by behavior patterns imposed upon entrants) do not ensure that behavior will be like that in a contestable market.

2G. Predecessors

While we believe that contestability analysis provides an important new approach to the analysis of market structure, conduct, and performance, our results clearly have important precursors, particularly in the single-product case. In this section, we cite some of the most noteworthy of these and point out fundamental distinctions in method even when their equilibrium results coincide with ours.

2G1. Perfect Competition: The Large-Numbers Case

The portion of the preceding analysis with which our critics are, perhaps, least likely to quarrel is the large-numbers case. Here, of course, our analysis yields precisely the same results as the classical, partial equilibrium theory of perfect competition as described by Robinson (1941/1952), Viner (1931/1952), Stigler (1957), and others.[21] However, our novel approach to the problem has, even here, offered us a new insight. The classical model employs as a crucial element in its mechanism the behavioral assumption that potential entrants evaluate the profitability of entry under the assumption that market prices will not change in response to their entry.

We see now that it is redundant to adopt the additional standard assumption that firms active in the market behave as price takers because they

---

[21] See Stigler and Boulding (1952) for a more complete selection of these classic works.

consider themselves to constitute an insignificant portion of the market. For we have shown that one reaches the same conclusions without it.

The classical models also provide us with a persuasive explanation of contestable behavior. The very observation that the output of potential entrants is small relative to the output of the market makes it reasonable to expect them to take prices as given. Notice that, in this large-numbers case, such behavior by entrants is not affected by the presence or absence of sunk costs. Thus, other things being equal, markets in which minimum efficient scale is small relative to industry output are most likely to be perfectly contestable, as one would expect.[22] Indeed, if they are perfectly competitive in the classical sense they must also be perfectly contestable.

2G2. Small Numbers and Oligopoly

Perhaps the most surprising and controversial conclusion that emerges from our analysis is the optimality of the performance of perfectly contestable markets containing even as few as two active producers. However, here too, others have been there before us. In addition to the oral tradition that, in Chicago, two is taken to be a large number, the most notable of similar results is found in the work of Bertrand (1883). There too, the oligopolistic (Nash) equilibrium attained when two or more *price*-setting firms have constant marginal costs involves price equal to marginal cost.

However, though the *result* corresponds precisely with that of our small-group case, the underlying forces in the Bertrand model are fundamentally quite distinct from ours. The basic distinction once again resides in the crucial role of potential entrants in our analysis, as contrasted with that of incumbents in Bertrand's. This has important implications for the robustness of the result when the values of the structural parameters change. As is well known, the Bertrand result is highly dependent upon the assumption that marginal costs are constant *and* equal to average cost throughout the relevant range.[23] If marginal cost begins to rise, or lies below average cost at (what might otherwise be) an equilibrium output, serious problems arise for the existence and uniqueness of equilibrium price and output.[24] Furthermore, the zero profit property tends to evaporate, and with it the optimality of the quantity of resources allocated to the industry.

---

[22] Recently, general equilibrium theorists have been engaged in research which reaches similar conclusions. See Sonnenschein (1980) and Mas-Colell (1980).

[23] Here, the relevant range extends through the output $Q(MC)/n$, where n is the (fixed) number of firms.

[24] Recent work by Grossman (1981) reveals how complicated the appropriately extended analysis becomes under more general cost and demand conditions.

In contrast, optimality is achieved in our model because the threat of potential entry *ensures* that market participants will, in fact, produce outputs falling within the range where marginal and average cost are equal and where average cost is locally constant. It will be recalled that our analysis does not have to assume, with Bertrand, that this holds everywhere in the relevant range. Rather, at most, we require the firms' average cost curve to be flat only within a narrowly defined range that is determined structurally.

Perhaps the most important difference between Bertrand's result and ours is that while his relies on an arbitrary assumption about the behavior of incumbents,[25] ours relies on an assumption about the behavior of entrants that is linked to objective conditions in the market—the degree of contestability. As we argue in Chapter 1, and analyze in a formal model in Chapter 10, if entry is perfectly reversible and frictionless, profit-seeking potential entrants who are rational will behave in the manner indicated. In addition, as discussed in Chapter 1, entrants who are sufficiently small or who operate under particular regulatory regimes will behave in the same fashion if they are rational.

Thus, there are objective structural market conditions that can be examined to determine the relevance of contestability in practice. And where these conditions do not hold even approximately, actual industry configurations can, nevertheless, be usefully compared to those that would result if the markets were structurally contestable. In sum, it is not arbitrary heroic assumptions, but the operation of the invisible hand in structurally contestable markets that leads to our results.

2G3. Natural Monopoly

In its natural monopoly analysis, our work has much in common with that of Demsetz (1968) (though he, too, had at least one precursor[26]). Certainly his conclusion in the single-product case is the same as ours—that equilibrium here requires $p = AC$. Demsetz argued that even in a natural monopoly market the standard model's monopoly price and monopoly profit need not materialize if potential entrants, with access to the same set of productive

---

[25] However, there may be circumstances under which Bertrand's behavioral assumptions are a rational result of objective structural conditions. Recent research on consistent oligopoly reaction functions seems to suggest some such circumstances. It has also been argued (see Stigler, 1968) that where direct negotiations between competing suppliers and individual potential customers are the rule, the Bertrand behavioral assumption is a natural one. Samuelson (1963, p. 79) suggests that for the case of globally constant returns to scale the horizontal supply curve of actual *or potential* competitors would be sufficient to induce a monopolist to behave competitively.

[26] See Chadwick (1859, p. 381).

techniques, are able to proffer bids to serve the entire market at a lower unit price. While the analysis is expressed in terms of bidding procedures and contracting costs, it is clear that the operative force in the Demsetz model, as in ours, is freedom of entry. In both models, if an incumbent were to try to set $p > AC$, he would be undercut by an entrant, so that with complete freedom of entry only average cost pricing is possible in equilibrium.

Our analysis goes beyond Demsetz, particularly in the multiproduct case. It also goes beyond his work by freeing itself from a situation in which the incumbent and potential entrants compete by bidding for the legal franchise that permits the firm to operate as a monopoly. Unlike our analysis, this bidding scheme is not readily generalizable to the multiproduct world. Moreover, we provide analyses of entry barriers, and their relation to sunk costs, which, we believe, cast additional light upon the substantive structural requirements of the requisite freedom of entry and provide the basis for further analysis of policy. Yet, in saying this, we do not mean to deny or in any way to belittle our debt to Demsetz and our other predecessors.

In concluding this chapter we sum up by reemphasizing the analytic role of perfectly contestable markets. A new analytic technique (sustainability theory) permits us to show more fully the power of truly free entry to extend the results of the classical theory of perfect competition into the unfamiliar terrain of monopoly and oligopoly. We turn next to the multiproduct case of the world of reality, and to some of the cost concepts necessary for its analysis. There, some of the novel features of our analysis will emerge more clearly.

3

Ray Behavior and Multiproduct Returns to Scale

In this chapter we embark upon our quest for multiproduct cost concepts that yield insights into multiproduct market structure analogous to those average cost offers us in the single-product case. The obvious stumbling block is that a multiproduct cost function possesses no natural scalar quantity over which costs may be "averaged." That is, we cannot construct a measure of the magnitude of multiproduct output without committing the sin of adding apples and oranges. One's first inclination when faced with this dilemma is to fix output *proportions* and consider the behavior of costs as the size of the resulting output bundle is varied. By doing so, one reverts, in essence, to the single-product case.[1] Only now, one deals with a *composite commodity*, and there is absolutely no difficulty in measuring percentage increases or decreases in its quantity. If the output of each item in the composite commodity rises k percent, the same can obviously be said for the quantity of the

---

[1] Empirical studies of the behavior of costs in multiproduct industries typically resort to this approach, attempting to hold output proportions fixed so that the parameters of a one-dimensional cost curve can be estimated. See, for example, Koenker's discussion of the fact that, for this reason, "the measurement of output is always a problem in empirical studies of transport." He proposes to get around the problem by inserting variables that characterize some of the main output differences, such as length of haul, and by using "a relatively homogeneous sample of firms" (Koenker, 1977, p. 5). For further discussion see Chapter 15, on methods for the estimation of cost functions.

composite good. This chapter confines itself to this fixed proportions analysis—to the case of composite goods.

3A. Ray Average Cost

One can also measure the absolute quantity of a composite good once one decides on the quantity of the bundle that will be assigned the value unity, thereby constituting the point of reference for measurement of the size of all other bundles with the same fixed proportions. This choice is completely arbitrary, and one decision on this matter may be preferable to another only in terms of convenience of calculation. It is essential that such a unit be chosen, whatever its magnitudes, for that choice permits us to define the average cost of the composite good as

$$RAC = C(ty^0)/t,$$

where y^0 is the unit bundle for a particular mixture of outputs—the arbitrary bundle assigned the value 1—and t is the number of units in the bundle $y = ty^0$.

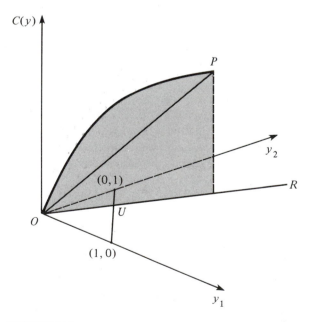

FIGURE 3A1

This average cost figure is referred to as *ray average cost* (RAC). The term refers to the geometry of the construct. In the two-product case, one considers the behavior of costs along a cross section of the total cost surface obtained by dropping a plane perpendicular to the y_1, y_2 plane along a ray such as OR in Figure 3A1. Ray average cost at any point P on that cross section is determined by the slope of a line from the origin to point P. It indicates how total costs vary as a function of the number of units moved in the "direction" (commodity bundle) indicated by OR. Without worrying for the moment about the units of output to be used in the averaging process, we can now formulate the following preliminary definition:

Definition 3A1 Ray average costs are strictly declining *at* $y = (y_1, \ldots, y_n)$ if there exists an $\varepsilon > 0$ such that $C(vy_1, \ldots, vy_n)/v < C(y)$ for all $1 < v < 1 + \varepsilon$, and $C(vy_1, \ldots, vy_n)/v > C(y)$ for all $1 - \varepsilon < v < 1$.

The intuitive interpretation is clear. Ray average cost is taken to decline whenever a small proportional change in output leads to a less than proportional change in total cost.

As already noted, we still must select a unit of output for the composite commodity in order to complete our definition of RAC. Because the choice of unit (like that for any product) is ultimately arbitrary, there are many conventions one can use for the purpose. For quantitative analysis we will find it useful to define our unit of distance along each ray in terms of the distance from the origin to the unit simplex *along that ray*. Here, the unit simplex is given by $U = \{y \geq 0 \,|\, \sum_{i=1}^{n} y_i = 1\}$. It is represented in Figure 3A1 by the 45° line with endpoints $(1, 0)$ and $(0, 1)$ on the axes. Thus, the unit bundle along each ray is defined to be the one such that the sum of the outputs of its component items, each measured in its own arbitrary units, is equal to unity.

Definition 3A2: Ray Average Cost The ray average cost of producing the output vector $y \neq 0$, denoted $RAC(y)$, is defined to be $C(y)/\sum_{i=1}^{n} y_i$. Ray average cost is said to be increasing (decreasing) at y if $RAC(ty)$ is an increasing (decreasing) function of the scalar t, at $t = 1$. Ray average cost is said to be minimized at y if $RAC(y) < RAC(ty)$, for all positive $t \neq 1$.

With the aid of this definition we can plot an RAC curve in the cross-section hyperplane erected on the ray that defines our composite commodity. This is done in Figure 3A2, which also shows the behavior of total cost along that ray, OR. We see that ray average cost and total cost have the usual relationships. Thus, they intersect at the unit output level y^0, and RAC reaches its minimum at the output $y = y^m$ at which the ray OT is tangent to the total cost surface in the hyperplane erected on OR.

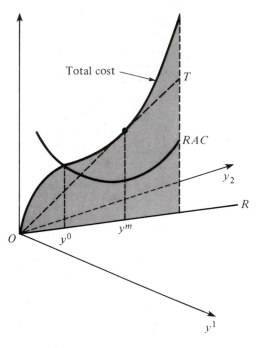

FIGURE 3A2

3B. Ray Average Cost and Multiproduct Scale Economies

The use of the raw sum of outputs as the denominator in the definition of RAC may still seem to smack of adding apples and oranges; but, since along any particular ray output proportions do not change, this convention is harmless. Furthermore, this convention makes it easy for us to relate RAC directly to a multiproduct generalization of our definition of S, the degree of scale economies:

Definition 3B1: Scale Economies   The degree of scale economies defined over the entire product set, $N = \{1, \ldots, n\}$, at y, is given by

$$S_N(y) = C(y)/y \cdot \nabla C(y) \equiv C(y) \bigg/ \sum_{i=1}^{n} y_i C_i(y),$$

where $C_i(y) \equiv \partial C(y)/\partial y_i$. Returns to scale are said to be increasing, constant, or decreasing as S_N is greater than, equal to, or less than unity, respectively.

It is easy to see that $S_N(y)$, the degree of scale economies over the entire product set, measured at y, is precisely the single-product measure of scale economies (discussed in Chapter 2) applied to the composite commodity y^0, where $y = ty^0$. This follows because the quantity of this composite commodity embodied in y is t, and because the marginal cost of the composite commodity is $dC(ty^0)/dt = \sum C_i(ty^0)y_i^0$. Consequently, $S_N(y)$ can be interpreted as the elasticity of the output of the relevant composite commodity with respect to the cost needed to produce it.

The relationship between returns to scale and the behavior of ray average cost is a little less obvious than the corresponding relationship in the scalar output case, but it is made clearer by the following Proposition, which relates S_N to the elasticity of RAC with respect to the output of the relevant composite commodity.

Proposition 3B1[2]   *Let e denote the elasticity of $RAC(ty)$ with respect to t, at the output point y. Then, at y, $S_N = 1/(1 + e)$.*

Corollary   *Returns to scale at the output point y are increasing, decreasing, or locally constant ($S_N > 1$, $S_N < 1$, $S_N = 1$, respectively) as the elasticity of RAC at y is negative, positive, or zero, respectively. Moreover, increasing or decreasing returns at y imply, respectively, that RAC is decreasing or increasing at y.*

Thus, we may interpret S_N as a measure of the percentage rate of decline or increase of RAC with respect to output. Since, as has been emphasized before, the unit of measurement of the denominator of RAC is inherently arbitrary, it should not be surprising that we have found it necessary to define S_N, this measure of the behavior of RAC, in terms of its elasticity, rather than its derivative. However, we note that the converse of the last part of the corollary is *not* true for the same reason as in the scalar case. For example, decreasing RAC at y does not imply that there must be increasing returns at y: $RAC(ty)$ may be strictly declining at $t = 1$ without having a negative derivative there.

Because of the central role played in our analysis by ray average costs and by the degree of scale economies, it is worth probing more deeply into their behavior. Section 3C relates these attributes of the multiproduct or scalar output cost function to the underlying set of production techniques.

---

[2] *Proof* $RAC(ty) = C(ty)/t\sum y_i.$

$$dRAC(ty)/dt = \frac{1}{(t\sum y_i)^2}\left[(t\sum y_i)y \cdot \nabla C(ty) - C(ty)\sum y_i\right]$$

$$e = \frac{dRAC(ty)}{dt}\frac{t}{RAC(ty)} = \frac{1}{C(ty)}\left[ty \cdot \nabla C(ty) - C(ty)\right] = (1/S_N(ty)) - 1.$$

Thus, at $t = 1$, $S_N(y) = 1/(e + 1)$.

This way of looking at the matter permits qualitative information on input–output relationships to be used to form judgments about returns to scale. It also serves as the first of several sections throughout the sequel that suggest how empirical studies of production can be designed to provide information about market structure through the light they can shed upon the characteristics of the cost function.

3C. The Multiproduct Production Set and Returns to Scale[3]

To show how our cost attributes are related to technology, we must first describe the production conditions which are considered. We deal with cases involving a multiplicity of inputs whose quantities are represented by the vector $x = (x_1, \ldots, x_m)$ together with a multiplicity of outputs represented by the vector $y = (y_1, \ldots, y_n)$. The most fundamental characterization of the product relationships is provided by the production-possibility set T, which in its simplest form is merely a list of *possible* combinations of inputs and outputs. More formally,

(3C1) $T = \{(x, y) \mid y \text{ can be produced from } x\}$.

In the scalar output case, a more concrete representation can be formulated using the simple production function $y = F(x)$. In that case, assuming free disposal, the production-possibility set can be written as $\{(x, y) \mid y \leq F(x)\}$.

While T, as defined by (3C1), has a straightforward intuitive interpretation, more must be assumed about it to facilitate mathematical analysis. We begin with a weak regularity condition that is commonly used:

R1:   Input vectors x are drawn from the compact set $X \subset R_+^m$ and output vectors y from the compact set $Y \subset R_+^n$. The production possibility set T is a nonempty closed subset of $X \times Y$. Furthermore
   (i) $(0, y) \in T$ iff $y = 0$
   (ii) $(x, y) \in T$, $(x', y') \in X \times Y$, $x' \geq x$, and $y' \leq y$ imply that $(x', y') \in T$.

This weak assumption can be taken, intuitively, to mean that there exist efficient input–output combinations, that is, combinations such that a further increase in any output requires either a decrease in some other outputs or an increase in some input quantities (actually, the premise is sufficient but not necessary for the existence of such efficient vectors). It stipulates that T includes its limit points, and that inputs and outputs can be drawn from

---

[3] This section is fairly technical and may be skipped by readers interested primarily in application of the theory to industrial organization.

bounded sets. Proviso (i) of the assumption merely asserts that if input use is zero, outputs must be zero. Finally, proviso (ii) of the assumption assures us that the input–output relationship is weakly monotonic; that is, an increase in input use will either permit an increase in outputs or, at worst, permit output quantities to remain unchanged (this premise has been referred to as the free disposal condition for unwanted inputs).

Regularity condition R1 is sufficient (see McFadden, 1978) for the existence of a continuous production transformation function $\phi(x, y)$, increasing in x and decreasing in y, such that

$$(3C2) \qquad \phi(x, y) \geq 0 \quad \text{iff} \quad (x, y) \in T.$$

The production transformation function provides a convenient representation of the set of available multiproduct techniques in the same way that a production function represents a set of techniques for the production of a single output. Thus, for example, if $y = F(x)$, then $\phi(x, y) \equiv F(x) - y$ satisfies (3C2).

Given a vector of parametric input prices $w > 0$, one can then define the multiproduct cost function:

$$(3C3) \qquad C(y, w) = \min_{x} \{w \cdot x | (x, y) \in T\} = w \cdot x^*(y, w).$$

Here $x^*(y, w)$ is some vector of input levels that minimizes the cost of producing y at input prices w. Regularity condition R1 is necessary and sufficient (see McFadden, 1978) for the existence of such a cost function that is positive for positive outputs, weakly increasing in output quantities and input prices, and which satisfies $C(0) = 0$.

Since the multiproduct cost function is our primary analytic construct, we shall find it useful (and sometimes necessary) to impose some additional "smoothness" properties upon it:

R2: For all i, if $y_i > 0$, $C(y, w)$ has a partial derivative with respect to y_i, denoted by C_i.

This simply asserts that a marginal cost exists for any output wherever the quantity of that output is positive. We do not assume a global differentiability property because it is undesirable to rule out the possibility that fixed or start-up costs will cause jump-discontinuities in C at the axes, that is, where some or all of the $y_i = 0$. For example, one such cost function is given by

$$C(y_1, y_2, w) = \begin{cases} F^J + c_1 y_1 + c_2 y_2 & y_1 > 0, y_2 > 0 \\ F^1 + c_1 y_1 & y_1 > 0, y_2 = 0 \\ F^2 + c_2 y_2 & y_1 = 0, y_2 > 0 \end{cases}$$

$$0 < F^1, F^2 < F^J, \text{ and } C(0, 0, w) = 0$$

Here, C exhibits start-up costs when either output is raised from zero to a strictly positive amount. Note, however, that we are dealing with a long-run cost function for which $C(0, w) = 0$, a property that follows immediately from assumption R1(i).

Our formulation of the cost function assumes that the firm faces a vector of *constant* input prices that are unaffected by its actions. Therefore, we shall simplify our notation by writing $C(y, w)$ as $C(y)$. It should be pointed out, however, that the assumption of input price-taking behavior on the part of firms is not essential for our analysis. Whether $C(y)$ is taken merely as a shorthand expression for $C(y, w)$ or as the reduced form of some $C(y, w(y))$ does not directly affect the logic of our inquiries into the determinants of market structure, though it can affect our conclusions. For example, it must be recognized that if the input prices faced by a firm decline (rise) with its size, that will tend to increase (decrease) the likelihood that it is a natural monopoly.[4] More will be said on this subject in Chapter 6.

Our primary interest here is the relationship between costs and output. Before embarking upon a discussion of the details of this relationship, two caveats must be mentioned: (1) If input prices are determined endogenously, our technological duality relationships require modification; (2) if such endogeneity reflects pecuniary, rather than real, advantages of large purchases (such as those yielded by monopsony power), $C(y)$ cannot be relied upon in an analysis of socially optimal industry structure because the structure that minimizes $C(y)$ need not minimize social cost.

Finally, we shall sometimes need this last and strongest regularity condition:

> R3: T is representable by a production transformation function, like that in (3C2), which is
> (a) continuously differentiable in x, and
> (b) continuously differentiable in y_i, for $y_i > 0$, at points (x, y), where x is cost-efficient for y.

---

[4] Moreover, as is shown in detail in Chapter 6, *relative* input prices can also affect industry structure. For example, where capital is cheap relative to labor, single-firm production may be cheapest, while the reverse may be true where low wages call for production techniques that are not capital-intensive. Therefore, the appropriate test to determine whether an industry is *now* a natural monopoly must indicate whether it is the least costly market form at (or in the neighborhood of) *current* input prices. To require this to be true for all input prices would be far too restrictive. Thus we deliberately avoid most of those areas of modern production theory that examine the *duality* of production and cost relationships under the assumption that the latter hold for *all* factor prices. See Fuss and McFadden (1978) and Blackorby, Primont, and Russell (1978) for thorough discussions of these issues. However, there is a gap in the literature that we wish to fill. The duality between superadditivity of T and subadditivity of C has been derived only for cases in which both properties may hold only weakly. Since strict subadditivity of costs plays an important role in much of our analysis, the dual conditions for the set of production techniques are derived in Appendix I.

This is a strong regularity condition implying smoothness in the production techniques that are used. It implies that isoquants have no corners, and it is sufficient for R2. It can be shown that, given R1, the existence of single-valued, continuous input demand functions that have a one-to-one relationship with relative factor prices suffices for R3(a). Furthermore, the existence of continuous marginal costs, together with R3(a), suffices for R3(b). Thus, while R3 is strong, it is a consequence of standard and plausible economic hypotheses.

Now, let us define a measure of scale effects directly from the production set:

Definition 3C1   Let $(x, y) \in T$. Then, at (x, y),

$$\tilde{S}_N \equiv \sup\{r \,|\, \exists \delta > 1 \text{ such that } (\lambda x, \lambda^r y) \in T \text{ for } 1 \leq \lambda \leq \delta\}.$$

$\tilde{S}_N$ indicates the maximal proportionate growth rate of outputs along their ray, as all inputs are expanded proportionally. If $\tilde{S}_N > 1$, then a proportional expansion of all inputs can yield a larger proportional expansion of all outputs. Further, if the latter holds, then $\tilde{S}_N \geq 1$. More precisely, $\tilde{S}_N$ behaves as the local degree of homogeneity of the production set. In fact, if the production set is globally homogeneous of degree α, in that

$$(x, y) \in T \Rightarrow (\lambda x, \lambda^\alpha y) \in T$$

(this is the multiproduct analog of homogeneous production functions), then for all $(x, y) \in T$, $\tilde{S}_N = \alpha$. Yet, $\tilde{S}_N$ is well defined for *any* production set, whether or not it has the strong property of homogeneity. We can now relate $\tilde{S}_N$ to the cost function and S_N, the degree of scale economies.

Proposition 3C1   Given R1 and R3,

$$S_N = C(y, w)/y \cdot \nabla C(y, w) = \tilde{S}_N,$$

where $\tilde{S}_N$ is defined at $(x^*(y, w), y) \in T$. (*The proof of this relationship is provided in Appendix II to this chapter.*)

Proposition 3C1 relates the degree of scale economies directly to the properties of the production set. It should be noted that $C(y, w)/y \cdot \nabla C(y, w)$ is a property of a *point* in output space, given factor prices, while $\tilde{S}_N$ appeared to be a property that related only to ray expansions of outputs. This shows that behavior along rays has fundamental implications for the behavior of costs throughout the range of outputs. This may also be inferred from Proposition 3B1, which showed the equality of S_N and $1/(e + 1)$, where e is the elasticity of RAC with respect to movements of outputs along a ray.

Proposition 3C1 permits one to see the interconnections between multi-product scale economies and the feasible combinations of inputs and output quantities. The next result shows explicitly how S_N can be "read off" from a production transformation function.

Proposition 3C2   *Given R1 and R3,*

$$S_N = - \sum_{i=1}^{m} x_i^* \frac{\partial \phi}{\partial x_i} \Big/ \sum_{j=1}^{n} y_j \frac{\partial \phi}{\partial y_j}$$

where ϕ is as defined in (3C2), and where the derivatives of ϕ are evaluated at $(x^*(y, w), y)$.

Note that this result (which is also proved in Appendix II) generalizes and deepens the standard result[5] that, with a differentiable scalar output production function $F(x)$, the "function coefficient," or "scale elasticity,"

$$\sum \frac{\partial F(x)}{\partial x_i} \frac{x_i}{F(x)} \equiv \varepsilon,$$

equals the ratio of average to marginal cost. Proposition 3C1 showed that ε, like S_N, can be related directly to the production set. Proposition 3C2 shows that S_N, like the ratio of average to marginal cost, which is S_N for a scalar output, can be calculated directly from a functional representation of T whose parameter values can be estimated econometrically.

Finally, this approach permits a substantial generalization of the standard result that for a scalar output produced in accord with a production function homogeneous of degree α, the cost function is homogeneous in output of degree $1/\alpha$.

Proposition 3C3   *If $C(y, w)$ is differentiable at (y, w), then*

$$S_N = \sup\{r | \exists \hat{\delta} > 1 \text{ such that } C(\lambda y, w) \leq \lambda^{1/r} C(y, w) \text{ for } 1 \leq \lambda \leq \hat{\delta}\}.$$

This result, which is proved in Appendix II, asserts that any differentiable cost function, whatever the number of outputs involved, and whether or not it derives from a homogeneous production process, has a local degree of homogeneity that is the reciprocal of S_N. And, as shown above, under our regularity conditions S_N is the local degree of homogeneity of the production set.

The preceding propositions are based on the assumption that the multi-product firm has no influence over input prices. However, as indicated

---

[5] See, for example, Ferguson (1969).

earlier, the analysis and use of the multiproduct cost function can readily be extended to cases in which the unit price faced by the firm for each of its inputs depends on the quantity of an input that it purchases; that is, where $w_i = w_i(x_i)$, the price of input i when x_i is the quantity used. While the equivalence between S_N and $\tilde{S}_N$ demonstrated under our regularity conditions no longer pertains, there remains a precise relationship between the structural returns to scale properties of the available techniques and the degree of scale economies relevant at the firm level. Extending a result of Ferguson (1969) to our multi-output case, we obtain, under our regularity conditions

$$(3C4) \qquad S_N = \tilde{S}_N \bigg/ \left(1 + \sum_{i=1}^{m} \tilde{e}_i \eta_i\right), \qquad \tilde{e}_i > -1,$$

where $\tilde{e}_i = x_i w_i'(x_i)/w_i(x_i)$, the elasticity of the ith input price schedule facing the firm, and $\eta_i = x_i w_i(x_i)/(\sum_{j=1}^{m} x_j w_j(x_j))$, the cost share of the ith input. If the $\tilde{e}_i$'s are negative, then $S_N > \tilde{S}_N$, and inversely.

Since our analyses of industry equilibrium are based on the properties of the *firms'* multiproduct cost function, the effects in (3C4) do not alter our *positive* theory of industry structure. However, to the extent that the input supply curves faced by the firm are less than perfectly elastic as a result of *pecuniary* rather than *real* economies (or diseconomies), our *normative* conclusions must be modified appropriately.

3D. Ray Average Cost, Returns to Scale, and Market Structure

The concepts described so far in this chapter play a central role in our analysis of market structure in the rest of this book. However, as we shall see, by themselves ray average cost and returns to scale for the entire product set are tools that are far from sufficiently powerful to relate market structure to technological conditions in multiproduct industries. To supplement these tools, we shall have to explore, in Chapter 4, some new cost concepts with no analogs in the scalar output case.

However, before we embark on this task, let us note that in multiproduct industries there is as much (or as little) reason as in the single-product case to expect the (ray) average cost curve to be U-shaped, as depicted in Figure 3A2. After all, as we have seen, *RAC* can be interpreted to refer to a single (composite) commodity so that the stories behind the U-shaped and flat-bottomed curves apply as much to the one case as to the other.[6]

---

[6] Indeed, it may not even be possible in principle to distinguish clearly between a simple and a composite commodity. Is an automobile a simple commodity or a composite good composed of tires, motor, chassis, and so forth? Surely, where proportions are held constant, nothing substantive is obtained by distinguishing the one from the other.

From the observation that cost behavior along a ray is analytically equivalent to that in the scalar output case, a number of conclusions from the single-output analysis can immediately be extended to ray behavior. Thus, as in scalar output cases, economies of scale imply that $S_N > 1$, $C(y) > y \cdot \nabla C(y)$, and the revenues from marginal cost pricing cannot cover production costs. Moreover, the larger is S, the larger the shortfall must be. Consequently, if firms must operate at output levels at which scale economies are not exhausted, because of the limited size of the market relative to the scale of operation at which RAC reaches its minimum, then perfect competition with prices equal to marginal costs cannot be viable financially. Of course, unlike the scalar output case, there will not be just one minimum efficient scale level of operation. Instead, because there is a multiplicity of outputs, each different output mix may well have its own minimum efficient scale at which scale economies are exhausted and $S_N = 1$.

The preceding observation immediately leads us to another construct that plays a crucial role in the analysis of the number of firms in an industry. It has just been implied that this number of firms is related to the ratio between the output of the industry and the output at which RAC reaches its minimum point. We have also observed that the distance of this minimum point from the origin may well vary and perhaps vary substantially from one ray to another.

For expository convenience, let us assume for the moment that the RAC curves are, in fact, all U-shaped. Then, along each ray there must exist a unique point at which RAC attains its minimum (though that point may sometimes lie outside the range of relevant output vectors). Let us use the "M locus" to designate the set of all output vectors that minimize RAC

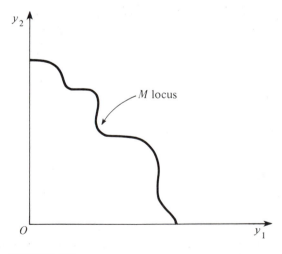

FIGURE 3D1

along their own rays. Thus, in the case of two outputs, M will be a locus, perhaps shaped irregularly, like that depicted in Figure 3D1. All we can say about the shape of M so far is that it must extend from one axis to the next, containing exactly one point in common with each ray that can be drawn in the figure. In the case of three products, M will, of course, be a surface. In a multi-output world, the position of the M locus plays the same role as y_m, the level of output with the minimal average cost, does in the scalar output case. As at y_m, there are locally constant returns to scale at every point in M. (Once more, the converse does not hold in general because even if along each ray RAC has a unique global minimum and is strictly increasing or decreasing elsewhere, a ray may contain inflection points of the cost surface at which $S_N = 1$.)

Later, we study in some detail the role that the M locus plays in determining the number of firms that minimizes the cost of producing a given industry vector of outputs, and this helps us to determine where that industry lies in the spectrum between a naturally monopolistic and a naturally competitive organization. We also see below how complementarity and substitutability of the industry's outputs affect the shape of the M locus. Ultimately, we examine, for the multiproduct case (as we did in Chapter 2 for the single-product case) some of the implications of the empirically based assumption that RAC curves are "flat-bottomed," so that the cost function exhibits constant returns to scale over a *segment* of each ray.

3E. Other Properties of Ray Cross Sections

While RAC and its scalar output analog, average cost, share many properties, the former by itself is a tool of relatively limited power for the analysis of multiproduct market structure. For this reason there is relatively little to be gained by formulating and proving the obvious analog of Proposition 2A1. For this purpose one could merely define such concepts as ray concavity and ray subadditivity in the obvious way, concluding, as in Proposition 2A1, that ray concavity implies declining ray average cost which in turn implies ray subadditivity, while the converse implications do not hold in general. However, these conclusions are of limited interest since ray subadditivity is tantamount to multiproduct natural monopoly *only* if all firms are somehow constrained to produce their outputs in *precisely* the *same* proportions, that is, in the proportions actually purchased on the market. Where, as is generally the case, no such restriction applies, even if a single firm is the least costly producer of outputs in the proportions purchased by the market, it still may be true, for example, that several more specialized firms may produce the equilibrium output vector of the industry still more cheaply. A concrete example will confirm this possibility.

Thus, consider the cost function $C(y_1, y_2) = (y_1^{1/3} + y_2^{1/3})^2$. This cost function exhibits decreasing ray average costs everywhere since $C(vy) = v^{2/3}C(y) < vC(y)$, for $v > 1$. Thus, by obvious analogy with Proposition 2A1, this cost function is subadditive along every ray—no two or more firms can produce as cheaply as one, if they all offer the same output proportions. However, a single firm producing output vector $(1, 1)$ would incur a cost of 4, while if this output vector were divided between two firms producing $(1, 0)$ and $(0, 1)$, respectively, total cost would be reduced since $4 = C(1, 1) > C(0, 1) + C(1, 0) = 2$. We soon see why a simplistic argument goes astray in such a case, and we show below that such complications are the essence of the multiproduct case rather than mere pathological exceptions. In any event, examples such as this reveal the need for multiproduct concepts that go beyond the ray-specific properties of the cost function. Such additional concepts constitute the subject of the next chapter.

APPENDIX I

The Duality of Strict Subadditivity

We adopt the framework of Uzawa (1964). Let the *input requirement set* be given by

$$A(y) \equiv \{x \,|\, (x, y) \in T\}.$$

This is the set of input bundles from which it is *possible* to produce the output y. Uzawa assumed (a) R1 and monotonicity and (b) that $A(y)$ is convex, closed, and bounded away from 0 for $y \neq 0$. He proved that

$$(1) \qquad A(y) = \{x \in R^m_+ \,|\, w \cdot x \geq C(y, w), \forall w \in R^m_+\};$$

that is, the input requirement set consists of precisely those input bundles that are at least as expensive as the cost-minimal bundle.

Let $A^0(y) \equiv \{x \,|\, (x, y) \in T, \exists x' \leq x, x' \neq x, (x', y) \in T\}$ denote the technically *inefficient* portion of $A(y)$. The efficient portion is then $\bar{A}(y) \equiv A(y) - A^0(y)$. $\bar{A}(y)$ is comprised of boundary points of the convex set $A(y)$. Therefore, for any $x \in \bar{A}(y)$ there is a separating hyperplane, $p \cdot z = k$, such that $p \neq 0$, $p \cdot x = k$, and $p \cdot z \geq k \, \forall z \in A(y)$. Monotonicity ensures that $p \geq 0$. However, we need a stronger additional assumption (c):

$$\forall x \in \bar{A}(y), \exists p > 0 \text{ such that } p \cdot z \geq k = p \cdot x, \forall z \in A(y).$$

For a sufficiently smooth technology (e.g., one satisfying R3), this assumption would merely require that all marginal products be positive and finite. In general, it implies that no *technically* efficient input bundle requires that some input prices be zero in order for it to be *cost*-efficient. This rules out technologies having points at which isoquants are horizontal. Thus, in Figure 1, the only hyperplane that supports $A(y)$ at point B is $x_2 = 0$ and condition (c) is violated.

We now can prove the following result:

Lemma *Assumptions (a), (b), and (c) imply that* $A^0(y) = \{x \in R^m_+ \,|\, w \cdot x > C(y, w), \forall w \in R^m_{++}\} \equiv \psi(y)$.

61

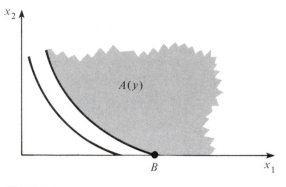

FIGURE 1

Proof   (i) $x \in A^0(y)$ implies $\exists x' \le x, x' \ne x$, such that $x' \in A(y)$. Thus, for $w \in R^m_{++}$, $C(y, w) \le w \cdot x' < w \cdot x$, so $x \in \psi(y)$.

(ii) Suppose $x \in \psi(y)$. Since $C(y, w)$ is continuous in w, $w \cdot x > C(y, w)$ $\forall w \in R^m_{++}$ implies that $w \cdot x \ge C(y, w)$ $\forall w \in R^m_+$. Thus from (1), $x \in A(y)$. It now suffices to show that $x \notin \bar{A}(y)$. Suppose $x \in \bar{A}(y)$. Then by (c), $\exists p > 0$ such that $p \cdot z \ge p \cdot x$ $\forall z \in A(y)$. Hence, $C(y, p) = p \cdot x$, which contradicts $x \in \psi(y)$.   Q.E.D.

The Lemma establishes conditions under which the *technically* inefficient input bundles are always *cost*-inefficient as well, and conversely. Note that (c) is crucial for this result. In Figure 1, point B is technically efficient but not cost-efficient for any set of positive input prices.

We can now establish our major result:

Theorem   *Given Assumptions (a), (b), and (c), the multiproduct cost function is* strictly *subadditive for all positive factor prices if and only if the technology set is* strictly *superadditive. That is,*

$$
(2) \qquad\qquad C(y, w) + C(y', w) > C(y + y', w) \qquad \forall w \in R^m_{++}
$$

if and only if

$$
(3) \qquad\qquad x \in A(y) \text{ and } x' \in A(y') \text{ imply that } (x + x') \in A^0(y + y').
$$

Proof   Suppose (3). Let $C(y, w) = x^* \cdot w$ and $C(y', w) = x^{**} \cdot w$, where $x^* \in A(y)$ and $x^{**} \in A(y')$, $w > 0$. Then $x^* + x^{**} \in A^0(y + y')$, so that $\exists z \in A(y + y')$ with $z \le x^* + x^{**}$ and $z \ne x^* + x^{**}$. Consequently, $C(y + y', w) \le z \cdot w < (x^* + x^{**}) \cdot w = C(y, w) + C(y', w)$.

Suppose (2). Let $x \in A(y)$ and $x' \in A(y')$. Now $\forall w > 0$, $w \cdot x \ge C(y, w)$, and $w \cdot x' \ge C(y', w)$. Then, $w \cdot (x + x') \ge C(y, w) + C(y', w) > C(y + y', w)$, $\forall w > 0$. Thus, $(x + x') \in \psi(y + y')$ and, by our Lemma, $(x + x') \in A^0(y + y')$.   Q.E.D.

Thus, under plausible regularity conditions, costs are strictly subadditive for all factor prices if and only if aggregation of outputs makes possible a *strict* reduction in input requirements.

APPENDIX II

Proofs of Propositions 3C1 and 3C2

We work from the representation of T provided by (3C2). First we show that

(1)
$$\frac{-\sum_i x_i^* \frac{\partial \phi}{\partial x_i}}{\sum_j y_j \frac{\partial \phi}{\partial y_j}} = \frac{C(y, w)}{\sum_j y_j C_j(y, w)} \equiv S_N$$

The cost minimization problem is to minimize $\sum w_i x_i$, over x, subject to $\phi(x, y) \geq 0$ and $x_i \geq 0$. The Kuhn-Tucker necessary conditions, with γ the Lagrangean multiplier, are

(2)
$$w_i - \gamma \partial \phi / \partial x_i \geq 0,$$

(3)
$$x_i^*[w_i - \gamma \partial \phi / \partial x_i] = 0,$$

(4)
$$\phi(x^*, y) \geq 0, \qquad \gamma \geq 0,$$

and

(5)
$$\gamma \phi(x^*, y) = 0.$$

Summing (3) over all inputs and using the identity $C(y, w) = w \cdot x^*(y, w)$, yields

$$C(y, w) = \gamma \sum_i x_i^* \frac{\partial \phi}{\partial x_i}.$$

The Envelope Theorem[1] tells us that

$$C_j = -\gamma \partial \phi / \partial y_j,$$

_____

[1] See Silberberg (1974) or Samuelson (1963) for expositions of the use of the Envelope Theorem.

and (1) follows as long as $\gamma \neq 0$. If γ were 0, (3) would give $x_i^* = 0$, for all i, which with (4) and $y \neq 0$ would contradict R1. Also, $\gamma \neq 0$ and (5) yield

$$(6) \qquad \qquad \phi(x^*, y) = 0.$$

This completes the proof of Proposition 3C1.

Now we demonstrate that

$$(7) \qquad \qquad \tilde{S}_N = -\sum_i x_i^* \frac{\partial \phi}{\partial x_i} \Big/ \sum_j y_j \frac{\partial \phi}{\partial y_j}.$$

Let $R = \{r \mid \exists \delta > 1$ such that $(\lambda x^*, \lambda^r y) \in T$ for $1 \leq \lambda \leq \delta\}$. Suppose that $r \in R$. Define $H(\lambda) \equiv \phi(\lambda x^*, \lambda^r y)$. $H(1) = 0$, by (6). For $1 \leq \lambda \leq \delta$, $H(\lambda) \geq 0$ by (3C2). H is differentiable by R3 and so, by a fundamental property of derivatives (see Apostol, 1957, p. 91), $H'(1) \geq 0$:

$$H'(\lambda) = \sum_i x_i^* \frac{\partial \phi(\lambda x^*, \lambda^r y)}{\partial x_i} + r\lambda^{r-1} \sum_j y_j \frac{\partial \phi(\lambda x^*, \lambda^r y)}{\partial y_j}$$

$$H'(1) = \sum_i x_i^* \frac{\partial \phi}{\partial x_i} + r \sum_j y_j \frac{\partial \phi}{\partial y_j} \geq 0$$

$$\varepsilon \equiv -\sum_i x_i^* \frac{\partial \phi}{\partial x_i} \Big/ \sum_j y_j \frac{\partial \phi}{\partial y_j} \geq r.$$

Thus ε is an upper bound on R. $\tilde{S}_N \equiv \sup R$ implies that $\varepsilon \geq \tilde{S}_N$. Suppose that $\varepsilon > \tilde{S}_N$. Then $\varepsilon > \bar{r} > \tilde{S}_N$, with $\bar{r} \notin R$. Consider then $G(\lambda) \equiv \phi(\lambda x^*, \lambda^{\bar{r}} y)$. $G(1) = 0$. If it were the case that $G'(1) > 0$, there would exist a $\delta > 1$ such that $G(\lambda) \geq 0$ for $1 \leq \lambda \leq \delta$ and $\bar{r} \in R$. Hence $G'(1) \leq 0$. But, calculating as before, we see that it follows that $\varepsilon \leq \bar{r}$. So $\varepsilon > \tilde{S}_N$ yields a contradiction, $\varepsilon = \tilde{S}_N$, and (7) is established, completing the proof of Proposition 3C2.
$$\text{Q.E.D.}$$

Proof of Proposition 3C3

Let $\hat{R} = \{r \mid \exists \delta > 1$ such that $C(\lambda y, w) \leq \lambda^{1/r} C(y, w)$ for $1 \leq \lambda \leq \delta\}$. Suppose that $r \in \hat{R}$. Define $H(\lambda) \equiv \lambda^{1/r} C(y, w) - C(\lambda y, w)$. As in the above proof $H(1) = 0$, $H(\lambda) \geq 0$ for $1 \leq \lambda \leq \delta$, and hence $(1/r)C(y, w) - \sum y_j C_j(y, w) = H'(1) \geq 0$. Thus, $S_N \geq r$, it is an upper bound on $\hat{R}$, and $S_N \geq \sup \hat{R}$.

Suppose that $S_N > \sup \hat{R}$. Then, there is an $\bar{r} \notin \hat{R}$, with $C / \sum y_j C_j > \bar{r}$. Consider $G(\lambda) \equiv \lambda^{1/r} C(y, w) - C(\lambda y, w)$. If $G'(1) > 0$, $\bar{r}$ would be in $\hat{R}$. So $G'(1) \leq 0$, and it follows from a now familiar calculation that $C / \sum y_j C_j \leq \bar{r}$. This is a contradiction, and hence $S_N = \sup \hat{R}$.   Q.E.D.

4

Cost Concepts
Applicable to
Multiproduct Cases

The analysis in Chapter 3 dealt with some properties of cost functions for multi-output production. But, as in most of modern production theory, the analysis there proceeds in much the same way as that for a single output. This simplicity is achieved by focusing upon measures of cost variation as output proportions remain unchanged, permitting the resulting commodity bundle to be treated as a single (composite) good.

In this chapter, we focus on the new phenomena that arise from the inter-relations among goods in production. For example, the presence or absence of complementarity between outputs in production becomes a crucial matter, which has no counterpart in the scalar model. At the same time, we are forced to design new cost concepts that relate to nonproportionate changes in output. These constitute the subject of this chapter and will be seen, in succeeding chapters, to be crucial for the analysis of the structure of industries.

One useful way of thinking about these concepts is to consider them as devices for the exploration of the shape of the cost surface. In Chapter 3 we began this process by analyzing cross sections of the cost surface along *rays* (such as OR in Figure 4A1) in output space, along the floor of the cost surface diagram. In this chapter we will proceed by considering cross sections of the cost surface along portions of output space (such as ST) that are parallel to one axis or another, and others, like WV, that cut diagonally from one axis to another. Each of these yields concepts and relationships that play a substantial role in the analysis later in the book.

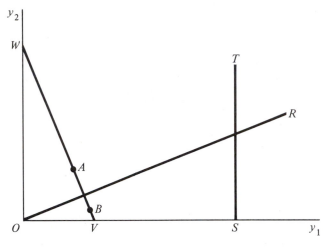

FIGURE 4A1

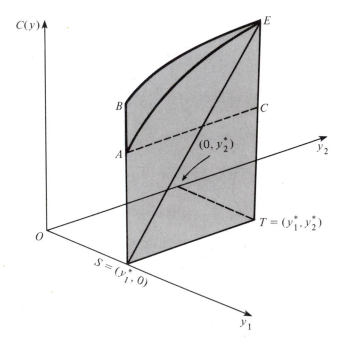

FIGURE 4A2

4A. Product-Specific Economies of Scale[1]

As was shown in Chapter 3, the concepts of RAC and multiproduct scale economies relate to *proportional changes* in the quantities in the entire product set. However, another analytically important way in which the magnitude of a firm's operations may change is through variation in the output of one product, holding the quantities of other products constant. To study the cost of such output variation it is useful to define the incremental cost of product i as the addition to the firm's total cost resulting from the given output of product i. That is, it is the firm's total cost with the given vector of outputs, minus what that total cost would be if production of good i were abandoned, all other output quantities remaining unchanged. More formally, we have

Definition 4A1   The *incremental cost* of the product $i \in N$ at y is

$$IC_i(y) = C(y) - C(y_{N-i})$$

where y_{N-i} is a vector with a zero component in place of y_i and components equal to those of y for the remaining products. The *average incremental cost*[2] of product i is then naturally defined as $AIC_i(y) \equiv IC_i(y)/y_i$.

Geometrically, this cost datum is found from the cross section of the cost surface that lies above the line in output space parallel to the axis for quantity i, for example, ST in Figure 4A2. Given output vector $T = (y_1^*, y_2^*)$, S is the corresponding output vector on the y_1 axis at which y_1 has been held the same as at point T, but y_2 has been reduced to zero. If product 2 has no output-specific fixed costs, then the total cost surface rises continuously above ST (curve AE). However, if some special fixed cost must be incurred to begin production of good 2 *as an addition* to the firm's line of other products, then the cross section of the cost surface will contain a vertical fixed cost segment

---

[1] This term seems to have been used first in Scherer *et al.* (1975) and Beckenstein (1975). Though similar in spirit to ours, their concept is not formulated in a manner convenient for our analysis.

[2] The term "average incremental cost" refers to the same magnitude as that (probably) referred to by those who use the term "average variable cost" (see, for example, Areeda and Turner, 1975, pp. 716–18). There is, however, some ambiguity in the latter term because it is not clear whether Areeda and Turner mean it to include capital costs. This tissue arises in particular when product i requires some fixed expenditure, F_i, which need not be undertaken if i were excluded from the product line. By Definition 4A1, F_i must be included in the average incremental cost, but we do not know how those who use the term average variable cost would deal with F_i in their definition.

AB, resulting in a jump discontinuity of $C(y)$ above the y_1 axis. Thus, the height *CE* in Figure 4A2 measures the *total incremental cost* of product 2 at output vector *T*, the addition to the firm's total cost resulting from a decision to add product 2 to its product line, raising its output from zero all the way to y_2^*. The average incremental cost, $AIC_2(y_1^*, y_2^*)$, is clearly given by the slope of the line from *A* to *E*.

It is also clear that, as Figure 4A2 is drawn, the average incremental costs of product 2 are declining with y_2, at least between 0 and y_2^*. This suggests, by analogy to the single-output case, the novel and useful concept of product-specific scale economies.

Definition 4A2: Product-Specific Returns to Scale   The degree of scale economies specific to product *i* at output vector *y* is given by

$$S_i(y) = IC_i(y)/y_i C_i \equiv AIC_i/(\partial C/\partial y_i).$$

Returns to the scale of product *i* at *y* are said to be increasing, decreasing, or constant as $S_i(y)$ is greater than, less than, or equal to unity, respectively.

Thus, the degree of scale economies specific to product *i* is measured by the ratio of the average incremental cost of the product to its marginal cost. As usual, if the marginal cost is less than the average incremental cost [as is true by definition when $S_i(y) > 1$], the latter has a negative derivative and will decline as y_i increases. So, in this case, the product-specific returns to scale will, indeed, be increasing. The explanations of the cases where $S_i(y) = 1$ and $S_i(y) < 1$ are, obviously, perfectly analogous.

This definition paves the way for a result that relates product-specific returns to scale to the financial viability of the sale of product *i* at a price equal to its marginal cost, *when product* i *is viewed as an increment to the firm's other outputs*. The obvious result states that the revenues ($y_i C_i$) collected from the sale of product *i* when its price is equal to its marginal cost will exceed, equal, or fall short of the incremental cost (IC_i) incurred by offering that product as S_i is less than, equal to, or greater than unity, respectively.

The preceding discussion indicates that the concept of product-specific scale economies has a natural and familiar interpretation based on the general concept of incremental cost. For $S_i(y)$ relates to the increment in the firm's total cost which results from the addition or total elimination of an entire product to or from the firm's set of products, given the magnitude of the output of each such item. It will also sometimes be useful to consider increments involving *subsets* of two or more products. Here we are faced once again with a problem of choice of units similar to that in the case of ray average cost. Fortunately, we can deal with this almost exactly as we did in the definition of *RAC*.

First we formulate the natural definition of the incremental cost of a set of products:

Definition 4A3 The incremental cost of the product set $T \subseteq N$ at y is given by $IC_T(y) = C(y) - C(y_{N-T})$, where y_{N-T} is a vector with zero components associated with the products in T and components equal in value to those of y for products in $N - T$. (Note that $y = y_T + y_{N-T}$.)

Definition 4A4 The *average incremental cost* of the product set T at y is $AIC_T \equiv IC_T(y)/\sum_{j \in T} y_j$. The average incremental cost of T is *decreasing* at y, $DAIC_T$, if $AIC_T(ty_T + y_{N-T})$ is a decreasing function of t at $t = 1$. Increasing average incremental cost is defined analogously.

For product sets including two or more items, AIC is analogous to ray average cost except that the pertinent "ray" does not go through the origin, and is instead limited to a subspace of R_+^n. This can be seen geometrically in Figure 4A3, which deals with the case of three products. There, given the magnitude of y_3, the incremental cost of a bundle with fixed proportions of y_1 and y_2 refers to the behavior of total cost when such bundles of y_1 and y_2 are added to the product line. The set of such bundles is described by a ray *in the plane* in which the magnitude of y_3 is held fixed. Two such loci are

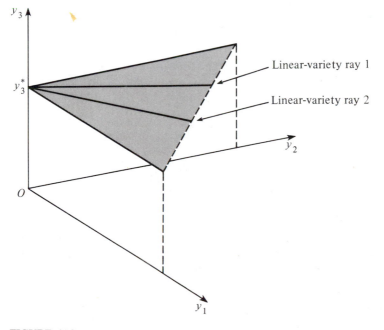

FIGURE 4A3

shown in the figure, and are labeled linear-variety ray 1 and linear-variety ray 2 (they are not, of course, ordinary rays of the three-dimensional diagram—hence the more complex designation, which is adapted from standard mathematical terminology).

We can now produce a natural multiproduct extension of our measure of product-specific returns to scale for the scalar case.

Definition 4A5   The degree of scale economies specific to the product set $T \subseteq N$ at y is given by $S_T(y) \equiv IC_T(y)/\sum_{j \in T} y_j C_j(y)$.

As is to be desired, this definition becomes identical to our multiproduct measure when $T = N$ and identical to the degree of scale economies in product i when $T = \{i\}$. Again, we can see directly that if there are increasing returns with respect to T at y, that is, $S_T(y) > 1$, the product set T is not financially compensatory under marginal cost pricing, since the returns collected ($\sum_{j \in T} y_j C_j(y)$) do not cover the product set's incremental costs. We can also state

Proposition 4A1 $S_T(y)$, the degree of product-specific economies of scale over T at y, is equal to $1/(1 + e_T)$, where e_T is the elasticity of average incremental cost of T at y.

Thus, if $S_T(y) > 1$, that is, if there are such scale economies at y, it follows that average incremental costs are strictly decreasing at y, $DAIC_T(y)$ (but not conversely). The proof of Proposition 4A1 is essentially the same as that of Proposition 3B1.

A basic relationship emerges by dividing the product set N into two disjoint subsets, T and $N - T$. One can then relate $S_N(y)$, the multiproduct degree of scale economies, to those specific to the subsets T and $N - T$. Using Definitions 4A3 and 4A4 we obtain

(4A1) $$S_N = \frac{\alpha_T S_T + (1 - \alpha_T) S_{N-T}}{(IC_T + IC_{N-T})/C}$$

where $\alpha_T = \sum_{j \in T} y_j C_j / \sum_{j \in N} y_j C_j$.

If the denominator of (4A1) were unity, the multiproduct degree of scale economies would merely be a weighted average of those of any subset and its complement.

If there were no production interdependencies between the commodities in T and those in $N - T$, the denominator in question would indeed be unity. To see this more clearly, we rewrite it, using Definition 4A1, as

$$\frac{C(y) - C(y_{N-T}) + C(y) - C(y_T)}{C(y)}$$

or

(4A2)
$$1 + \frac{C(y) - C(y_T) - C(y_{N-T})}{C(y)}.$$

If the production process is truly disjoint, the cost of producing the entire product set is exactly equal to the sum of the *stand-alone* costs of the subsets T and $N - T$ [i.e., $C(y_T)$ and $C(y_{N-T})$]. If, however, as one would expect, economies of joint production yield total costs *less* than the sum of stand-alone costs, then (4A2) will be less than unity. A discussion of the full implications of this conclusion must await the analysis of the next section.

4B. Economies of Scope

The multiproduct cost concepts described up to this point have dealt with the behavior of the cost hypersurface in several conveniently selected cross sections of cost and output space. However, we see in this section that there is a crucial cost concept that cannot be characterized directly in terms of a slice of the cost surface.

In addition to economies deriving from the size or scale of a firm's operations (concepts at least intuitively familiar to most economists), there is also the possibility that cost savings may result from simultaneous production of several different outputs in a single enterprise, as contrasted with their production in isolation, each by its own specialized firm. That is, there may exist economies resulting from the *scope* of the firm's operations.

Formally, economies of scope can be interpreted as a restricted form of subadditivity. One simply modifies Definition 2A1 (subadditivity) by restricting it to sets $y^1, \ldots, y^k$ of *orthogonal* nonnegative output vectors (vectors with no positive components in common), that is, such that $y^i \cdot y^j = 0, i \neq j$.

This apparently abstract concept merits such a precise, formal definition because it is a *necessary and sufficient condition for the existence of multi-product firms* in perfectly contestable markets. This is easy to see in the case of two products. If $C(y_1, y_2) < C(y_1, 0) + C(0, y_2)$, any market structure involving specialty firms would be unstable in that it would be profitable for them to merge. If the opposite inequality held, no arrangement involving multiproduct firms could be stable because separation would be profitable. Only in the equality case is the issue a matter of indifference.

The following is a general way of expressing the definition of the concept:

Definition 4B1 *Economies of Scope.* Let $P = \{T_1, \ldots, T_k\}$ denote a nontrivial partition of $S \subseteq N$. That is, $\bigcup_i T_i = S$, $T_i \cap T_j = \emptyset$ for $i \neq j$. $T_i \neq \emptyset$, and $k > 1$. There are economies of scope at y_S with respect to the partition P if

(4B1)
$$\sum_{i=1}^{k} C(y_{T_i}) > C(y_S)$$

There are said to be *weak economies of scope* if the inequality in (4B1) is weak rather than strict, and *diseconomies of scope* if the inequality is reversed. For example, if $N = \{1,2\}$, and $P = \{\{1\}, \{2\}\}$, the condition of economies of scope means that $C(y_1, y_2) < C(0, y_2) + C(y_1, 0)$.

Geometrically (Figure 4B1), the concept involves a comparison of $C(y_1^*, 0) + C(0, y_2^*)$, the sum of the heights of the cost surface over the corresponding points on the axes, with $C(y_1^*, y_2^*)$, the height of the cost surface at point (y_1^*, y_2^*) which is the vector sum of $(y_1^*, 0)$ and $(0, y_2^*)$. If $C(y_1^*, y_2^*)$ lies below the hyperplane OAB which goes through the origin and points $C(y_1^*, 0)$ and $C(0, y_2^*)$, then the condition for economies of scope is

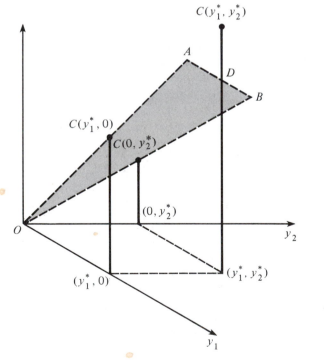

FIGURE 4B1

satisfied. Thus, in Figure 4B1 the height of D, the point on plane OAB above (y_1^*, y_2^*), must equal $C(y_1^*, 0) + C(0, y_2^*)$ since the hyperplane is described by $C = ay_1 + by_2$ for some constants a, b. Therefore, $C(y_1^*, 0) = ay_1^*$ and $C(0, y_2^*) = by_2^*$, and $C(y_1^*, y_2^*)$ must be less than $ay_1^* + by_2^*$ for economies of scope to hold.

Economies of scope, and some related measures which will be described in the remainder of this chapter, describe the basic phenomena that distinguish the multiproduct case from that of the single product. In the latter, we are used to an association between natural monopoly and economies of scale or declining average cost because, roughly speaking, the presence of such economies means that production of a given output by two or more firms will be more costly than its production by a single firm. A moment's thought will indicate that when the firm offers many products, even where ray average costs decline everywhere, the absence of economies of scope may nevertheless preclude natural monopoly. For in an industry that enjoys no economies of scope, a multiproduct firm can be broken up into several specialized firms without any increase in cost and, perhaps, even with some saving. This, in essence, suggests why economies of scope and the concepts related to it play so central a role in the analysis of multiproduct industry structure.

In order to study the relationships between the concept of economies of scope and our measures of the various types of economies of scale, we construct a measure of the magnitude of economies of scope.

Definition 4B2   The *degree of economies of scope* at y relative to the product set T is defined as $SC_T(y) \equiv [C(y_T) + C(y_{N-T}) - C(y)]/C(y)$.

Thus, the degree of economies of scope measures the relative increase in cost that would result from a splintering of the production of y into product lines T and $N - T$. Such a fragmentation of the firm increases, decreases, or leaves unchanged the total cost as SC_T is greater than, less than, or equal to zero, respectively.

It is easy to show that, if all products have positive incremental costs,[3] then $SC_T(y) < 1$. For, assume the contrary. Then, by the expression for $SC_T(y)$,

$$C(y_T) + C(y_{N-T}) - C(y) \geq C(y)$$

---

[3] While at first blush it may seem that every product must have positive incremental costs, further thought suggests that there may be exceptions. Thus suppose one wished to produce beef entirely without any hides or vice versa. The total cost may be interpreted to include the costs of disposing of the unwanted product, and these could make the incremental costs of the hides negative.

or

$$[C(y) - C(y_T)] + [C(y) - C(y_{N-T})] \leq 0.$$

Since the expressions in the square brackets are the incremental costs of $N - T$ and T, respectively, it follows that at least one of them must be zero or negative. Therefore, $SC_T(y)$ must be less than unity.

This permits us to see that integral role played by economies of scope in the relationship of product-specific to aggregate scale economies. By Definition 4A1.

$$IC_T(y) + IC_{N-T}(y) = C(y) + C(y) - C(y_T) - C(y_{N-T}).$$

Then, equation (4A1) can be rewritten, in view of Definition 4B2, as

(4B2) $$S_N(y) = \frac{\alpha_T S_T(y) + (1 - \alpha_T)S_{N-T}(y)}{1 - SC_T(y)},$$

where we note that our measure of economies of scope, $SC_T(y)$, appears in the denominator. Equation (4B2) indicates precisely the way in which economies of scope magnify the effects of product-specific scale economies in the determination of overall scale economies. In the absence of any economies or diseconomies of scope, $SC_T(y) = 0$ and S_N is a simple weighted sum of its component product-specific scale economies. However, if economies of scope are present, then $SC_T(y) > 0$, the denominator of (4B2) is less than one, and S_N is larger than the weighted sum of $S_T(y)$ and $S_{N-T}(y)$. Consequently, with economies of scope, $DAIC_T$ and $DAIC_{N-T}$ together imply increasing returns to scale. Even if product-specific returns to scale happen to be constant for both T and $N - T$, with scope economies there will nevertheless be increasing returns to scale overall. And, sufficiently strong economies of scope can impart scale economies to the entire product set even if there are diseconomies to the scale of individual products.

Although the notion of economies of scope seems to be intuitively attractive, judging by other writers, the concept, nevertheless, remains somewhat unfamiliar. It may therefore be helpful to relate it to a more familiar multiproduct cost concept.

Definition 4B3 A twice-differentiable multiproduct cost function exhibits *weak cost complementarities* over the product set N, up to y, if $\partial^2 C(\hat{y})/\partial y_i \partial y_j \equiv C_{ij}(\hat{y}) \leq 0$, $i \neq j$, for all $\hat{y}$ with $0 \leq \hat{y} \leq y$, with the inequality holding strictly over a set of nonzero measure.

Essentially, the presence of weak cost complementarities implies that the marginal cost of producing any one product decreases (weakly) with increases in the quantities of all other products. Weak cost complementarities were held to be a "normal" property of joint production by Sakai (1974). We can demonstrate that weak cost complementarities are a sufficient condition for economies of scope.

Proposition 4B1[4]   *A twice-differentiable multiproduct cost function which exhibits weak cost complementarities over N, up to y, exhibits economies of scope at y with respect to all partitions of N.*

With only a slight modification, this result can be strengthened to cover the cases in which product-specific fixed costs are present, that is, when $C(\cdot)$ is not differentiable along the relevant axes. Without loss of generality, any multiproduct cost function can be expressed as

$$(4B2) \qquad\qquad C(y) = F(S) + c(y),$$

where $S = \{i \in N \mid y_i > 0\}$.
   We can now state

Proposition 4B2   *If c(y) is a twice-differentiable function which exhibits weak cost complementarities over N, up to y, and if F is not superadditive—that is, $F(S) + F(T) \geq F(S \cup T) \; \forall S, T \subseteq N$—then the cost function exhibits economies of scope at y > 0 with respect to all partitions of N.*

In addition to widening the applicability of Proposition 4B1, Proposition 4B2 suggests that economies of scope can be present despite some degree of anticomplementarity ($c_{ij} > 0$), provided that $F(S)$ is sufficiently *subadditive*.

4C. Sources of Economies of Scope

It is clear from the definition of economies of scope that the presence of such economies creates an incentive for specialty firms to merge and become multiproduct firms. However, we have not, as yet, offered any explanation indicating why a multiproduct cost function may exhibit such economies. It seems incumbent upon us to do so, since, after all, the traditional view that there are advantages in *specialization* goes back at least 200 years to Adam Smith's discussion of the division of labor. A logical place to begin our search for sources of economies of scope is the Marshallian case of joint production. Intuition tells us that it must clearly be cheaper to produce pairs of items such as wheat and straw, wool and mutton, and beef and hides

---

[4] The results of Propositions 4B1 and 4B2 are proved in Appendix I.

in one enterprise rather than in two specialized firms. We will therefore construct a formal model of joint production and derive the relationship of that concept to economies of scope and cost complementarities.

We take the position that joint production, in the Marshallian sense, arises because some factors of production are *public inputs* in the sense that, once they are acquired for use in producing one good, they are available costlessly for use in the production of others.[5] Formally, we assume that there are n production processes

(4C1) $$y_i = f^i(z^i, K), \qquad i = 1, \ldots, n$$

where z^i is a vector of inputs directly assignable to the production of good i, and K is the public input. For our purpose it will be more convenient to employ an equivalent representation of the production technology which expresses the minimum attributable costs, V^i, of producing any good i as a function of its quantity, the quantity of the public input, and the vector of nonpublic input prices. That is,

$$V^i(y_i, K, w) = \min_{z^i} \{z^i \cdot w \,|\, f^i(z^i, K) \geq y_i\} \qquad i = 1, \ldots, n.$$

We assume that the public input is at least weakly productive in all uses. This is equivalent to

(4C2) $$V^i(y_i, K_1) \geq V^i(y_i, K_2) \text{ for } K_2 \geq K_1, i = 1, \ldots, n.$$

In addition, we make the assumption that the public input is strictly productive in each process in the weak sense that any positive amount of the public input yields a larger output than none at all. That is,

(4C3) $$V^i(y_i, K) < V^i(y_i, 0), \qquad \forall y_i, K > 0, i = 1, \ldots, n.$$

[A more familiar, but more restrictive, means of assuring that (4C3) holds is the assumption that K is an essential input, that is, that $f^i(z^i, 0) = 0$.] Finally, we assume for simplicity that as many units of the public input as desired are obtainable at a constant price β.

We are now able to relate this model of joint production with a public input to the concept of economies of scope.

---

[5] Perhaps the clearest example of the phenomenon occurs in the peak load pricing arena; see Clark (1923), Demsetz (1973), and Panzar (1976). Marshall's (1925) bucolic illustrations can, perhaps, be viewed in this light by considering the *animal* in question as a public input. Starrett (1977) provides a discussion of public inputs in a macro context.

Proposition 4C1 [6]   *The multiproduct minimum cost function $C(y)$ that is dual to a set of multiproduct production techniques employing a public input of nontrivial value exhibits economies of scope.*

This derivation of economies of scope from the public input model is quite general because its additional technological assumptions are minimal. The public input model yields additional and useful implications about the multiproduct cost function if somewhat stronger (but still plausible) restrictions are imposed on the pertinent production techniques.

Proposition 4C2   *A multiproduct cost function that is twice differentiable and dual to a multiproduct set of production techniques employing a strongly normal public input exhibits strong cost complementarities.*

The public input model has demonstrated one technological source of economies of scope. However, while the presence of a public input is a sufficient condition for the existence of economies of scope, it is certainly not necessary. Indeed, as shown by Proposition 4C2, with a modest strengthening of its assumptions the public input model exhibits the much stronger property of strong cost complementarity.

However, cases of joint production processes making use of public inputs do not seem sufficiently common to account for the near universality of multiproduct firms, many of which presumably enjoy economies of scope. There is another tradition in the literature which undertakes to explain the presence of economies of scope. This rests, ultimately, on the presence of inputs that are readily shared by the processes utilized to produce several different outputs. Often, this results because of indivisibilities or lumpiness in the plant of the productive enterprise. The following quotations clearly illustrate this view of the matter:

> [A]lmost every firm does produce a considerable range of different products. It does so largely because there are economies to be got from producing them together, and these economies consist largely in the fact that the different products require much the same overhead. (Hicks, 1935/1952, p. 372)

> It is a commonplace of business practice that the production and sales managers work hand in hand to devise new products that can be produced with the company's idle capacity . . . . What the firm has to sell is not a product, or even a line of products, but rather its capacity to produce. (Clemens, 1950/1958, p. 263)

In order to analyze the implications of this type of phenomenon more precisely, we have formulated a micro model of the production relationships

---

[6] The results of Propositions 4C1 and 4C2 are proved in Appendix I.

which involves, as before, n otherwise independent production processes that are capable of sharing the services of some productive inputs. For ease of exposition, we assume that there is only one such input, called "capital." Then the multiproduct (minimum) cost function, which embodies the least costly way of producing y_S, is obtained by solving the program

$$(4C4) \qquad C(y_S) \equiv \min_{k} \sum_{i \in S} V^i(y_i, k_i) + \psi(k, \beta),$$

where V^i represents the minimum variable cost of producing the output y_i using k_i units of capital *services*. The capital service cost function, $\psi(k, \beta)$, represents the cost of acquiring the requisite vector k of capital services, where β represents relevant factor prices. Finally, let $k^*(y_S)$ denote the argmin of (4C4), the cost-minimal vector of capital services required for the production of y_S. We simply assume here that $k^*(y)_i > 0$ for $y_i > 0$, at prevailing input prices, while $k^*(y)_i = 0$ for $y_i = 0$.

We shall focus on cases in which ψ is *strictly* subadditive in k. Another way to describe such situations is to say that "capital" is a *quasipublic* input, since its services can be shared by two or more product lines without *complete* congestion. The most extreme form occurs when, as in the preceding discussion, capital is a pure public input; that is $\psi(k, \beta) = \beta \max_i k_i$. On the other hand, when capital is a pure private input that is marketed competitively, $\psi(k, \beta) = \beta \sum_i k_i$, which is only weakly subadditive in k. Intermediate cases also fit well into this framework, hence the term quasipublic input. We are now in a position to state

Proposition 4C3   *For any nontrivial partition of N, there are economies (diseconomies) of scope if and only if ψ is strictly subadditive (superadditive) in the relevant range.*

   Proof   Let $\{T_1, \ldots, T_l\}$ be a nontrivial partition of N, and let $\hat{k} = \sum_j k^*(y_{T_j})$. It follows from (4C4) and the definition of $k^*(\cdot)$ that

$$(4C5) \qquad \sum_{i \in N} V^i(y_i, k_i^*(y)) + \psi(k^*(y)) = C(y) \le \sum_{i \in N} V^i(y_i, \hat{k}_i) + \psi(\hat{k})$$

$$(4C6) \qquad \sum_{i \in T_j} V^i(y_i, k_i^*(y)) + \psi(k^*(y)_{T_j}) \ge C(y_{T_j})$$

$$= \sum_{i \in T_j} V^i(y_i, k_i^*(y_{T_j})) + \psi(k^*(y_{T_j})).$$

Summing (4C6) over j and subtracting from (4C5) yields

$$(4C7) \qquad \psi(k^*(y)) - \sum_j \psi(k^*(y)_{T_j}) \le C(y) - \sum_j C(y_{T_j}) \le \psi(\hat{k}) - \sum_j \psi(k^*(y_{T_j})).$$

The conclusions follow since the leftmost (rightmost) term in (4C7) is positive (negative) if and only if ψ is strictly superadditive (subadditive) over the relevant range.   Q.E.D.

Thus, the micro model we have constructed illustrates the equivalence between the existence of quasipublic, sharable inputs and economies of scope. This discussion, of course, may not exhaust the possible sources of economies of scope. It has been intended to show only that they can be attributed to phenomena which have appeared before in the literature, and about which there need be little mystery.

4D. Trans-Ray Convexity

While the concept of economies of scope does not lend itself to complete description in terms of a cross section of the cost hypersurface, there is another indicator of complementarity in production which can be described in such terms and which is seen in succeeding chapters to be a powerful tool for the analysis of industry structure and the issue of potential entry. This concept, called "trans-ray convexity," relates to any cross section of the cost hypersurface that connects points on the output axes. Figures 4D1 and 4D2

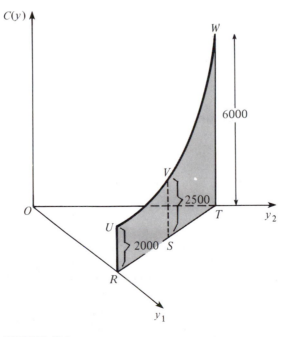

FIGURE 4D1

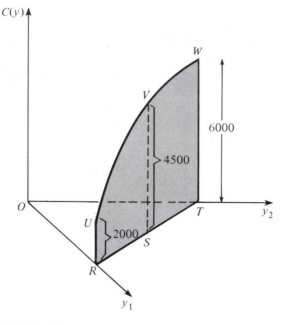

FIGURE 4D2

show two possible shapes of such a cross section taken above line segment RT. Here R is a point on the y_1 axis and T is some point on the y_2 axis. Since any such line segment RT must be perpendicular to some ray in the $y_1 y_2$ plane, the segment of the cost surface above RT is referred to as a trans-ray cross section.

This cross section is important for our purposes because it compares the costs of operation of specialized firms, corresponding to points R and T on the axes, with the costs of firms producing a weighted average of the outputs on the axes. Thus, if point R represents 100 units of output of y_1 alone, while point T represents 150 units of output of y_2 alone, point S, the midpoint of RT, represents the production of $(y_1, y_2) = (50, 75)$—exactly half as much y_1 as point R and half as much y_2 as point T. Thus, one may regard a move toward the center of RT as a decrease in the firm's degree of specialization.

Now, a comparison of Figures 4D1 and 4D2 will immediately suggest the relevance of the property of trans-ray convexity, which is satisfied in the cross section of Figure 4D1 and violated in Figure 4D2. In the former, the cross section curves downward toward the center, meaning that multiproduct production is relatively cheaper than specialized production. Specifically, we see that for specialized production, $C(R) = 2000$ and $C(T) = 6000$, while for the midpoint, $C(S) = 2500 < \frac{1}{2}C(R) + \frac{1}{2}C(T) = 4000$. What this suggests is that in this case a multiproduct firm will enjoy a cost advantage over specialized firms, and, in fact, trans-ray convexity turns out to be one of a set of conditions which together are sufficient for natural monopoly in a multiproduct industry. The situation is quite different in the circumstances illustrated in Figure 4D2. There, multiproduct production is at a relative cost disadvantage, as indicated by the upward curvature of the cross section toward the center of the diagram. Here, $C(S) = 4500$ and is thus greater than a weighted average of the costs of specialized production at R and T. Such a shape of the trans-ray cross section reduces the likelihood that the industry will be a natural monopoly, though we will see in Chapter 7 that, even so, subadditivity (natural monopoly) may still be possible. That is, trans-ray convexity is not a necessary condition for natural monopoly.

We now offer our formal definition of trans-ray convexity:

Definition 4D1: Trans-Ray Convexity   A cost function $C(y)$ is trans-ray convex through some point $y^* = (y_1^*, \ldots, y_n^*)$ if there exists any vector of positive constants $w_1, \ldots, w_n$ such that for every two output vectors $y^a = (y_1^a, \ldots, y_n^a)$ and $y^b = (y_1^b, \ldots, y_n^b)$ that lie on the hyperplane $\sum w_i y_i = w_0$ through point y^* (so that they satisfy $\sum w_i y_i^a = \sum w_i y_i^b = \sum w_i y_i^*$), for any k such that $0 < k < 1$ we have

(4D1) $$C[ky^a + (1 - k)y^b] \leq kC(y^a) + (1 - k)C(y^b).$$

This definition asserts that the trans-ray cross section above the line $\sum w_i y_i = w_0$ is convex like that in Figure 4D1. But there are two other features of this definition which are less obvious. First, the concept refers to a particular output vector, y. That is, the cost surface may be trans-ray convex at one point and yet not satisfy this condition at another point. It may, for example, have cross sections like that in Figure 4D1 at points near the origin while, at more distant points, its trans-ray cross sections may look like that in Figure 4D2.

A second important feature of the definition makes it a far weaker (less demanding) requirement than it may seem at first glance. Through output point y there are an infinity of different trans-ray cross sections. Thus, in Figure 4D3, every negatively sloping line segment through y corresponds to a trans-ray cross section through y. The concept of trans-ray convexity does *not* require that the cross section of the cost surface above each and every such negatively sloping line have the convex shape depicted in Figure 4D1. All that is required for the cost function to satisfy the definition at point y is that there exist at least *one* negatively sloping line through y for which the cross section of $C(y)$ is convex. If this is true, say, above RT in Figure 4D3, it does not matter if convexity is violated above $R'T'$, $R''T''$, and so forth. The one convex cross section above y is all we will need for our later results.

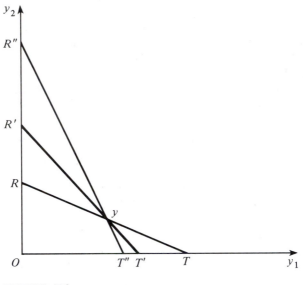

FIGURE 4D3

The economic content of trans-ray convexity is made clearer by examining the two-product case and taking $y^a = (y_1, 0)$ and $y^b = (0, y_2)$ in (4D1). While $ky^a + (1 - k)y^b$ includes a smaller quantity of each good, it involves simultaneous production of both items. Thus (4D1) implies that the cost savings offered by economies of scope outweigh the effects of any increasing returns to scale in production of the items individually. This property of trans-ray convexity shows up clearly in a characterization of the concept discussed later in Chapter 15. If the cost function in two outputs is twice-differentiable, (4D1) holds if and only if, on the pertinent trans-ray hyperplane, $w_1^2 C_{22} + w_2^2 C_{11} - 2w_1 w_2 C_{12} \geq 0$. It suffices for this that the incremental cost of product i be convex in y_i (which implies that average incremental cost of i is increasing) and that $C_{12} \leq 0$ (weak cost complementarity). Further, decreasing marginal costs ($C_{ii} < 0$), which suggest the presence of product-specific scale economies, can only be consistent with trans-ray convexity if C_{12} is sufficiently negative, that is, if cost complementarity is sufficiently strong.

There is one important class of cases in which the condition of trans-ray convexity is always violated everywhere because of the strength of product-specific returns to scale. In these cases, the cost function involves fixed costs incurred on account of particular products. Here, for a firm already producing y_1 to add y_2 to its product line, an outlay F_2 must be incurred, where the outlay does not vary with the magnitude of y_2. Then, F_2 is a product-specific fixed cost for good 2. It is easy to see that such product-specific fixed costs preclude trans-ray convexity, even if this condition would other-

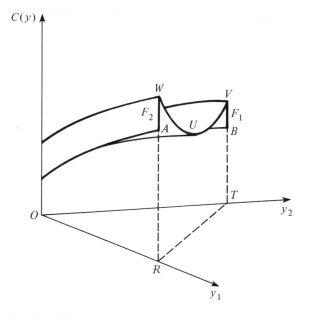

wise have been satisfied. Figure 4D4 shows such a "transylvanian" cost function. Here WUV, the interior portion of the cross section above RT, is clearly convex. But, since production of product 1 incurs a fixed cost, $F_1 = BV$, above the y_2 axis the cross section jumps vertically (and discontinuously) from B to V. Similarly, because of product 2's fixed cost, $F_2 = AW$, there is a jump from A to W above the y_1 axis. The resulting batlike wings (hence the term transylvanian) of the cost surface yield trans-ray cross sections like $AWUVB$, which obviously violate the convexity requirement.

4E. Iso-Cost Contours

The last type of cross section of the cost surface we will consider is one parallel to the y_1, y_2 plane (Figure 4E1). This, of course, yields the iso-cost surfaces (cost indifference curves) familiar from elementary microeconomics. We will employ one attribute that characterizes the shape of some such iso-cost surfaces:

Definition 4E1: Quasiconvexity   A cost function $C(y)$ is said to be (strictly) quasiconvex at y^0 if the set $\{y \mid C(y) \leq C(y^0)\}$ is a (strictly) convex set. That is, in the two-output case, the cost function is strictly quasiconvex if all of the cost indifference curves are everywhere strictly concave to the origin.

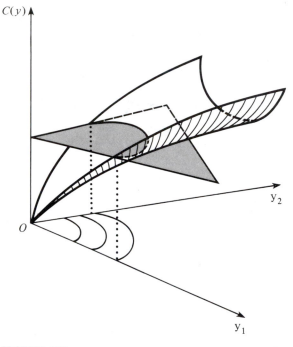

FIGURE 4E1

Thus, as depicted on the floor of Figure 4E1, the iso-cost surfaces have the shape familiar from the "guns-and-butter" product transformation curves of elementary economics.[7]

In one of the basic theorems proved in this book, quasiconvexity can be substituted for trans-ray convexity and vice versa. That is, in that case the two characteristics serve the same analytic purpose.

There is some reason to expect that the two attributes will normally accompany each other. The reason is suggested in Figure 4E2, which shows a set of iso-cost contours shaped as quasiconvexity requires. Then, along the trans-ray cross section *SR* total cost goes lower and lower as one moves from either end toward *T*, the point of tangency with an iso-cost curve. This suggests that the trans-ray cross section may be U-shaped, with its minimum point above *T*, and that there may also be trans-ray convexity. However, while this argument is suggestive, it is not generally valid since the trans-ray

---

[7] While a convex cost function is, of course, also quasiconvex, the converse need not hold. This is obviously the same as the relationship between concavity and quasiconcavity of production and utility functions.

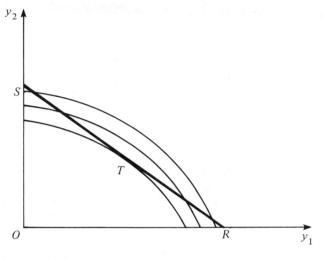

FIGURE 4E2

cross section corresponding to Figure 4E2 can have concave segments on either side of T, even though sloping downward toward T. Indeed, as is shown in Appendix II, quasiconvexity does not imply trans-ray convexity, nor does the latter imply the former. Thus, while related, neither attribute is formally stronger than the other.

Where either premise will do, the choice between the assumptions of trans-ray convexity and quasiconvexity must be made on grounds of economic interpretation rather than mathematical power. While quasiconvexity is the more familiar of the two concepts because it has been used explicitly for other purposes by economists for some years, and implicitly far longer, we will generally choose to work with the concept of trans-ray convexity. In part, this is because there are several key propositions which, at least so far, have only been proved with the aid of the latter concept. Perhaps more important, the straightforward economic meaning of trans-ray convexity in terms of the complementarity benefits of simultaneous production of several outputs makes it easier to understand the role it plays in our analysis.

Finally, we mention a characteristic related to, but weaker than quasiconvexity which will also be useful to us later:

Definition 4E2 An iso-cost surface has a support at y^0 if there exists a strictly positive vector v such that every point in the hyperplane $H = \{y \in R^n_+ | v \cdot y = v \cdot y^0\}$ is more costly to produce than y^0; that is, $C(y) \geq C(y^0), \forall y \in H$.

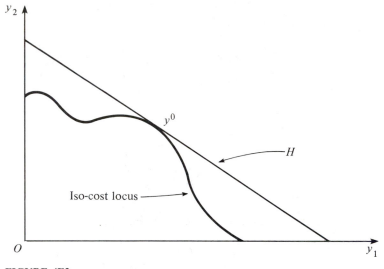

In the case where the iso-cost locus is continuous and y^0 does not lie on the axes, a support is simply a line (hyperplane) tangent to the iso-cost locus at y^0 which lies above that locus everywhere except the point of tangency. That is, if the iso-cost locus crosses the tangent to it at y^0 anywhere, then the locus possesses no support at y^0 and so does not satisfy the requirements of Definition 4E2.

As illustrated in Figure 4E3, while all quasiconvex functions possess such a support, quasiconvexity is not necessary for this condition to be satisfied.

4F. Trans-Ray Supportability

While the preceding multiproduct cost concepts have the advantages of familiarity, intuitive clarity, or ready adaptation to geometric exposition, it turns out that there is a weaker concept of cost complementarity that serves equally well for our most basic results on subadditivity of costs and the sustainability of industry configurations.

Definition 4F1: Trans-Ray Supportability   A cost function $C(y)$ is trans-ray supportable at y^0 if there exists at least one trans-ray direction above which the cost surface is supportable, that is, if there is a trans-ray hyperplane in output space, $H_{TR} \equiv \{y \geq 0 \,|\, w \cdot y = w \cdot y^0\}$, with $w > 0$, such that there is a constant v_0 and a vector v with the properties that $C(y) \geq v_0 + v \cdot y$ for all $y \in H_{TR}$ and $C(y^0) = v_0 + v \cdot y^0$.

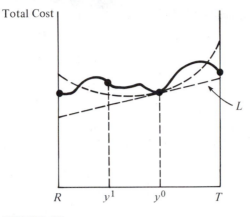

FIGURE 4F1

While powerful, this condition is not easy to interpret intuitively. It can, however, be made clearer with the aid of a diagram (Figure 4F1) in which the vertical axis measures total cost and the horizontal axis coincides with the base RT of the trans-ray slice in Figure 4D1. Consider the point y^0 on this ray. If the cost function were (strictly) trans-ray convex in the cross section in question, it would assume the shape of the dashed curve and would, by definition, have a support, as indicated by the dashed line L tangent to it at y^0. Observe, however, that the cost function $\tilde{C}(y)$ associated with the solid curve is also supported by L at y^0 although it is clearly *not* trans-ray convex. On the other hand, $\tilde{C}(y)$ cannot be supported at y^1 because it is simply impossible to construct a line through $(y^1, \tilde{C}(y^1))$ which does not also cut $\tilde{C}(y)$ somewhere else. Thus, $\tilde{C}(y)$ is trans-ray supportable over RT at y^0 but not at y^1.

We are now in a position to state two important results (proved in Appendix I) concerning this cost concept.

Proposition 4F1 *If the iso-cost surface has a support at y^0, then $C(y)$ is trans-ray supportable there. However, supportability is guaranteed only with respect to the trans-ray hyperplane defined by $w = \nabla C(y^0)$.*

Corollary *If the cost function is quasiconvex at y^0, it is trans-ray supportable there. (This follows immediately from Definitions 4E1 and 4F1. The convex set $\{y \geq 0 \,|\, C(y) \leq C(y^0)\}$ is guaranteed to have a support.)*

Proposition 4F2 *If $C(y)$ is trans-ray convex at y^0 along some trans-ray hyperplane, it is trans-ray supportable there over that hyperplane. (The logic of this result is evident from our discussion of Figure 4F1.)*

Thus, both quasiconvexity and trans-ray convexity imply trans-ray supportability, which is a weaker and more general condition than either of those.

4G. Epilogue

This, then, completes our description of the building blocks required in our analysis of multiproduct market structures. We are now in a position to proceed in the next several chapters to describe how the cost-minimizing number of firms in a multiproduct industry can be determined, and then to determine when an industry will be a natural monopoly, and the degree to which potential entry constrains even a natural monopoly.

APPENDIX I

Proof of Proposition 4B1

Since any partition of N can be obtained via a series of binary partitions, it suffices to demonstrate the result for the partition T, $N - T$, $N \neq T \neq \emptyset$. We can rearrange terms so that the condition to be demonstrated is

(1) $$[C(y_T + y_{N-T}) - C(y_T)] - [C(y_{N-T}) - C(0)] < 0.$$

The first bracketed term can be written as

$$\int_\Gamma \sum_{i \in N-T} C_i(y_T + x_{N-T}) dx_i$$

and the second bracketed term as

$$\int_\Gamma \sum_{i \in N-T} C_i(x_{N-T}) dx_i,$$

where Γ is any smooth *monotonic* arc from 0 to y_{N-T}. Since these are line integrals along the common path Γ, their difference can be written as

(2) $$\int_\Gamma \sum_{i \in N-T} [C_i(y_T + x_{N-T}) - C_i(x_{N-T})] dx_i$$

$$= \int_\Gamma \sum_{i \in N-T} \left[\int_\Lambda \sum_{j \in T} C_{ij}(z_T + x_{N-T}) dz_j \right] dx_i < 0,$$

where Λ is a smooth monotonic arc from 0 to y_T.   Q.E.D.

Proof of Proposition 4B2

The proof is the same as that of Proposition 4B1, with $c(\cdot)$ replacing $C(\cdot)$, so that equations (1) and (2) also contain the expression $F(N) - F(T) - F(N - T)$. This term is nonpositive by hypothesis.   Q.E.D.

Proof of Proposition 4C1

Under our assumptions, a specialized firm which produces at minimum cost any subset of products T_j with output y_{T_j} must solve the following program:

(3)
$$\min_{K} \sum_{i \in T_j} V^i(y_i, K) + \beta K.$$

Let $K^*_{T_j}$ be the solution of this program. Then, one property of the multi-product minimum cost function is that

(4)
$$C(y_{T_j}) = \sum_{i \in T_j} V^i(y_i, K^*_{T_j}) + \beta K^*_{T_j}.$$

Letting $\{T_1, \ldots, T_k\}$ constitute a nontrivial partition of N, we define this feasible cost function:

(5)
$$\bar{C}(y) \equiv \sum_{j=1}^{k} \sum_{i \in T_j} V^i(y_i, \bar{K}) + \beta \bar{K}$$

where

(6)
$$\bar{K} = \max_{j} K^*_{T_j} \qquad j = 1, \ldots, k.$$

Then we have

(7)
$$\bar{C}(y) - \sum_{j=1}^{k} C(y_{T_j}) = \sum_{j=1}^{k} \sum_{i \in T_j} [V^i(y_i, \bar{K}) - V^i(y_i, K^*_{T_j})]$$
$$+ \beta \left[\bar{K} - \sum_{j=1}^{k} K^*_{T_j} \right].$$

Under our assumptions, both terms on the right-hand side of (7) are non-positive, with at least one strictly negative. Because $C(y)$ is defined to be the *minimum* cost function, we know that $C(y) \le \bar{C}(y)$ and, hence,

(8)
$$C(y) - \sum_{j=1}^{k} C(y_{T_j}) \le \bar{C}(y) - \sum_{j=1}^{k} C(y_{T_j}) < 0.$$

For $y > 0$, the inequality in (8) is our definition of economies of scope.
                                                                                                     Q.E.D.

Proof of Proposition 4C2

By definition,

(9)
$$C(y) \equiv \min_{K} \sum_{k=1}^{n} V^k(y_k, K) + \beta K.$$

Necessary and sufficient conditions for a regular interior optimum for the program (9) are given by

(10)
$$\sum_{k=1}^{n} V_K^k + \beta = 0$$

(11)
$$\sum_{k=1}^{n} V_{KK}^k > 0$$

where subscripts indicate partial differentiation. Let $K^*(y)$ be a solution of these equations. Substituting K^* into (9) and differentiating with respect to y_i yields (via the Envelope Theorem)

(12)
$$C_i = V_y^i(y_i, K^*).$$

Differentiation with respect to y_j yields

(13)
$$C_{ij} = V_{yK}^i \frac{\partial K^*}{\partial y_j} \qquad j \neq i.$$

Standard comparative statics analysis of equations (9) and (10) yields

(14)
$$\frac{\partial K^*}{\partial y_j} = -V_{yK}^j \Bigg/ \sum_{k=1}^{n} V_{KK}^k.$$

Since strong normality implies $V_{yK}^i < 0$, $i = 1, \ldots, n$, we have, upon substituting (14) into (13),

$$C_{ij} = -V_{yK}^i V_{yK}^j \Bigg/ \sum_{k=1}^{n} V_{KK}^k < 0 \qquad i \neq j. \qquad \text{Q.E.D.}$$

Proof of Proposition 4F1

For simplicity, let us assume that C is continuously differentiable at y^0 so that the iso-cost surface is supported by the gradient hyperplane. This ensures that $C(y) \leq C(y^0)$ implies $\nabla C(y^0) \cdot y \leq \nabla C(y^0) \cdot y^0$. Forming the contrapositive of this relation, we find that $\nabla C(y^0) \cdot y > \nabla C(y^0) \cdot y^0$ implies $C(y) > C(y^0)$. Then, by continuity, it must be the case that $\nabla C(y^0) \cdot y = \nabla C(y^0) \cdot y^0$ implies $C(y) \geq C(y^0)$. Now let $w = \nabla C(y^0)$, $v_i = S_N(y^0)C_i(y^0)$, and $v_0 = 0$. Since $C(y^0) \equiv S_N(y^0)\nabla C(y^0) \cdot y^0$, clearly $w \cdot y = w \cdot y^0$ implies $C(y) \geq C(y^0) = S_N(y^0)w \cdot y^0 = S_N(y^0)w \cdot y = v \cdot y$. Q.E.D.

Proof of Proposition 4F2

$C(y)$ is convex on the hyperplane $w \cdot y = w \cdot y^0$, $w > 0$. Consider the set $X = \{(x, y) \mid w \cdot y = w \cdot y^0, y \geq 0, x \geq C(y)\}$.

First we show that X is a convex set. Let (x', y') and (x'', y'') be elements of X and consider $(\lambda x' + (1 - \lambda)x'', \lambda y' + (1 - \lambda)y'')$, $\lambda \in (0, 1)$. Clearly,

$$w \cdot [\lambda y' + (1 - \lambda)y''] = \lambda w \cdot y' + (1 - \lambda)w \cdot y'' = \lambda w \cdot y^0 + (1 - \lambda)w \cdot y^0$$
$$= w \cdot y^0,$$

and $\lambda y' + (1 - \lambda)y'' \geq 0$ by construction. Now, by definition, $\lambda x' + (1 - \lambda)x'' \geq \lambda C(y') + (1 - \lambda)C(y'') \geq C(\lambda y' + (1 - \lambda)y'')$, by trans-ray convexity. So $\lambda(x', y') + (1 - \lambda)(x'', y'') \in X$ and X is convex.

Since $(C(y^0), y^0)$ is a boundary point of X, there is a separating hyperplane through it. That is, there exist numbers $b_0, b_1, \ldots, b_n$ such that

$$(x, y) \in X \text{ implies that } b_0 x + b \cdot y \geq b_0 C(y^0) + b \cdot y^0.$$

We know that $b_0 > 0$ because, for $y > 0$ and $(x, y) \in X$, $(0, y) \notin X$ implies that $b_0 \neq 0$, and $(\infty, y) \in X$ implies that $b_0 \geq 0$. Let $y \geq 0$ and $w \cdot y = w \cdot y^0$. Then $(C(y), y) \in X$ so that $b_0 C(y) + b \cdot y \geq b_0 C(y^0) + by^0$, or $C(y) \geq (-b/b_0) \cdot y + C(y^0) + (b/b_0) \cdot y^0$. Hence, letting $v = (-b/b_0)$ and $v_0 = C(y^0) + (b/b_0) \cdot y^0$ would complete the proof, if $v \cdot y^0 + v_0 = C(y^0)$ held. This is in fact the case since

$$(-b/b_0)y^0 + [C(y^0) + (b/b_0)y^0] = C(y^0). \qquad \text{Q.E.D.}$$

APPENDIX II

Quasiconvexity versus Trans-Ray Convexity of Cost Functions[1]

Figure 4E2 illustrated why we would normally expect a quasiconvex cost function (with concave iso-cost contours) to be trans-ray convex, and vice versa. Here we will provide two examples that show that the two concepts are not equivalent and that neither implies the other. The first cost function is quasiconvex, but not trans-ray convex. The second cost function is trans-ray convex, but not quasiconvex. Both have declining ray average costs.

Consider the cost function in two products given by

$$(1) \qquad C(y_1, y_2) = \begin{cases} 10 + \max(y_1, y_2) & \text{if } \max(y_1, y_2) \leq 1 \\ 11 & \text{if } 1 \leq \max(y_1, y_2) \leq 2 \\ 9 + \max(y_1, y_2) & \text{if } 2 \leq \max(y_1, y_2) \end{cases}$$

which is illustrated in Figure 1a.

All iso-cost loci are squares, that is, concave to the origin, so that the cost function is quasiconvex by definition. But it is not trans-ray convex everywhere. For example, all trans-ray cross sections through the point $(.75, .75)$ have the same sort of nonconvexity as that for the trans-ray $y_1 + y_2 = 1.5$, depicted in Figure 1b.

The second example is provided by the cost function

$$(2) \quad C(y_1, y_2) = \begin{cases} 10 + \frac{1}{2}(y_1 + y_2) + \frac{1}{2}|y_1 - y_2| & 0 \leq y_1 + y_2 \leq 1.5 \\ 10.75 + \frac{1}{2}|y_1 - y_2| & 1.5 \leq y_1 + y_2 \leq 2.5 \\ 9.5 + \frac{1}{2}(y_1 + y_2) + \frac{1}{2}|y_1 - y_2| & 2.5 \leq y_1 + y_2 \end{cases}$$

---

[1] We are indebted to Dietrich Fischer for the materials in Appendix II.

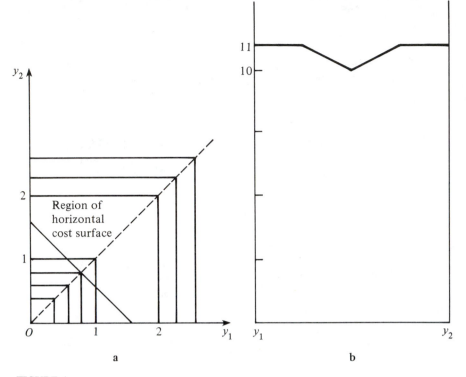

FIGURE 1
Horizontal and trans-ray cross sections of cost function (1) which is quasiconvex but not trans-ray convex.

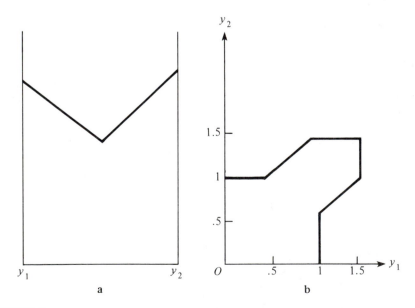

FIGURE 2
Trans-ray and horizontal cross sections of cost function (2) which is trans-ray convex but not quasiconvex.

Each cross section of this cost function along a trans-ray perpendicular to the 45° line has the shape shown in Figure 2a; that is, it is trans-ray convex. But an intersection with a horizontal plane at the level $C(y_1, y_2) = 11$ has the shape shown in Figure 2b, which is clearly not concave to the origin. Therefore, this cost function is not quasiconvex.

We see therefore that quasiconvexity is neither a necessary nor a sufficient condition for trans-ray convexity.

5

The Cost-Minimizing Industry Structure[1]

In this chapter we begin our study of the determination of the market structure of an industry by examining the number of firms that is efficient in the industry. For this purpose we relate the cost functions of the industry's firms to the market demands for the industry's products to determine the structural requirements of efficiency. If the number of firms needed for efficient production of the relevant quantities of industry output is very large, then the industry may be "naturally competitive." If the relevant cost-minimizing number of firms is one, then the industry is a natural monopoly. And if neither of these cases holds so that the cost-minimizing number of firms for relevant output levels is at some intermediate level, then the industry can be termed a natural oligopoly.

We proceed by describing a general analytic procedure that permits one to determine whether an industry is naturally competitive, a natural monopoly, or naturally oligopolistic on the basis of complete information about the market demands for the industry's outputs and the cost functions of the industry's firms. Then, recognizing that such complete information is unlikely to be available to the analyst, we provide procedures that require far less information, but that can nevertheless indicate the likely form of an industry's structure.

---

[1] This chapter contains a number of major contributions by Dietrich Fischer, who was coauthor of the article (Baumol and Fischer, 1978) on which much of this chapter is based.

We first describe versions of these procedures for the analysis of industries composed of firms that all produce a single, homogeneous product. Here, we are able to extend and make considerably more precise some standard insights and results that were discussed earlier.

Then, armed with the cost concepts and tools introduced in Chapters 3 and 4, we turn to the far richer and more realistic case of the multiproduct industry. Here we will see, surprisingly, that the methods and results analogous to those for single-product industries are not valid. Instead, new techniques are required with which we are able to derive bounds on the number of firms that can produce any given bundle of different outputs at least total industry cost. The bounds are based on a minimum of information about the cost function of each firm—namely, the set of outputs at which scale economies are exhausted (the M locus). And examples show that the bounds cannot be uniformly improved.

We find that the bounds, and therefore the range of feasible industry structures, are quite sensitive to the shape of the M locus. Consequently, we examine some of the characteristics of the industry that influence that shape. This analysis permits some general inferences about the determinants of market structures of industries.

We provide procedures that determine the set of industry outputs relevant for the determination of the bounds upon the efficient number of firms. Each of these procedures is related to, but requires less information than, the one which is theoretically ideal but whose use requires complete information. We conclude by noting the relationships between these tests of market structure and the analyses of natural monopoly and perfectly competitive industry structures discussed in Chapters 8 and 9.

While the analysis throughout the body of this chapter concerns the standard cases of U-shaped average cost and ray average cost curves, the case of flat-bottomed average cost curves, critical to Chapters 2 and 11, is treated in Appendix IV. Appendix II provides additional details on the cost-minimizing distribution of production among the firms in a single-output industry. Appendices I and III contain proofs of key propositions.

5A. Feasible and Efficient Industry Configurations

In this section, we define an analytic procedure that relies upon complete information on the economic structure of an industry to determine the range of efficient market structures. Our only requirement is that any such structure be feasible financially: that it yield nonnegative economic profit to each firm. Of course, as indicated in Chapter 1, profitability and efficiency may be considered to be implications of various sets of more basic postulates about the behavior of firms and of entrepreneurs. Throughout the remainder of this volume we shall work from such behavioral assumptions to derive far more detailed characterizations of firm and industry performance. Nevertheless, here, we confine ourselves to the implications of financial feasibility and industry efficiency because we find that these simple requirements alone have surprisingly rich implications.

The structural economic data with which we work are the industry inverse demand function, denoted by the vector function $P(y)$ and the cost function $C(y)$ of a firm. We proceed throughout this chapter under the assumption that all firms, extant or potential, have available to them the same input prices and the same production techniques representing the best current practice, and which are, together, dual to $C(y)$. We first construct the *industry cost function* from the cost function that pertains to each firm.

Definition 5A1   The industry cost function, denoted $C^I(y^I)$, gives the total cost to the industry of producing the output vector y^I when these outputs are apportioned in a cost-minimizing fashion among a cost-minimizing number of cost-minimizing firms. Thus,

$$(5A1) \quad C^I(y^I) = \min_{m, y^1, \ldots, y^m} \sum_{j=1}^m C(y^j), \quad \text{where } \sum_{j=1}^m y^j = y^I \text{ and } y^j \geq 0.$$

Definition 5A2   The set of cost-minimizing numbers of firms for the production of the industry output vector y^I is denoted by $m(y^I)$. It is the set of values of m which permit the minimum in (5A1) to be attained.

A feasible and efficient industry configuration is a set of output vectors, one for each firm in the industry, which constitutes an efficient (cost-minimizing) apportionment of the total industry output and which permits each of the firms to cover its production cost at the prices which induce the market to demand the given industry output vector. Restating this formally in terms of the industry cost function:

Definition 5A3 $\{y^1, \ldots, y^m\}$ is a feasible and efficient industry configuration if and only if $y^j \not\equiv 0, j = 1, \ldots, m$

$$\sum_{j=1}^m C(y^j) = C^I(y^I) \quad \text{where } y^I = \sum_{j=1}^m y^j \quad \text{(efficiency)}$$

and

$$P(y^I) \cdot y^j \geq C(y^j) \quad \text{for } j = 1, \ldots, m \quad \text{(feasibility)}.$$

This definition links the postulate of efficient industry production by financially viable firms with the economic data comprised of the market (inverse) demand function for the industry's outputs and the best-practice cost function of a firm.

Associated with each feasible and efficient industry configuration is a total industry output vector y^I and the number of firms m which, in total, produce that output in the configuration. Of course, that is a cost-minimizing number of firms for the production of y^I. We let $\tilde{Y}$ and $\underline{R}$ denote the sets of industry output vectors and numbers of firms, respectively, that are associated with some feasible and efficient industry configuration. Formally,

(5A2) $\tilde{Y} \equiv \{y^I | \exists$ a feasible and efficient industry configuration

$$\{y^1, \ldots, y^m\} \text{ with } \sum_{j=1}^{m} y^j = y^I\}.$$

(5A3) $\underline{R} \equiv \{m | \exists$ a feasible and efficient industry configuration

$$\{y^1, \ldots, y^m\}\}.$$

The shaded region in Figure 5A1 depicts the set $\tilde{Y}$ for a two-product industry. The subsets of $\tilde{Y}$ that are separated by dashed lines contain industry output vectors that are each produced efficiently by the given number of firms. Thus, for example, the points in the subregion $m = 4$ constitute the set of output combinations produced most cheaply by four firms. In the figure, this number of firms ranges from one to five, so that $\underline{R} = \{1, 2, 3, 4, 5\}$. Here, the industry appears to be a natural oligopoly, because of the intermediate numbers of firms in the feasible and efficient industry configurations.

In general, it is the set $\underline{R}$ that indicates the range of feasible and efficient market structures. If $\underline{R} = \{1\}$, then the industry is a natural monopoly. If

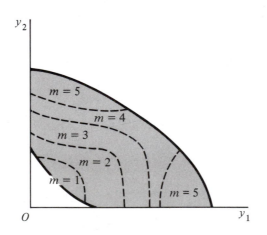

FIGURE 5A1

$\underline{R}$ includes numbers sufficiently large, then the industry may be[2] naturally competitive. And, if $\underline{R}$ contains and is bounded by an intermediate number, then the industry is a natural oligopoly. Thus, the focus of much of this chapter is upon methods which enable us to glean information about the set $\underline{R}$.

5B. The Two-Part Approach to Determination of Efficient Structure

While Definition 5A3 provides a precise specification of the set $\underline{R}$, the range of feasible and efficient numbers of firms, its heavy information requirements considerably impede its use. Consequently, we will seek to design methods that require much less information to bound $\underline{R}$. To describe our general approach, we must first distinguish two basic influences upon $\underline{R}$.

The number of firms that can produce an industry's output at lowest cost clearly depends on the nature of its production techniques. We are used to thinking of industries which are highly capital-intensive and which use large, indivisible pieces of equipment as naturally oligopolistic or monopolistic, while industries which are very labor-intensive and use very divisible equipment are more likely to be competitive. Thus, technological information, as reflected in the nature of the firm's cost function, will play a key role in our analysis. The issue here is how to proceed from relatively limited information about that cost function to helpful conclusions about the cost-minimizing number of firms for the industry.

There is also a second crucial influence, the size of the elements in the industry's output vector. An industry may, for example, call for only a small number of firms if demand for its product is very limited, but with a sharp upward shift in demand the number of firms required for efficiency may increase commensurately. Thus, if we have little or no information about the relevant range of industry output vectors, we may be left with an unfortunately broad range of possibilities for the cost-minimizing number of firms. If all we can conclude is that the optimal number of firms lies somewhere between 3 and 30,000, we will have learned little, if anything, about the industry's structure.

Thus, our procedure encompasses two main elements: (i) means to narrow the relevant range of industry outputs as far as is possible. For example, we will rule out some output vectors for which the demand and cost functions indicate that potential revenues cannot equal the cost. Having thus restricted the region of relevant outputs, we will provide (ii) bounds on

---

[2] Even if the most efficient industry arrangement calls for a very large number of firms, it need not be naturally competitive. For those firms need not be equal in size. An industry composed of 100,000 firms will not necessarily be competitive if half its assets are possessed by a single enterprise.

the least costly numbers of firms for the outputs within this region. The analytic techniques for the first step are described in Section 5C, and those for the second step are described in Sections 5D–5G. However, before proceeding, let us examine the logical underpinnings of this approach.

Proposition 5B1   *If it is possible to establish upper and lower bounds upon the numbers of firms needed to produce industry outputs efficiently for some set of output vectors that includes all the industry outputs associated with efficient and feasible industry configurations, then the number of firms in any feasible and efficient configuration will lie between these bounds. Formally, if $\underline{Y}$ is some set of industry outputs, let $m(\underline{Y}) \equiv \{x \mid x \in m(y) \text{ for some } y \in \underline{Y}\}$. Then, if $\underline{Y} \supseteq \tilde{Y}$, it follows that $\underline{R} \subseteq m(\underline{Y})$. Consequently, any upper and lower bounds, $\bar{m}$ and $\underline{m}$ respectively, on $m(\underline{Y})$ are also upper and lower bounds on $\underline{R}$.*

   Proof   If $m \in \underline{R}$, there is a feasible and efficient industry configuration $\{y^1, \dots, y^m\}$. Then $m \in m(y^I)$, where $y^I = \sum_{j=1}^{m} y^j$, and, hence, $m \in m(\tilde{Y})$. Thus, $\underline{R} \subseteq m(\tilde{Y})$. $\underline{Y} \supseteq \tilde{Y}$ implies that $m(\tilde{Y}) \subseteq m(\underline{Y})$. The Proposition follows.

   Proposition 5B1 makes clear the criterion that can be used in the search for a relevant region of industry outputs. We require a region Y that is sure to include all industry outputs associated with feasible and efficient industry configurations. Once we have found such a larger region, if we can find bounds for the cost-minimizing number of firms for any of Y's outputs, we can be sure that these same bounds also hold for the number of firms in the subregion of $\underline{Y}$ which is composed of all outputs that can be produced efficiently by financially viable firms.

5C. Relevant Sets of Industry Outputs

   The first method of establishing a relevant set of industry outputs makes use of the market inverse demand function and the industry cost function to determine the output vectors which can conceivably yield a total industry revenue that equals or exceeds total industry cost. For, clearly, any output vector that does not satisfy this requirement is not feasible financially. Formally,

Proposition 5C1   *Any industry output vector produced by an efficient and feasible industry configuration must earn market revenues at least as large as industry costs. That is,*[3]

(5C1) $$T \equiv \{y^I \mid P(y^I) \cdot y^I \geq C^I(y^I)\} \supseteq \tilde{Y}.$$

---

   [3] *Proof*   For any $y^I \in \tilde{Y}$, there is a set of firms' outputs, $\{y^1, \dots, y^m\}$, with $\sum_{j=1}^{m} y^j = y^I$, $\sum_{j=1}^{m} C(y^j) = C^I(y^I)$, and $P(y^I) \cdot y^j \geq C(y^j)$. Adding these inequalities gives $\sum_{j=1}^{m} P(y^I) \cdot y^j \geq \sum_{j=1}^{m} C(y^j)$. Thus, $P(y^I) \sum_{j=1}^{m} y^j = P(y^I) \cdot y^I \geq C^I(y^I) = \sum_{j=1}^{m} C(y^j)$, as required.

It should be noted that the set T of potentially profitable outputs may not be identical with the set $\tilde{Y}$ of outputs that can be produced in a manner that is efficient and financially feasible. In particular, T may include some industry output vectors which, although they earn enough market revenue in total to cover total industry cost, do not yield prices that permit each and every firm in the efficient apportionment to cover its own cost. Yet, Proposition 5C1 assures us that $\tilde{Y}$ must be a subset of T. Thus, the set T does satisfy the criterion established in Proposition 5B1 to test whether it can serve as a relevant region of industry outputs.

The construction of the set T of outputs for which industry revenues equal or exceed industry costs is illustrated in Figure 5C1. T is the shaded projection on the floor of the diagram of the region where the total revenue dome lies on or above the total cost surface. Here, total revenue is determined from industry output via the calculation $P(y^I) \cdot y^I$.

The construction of T in this formal manner requires knowledge of the industry cost function, which itself must be deduced from the entire firm cost function via the cost-minimization calculations described in Definition 5A1. Since these same calculations determine precisely the number of firms that produces each of the various industry output vectors efficiently, this analytic approach makes it unnecessary to find bounds for the numbers $m(y^I)$. But, on the other hand, such a formal construction of both T and $m(T)$ requires a great deal of information.

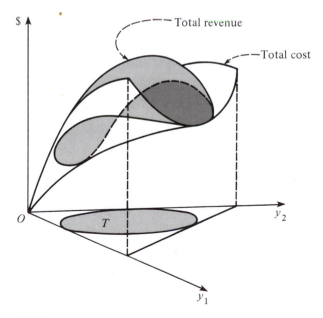

FIGURE 5C1

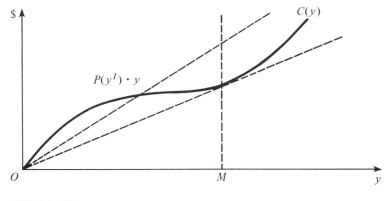

FIGURE 5C2

Nevertheless, the conceptual definition of the set of potentially profitable outputs T, and the result that $T \supseteq \tilde{Y}$, may permit an analyst to select an acceptable relevant region using methods that are less formal and information that is less complete. In particular, it is useful to design methods to bound the numbers in $\underline{R}$ that simply begin by taking a relevant region of industry outputs as a datum. Methods of this sort, which we describe in Sections 5D to 5G, require neither information about the industry cost function nor complete knowledge of the cost functions of firms. Instead, the methods rely only on the M locus, the set of outputs at which a firm's returns to scale are exhausted.

But first we will describe an alternative and compatible procedure for the determination of a relevant region of industry outputs that also relies on information about the M locus. In particular, the only cost information required by this mehod is that pertaining to the outputs of the firm that comprise the M locus.

This approach is most easily described in the single-product case, with which we begin our discussion. With such a scalar output, for y^I to be a relevant industry output quantity, $P(y^I)$ must be sufficiently large to permit a firm to cover its costs. If so, $P(y^I)$ must certainly be large enough to permit a firm operating *at minimum average cost* to break even, or do better. So, as shown in Figure 5C2, $P(y^I)$ must define a pseudorevenue hyperplane $P(y^I) \cdot y$ that lies at or above the cost of a firm at M, where average cost is at its minimum. This is the criterion we will now use to determine whether a given industry output is relevant for our analysis.

Figure 5C3 exhibits this average-cost criterion more explicitly and shows how the admissable prices translate into a region of relevant industry outputs $\underline{\hat{Y}}$. Here, if the firm's average cost curve is either *amb* or *smt*, the set $\underline{\hat{Y}}$ is composed of those values of y^I satisfying $0 < y^I \leq r$, since at every such output the demand curve DD lies at or above the horizontal at height Mm,

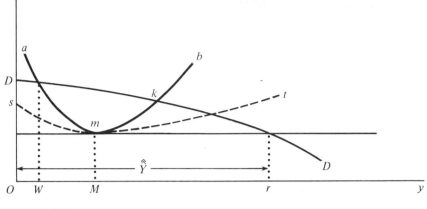

FIGURE 5C3

the minimum value of average cost. This cannot be compared directly with T, the set of potentially profitable outputs for the industry, because the average cost curves shown are those for the firm, not for the industry. Yet it is easy to see that if the firm's average cost curve is amb, then the set T excludes outputs smaller than w. For such industry output levels, the market price lies below the average costs of any firm that could produce some or all of the industry's output. However, outputs smaller than w cannot be excluded from $\hat{\hat{Y}}$, because this relevant region is defined solely in terms of the minimum level of average cost. These outputs would be included in T if the average cost curve were smt rather than amb, and both these average cost curves define the same set $\hat{\hat{Y}}$.

When we deal with a multiproduct industry, generalization of this procedure involves some increase in complexity. For each candidate industry output y^I, we first define a set $\psi(y^I)$ which is the set of points in M (the locus of outputs which minimize ray average costs) at which $P(y^I)$ yields revenues sufficient to cover costs of production. That is, it is the set of points in M which can be produced profitably at the vector of prices that the market is prepared to pay for y^I. If $\psi(y^I)$ is empty, it means that *at those prices*, $P(y^I)$, there exists no ray, that is no mix of the industry's outputs, which can be produced profitably even if that mix is produced at minimum ray average cost. Thus, if $\psi(y^I)$ is empty, y^I is clearly not financially feasible. Hence, we can restrict our attention to those industry outputs y^I for which $\psi(y^I)$ is not empty. The set of y^I for which $\psi(y^I)$ is not empty will be labeled $\hat{Y}$.

In the single-product case, M is just a point—the output corresponding to the minimum point on the average cost curve. So, if $\psi(y)$ is not empty, it *is* the set M. The construction of ψ in the multiproduct case is depicted in Figure 5C4. Suppose y^a is the industry output under consideration and that the corresponding point on the total revenue hypersurface $y \cdot P(y)$ is A.

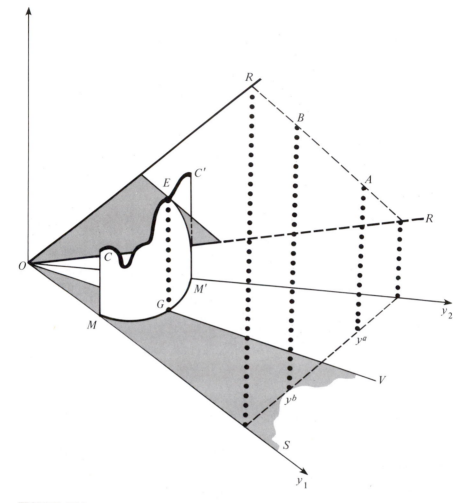

FIGURE 5C4

Let ORR be the hyperplane, passing through the origin and through point A, whose "height" above y is $y \cdot P(y^a)$. This is termed the pseudorevenue hyperplane through y^a. Let MM' be the M locus of points of minimum ray average cost, and CC' the corresponding total costs. Then CE is the portion of this "total cost cross section" which lies below ORR. Consequently, $\psi(y^a)$ is MG, the set of points on the M locus whose costs are covered at prices $P(y^a)$. Since we see that $\psi(y^a)$ is not empty, it follows that y^a is in the set $\hat{Y}$.

We can further narrow the set of candidate industry outputs to which we restrict our attention. If industry output is y^I, only a firm operating on a ray that includes a point in $\psi(y^I)$ can possibly cover its costs. For, at any other output level on that ray, a firm's average cost will be higher than at the corresponding point in $\psi(y^I)$, while its average revenue will still be the same as at the point in $\psi(y^I)$. Thus, any feasible industry output must be the sum of outputs by the firms that compose the industry represented by points on the rays that intersect ψ. That is, the industry output must be some linear combination, $\sum w_i y^i$, $w_i > 0$, of output vectors y^i in $\psi(y^I)$. Now, the graph of the set of all such linear combinations, $\sum w_i y^i$, is the convex cone generated by the portion of M, $\psi(y^I)$. That is, it is the cone composed of all rays through $\psi(y^I)$, together with the rays that lie "between" them. This convex cone is denoted by Cone $[\psi(y^I)]$.

In the single-product case, if $\psi(y^I)$ is not empty, then Cone $[\psi(y^I)]$ is the entire real line, for example, the horizontal axis to the right of the origin in Figure 5C3. The multiproduct case is again illustrated in Figure 5C4. Here, with MG depicting $\psi(y^a)$, the set of points in MM' at which cost lies on or below ORR, Cone $[\psi(y^a)]$ is the shaded region OSV. Note that here $\psi(y^a) = MG$ is a proper subset of the M locus. More important, we note that point y^a happens not to lie in Cone $[\psi(y^a)]$. But, as we have just seen, even though $\psi(y^a)$ is not empty, the fact that y^a does not lie in Cone $[\psi(y^a)]$ means that y^a cannot possibly be apportioned among firms able to cover their costs at the prices $P(y^a)$ which the market is willing to pay.

For this reason, we may choose to consider as candidate industry outputs the more restricted set $\hat{Y}$ of those output vectors y^I with two attributes: (i) $\psi(y^I)$ is not empty, and (ii) y^I lies in Cone $[\psi(y^I)]$. Again, referring to Figure 5C4, suppose now that the pseudorevenue hyperplane ORR corresponds to the candidate industry output vector y^b (instead of y^a, as before), so that $\psi(y^b)$ is still MG. Thus $\psi(y^b)$ is not empty. Moreover, a glance at the diagram confirms that y^b lies in Cone $[\psi(y^b)]$. This indicates that y^b would be included in $\hat{Y}$ and therefore would be an appropriate candidate for consideration as an industry output.

Let us now prove more formally that sets $\underline{Y}$ and $\hat{Y}$ include all feasible and efficient industry output vectors and therefore include all candidate industry outputs. We have

Proposition 5C2 *Suppose that on each output ray, a firm's ray average cost achieves a minimum. For any industry output vector y^I, define the following subset of M, the locus of outputs that achieve these minima.*

(5C2) $$\psi(y^I) \equiv \{y \in M \mid P(y^I) \cdot y \geq C(y)\}.$$

Let Cone $[\psi(y^I)]$ *denote the positive convex cone generated by the points in the set $\psi(y^I)$. That is,*

(5C3) Cone $[\psi(y^I)] \equiv \left\{ y \mid y = \sum_{i=1}^{k} \alpha_i q^i, \text{ where } \alpha_i > 0 \text{ and } q^i \in \psi(y^I) \right\}.$

Then, we can use this to define the set $\hat{\tilde{Y}}$ as in (5C4) and show[4] that it includes the entire set $\tilde{Y}$ of potentially efficient and feasible vectors of industry outputs. That is,

(5C4) $\hat{\tilde{Y}} \equiv \{ y^I \,|\, \psi(y^I) \text{ is not empty and } y^I \in \text{Cone } [\psi(y^I)] \} \supseteq \tilde{Y}.$

As a less restrictive alternative for $\hat{\tilde{Y}}$ we may adopt $\hat{Y}$ as defined in (5C5) and prove that it, too, must include $\tilde{Y}$, that is,

(5C5) $\hat{Y} \equiv \{ y^I \,|\, \psi(y^I) \text{ is not empty} \} \supseteq \tilde{Y}.$

5D. Efficient Numbers of Firms in Single-Product Industries

Economists have often drawn inferences about market structure in a single-product industry from data about market demand and the cost function of its firms.[5] In particular, the presence or absence of natural monopoly or of viable perfect competition, the polar cases of the spectrum of market structures, have been considered testable in this fashion. An industry has been said to be a natural monopoly if, as in Figure 5D1, the market's inverse demand curve crosses the firms' average cost curve in the region of increasing returns to scale, that is, the downward sloping portion of the average cost curve.

Similarly, Figure 5D2 depicts the standard test for the viability of perfect competition. Perfect competition is said to be viable if the market's inverse demand curve crosses the horizontal drawn at the competitive price p_c at a quantity y_c^I that is large compared with the output which minimizes the average cost of each firm. Further, if y_c^I is not a large multiple of M and if the natural-monopoly criterion illustrated in Figure 5D1 is not satisfied, then the industry is, in standard practice, considered a natural candidate for oligopoly status.

In this section, we describe a procedure for the analysis of the market structure of a single-product industry that makes precise and unifies these

---

[4] *Proof*   Let $y^I \in \tilde{Y}$. Then there exists a feasible and efficient industry configuration, $\{y^1, \ldots, y^m\}$, with $\sum_{j=1}^{m} y^j = y^I$. By definition, $P(y^I) \cdot y^j \geq C(y^j)$, for $j = 1, \ldots, m$. By assumption, there exists a $t_j > 0$ such that $t_j y^j \in M$, for $j = 1, \ldots, m$. Then,

$$P(y^I) \cdot (t_j y^j) \geq t_j C(y^j) \geq C(t_j y^j), \qquad j = 1, \ldots, m,$$

where the second inequality follows from the definition of M as the locus of outputs that minimize ray average cost. From this and (5C2), $t_j y^j \in \psi(y^I)$, so that $\psi(y^I)$ is not empty. And, since

$$y^I = \sum_{j=1}^{m} y^j = \sum_{j=1}^{m} (1/t_j)(t_j y^j), \qquad \text{with } t_j y^j \in \psi(y^I),$$

it follows that $y^I \in \text{Cone } [\psi(y^I)]$. Thus, (5C4) has been established, and (5C5) follows immediately.

[5] See, for example, Samuelson (1973, p. 484).

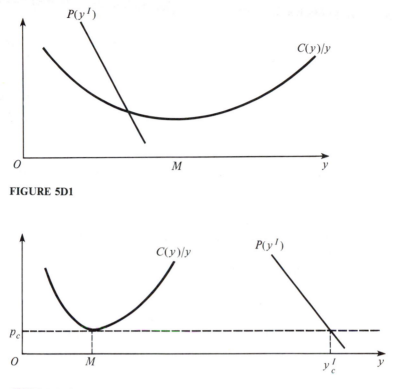

FIGURE 5D1

FIGURE 5D2

standard tests. This procedure requires cost information only about the output level that minimizes average cost (the M locus). The procedure builds on the results of Sections 5B and 5C, and it rests on the following powerful theorem.[6]

Proposition 5D1   *Assume that the average cost function $C(y)/y$ has a unique minimum at y^M, is strictly decreasing for $0 < y < y^M$, and is strictly increasing for $y > y^M$. Then the cost-minimizing number of firms for the production of the industry output y^I, denoted by $m(y^I)$, is exactly y^I/y^M if that number is an integer. In this case, $C^I(y^I) = (y^I/y^M)C(y^M)$. If y^I/y^M is not an integer, then $m(y^I)$ is either the integer just smaller or the integer just larger than y^I/y^M. Formally,*

(5D1) $$\lfloor y^I/y^M \rfloor \leq m(y^I) \leq \lceil y^I/y^M \rceil.$$

---

[6] This result has also been proved by Ginsberg (1974) using more restrictive conditions equivalent to assuming a cost function which is concave up to y^M and convex thereafter.

Here, $\lfloor x \rfloor$ denotes the largest integer which does not exceed x, and $\lceil x \rceil$ denotes the smallest integer which is not less than x. For example, $\lfloor 17.5 \rfloor = 17$, $\lceil 17.5 \rceil = 18$, $\lfloor 17 \rfloor = \lceil 17 \rceil = 17$. This notation will be helpful throughout the discussion that follows.

The proof of Proposition 5D1 is provided in Appendix I, but its basic idea is straightforward. To establish the upper bound on $m(y^I)$ given in (5D1), consider an allocation of y^I among k firms, where $k > \lceil y^I/y^M \rceil$. The average output of these firms, y^I/k, is less than y^M, and, moreover, the industry output averaged over one fewer firm, $y^I/(k-1)$, is still not greater than y^M. Consequently, it is possible to construct an allocation of the industry output among $k-1$ firms that is cheaper than the given allocation among k firms. This is accomplished by eliminating the smallest firm, which, by the assumption about the shape of the average cost curve, has the highest average cost of the firms producing less than y^M. The output of this firm is reallocated among the remaining $k-1$ firms in such a way that every firm is guaranteed to have average costs no higher than before, and at least one firm is assured of having a lower average cost. Consequently, the original allocation among k firms cannot have minimized cost, and therefore $m(y^I) \leq \lceil y^I/y^M \rceil$. An analogous argument involving the introduction, rather than the elimination, of a firm can be used to derive the lower bound on $m(y^I)$.

An example suffices to show that the assumption about the shape of the average cost curve used in Proposition 5D1 cannot be dispensed with.

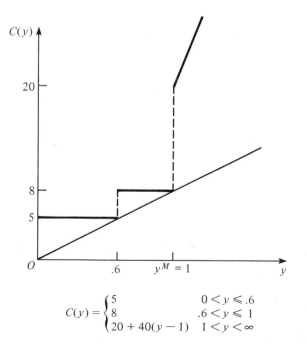

$$C(y) = \begin{cases} 5 & 0 < y \leq .6 \\ 8 & .6 < y \leq 1 \\ 20 + 40(y-1) & 1 < y < \infty \end{cases}$$

FIGURE 5D3

Figure 5D3 shows a cost function that has a unique minimum of average cost at $y^M = 1$, has rising average cost to the right of y^M, and is subadditive to the left of y^M. But, average costs are not declining for all y less than y^M, so that the assumption of Proposition 5D1 does not hold. An industry output of $1.8 = y^I$ would require $m(y^I)$ to be either 1 or 2, if (5D1) were valid for this case. However, a straightforward calculation shows that $y^I = 1.8$ is produced most efficiently by three firms, each producing .6 units.

Now we can describe the procedure we use to investigate potential market structures using only cost information about the output at which scale economies are exhausted.

Proposition 5D2[7]   *Assume that the average cost function $C(y)/y$ has a unique minimum at y^M, is strictly decreasing for $0 < y < y^M$, and is strictly increasing for $y > y^M$. Then,*

(5D2) $\qquad\qquad \lceil Q[C(y^M)/y^M]/y^M \rceil$ *is an upper bound for $\underline{R}$,*

where $Q(p)$ is the scalar market demand when the market price is p. Further,

(5D3) $\qquad$ *if $\lfloor Q[C(y^M)/y^M]/y^M \rfloor > 0$, then $\lfloor Q[C(y^M)/y^M]/y^M \rfloor \in \underline{R}$.*

This result generalizes and confirms the validity of the standard procedures for investigation of the structure of a single-product industry, while only requiring cost data about M. As illustrated in Figures 5D4 and 5D5, it instructs the analyst to find the value of industry demand at which the market price is equal to the firms' minimum average cost, $C(y^M)/y^M$. One then divides that industry demand by y^M, the output of the firm that minimizes its average cost, to find the critical number $Q[C(y^M)/y^M]/y^M$, which we shall call m^*. The proposition establishes that no feasible and efficient industry configuration has more than $\lceil m^* \rceil$ firms and, equally important, that there is such a configuration with $\lfloor m^* \rfloor$ firms.

Consequently, if m^* is large, the industry is likely to be naturally competitive because there are feasible and efficient configurations with as many as $\lfloor m^* \rfloor$ firms. If, at the other extreme, $0 < m^* < 1$, as in Figure 5D5, the industry

---

[7] *Proof*   Working from Proposition 5C2 and definitions (5C2) and (5C5), since the M locus is the single point y^M, y^I is in $\hat{Y}$ if and only if $P(y^I)y^M \geq C(y^M)$, iff $P(y^I) \geq C(y^M)/y^M$, and iff $y^I \leq Q[C(y^M)/y^M]$. Hence, as represented in Figures 5D4 and 5D5, $\hat{Y}$ is the set of y's between 0 and $Q[C(y^M)/y^M]$. Then, by (5D1), for $y^I \in \hat{Y}$, $m(y^I) \leq \lceil Q[C(y^M)/y^M]/y^M \rceil$. Finally (5D2) follows from (5C5) and Proposition 5C1. That (5D3) holds follows because, if

$$\lfloor Q[C(y^M)/y^M]/y^M \rfloor > 0,$$

then $x \equiv y^M \lfloor Q[C(y^M)/y^M]/y^M \rfloor \leq Q[C(y^M)/y^M]$, so that $P(x) \geq C(y^M)/y^M$, and $P(x)y^M \geq C(y^M)$. $C^I(x) = \lfloor Q[C(y^M)/y^M]/y^M \rfloor C(y^M)$ by Proposition 5D1, so that $\lfloor Q[C(y^M)/y^M]/y^M \rfloor$ firms, each producing y^M, is a feasible and efficient industry configuration. Then (5D3) follows by definition of $\underline{R}$.

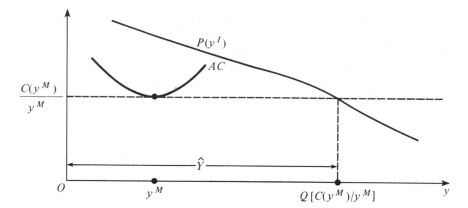

FIGURE 5D4

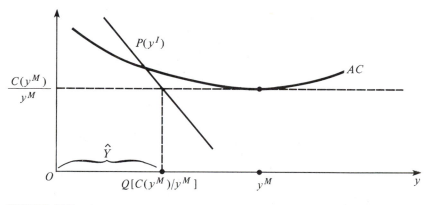

FIGURE 5D5

is a natural monopoly because there are no feasible and efficient configurations involving more than one firm. [However, it must be noted that there is no assurance in this case that there are *any* feasible configurations. It depends on whether $P(y^I)$ crosses $AC(y^I)$, and this cannot be deduced from information about only M.] And, if m^* is intermediate between these extremes, not only can we infer that the industry is a natural oligopoly, but we know where in the spectrum of market structures it lies. We know that the industry cannot contain more than $\lceil m^* \rceil$ firms in a feasible and efficient configuration, and we know that it can contain $\lfloor m^* \rfloor$ firms.[8]

---

[8] This completes the discussion of the cost-minimizing *number* of firms in the single-product case. However, something is also known about the way in which industry output must be divided among firms of the single-product industry. This is discussed in Appendix II.

5E. Efficient Numbers of Firms in Multiproduct Industries: Preliminary Exposition

As is generally recognized, in the multiproduct case it is not easy to pinpoint the set of products that delineate the industry. In economic theory an industry is usually defined to offer a set of products that are closely related (e.g., they may be substitutes) in consumption but not necessarily in production (e.g., pens, pencils, and typewriters). Alternatively, in accord with business usage, it may involve outputs closely related in production but not necessarily in consumption (e.g., automobiles and trucks). We shall see that the choice between these two criteria does make some difference for our analysis. For now, we therefore simply take the industry to be composed of all suppliers of all products that are sufficiently closely related on one or the other of these criteria.

Having seen in Proposition 5D2 the powerful analysis of the number of firms that minimizes industry cost made possible by the assumption that average costs first decline, achieve their minimum, and then rise, we will now explore the consequences of the analogous premise about the costs of multi-output production. Thus, throughout this section we employ

Assumption 5E1   *For each ray in the space of outputs, ray average cost achieves a unique minimum. For any point y in M, the set of outputs at which these minima are achieved, ray average costs are decreasing at ty, $0 < t < 1$, and ray average costs are increasing at ty, for $t > 1$.*

As mentioned in Chapter 3, in multi-output production, unlike the scalar output case, increasing returns to scale and declining ray average costs do not suffice for cost subadditivity. Consequently, hoping to establish results that approach the power of those proven for single-product industries, we here *assume* that cost subadditivity holds throughout a portion of the region of outputs that exhibit nondecreasing returns to scale, the region "inside" and on the M locus.

Since the M locus can sometimes be very irregular, it is useful to simplify the shape with which we deal by enclosing the M locus in a region whose shape is more regular. For this purpose, we form the smallest closed convex region which contains the entire M locus. That region is called the *convex closure* of M, and we use the symbol $[\ddot{M}]$ to denote it.

We need only assume that cost subadditivity holds throughout the region "inside" the convex closure of M. This region, denoted by $[\ddot{M}]^-$, is defined formally as

(5E1) $[\ddot{M}]^- \equiv \{y \geq 0 \,|\, \lambda y \in [\ddot{M}] \text{ for some } \lambda \geq 1,$

$\text{and } \lambda < 1 \text{ implies } \lambda y \notin [\ddot{M}]\}.$

Assumption 5E2   *The firm's cost function $C(y)$ is strictly subadditive in $[\ddot{M}]^-$. That is, if y^1, y^2, and $y^1 + y^2$ are all in $[\ddot{M}]^-$, then $C(y^1 + y^2) < C(y^1) + C(y^2)$.*

It must be noted here that this subadditivity over $[\ddot{M}]^{-}$ can be derived from Assumption 5E1, together with one of a variety of additional cost conditions. Because this subject is treated in detail in Chapter 7 (see also ten Raa, 1980), here we shall simply adopt subadditivity as an assumption.

Given Assumptions 5E1 and 5E2, one might expect a result analogous to Proposition 5D2 to hold, where the relevant standard of comparison is an output point on the M locus. In terms of Figure 5E1 one might expect that $m(y^I)$ is bounded between $\lceil 1/t \rceil$ and $\lfloor 1/t \rfloor$, where ty^I is the output vector that minimizes ray average cost along the ray on which y^I lies. Unfortunately, this is not the case.

Consider the cost function given by

(5E2) $C(y_1, y_2) = \begin{cases} 1 + 2\min(y_1, y_2), & \text{if } y_1 + y_2 \leq 1 \\ (1 + \varepsilon)(y_1 + y_2) - \varepsilon + 2\min(y_1, y_2), & \text{if } y_1 + y_2 > 1 \end{cases}$

where ε is a nonnegative parameter. Figure 5E2 displays this cost function. Here, $M = \{y \mid y_1 + y_2 = 1, y_1 \geq 0, y_2 \geq 0\}$, and Assumption 5E1 is satisfied; that is, ray average costs are U-shaped on each ray. $C(y)$ is subadditive inside the M locus, since, in this region, $1 \leq C(y) \leq 2$ for single-firm production while the cost of production by k firms is greater than or equal to k. Outside the M locus there are diseconomies of scope at some output points. Calculation shows that for $0 < \varepsilon < \delta$, and δ small, $m = 2$ when industry output is $(2 - \delta, 2 - \delta)$ because, in this case, ray average costs increase very slowly past the M locus so that industry costs are minimized with two specialized firms, each producing $2 - \delta$ units. Yet, for δ small, the point

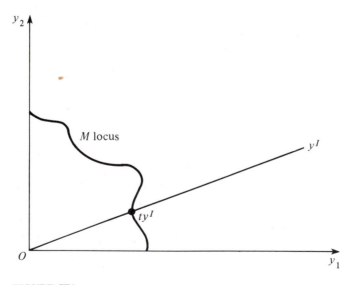

FIGURE 5E1

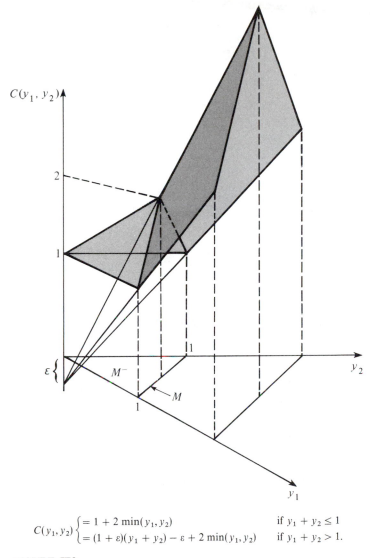

$$C(y_1, y_2) \begin{cases} = 1 + 2 \min(y_1, y_2) & \text{if } y_1 + y_2 \leq 1 \\ = (1 + \varepsilon)(y_1 + y_2) - \varepsilon + 2 \min(y_1, y_2) & \text{if } y_1 + y_2 > 1. \end{cases}$$

FIGURE 5E2

$y^I = (2 - \delta, 2 - \delta)$ is almost a quadruple of the point $(\frac{1}{2}, \frac{1}{2})$ which minimizes RAC along the relevant ray. Similarly, for ε large, and for small $\delta > 0$, ray average costs increase rapidly past M so that when $y^I = (1 + \delta, 1 + \delta)$, $m = 4$.[9] Yet, the point $y^I = (1 + \delta, 1 + \delta)$ is only a bit more than a double of the point $(\frac{1}{2}, \frac{1}{2})$ which minimizes RAC along the relevant ray.

---

[9] In an optimal allocation, two firms will produce $((1 + \delta)/2, 0)$, and two firms $(0, (1 + \delta)/2)$.

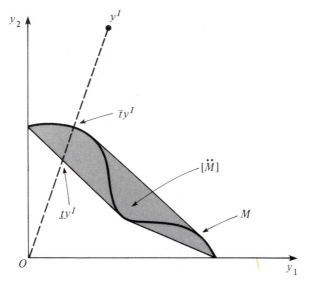

FIGURE 5E3

These examples show that, under only Assumptions 5E1 and 5E2, bounds on $m(y^I)$ as tight as those given by Proposition 5D2 do not exist for multi-output production. However, under these assumptions we can derive looser general bounds on $m(y^I)$ that require information about only the location of the M locus. And, these bounds are shown to be as tight as possible in some cases, so that there are no other bounds that are uniformly better.

The determination of the bounds on the optimal number of firms for the industry rests on constructions that are essentially geometric.

In Figure 5E3 the heavy curve is the M locus, and the shaded region is its convex closure $[\ddot{M}]$. The point $\bar{t}y^I$ is the furthest point from the origin on the ray containing y^I that is in the set $[\ddot{M}]$, and $\underline{t}y^I$ is the closest point to the origin on that ray that is in the set $[\ddot{M}]$. In the case depicted, $\bar{t}y^I$ is on the M locus itself, but generally, neither $\bar{t}y^I$ nor $\underline{t}y^I$ need be in M.

5F. The Basic Theorems for the Multiproduct Case

We can now relate the concepts we have just discussed to the optimal number of firms for the industry.

Proposition 5F1 *Let* $[\ddot{M}]$ *denote the convex closure of the set M, and for a given industry output vector y^I, let*

$$\bar{t} \equiv \max\{t \mid ty^I \in [\ddot{M}]\}$$
$$\underline{t} \equiv \min\{t \mid ty^I \in [\ddot{M}]\}.$$

Then, Assumption 5E1 implies that

(5F1) $m(y^I) > 1/2\bar{t}$ *and* $m(y^I) \geq \lceil 1/2\bar{t} \rceil$.

And, Assumptions 5E1 and 5E2 together imply that

(5F2) $m(y^I) = 1$ *if* $\underline{t} \geq 1$.

Otherwise, $1 \leq m(y^I) < 2/\underline{t}$ and $1 \leq m(y^I) \leq \lfloor 2/\underline{t} \rfloor$.

A rigorous proof of this proposition is left to Appendix III. But here we can discuss its logic intuitively. Roughly speaking, the proposition tells us that the cost-minimizing number of firms in the industry must be such that the "average-sized" firm is sufficiently close to the M locus. Specifically, the requirement that the number of firms $m(y^I) > 1/2\bar{t}$ is clearly tantamount to the requirement that the average output $y^I/m(y^I)$ be less than $2\bar{t}y^I$. Similarly, the condition for the multifirm case that $m(y^I) < 2/\underline{t}$ amounts to the requirement that the average firm's output satisfy $y^I/m(y^I) > \underline{t}y^I/2$.

Let us see the logic behind the first of the inequalities. Figure 5F1, which is based on the preceding diagram, shows H, the supporting hyperplane at

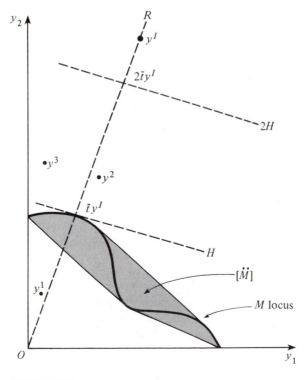

FIGURE 5F1

point $\bar{t}y^I$ of $[\ddot{M}]$, the (shaded) convex closure of the M locus. It also shows the hyperplane $2H$ which is parallel to H and lies twice as far from the origin as H.

Now, *any* firm i in the cost-minimizing solution must have an output y^i which lies below hyperplane $2H$. For, otherwise, y^i could be divided in half and produced by two different firms at lower cost, since $y^i/2$ must then still lie above the M locus, but closer to that locus than y^i. Because each firm's output must lie below $2H$ (three such points are shown in the figure), the same must be true of the point representing the "average" output $\sum y^i/m(y^I) = y^I/m(y^I)$. But $y^I/m(y^I)$ must also lie on the ray R through industry output y^I. Since the intersection of R with $2H$ is $2\bar{t}y^I$, it follows that $y^I/m(y^I) < 2\bar{t}y^I$ or $m(y^I) > 1/2\bar{t}$. This, then, is how one derives the upper bound on the size of the average firm, and the corresponding lower bound on the cost-minimizing number of firms in the industry.

The logic underlying the other bound is similar. Figure 5F2 shows L, the supporting hyperplane to $[\ddot{M}]$ at $\underline{t}y^I$, and $L/2$, the parallel hyperplane which lies midway between L and the origin. By the assumed subadditivity of the

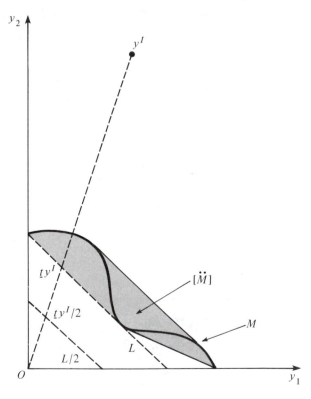

FIGURE 5F2

cost function in the region below the set $[\ddot{M}]$, no two firms in a cost-mini-mizing configuration can have outputs y^i and y^j both below $L/2$. For other-wise, their (vector) sum would lie below L and, hence, below $[\ddot{M}]$. Thus, the larger combined firm could produce the same total output $y^i + y^j$ at a lower total cost than it can be done by firms i and j. The similarity of the remainder of the argument to that in the preceding paragraph should be clear. The formal proof of our proposition essentially replicates the steps which have just been described in our intuitive argument.

Unfortunately, the range of values of $m(y^I)$ permitted by the preceding result is still rather broad. This is true even when the M locus happens to assume a shape that tends to minimize this range. It is clear that the upper and lower bounds on $m(y^I)$ given by Proposition 5F1 are closest together when M happens to be a hyperplane because then M is itself a convex set, so that $M = [\ddot{M}]$ and $\bar{t} = \underline{t} \equiv t^*$. However even in this case, (5F1) and (5F2) together give $\lceil 1/2t^* \rceil \le m(y^I) \le \lfloor 2/t^* \rfloor$. This seems to be a very broad range for $m(y^I)$ to lie in when it is compared with the range established earlier for the scalar output case. There, it will be recalled, the corresponding relation-ship was $\lfloor 1/t^* \rfloor \le m(y^I) \le \lceil 1/t^* \rceil$. Yet, we have this result:

Proposition 5F2 *There exist cases in which the lower bound of (5F1) is realized and cases in which the upper bound of (5F2) is realized.*

Cases needed to establish the theorem are provided by the example given in (5E2) and depicted in Figure 5E2. There, it will be recalled, for $0 < \varepsilon < \delta$ and δ small, $m(2 - \delta, 2 - \delta) = 2$. For $y^I = (2 - \delta, 2 - \delta)$, $\bar{t}(2 - \delta) = \frac{1}{2}$ because the point $(\frac{1}{2}, \frac{1}{2})$ minimizes RAC along the ray through $(2 - \delta, 2 - \delta)$, and because the M locus is a hyperplane. Then $\lceil 1/2\bar{t} \rceil = \lceil 2 - \delta \rceil = 2$, and the lower bound of (5F1) is equal to $m(y^I)$.

It will be recalled also that for large ε and small $\delta > 0$, $m(1 + \delta, 1 + \delta) = 4$. Here, $\underline{t}(1 + \delta) = \frac{1}{2}$ because M is still the same hyperplane through $(\frac{1}{2}, \frac{1}{2})$. Then $\lfloor 2/\underline{t} \rfloor = \lfloor 4 + 4\delta \rfloor = 4$, and the upper bound of (5F2) is equal to $m(y^I)$.

Q.E.D.

Thus, the bounds given in Proposition 5F1 cannot be uniformly im-proved, despite the very limited information on which they are based.

We now have the means to provide the full specifications of our informa-tion-economizing approach to the determination of efficient and feasible structures for multiproduct industries. Let $\underline{Y}$ be the relevant region of industry outputs, as determined by one of the techniques described earlier in Section 5C. Let $\bar{m}$ and $\underline{m}$, respectively, be the largest of the upper bounds and the smallest of the lower bounds given by the application of Proposition 5F1 to the various output vectors in the set $\underline{Y}$. Then, it follows from Proposi-tion 5B1 that $\bar{m}$ and $\underline{m}$ are bounds on the set $\underline{R}$. That is, the number of firms in any efficient and feasible industry configuration must lie between $\bar{m}$ and $\underline{m}$. Consequently, if $\underline{m}$ is large, then efficient and feasible industry operation

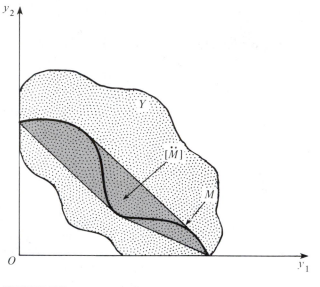

FIGURE 5F3

must involve a large number of firms. And, if $\bar{m}$ is small, then only a small number of firms can comprise a configuration that is feasible and efficient.

The representation of the bounds on $m(y^I)$ that is given by Proposition 5F1 offers considerable insight into the basic economic determinants of the sizes of $\bar{m}$ and $\underline{m}$. If, as depicted in Figure 5F3, the relevant region of industry outputs $\underline{Y}$ lies close to the convex closure of the M locus, along the rays through $\underline{Y}$, then $\bar{m}$ will be small. Intuitively, this is so because in this case just a few firms of approximately cost-minimizing size can produce any relevant industry output. This is the analog of the single-product case in which every industry output in the relevant range is close to y^M, the cost-minimizing output level for the firm.

The extreme case is that in which the relevant region, $\underline{Y}$, lies entirely in $[\ddot{M}]^-$, inside the M locus. Then, by Assumption 5E2, the industry is a natural monopoly. On the other hand, if as depicted in Figure 5F4, $\underline{Y}$ were to lie very far from $[\dot{M}]$ along the rays through $\underline{Y}$, then $\underline{m}$ would be large.

The reason for this relationship is clear. The larger the potential demand relative to the cost-minimizing firm size, as dictated by the available techniques, the larger the number of firms that will be optimal. Specifically, it implies that, given industry technology, *monopoly will tend to be promoted by small-scale industry operations*, that is, by low output demands (at prices that cover costs) relative to the cost-minimizing scale of the individual firm. This relationship also suggests that the evolution of the cost-minimizing market form of an expanding industry with the passage of time is determined by a race between the growth in demand for its products and the evolution

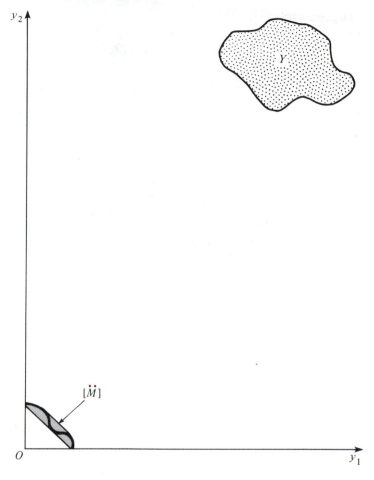

FIGURE 5F4

of its production techniques. We may expect that expansion in industry demand over time will call for increasing competitiveness unless innovation leads to increases in the cost-minimizing size of firms that are at least proportionate to the expansion in demands.

5G. The Large-Numbers Case and Pure Competition

It is tempting to conclude from a large value of m that, because efficient and feasible industry operation requires a large number of firms, the industry must be naturally competitive. This would certainly be a valid inference in an industry with only a single product. However, even if a multiproduct industry contains a large number of firms, it is possible that some particular good, or

some particular subset of goods in the industry's product line is actually
produced by only a small number of firms.

In fact, later, in Chapter 11, we will deal with an analytically significant
case in which this can be expected to occur. There, we consider a good whose
production involves product-specific scale economies globally, and which is
linked by economies of scope to the production of a second good whose
product-specific scale economies are exhausted at a small scale. In such
cases, efficient industry operation may well require each of a great number of
firms to produce some of the second good. Taken by itself, this product may
be said to be supplied in a naturally competitive fashion. However, efficiency
will also require that the first good be supplied by only one firm. Thus, an
industry of this kind is not naturally competitive because one of its firms is
the sole supplier of the first product. Yet, in such a case, all relevant industry
outputs may lie far along their rays from the convex closure of M.

Consequently, the tools that have been described so far are not sufficiently
powerful to determine conclusively when an industry is naturally competi-

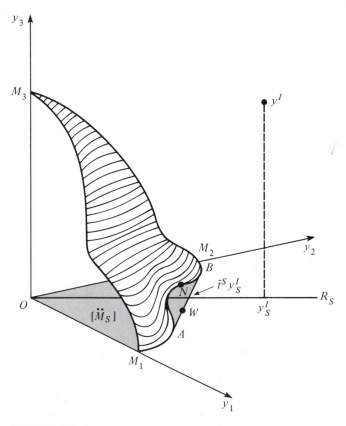

FIGURE 5G1

tive. While a small value of $\bar{m}$ indicates that an industry is not naturally competitive, a large value of $\underline{m}$ leaves open the possibility that some of the industry's outputs are most efficiently supplied by a number of firms too small to ensure pure competition. However, the following result remedies this deficiency by providing a lower bound on the number of firms in an efficient industry configuration that produce positive quantities of some of the goods in any given subset S. This lower bound is derived by the same method that was prescribed by Proposition 5F1, but it is now applied in the subspace corresponding to the subset S, making use of the projections of y^I and the M locus in this subspace.

Formally, let $m_S(y^I)$ denote the number of firms that produce positive quantities of some of the goods in the subset S in an efficient configuration of the industry with total output y^I. For any vector $y \in \mathbf{R}^n$, and for $S \subseteq \{1, 2, \ldots, n\}$, let y_S denote the projection of y in the subspace corresponding to the subset S. That is, the ith component of y_S is y_i for $i \in S$, and is 0 for $i \notin S$. Finally, let M_S denote the projection of the M locus. That is,

$$M_S \equiv \{ y_S \in \mathbf{R}^n | y \in M \}.$$

Proposition 5G1[10]   *Assumption 5E1 implies that*

(5G1) $$m_S(y^I) > \frac{1}{2\bar{t}^S} \quad and \quad m_S(y^I) \geq \left\lceil \frac{1}{2\bar{t}^S} \right\rceil$$

where

(5G2) $$\bar{t}^S \equiv \max\{t \,|\, ty_S^I \in [\ddot{M}_S]\}.$$

The construction of $m_S(y^I)$ prescribed by the proposition is depicted for a three-product case in Figure 5G1, where $S = \{1, 2\}$ and y^I entails production of goods 1, 2, and 3. Here y_S^I is the projection of y^I onto the floor of the diagram (y_1, y_2 space), and region OM_1NM_2 is the projection of the three-dimensional M locus, $M_1M_2M_3$. The convex closure $[\ddot{M}_S]$ of that projection is the shaded region OM_1WM_2, which is the projection of the M locus together with the addition of the area $AWBN$. Next, we draw in ray R_S through y_S^I and find point $\bar{t}^S y_S^I$, the outermost point common to ray R_S and $[\ddot{M}_S]$. Then $1/2\bar{t}^S$ provides the desired lower bound on the number of

---

[10] *Proof*   The point $\bar{t}^S y_S^I$ is constructed to lie on the outside boundary of the convex set $[\ddot{M}_S]$. Then, let $\bar{h}^S$ and $\bar{k}^S$ define a separating hyperplane with $\bar{h}^S \bar{t}^S y_S^I = \bar{k}^S$ and $\bar{h}^S \cdot y \leq \bar{k}^S$ for $y \in [\ddot{M}_S]$. As in the proof of Proposition 5F1, $\bar{k}^S > 0$. Let $\{y^1, \ldots, y^m\}$ be an industry cost-minimizing partition of y^I, and let $t^i y^i \in M$. As in the proof of Proposition 5F1, it follows that $t^i > \frac{1}{2}$. Then, $t^i y_S^i \in M_S \subseteq [\ddot{M}_S]$ so that $\bar{h}^S t^i y_S^i \leq \bar{k}^S$, $\bar{h}^S y_S^i \leq \bar{k}^S/t^i < 2\bar{k}^S$, and $\bar{h}^S y_S^i < 2\bar{k}^S$. A summation of these inequalities over $\{i | y_S^i \neq 0\}$ omits only firms producing none of the goods in S, and hence yields $\bar{h}^S y_S^I < 2\bar{k}^S m_S(y^I)$. Thus, $m_S(y^I) > \bar{h}^S y_S^I/2\bar{k}^S = 1/2\bar{t}^S$, as required.   Q.E.D.

firms producing goods in S. Proposition 5G1 provides a lower bound on the number of firms that produce at least one of the goods in the set S in efficient industry configurations. This can be useful in assessing the degree of competition in the industry if the goods that comprise S are close substitutes in demand. In this case, the potential market power of a firm marketing one of the goods in S can be presumed to be constrained effectively if there are many other firms supplying any of the goods in S.

However, to determine whether we are dealing with a case of pure competition, we require a lower bound upon the numbers of firms producing *each* of the industry's goods. To test for this, we can repeat the procedure that has just been described, this time letting S contain just one product at a time. That is, to determine whether a large number of firms is required to produce some good i efficiently, we define S to contain only i and use Proposition 5G1 to determine the lower bound on the number of firms.

With only one product contained in S, we can simplify our notation to write for the number of firms producing i, $m_i(y^I) \equiv m_S(y^I)$, where $S = \{i\}$, and we have this special version of Proposition 5G1.

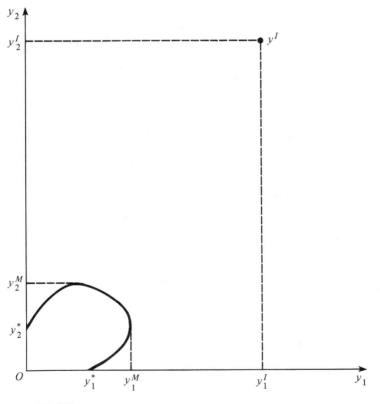

FIGURE 5G2

Corollary   *Assumption 5E1 implies that*

(5G3) $$m_i(y^I) > y_i^I/2y_i^M \quad and \quad m_i(y^I) \geq \lceil y_i^I/2y_i^M \rceil$$

where $y_i^M = \max\{y_i \mid y \in M\}$.

Figure 5G2 depicts this construction where y^I entails production of goods 1 and 2. It should be noted that in this case if y_1 were produced entirely alone, the output that minimized its average cost would be y_1^*. But for a producer of both goods, the largest output of y_1 that lies on the M locus is y_1^M, which is considerably greater than y_1^*. A similar relationship holds for y_2. This is so because a firm engaged in joint production in this industry has a larger efficient scale. It is essential to note that it is y_1^M and *not* y_1^* that must be used in (5G3) to determine the minimum number of firms engaged in the production of commodity 1. Criterion (5G3) tells us that efficient

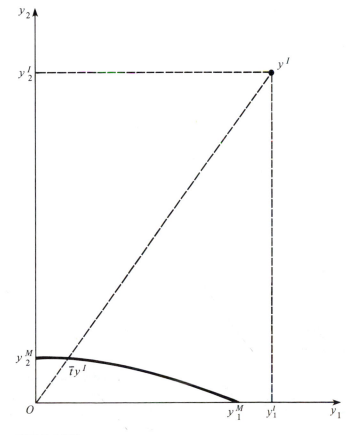

FIGURE 5G3

production requires good 1 to be produced by a number of firms equal to at least half the ratio of industry output y_1^I to y_1^M.

Figure 5G3 represents a case in which the ratio y_2^I/y_2^M is much larger than y_1^I/y_1^M, so that $m_2(y^I)$, the number of firms producing good 2, has a relatively large lower bound, while $m_1(y^I)$ has a small one. Here, industry efficiency may require many firms to produce good 2 but only one firm to produce good 1. Of course, that firm may well also produce some of good 2.

Now we are finally equipped with tools sufficiently powerful to identify categorically cases in which an industry is naturally competitive. This will be true if each component of y^I, y_i^I, for all y^I in the relevant region, is very large relative to y_i^M. This condition assures us that all efficient and feasible industry configurations involve large numbers of firms producing each and every good supplied by the industry. Chapter 9 will provide a detailed analysis of the equilibria that characterize multiproduct industries that are naturally competitive.

5H. Determinants of the Form of the M Locus

This chapter has shown the critical role played by the M locus in determining the range of market structures compatible with efficiency and financial feasibility of the industry's operations. In particular, Propositions 5F1 and 5G1 show clearly the significance of the economic and analytic implications of the shape of the M locus. Thus, it is useful to examine the economic influences that determine the location and shape of M. The relationships between factor prices and M will be investigated later, in Chapter 6. Here, we focus upon the influence of technological characteristics of the industry and upon relationships among the products that serve to define the industry.

It is helpful to consider the M locus to be anchored at the points $(0, \ldots, 0, y_i^*, 0, \ldots, 0)$ where it intersects the axes. At such a point, scale economies are exhausted in *stand-alone production* of good i, whose output is represented on the corresponding axis. At each such point the average and marginal cost of the nonzero output, y_i, are equal, so that $C(0, \ldots, 0, y_i^*, 0, \ldots, 0) = y_i^* C_i(0, \ldots, 0, y_i^*, 0, \ldots, 0)$.

We will now examine several cases involving, respectively, independence, complementarity, and substitutability in production of the goods in question, to determine how such relationships affect the shape of the M locus.

Case 1: Independence in Production. The simplest benchmark case is that in which there are neither economies nor diseconomies of scope, so that productive inputs are completely specialized by product and there is strong independence among the costs of the products. Here, $C(y) = \sum_i C(0, \ldots, 0, y_i, 0, \ldots, 0)$, so that $C(y^*) = \sum_i y_i^* C_i(0, \ldots, 0, y_i^*, 0, \ldots, 0) = \sum_i y_i^* C_i(y^*)$. Thus $C(y)$ must be (locally) linearly homogeneous at y^* and RAC must be

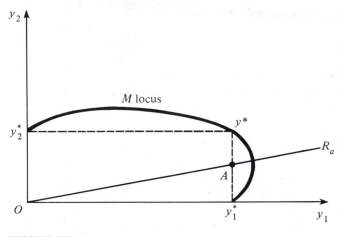

FIGURE 5H1

constant there. Consequently, in this case, the M locus must pass through y^*, which entails production of that quantity of each of the goods that minimizes its own stand-alone average cost. This suggests that the M locus will have the shape of an inverted L. Surprisingly, however, despite the independence in production of the goods, the M locus must bow outward as in Figure 5H1 because, as discussed in Chapter 4, in the absence of economies of scope, the index of overall scale economies S_N is a weighted average of S_1 and S_2, the indices of scale economies of products 1 and 2, respectively. Thus, at an intermediate point, such as point A on the vertical line segment $y_1^* y^*$ in Figure 5H1, economies of scale in the production of good 2 will not yet be exhausted, so that there $S_2 > 1$ and, even though $S_1 = 1$, S_N will also exceed 1. Only by going further than point A along its ray R_a does one reach the point where $S_N = 1$, that is, the point at which R_a and the M locus intersect.

We can show[11] that where there are economies (diseconomies) of scope, y^* will be outside (inside) the M locus, and it will be further outside (inside)

[11] *Proof*   By a bit of algebraic manipulation it follows that

$$C(y^*) - \sum_i C_i(y^*)y_i^* = C(y^*) - \sum_i \frac{C_i(y^*)C(0, \ldots, 0, y_i^*, 0, \ldots, 0)}{C_i(0, \ldots, 0, y_i^*, 0, \ldots, 0)}.$$

Then, if there is (almost) total independence among the variable costs of the products, so that $C_i(y^*) \approx C_i(0, \ldots, 0, y_i^*, 0, \ldots, 0)$, it follows that

$$C(y^*) - \sum_i C_i(y^*)y_i^* \approx C(y^*) - \sum_i C(0, \ldots, 0, y_i^*, 0, \ldots, 0).$$

Since $C(y^*) - \sum C(0, \ldots, 0, y_i^*, 0, \ldots, 0)$ is negative and greater in absolute value the larger the economies of scope, and $C(y^*) - \sum C_i(y^*)y_i^*$ increases with the degree of economies of scale, the result follows.

the M locus the greater are the economies (diseconomies) of scope that inhere in the part of costs that are fixed with respect to the scale of outputs but that vary with the set of goods produced.[12] The intuitive explanation is that, for a given value of y^*, by definition, economies of scope of this sort mean that fixed cost will be lower relative to output in a firm that produces the goods jointly, and lower fixed cost implies that (*ceteris paribus*) economies of scale will be exhausted at lower levels of output.

Case 2: Perfect Complementarity. At the opposite extreme from the case of strong independence of the costs of the different products, in which all inputs are completely specialized by use, is the case of perfect complementarity. In this case, all inputs are shared without congestion among the processes used, so that once an input is purchased to produce one good i, that input is readily usable in the production of the others without loss in its capacity to produce i. Such a function may characterize the total cost when the same capital facilities are used to supply a given service to customers at different times. Here, the maximum value of y_i is the peak load, and the other components of y are the off-peak loads. If the stand-alone production techniques are the same for all goods, $C(y) = \max_i C(0, \ldots, 0, y_i, 0, \ldots, 0)$. That is, the firm must purchase input quantities sufficient to produce its largest output, and all its other goods can be produced without additional cost. It follows that an output vector y will minimize ray average cost if its largest component is at the level that minimizes its own average stand-alone cost; i.e., $y \in M$ iff $C(0, \ldots, 0, y_i^*, 0, \ldots, 0) = \max_j C(0, \ldots, 0, y_j, 0, \ldots, 0)$.

---

[12] These costs fixed with respect to scale enter the matter in the following way. Complete independence clearly requires a cost function of the form $C^* = \sum f_i(y_i)$. However, it is easily seen that our argument also holds for a more general class of cost functions, $C = \sum f_i(y_i) + F(S)$ where S is the set of products offered by the firm. It is clear that the value of F does not vary with the magnitude of the outputs of particular products and that it is compatible with economies of scope. Now when S consists of the single product i, at y_i^* marginal cost pricing just covers total cost so that

(a) $f_i'(y_i^*)y_i^* = F(i) + f_i(y_i^*).$

However, when S consists of the entire set of products, N, we have

(b) $C(y^*) - \sum y_i^* C_i(y^*) = \sum f_i(y_i^*) + F(N) - \sum y_i^* f_i'(y_i^*)$

or, in view of (a), it follows that (b) becomes

$$C(y^*) - \sum y_i^* C_i(y^*) = F(N) - \sum F(i).$$

With economies of scope the right-hand side is negative, and more so the greater the economies of scope. Thus, total cost increasingly falls short of the revenues from marginal cost pricing and the greater are the diseconomies of scale.

Therefore, again letting y^* be the vector of the output quantities y_i^*, each of which minimizes its stand-alone average cost, once more we have $y^* \in M$. This time, however, the M locus must be rectangular, like that depicted by the dashed lines in Figure 5H1. These features of the M locus are unaltered if, in addition to the shared inputs, there are specialized inputs required in fixed proportion to the output quantities. In that case,

$$C(y) = \sum_i c_i y_i + f\left[\max_i (y_i)\right].$$

Irrespective of the global behavior of costs, the M locus will curve outward from its points on the axes, $(0, \ldots, 0, y_i^*, 0, \ldots, 0)$, as in Figure 5G2, if there are cost complementarities in the vicinities of these points.[13] Consequently, when there are cost complementarities, $y_i^M > y_i^*$, where, as before, y_i^M is the largest output of good i corresponding to a point on the M locus. Then, the entire M locus must be utilized in applying Proposition 5G1. That is, the calculations of the lower bounds on the numbers of firms producing each individual good must take joint production into account and not be based solely on y_i^*, the output of good i that minimizes its stand-alone average cost. We also conclude that where cost complementarities are present, the M locus cannot be convex to the origin.

Case 3: Perfect Transferability of Inputs. However, the M locus need not be strictly concave to the origin in the way depicted in Figures 5G2 and 5H1. In particular, M will take the form of a hyperplane where all inputs are perfectly transferable from one output to another. If they can be transferred from production of y_1 to that of y_2, and the ratio of y_2/y_1 thereby increased steadily without any diminishing returns, and if the same holds for all other pairs of products, then the cost function must have the form $C = \phi[y_1 + \sum_{i=2}^n a_i y_i]$. In this case, the M locus is the hyperplane $\{y \in \mathbf{R}_+^n \mid y_1 + \sum_{i=2}^n a_i y_i = k\}$, where $k\phi'(k) = \phi(k)$.

Here, $\phi(t)$ is the cost of producing t units of the first good alone, and $1/a_i$ units of good i can be substituted costlessly for one unit of good 1. $\phi(t)$ can be regarded as the cost of operating the basic production process, and

---

[13] *Proof* The M locus is defined implicitly by the equation $C(y) - \sum C_i(y)y_i = 0$. Total differentiation gives $\sum_i dy_i \sum_j C_{ij}(y)y_j = 0$. At $(0, \ldots, 0, y_i^*, 0, \ldots, 0)$, then,

$$C_{ii}\, dy_i + \sum_{j \neq i} C_{ij}\, dy_j = 0$$

characterizes the shape of the M locus. $C_{ii} > 0$ follows from the fact that $C(0, \ldots, 0, y_i, 0, \ldots, 0)/y_i$ is minimized at y_i^*. And cost complementarities mean that $C_{ij} < 0$, $i \neq j$. Hence, as the M locus moves away from $(0, \ldots, 0, y_i^*, 0, \ldots, 0)$, it must be true that $dy_i > 0$, since $dy_j > 0$ for some $j \neq i$ and $dy_j \geq 0$ for all $j \neq i$.

the average cost of the process is minimized at the level of output $y_1 = k$. From this viewpoint, the perfect transferability of the inputs means that the output goods are perfect substitutes in production.

Somewhat more generally, the M locus is a hyperplane for the cost function

$$C(y) = \sum c_i y_i + \phi \left[y_1 + \sum_{i=2}^{n} a_i y_i \right].$$

Here, the outputs require the provision of specialized inputs that cost c_i per unit. But, at the same time, they require operation of the process that incurs the cost ϕ in which they are perfect substitutes. Thus, the M locus must be a hyperplane when the outputs are perfect substitutes in that portion of the production process with scale effects, that is, where the capacity of that portion of the production process is perfectly transferable among outputs.[14]

Case 4: M Loci Convex to the Origin. However, if substantial coordination and administrative costs are incurred by firms, and if these costs grow with both the size and diversity of outputs, then even if physical capacity is perfectly transferable, we may expect the M locus to curve in toward the origin. For, in such a case, multiproduct firms may grow more costly to manage at smaller scales of output than do specialized enterprises. (This may be the reason larger farms seem to tend toward specialization even where the land is equally suited to several crops.)

5I. The Shape of the M Locus and the Strength of Its Implications

The case of perfectly substitutable outputs is well adapted to analyses based on Propositions 5F1 and 5G1. Here, because M is a hyperplane, $[\ddot{M}] = M$ and $\bar{t} = \underline{t}$ so that the bounds on $m(y^I)$ given in (5F1) and (5F2) are as close together as possible. Furthermore, we necessarily have $y_i^* = y_i^M$ in this case, because, unlike the case depicted in Figures 5G2 and 5H1, the M locus does not bend back (it does not have a positively sloping segment followed by a negatively sloping segment). Therefore, the output yielding

---

[14] It is interesting to note that such technological conditions involve economies of scope for combinations of outputs that fail to exhaust the scale economies of the basic process. To see this, we note that in this case (with $a_1 = 1$) $\sum C(0, \ldots, 0, y_i, 0, \ldots, 0) - C(y) = \sum \phi(a_i y_i) - \phi(\sum a_i y_i)$, so that there will be economies of scope iff there is natural monopoly in the scalar output process. As we have seen, a sufficient condition is that, when producing all of the y_i, economies of scale not be exhausted. On the other hand, for levels of the output combination that push the firm sufficiently far into the region of decreasing returns to violate the requirements of natural monopoly, there will be diseconomies of scope.

the minimum average cost in stand-alone production can be used in calculating the lower bounds on $m_i(y^I)$ given in (5G3). When M curves in
toward the origin, it is still true that $y_i^* = y_i^M$. However, here, a wedge is
driven between $\bar{t}$ and $\underline{t}$ so that the bounds given by Proposition 5F1 encompass a wider range for $m(y^I)$ and thereby provide less information.

An M locus that curves out from the axes and is concave to the origin
causes each y_i^M to be larger than y_i^* and the lower bounds on each $m_i(y^I)$
to be smaller and thus less informative. This shape also leads to divergences
between $\bar{t}$ and $\underline{t}$ that grow with both the curvature of the M locus and with
the number of products in the industry.[15] As we have seen, such an M locus
can be the result of cost complementarities or of inputs shared without
congestion by the various outputs' production processes, of economies of
scope in set-up costs, or of cost independence. In this last case, it must be
demand relationships that serve to define the industry, while, in the other
cases, the production relationships among the goods alone may suffice for
the industry's definition.

---

[15] This can be seen most easily in the case where the M locus is rectangular as in Figure 5I1
and 5I2, which depict the two- and three-product cases, respectively. In Figure 5I1 the M locus
is BAK and the lower bound of its convex closure is the diagonal hyperplane L, that is, BL_aK.
To make our point most clearly, let us assume that, by coincidence, y^I lies on the ray R_a through
point A, the corner point of the M locus. Then $\underline{t}y^I = L_a$, the point at the intersection between
L and R_a. Now, L_a is clearly exactly halfway between the origin and A. But in the three-
dimensional case in Figure 5I2 it is easy to show that L_a is $\frac{2}{3}$ of the distance from A to the origin.
That is, as one more item is added to the industry's product line, the zone of ignorance about
the optimal number of firms represented by the distance between $\underline{t}y^I$ and $\bar{t}y^I$ has increased
from $\frac{1}{2}OA$ to $\frac{2}{3}OA$. More generally, we can show that in the N product case this distance is
$((N-1)/N)OA$ and so asymptotically approaches OA as N increases.

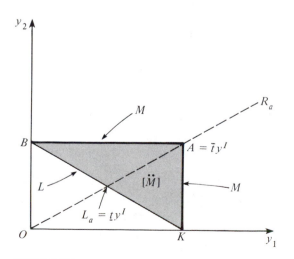

FIGURE 5I1

It is to be expected that these different cases will have very different implications for the structure of a multiproduct industry, yet we have found that they can all yield an M locus of the same general shape and position.

Footnote 15 (continued)

Proof   We can denote the relevant points as

$$A = (a_1, a_2, \ldots, a_n)$$

$$L_a = (ka_1, ka_2, \ldots, ka_n), \qquad k < 1$$

$$\left.\begin{array}{l} y_1^* = (a_1, 0, \ldots, 0) \\ y_2^* = (0, a_2, \ldots, 0) \\ \quad\cdots\cdots\cdots \\ y_n^* = (0, 0, \ldots, a_n) \end{array}\right\} \text{the intercepts of the } M \text{ locus.}$$

Our objective is to show $OL_a/OA = k = 1/n$. The equation of hyperplane L can be written

$$w_1 y_1 + \cdots + w_n y_n = w \qquad \text{(constant).}$$

Since points $y_1^*, \ldots, y_n^*$ and L_a all lie on L, they must all satisfy the preceding equation so that (by direct substitution)

$$w_1 a_1 = w_2 a_2 = \cdots = w_n a_n = \sum w_i k a_i = w.$$

Summing the $w_i a_i$ and bringing the k out of the summation we have

$$\sum w_i a_i = nk \sum w_i a_i \quad \text{or} \quad k = 1/n,$$

as was to be shown.

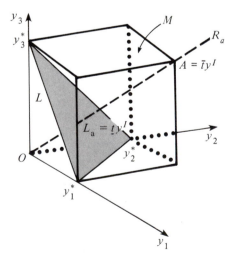

FIGURE 5I2

This is the fundamental reason why bounds that are determined only from information about the M locus must permit a wide range of values for $m(y^I)$ and $m_i(y^I)$ when the M locus curves outward from the axes.

Succeeding chapters will build upon the results we have described here, offering analyses of market structures that are based on significant technological information going beyond the information on returns to scale embedded in the M locus. After the discussion in Chapter 6 on the effects of input prices, in Chapters 7 and 8 we explain the roles played by trans-ray cost properties in multiproduct natural monopoly. Then, in Chapter 9, we proceed with this program by characterizing multiproduct competitive industries. As we have seen here, such a market form cannot occur if $\bar{m}$ is small. And perfect competition can be hoped for in an industry that is shown to be naturally competitive by large lower bounds on each of the $m_i(y^I)$, for all relevant y^I. Yet, a more complete analysis of multiproduct competitive equilibria, like the analysis of natural monopoly, must make use of the trans-ray cost concepts that were introduced in Chapter 4, and which go beyond the returns-to-scale phenomenon whose implications were explored here.

APPENDIX I

Proof of Proposition 5D1[1]

(i) Suppose that $y^I = my^M$ and that m happens to be an integer. To show that $C^I(y^I) = mC(y^M)$ and that exactly m firms, each producing y^M, are required for industry cost minimization, let $\{y^1, \ldots, y^k\}$ be any other partition of y^I. Its total cost is $\sum C(y^i) = \sum y^i[C(y^i)/y^i] > \sum y^i[C(y^M)/y^M] = my^M C(y^M)/y^M = mC(y^M)$. Thus, $mC(y^M)$ is the minimum industry cost of producing y^I, and m firms producing y^M is the unique configuration that is efficient.

(ii) To show that $m(y^I) \geq \lfloor y^I/y^M \rfloor$, suppose instead that the efficient partition of y^I is $\{y^1, \ldots, y^m\}$, with $0 < m < \lfloor y^I/y^M \rfloor$. For convenience in notation, let $y^1 \leq y^2 \leq \cdots \leq y^m$. By the hypothesis, $y^M < y^i$ for some i. Let y^{k+1} be the smallest such member of the partition.

First we establish that $\sum_{i=k+1}^{m}(y^i - y^M) \geq y^M$. Otherwise, $y^I = \sum_{i=1}^{m} y^i \equiv \sum_{i=1}^{k} y^i + \sum_{i=k+1}^{m}(y^i - y^M) + (m-k)y^M < ky^M + y^M + (m-k)y^M = (m+1)y^M$. Then, $m + 1 > y^I/y^M$, which, because m is an integer, contradicts the assumption $0 < m < \lfloor y^I/y^M \rfloor$.

Because $y^i > y^M$ for $i > k$, and because $\sum_{i=k+1}^{m}(y^i - y^M) \geq y^M$, we can choose $\Delta y^i > 0$ for $i = k+1, k+2, \ldots, m$ so that $\sum_{i=k+1}^{m}\Delta y^i = y^M$, $\Delta y^i \leq y^i - y^M$, and $y^i - \Delta y^i \geq y^M$. We show now that removal of Δy^i from the output of the ith firm, for $i > k$, together with the creation of a new firm that produces $\sum_{i=k+1}^{m}\Delta y^i = y^M$, reduces the total industry cost of producing y^I. This will constitute the requisite contradiction. The new total industry cost is

$$
\sum_{i=1}^{k} C(y^i) + \sum_{i=k+1}^{m} C(y^i - \Delta y^i) + C(y^M)
$$

$$
= \sum_{i=1}^{k} C(y^i) + \sum_{i=k+1}^{m} y^i[C(y^i - \Delta y^i)/(y^i - \Delta y^i)]
$$

$$
+ \sum_{i=k+1}^{m} \Delta y^i[(C(y^M)/y^M) - (C(y^i - \Delta y^i)/(y^i - \Delta y^i))] < \sum_{i=1}^{k} C(y^i)
$$

$$
+ \sum_{i=k+1}^{m} y^i[C(y^i)/y^i] = \sum_{i=1}^{m} C(y^i).
$$

---

[1] The first proof of this result appeared in Baumol and Fischer (1975); the treatment here is based on a proof by David Romer (personal communication).

The inequality follows from $C(y^M)/y^M \leq C(y^i - \Delta y^i)/(y^i - \Delta y^i)$ and from $C(y^i - \Delta y^i)/(y^i - \Delta y^i) < C(y^i)/y^i$, for $i > k$, which is a consequence of the stipulations that $y^i > y^i - \Delta y^i \geq y^M$ and that $C(y)/y$ is an increasing function of y for $y \geq y^M$.

(iii) To show that $m(y^I) \leq \lceil y^I/y^M \rceil$, suppose instead that the efficient partition of y^I is $\{y^1, \ldots, y^m\}$, with $m > \lceil y^I/y^M \rceil$. For convenience in notation, let $0 < y^1 \leq y^2 \leq \cdots \leq y^m$. By the hypothesis, $y^M > y^i$ for some i. Let y^k be the largest such member of the partition.

First, we show that $\sum_{i=2}^{k} (y^M - y^i) \geq y^1$. Otherwise,

$$y^I = \sum_{i=1}^{m} y^i = y^1 - \sum_{i=2}^{k} (y^M - y^i) + (k-1)y^M + \sum_{i=k+1}^{m} y^i > (k-1)y^M$$

$$+ \sum_{i=k+1}^{m} y^i \geq (k-1)y^M + (m-k)y^M = (m-1)y^M.$$

Then, $m - 1 < y^I/y^M$, which, because m is an integer, contradicts $m > \lceil y^I/y^M \rceil$.

Because $y^i < y^M$, for $i \leq k$, and because $\sum_{i=2}^{k} (y^M - y^i) \geq y^1$, we can choose $\Delta y^i > 0$ for $i = 2, 3, \ldots, k$ so that $y^i + \Delta y^i \leq y^M$ and $\sum_{i=2}^{k} \Delta y^i = y^1$. We now show that industry cost can be decreased by eliminating firm 1 and redistributing its output in quantities Δy^i among firms i, $i = 2, 3, \ldots, k$. This result will provide the contradiction we require. The new industry cost is

$$\sum_{i=2}^{k} C(y^i + \Delta y^i) + \sum_{i=k+1}^{m} C(y^i)$$

$$= \sum_{i=2}^{k} y^i [C(y^i + \Delta y^i)/(y^i + \Delta y^i)]$$

$$+ \sum_{i=2}^{k} \Delta y_i [[C(y^i + \Delta y^i)/(y^i + \Delta y^i)] - C(y^1)/y^1] + C(y^1)$$

$$+ \sum_{i=k+1}^{m} C(y^i) < C(y^1) + \sum_{i=2}^{k} y^i [C(y^i)/y^i] + \sum_{i=k+1}^{m} C(y^i) = \sum_{i=1}^{m} C(y^i).$$

The inequality follows from $[C(y^i + \Delta y^i)/(y^i + \Delta y^i)] < C(y^i)/y^i$ and $[C(y^i + \Delta y^i)/(y^i + \Delta y^i)] < C(y^1)/y^1$, for $i = 2, \ldots, k$, which hold because $C(y)/y$ is a decreasing function of y for $y < y^M$.   Q.E.D.

The Cost-Minimizing Distribution of Output among Firms

Proposition 5D1 has provided bounds on the optimal number of firms, but it does not tell us anything about the manner in which a given industry output should, optimally, be divided among a given (i.e., the optimal) number of firms. This is the subject treated here, for the case of a single-product industry with U-shaped average cost curves for its firms.

Here intuition is particularly likely to prove misleading. It is tempting to argue that an optimal solution will involve approximately equal outputs by all firms and that it will not permit some firms to produce outputs exceeding and others to produce outputs falling short of the cost-minimizing quantity y^M. For example, suppose firms A and B produce outputs y_a and y_b with $y_a < y^M < y_b$. Let $y_a + \delta \leq y^M$ and $y^M + \delta = y_b$. If firm A's output is increased to $y_a + \delta$ and firm B's output is reduced to $y_b - \delta$, then total industry output will obviously be unchanged. Yet both firms must have lower average costs than before (with a U-shaped average cost curve) since their outputs will both have moved closer to y^M. Thus, it would seem that the original outputs y_a and y_b could not have been cost-minimizing. More generally, it would seem that, if we start with one firm producing less than y^M and with another producing more than y^M, it must always be possible to reduce the industry's total production cost by transferring some of the larger firm's output to the smaller producer, thus lowering both their average costs. However, as we will show now by a concrete example, this conclusion is fallacious. The reason, it may be noted, is that firm A's average cost may well be considerably higher than that of firm B both before and after the transfer takes place. Hence, the reassignment of output may involve a move from a low-cost to a high-cost producer. Thus, even if both firms' average costs are reduced in the process, the net result may be an addition to total cost.

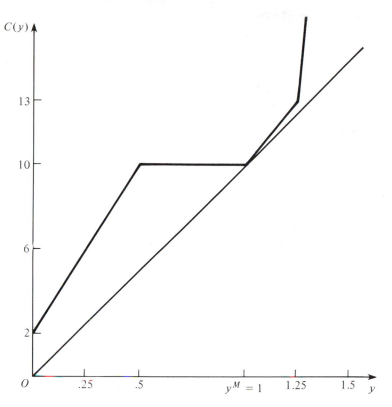

FIGURE 1

This is proved by the following illustrative cost function which is depicted in Figure 1,

$$C(y) = \begin{cases} 2 + 16y & 0 < y \le 0.5 \\ 10 & 0.5 \le y \le 1 \\ 10 + 12(y - 1) & 1 \le y \le 1.25 \\ 13 + 100(y - 1.25) & 1.25 \le y \end{cases}$$

This cost function obviously has its minimum average cost at $y^M = 1$.

Average cost strictly declines in the interval $0 < y < 1$, and strictly increases in the interval $1 < y < \infty$. Thus, $C(y)$ satisfies the assumptions of Proposition 5D1.

Now, consider the industry output $y^I = 1.5$. By Proposition 5D1 the optimal number of firms is either 1 or 2. It is clearly not optimal to have a

Table 1

Total Cost of Output y^I as a Function of the Output y_1 of the First Firm

y_1	0+	0.25	0.50	1	1.25	1.5−
$C(y_1) + C(y_2)$	40	19	20	20	19	40

single firm produce $y^I = 1.5$, because for $y > 1.25$, $C(y)$ increases very sharply. Thus, it is cheapest to use two firms producing amounts y_1 and $y_2 = 1.5 - y_1$, where both y_1 and y_2 are ≤ 1.25. Table 1 gives the total cost $C(y_1) + C(y_2)$ for all corner values of y_1. For values of y_1 not in the table, $C(y_1) + C(y_2)$ can be interpolated linearly.[1]

From Table 1 it is seen immediately that the optimal allocation of the total output $y^I = 1.5$ assigns 0.25 units to one firm, and 1.25 units to the other. Thus, in an optimal allocation it may indeed be possible for some firms to produce more than y^M, and others less than y^M. We are not prepared to conjecture how frequently such cases might be expected to arise. However, the intuitive argument that attempts to rule out such cases is not entirely fallacious, as is shown by

Proposition 1 *Let $C(y)$ be a cost function satisfying the hypotheses of Proposition 5D1, y^I be a given industry output and m be a given number of firms. If $(y^1, \ldots, y^m)$, with $y^i > 0$ $(i = 1, \ldots, m)$ and $\sum y^i = y^I$, is an optimal allocation of the industry output among the m firms, then*

$$y^M - y^i > \sum_{y^j > y^M} (y^j - y^M) \qquad for\ all\ y^i < y^M.$$

In discussing this result it will be convenient to follow Chamberlin (1962) by referring to a positive difference $z_i = y^M - y^i$ as the *excess capacity* of firm i. Similarly, we will refer to a positive difference $w_j = y^j - y^M$ as the *supracapacity output* of firm j. In these terms, Proposition 1 asserts that if any firm has excess capacity in an optimal arrangement, then the excess capacity *of that firm alone* must exceed the sum of the supracapacity outputs of all firms in the industry. In other words, in an optimal arrangement where there are both excess capacities and supracapacity outputs, the former will tend to outweigh the latter heavily.

Proof of Proposition 1 Suppose the contrary. Then, letting $0 < y^1 \leq y^2 \leq \cdots \leq y^{r-1} < y^M \leq y^r \leq \cdots \leq y^m$, it must be true that

$$z_{r-1} \equiv y^M - y^{r-1} \leq \sum_{j=r}^{m} (y^j - y^M) \equiv \sum_{j=r}^{m} w_j.$$

---

[1] The value corresponding to $y_1 = 0+$ means the limit as y_1 approaches 0 from above. The value for $y_1 = 1.5-$ means the limit as y_1 approaches 1.5 from below.

But then we can show that a cheaper way of producing the same total output can be constructed as follows: Let $x^{r-1} = y^M$, $x^i = y^i$ for $i < r - 1$, and $x^j = y^j - w_j + \delta_j \geq y^M$ for $j \geq r$, where the δ_j are any numbers chosen so that

$$w_j \geq \delta_j \geq 0 \quad \text{and} \quad \sum_{j=r}^{m} (w_j - \delta_j) = z_{r-1}.$$

In other words, the new allocation of industry output, $x^1, \ldots, x^m$, involves the expansion of the output of firm $r - 1$ to the level y^M, with offsetting reductions in the outputs of the firms with supracapacity, not reducing any of their outputs below y^M. To show that the new allocation incurs a smaller total cost than the old, it suffices to consider the total costs of the firms whose outputs have been altered:

$$C(x^{r-1}) + \sum_{j=r}^{m} C(x^j) = y^M AC(y^M) + \sum_{j=r}^{m} x^j AC(x^j)$$

$$= y^{r-1} AC(y^M) + z_{r-1} AC(y^M) + \sum_{j=r}^{m} x^j AC(x^j)$$

$$< y^{r-1} AC(y^{r-1}) + \sum_{j=r}^{m} (w_j - \delta_j) AC(y^j) + \sum_{j=r}^{m} x^j AC(y^j)$$

$$= y^{r-1} AC(y^{r-1}) + \sum_{j=r}^{m} y^j AC(y^j) = C(y^{r-1}) + \sum_{j=r}^{m} C(y^j).$$

Here, the inequality follows from the relationships $AC(y) > AC(y^M)$ for $y \neq y^M$, $z_{r-1} = \sum_{j=r}^{m} (w_j - \delta_j)$, $w_j - \delta_j \geq 0$, and $AC(x^j) \leq AC(y^j)$ for $j \geq r$. This last relationship holds because $y^M \leq x^j \leq y^j$ and because, by assumption, $AC(y)$ is increasing in y, for $y \geq y^M$. Consequently, the hypothesized allocation cannot be cost-minimizing, and we have the requisite contradiction.   Q.E.D.

The asymmetry of this result—its bias toward a preponderance of excess capacity over supracapacity output calls for some comment. When some firms produce at less than capacity and others exceed the capacity level y^M, it is natural, as we saw earlier, to *consider* the transfer of some output from the latter to the former in order to reduce both firms' average costs. We have noted that this may conceivably turn out to increase total cost for the industry if the firms producing below capacity have much higher average costs than those producing above capacity so that the transfer we are considering constitutes a shift from a low-cost to a high-cost producer; and in that case, as was observed before, optimality may require the presence of some firms with overcapacity and some firms with undercapacity outputs. However, Proposition 1 tells us that this cannot possibly happen if any one excess capacity is less than the total of the supracapacity outputs. For then there is enough output available for the transfer to an undercapacity firm to raise its output *all the way to* y^M—the output of minimum average cost. *Such a transfer must then always be a shift of output to a firm with lower* (indeed, minimal) *average cost*, and the resulting output reassignment will therefore inevitably reduce total industry cost. Thus, where excess capacities are relatively small, it will always pay to eliminate them, because the transfer required for the purpose will both reduce the average costs of all affected firms *and* shift output from a nonminimum-cost to a minimum-cost producer. Only where each quantity of excess capacity is relatively high compared to the total supracapacity output may such transfers fail to reduce total industry cost.

There is a noteworthy class of cases in which excess capacities and supracapacity outputs will never simultaneously be optimal—indeed, in which it will be optimal for all firms to produce the same quantity. We have

Proposition 2   *Let $C(y)$ be a cost function which is strictly convex for $0 < y < \infty$. Then the cheapest way to produce a given industry output y^I using m firms is to have each firm produce the same amount, y^I/m. This is true, in particular, if m is the optimal number of firms.*

Proof   Suppose the contrary. Then there would exist a pair of firms, say firms 1 and 2, which produce different positive amounts, $y^1 \neq y^2$. Consequently, by strict convexity of $C(y)$,

$$C\left(\frac{y_1 + y_2}{2}\right) < \tfrac{1}{2}(C(y_1) + C(y_2)).$$

Then, a cheaper allocation of the output y^I is obtained immediately by letting firms 1 and 2 split their joint output in half. Thus, any allocation in which all firms do not produce the same amount cannot be optimal.   Q.E.D.

Note that for $C(y)$ to be convex it is not sufficient that $AC(y)$ be convex.[2] Even when $AC(y)$ is U-shaped, as is frequently assumed, $C(y)$ typically has an S shape. This may make it economical to have some portion of the output supplied by a very small firm, even at a higher average cost.

Corollary (i)   *If $C(y)$ is a cost function satisfying the assumptions of Proposition 5D1 and is strictly convex* in the interval $0 < y < 2y^M$, *then in the optimal allocation all firms produce equal amounts.*

---

[2] Indeed, assuming differentiability, convexity of the total cost curve is tantamount to the assertion that marginal costs are rising throughout, which—as is well known—is not the case when the AC curve is U-shaped and there are no fixed costs.

Proof   In the globally optimal allocation of the output y^I among any number of firms, no firm can produce an amount greater than or equal to $2y^M$, since otherwise it would be cheaper to split up this firm into two smaller firms producing equal amounts. Thus, the requirement of convexity is only needed in the interval $0 < y < 2y^M$.   Q.E.D.

Figure 2 shows an example of a cost function which satisfies the assumptions of Corollary (i).

Corollary (ii)   *If $C(y)$ is strictly convex in the interval $0 < y < 2y^M$ and satisfies the assumptions of Proposition 5D1, then the optimal number of firms, $m(y^I)$, is exactly $m' = \lfloor y^I/y^M \rfloor$ if*

$$AC(y^I/m') < AC(y^I/(m' + 1)).$$

It is exactly $m'' = \lceil y^I/y^M \rceil$ if

$$AC(y^I/m'') < AC(y^I/(m'' - 1)).$$

It is either $m' = \lfloor y^I/y^M \rfloor$ or $m'' = \lceil y^I/y^M \rceil$ if

$$AC(y^I/m') = AC(y^I/m'').$$

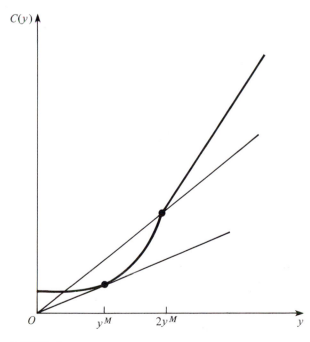

FIGURE 2

Proof   Since by Corollary (i) all firms must produce equal amounts in the optimal allocation, we simply have to determine for which number of firms m the *average* cost $AC(y^I/m)$ is minimal. That is, since the total industry cost is

$$m \cdot (y^I/m)AC(y^I/m) = y^I AC(y^I/m),$$

with y^I given, total cost will be minimized for the m that minimizes AC. Proposition 5D1 then narrows the search to the two values of m specified in the corollary.   Q.E.D.

APPENDIX III

Proof of Proposition 5F1

We may take y^I to be strictly positive, without loss of generality. If some components of y^I were zero, this proof could be applied, unchanged, in the nonnegative orthant of the subspace pertaining to the positive components of y^I.

As in the statement of the proposition, the points $\bar{t}y^I$ and $\underline{t}y^I$ are constructed to be on the boundary of the convex set $[\ddot{M}]$, so that there are separating hyperplanes that pass through these points. That is, there exist nonzero vectors $\bar{h}$ and $\underline{h}$, and scalars $\bar{k}$ and $\underline{k}$ such that

(1)
$$\bar{h} \cdot \bar{t}y^I = \bar{k}, \qquad \underline{h} \cdot \underline{t}y^I = \underline{k}$$

and

(2)
$$y \in [\ddot{M}] \text{ implies that } \bar{h} \cdot y \le \bar{k} \text{ and } \underline{h} \cdot y \ge \underline{k}.$$

Before proceeding with the portion of the proof that was explained in the text, it is necessary to establish this preliminary result:

Lemma $\bar{k}$ *and* $\underline{k}$ *can be chosen to be positive numbers.*

Proof First we show that $\bar{k} \ne 0 \ne \underline{k}$. By Assumption 5E1, for each i there exists a $y_i^* > 0$ such that

(3)
$$(0, \dots, 0, y_i^*, 0, \dots, 0) \in M \subseteq [\ddot{M}].$$

Suppose $\bar{k} = 0$. Then, by (2), $\bar{h}_i y_i^* \le 0$, so that $\bar{h}_i \le 0$ for all i. Since $\bar{h} \ne 0$ and $y^I > 0$, it follows that $\bar{h} \cdot \bar{t}y^I < 0$. This contradicts (1). The argument for $\underline{k}$ is analogous.

Clearly, if the set $[\ddot{M}]$ is itself a portion of a hyperplane, then the two separating hyperplanes coincide with it, $\bar{t} = \underline{t}$, and the parameters of the hyperplanes can be chosen so that $\bar{h} = \underline{h}$ and $\underline{k} = \bar{k} \ne 0$. Both $\underline{k}$ and $\bar{k}$ can be chosen positive because, if they were negative, neither (1) nor (2) would be falsified (in this case) by replacing $\bar{k}$, $\underline{k}$, $\bar{h}$, and $\underline{h}$ with their negatives.

If $\bar{t}$ exceeded $\underline{t}$, it would follow that $\bar{k}$ and $\underline{k}$ were positive. The argument for $\bar{k}$ is this: $\bar{h} \cdot \bar{t}y^I = \bar{k}$ and $\bar{h} \cdot \underline{t}y^I \leq \bar{k}$ since $\underline{t}y^I \in [\ddot{M}]$. Then, since $\bar{t} \neq \underline{t}$, $\bar{h} \cdot \underline{t}y^I < \bar{k}$. Hence $(\bar{t} - \underline{t})\bar{h}y^I > 0$, so that $\bar{h}y^I > 0$. Since $\bar{h}y^I = \bar{k}/\bar{t}$, $\bar{k}$ must be positive The argument for $\underline{k}$ is analogous.

To prove the Lemma, it remains to show that $\bar{t} > \underline{t}$ for $[\ddot{M}]$ not a portion of a hyperplane. By (3), $[\ddot{M}]$ includes the convex hull of the points $(0, \ldots, 0, y_i^*, 0, \ldots, 0)$, $i = 1, 2, \ldots, n$. This is the hyperplane segment $A = \{y \geq 0 \mid a \cdot y = 1\}$, where $a_i = 1/y_i^*$. Suppose that $\bar{t} = \underline{t}$. Then $ty^I \in [\ddot{M}]$ must imply that $t = \bar{t} = \underline{t}$. Since $(1/a \cdot y^I)y^I \in A \subseteq [\ddot{M}]$, it must be true that $\bar{t} = \underline{t} = (1/a \cdot y^I)$. Now, with $[\ddot{M}]$ not a hyperplane segment, there must exist $x \in [\ddot{M}]$ and $x \notin A$. Choose $\delta \in \mathbf{R}_{++}$ small enough that $y^I - \delta x > 0$. Let $t_A = 1/a \cdot (y^I - \delta x)$, so that $t_A(y^I - \delta x) \in A \subseteq [\ddot{M}]$. By convexity,

(4) $\theta x + (1 - \theta)t_A(y^I - \delta x) \in [\ddot{M}]$ for $0 < \theta < 1$.

Also, since $x \notin A$,

$$\theta x + (1 - \theta)t_A(y^I - \delta x) \notin A.$$

But, for $\theta = \delta t_A/(1 + \delta t_A)$, the expression in (4) is $[t_A/(1 + \delta t_A)]y^I$. Thus, we have constructed a point of the form ty^I which is in $[\ddot{M}]$, by (4), and which is not in A. This is the requisite contradiction of $\bar{t} = \underline{t}$, and the proof of the Lemma is thereby completed.

Now we can establish the lower bounds in (5F1). Let $\{y^1, y^2, \ldots, y^m\}$ be an industry cost-minimizing partition of y^I, and let $t^i y^i$ be the point on the ray through y^i that is on the M locus. Each number t^i must be greater than $\frac{1}{2}$, because otherwise the point $y^i/2$ would be on or past the M locus and then, because of the assumption that RAC increases past the M locus, it would follow that $C(y^i) > 2C(y^i/2)$. Such a relationship would exclude y^i from the industry cost-minimizing partition.

Each point $t^i y^i \in M \subseteq [\ddot{M}]$, by construction, so that $\bar{h} \cdot t^i y^i \leq \bar{k}$ by (2). But $t^i > \frac{1}{2}$, so $\frac{1}{2}\bar{h} \cdot y^i < \bar{k}$. Summing this inequality over the m values of y^i gives $m\bar{k} > \frac{1}{2} \sum_i \bar{h} \cdot y^i = \frac{1}{2}\bar{h} \cdot y^I = (1/2\bar{t})\bar{k}$. The relationships in (5F1) follow by division by $\bar{k} > 0$, recognizing that m must be an integer.

Now we prove the relationships in (5F2) under Assumptions 5E1 and 5E2. The assumed subadditivity holds over the set of points inside the separating hyperplane through $\underline{t}y^I$, since $\{y \geq 0 \mid \underline{h} \cdot y \leq \underline{k}\} \subseteq [\ddot{M}]^-$. This follows because $\underline{h} \cdot y \leq \underline{k}$ and $\lambda y \in [\ddot{M}]$ together imply, by (2), that $\lambda \geq 1$.

As before, let $\{y^1, \ldots, y^m\}$ be an industry cost-minimizing partition of y^I, and let $t^i y^i \in M$. Then, by (2), $\underline{h} \cdot t^i y^i \geq \underline{k} > 0$, so $\underline{h} \cdot y^i > 0$. For $i \neq j$, it is necessarily true that $\underline{h} \cdot y^i + \underline{h} \cdot y^j > \underline{k}$. If not, that is, if $\underline{h} \cdot y^i + \underline{h} \cdot y^j \leq \underline{k}$, it would follow from $\underline{h} \cdot y^i > 0$ and $\underline{h} \cdot y^j > 0$ that $\underline{h} \cdot y^i < \underline{k}$ and $\underline{h} \cdot y^j < \underline{k}$. Then, y^i, y^j, and $y^i + y^j$ would all be in $[\ddot{M}]^-$, so that, by Assumption 5E2,

$C(y^i) + C(y^j) > C(y^i + y^j)$. The relationships would exclude y^i and y^j from the industry cost-minimizing partition.

Thus, if m were larger than one, we would have $\underline{h} \cdot y^1 + \underline{h} \cdot y^2 > \underline{k}$, $\underline{h} \cdot y^2 + \underline{h} \cdot y^3 > \underline{k}, \ldots, \underline{h}y^{m-1} + \underline{h}y^m > \underline{k}$, and $\underline{h} \cdot y^1 + \underline{h} \cdot y^m > \underline{k}$. Addition of these m inequalities yields $2 \sum_{i=1}^{m} \underline{h} \cdot y^i > m\underline{k}$. Since $\sum_{i=1}^{m} y^i = y^I$, $\underline{h} \cdot y^I = \underline{k}/\underline{t}$, and $\underline{k} > 0$, rearrangement gives $m < 2/\underline{t}$, provided that $m > 1$. Also, if $\underline{t} \geq 1$, $y^I \in [\ddot{M}]^-$ so that $m = 1$ by Assumption 5E2. Consequently, the relationships given in (5F2) follow.   Q.E.D.

APPENDIX IV

Flat-Bottomed Ray Average Cost Curves

In this appendix, we show how the bounds of Propositions 5F1 and 5G1 on the industry cost-minimizing numbers of firms can be generalized to apply to cases in which ray average cost curves have flat bottoms. In such cases, the minimum level of ray average cost along a given ray is achieved not at a unique output vector, but instead at every output vector in an interval along the ray. Then, Assumption 5E1 is replaced by the following assumption, which plays an important role in Chapter 11 as the definition (Definition 11D1) of flat-bottomed RAC:

Assumption 1   *On each ray in the space of outputs there is a point y_m and a positive number k such that RAC is decreasing up to y_m, is constant between y_m and $(1 + k)y_m$, and is increasing at ty_m for $t > (1 + k)$. These numbers k may be different for different rays. Denote by $\underline{M}$ the set of points of minimum efficient scale y_m and by $\bar{M}$ the set of points $(1 + k)y_m$ of maximum efficient scale beyond which RAC begins to increase. Here, the M locus is the set of all output vectors at which there are locally constant returns to scale: the set of points on each ray that lie between $\underline{M}$ and $\bar{M}$.*

Before turning to the formal statements and proofs of the generalizations of Propositions 5F1 and 5G1, we can indicate the heart of their logic. Let y^i be the output vector of any of the firms in an industry cost-minimizing configuration, let $y_m \in \underline{M}$ be the point on its ray with minimum efficient scale, and let $(1 + k)y_m \in \bar{M}$ be the point on its ray with maximum efficient scale. If $(1 + k) < 2$, it must be true that $y^i < 2y_m$. Here, as when ray average cost curves are U-shaped, $y^i \geq 2y_m$ would imply that y^i could be produced at a smaller total cost by two firms, each producing $y^i/2$. This would then contradict the posited role of y^i in an industry cost-minimizing configuration.

However, matters are different if $(1 + k) \geq 2$. Here, as was explained in Chapter 2 for single-output industries, any output on the ray beyond minimum efficient scale can be apportioned among firms that each produce at minimum *RAC*. Hence, in this case, y^i cannot exceed $(1 + k)y_m$ because, if it did, y^i could be produced at a smaller total cost by two or more firms, each producing at minimum *RAC*. It is possible here that $y^i \geq 2y_m$ because, along the ray in question, the M locus extends out to $(1 + k)y_m$ and $(1 + k)$ may exceed 2. Nevertheless, industry-wide cost efficiency does not require y^i to be that large. For $(1 + k) \geq 2$, any total output on the ray larger than y_m can be apportioned in a cost-minimizing manner among firms that each produce output vectors between y_m and $2y_m$. Thus, for any industry cost-minimizing configuration with $(1 + k)y_m \geq y^i \geq 2y_m$, there is another configuration producing the same total output at the same total cost that entails no firm as large as $2y_m$.

To summarize, then, y^i need not exceed $2y_m$, and it cannot exceed the greater of $2y_m$ and $(1 + k)y_m$. Consequently, the analysis to follow will make heavy use of the outer envelope of the loci $\bar{M}$ and twice $\underline{M}$. This set, pictured as the dark curve in Figure 1, is defined formally as

(1) $$L = \{\, y | \max(\bar{t}, 2\underline{t}) = 1, \bar{t}y \in \bar{M}, \underline{t}y \in \underline{M} \,\}.$$

Proposition 1   *For a given industry output vector* y^I, *let*

$$\bar{t} \equiv \max\{t | ty^I \in [\ddot{M}]\},$$

$$\underline{t} \equiv \min\{t | ty^I \in [\ddot{\underline{M}}]\},$$

$$\bar{\bar{t}} \equiv \max\{t | ty^I \in [\ddot{L}]\}.$$

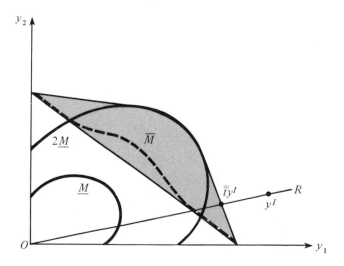

FIGURE 1

Then, Assumption 1 implies that

(2) $$m(y^I) \geq 1/\bar{\bar{t}}.$$

(3) $$1/2\bar{t} \geq 1/\bar{\bar{t}}.$$

Assumption 1 also implies that there exists an industry cost-minimizing configuration with m firms such that

(4) $$m > 1/2\bar{t} \quad \text{and} \quad m \geq \lceil 1/2\bar{t} \rceil.$$

Assumptions 1 and 5E2 together imply that

(5) $$m(y^I) = 1 \quad \text{if} \quad \underline{t} \geq 1.$$

Otherwise,

$$1 \leq m(y^I) < 2/\underline{t} \quad \text{and} \quad 1 \leq m(y^I) \leq \lfloor 2/\underline{t} \rfloor.$$

Proof Much of the requisite proof is analogous to that of Proposition 5F1 presented in Appendix III. The proof of (5) is identical to that of (5F2). To prove (2), (3), and (4), as before we may take $y^I > 0$. As before, there are hyperplanes through the points $\bar{t}y^I$ and $\bar{\bar{t}}y^I$ that separate, respectively, the convex sets $[\ddot{\underline{M}}]$ and $[\ddot{L}]$. That is, there exist nonzero vectors $\bar{h}$ and $\bar{\bar{h}}$, and scalars $\bar{k}$ and $\bar{\bar{k}}$ such that

(6) $$\bar{h} \cdot \bar{t}y^I = \bar{k}, \qquad \bar{\bar{h}} \cdot \bar{\bar{t}}y^I = \bar{\bar{k}},$$

(7) $$y \in [\ddot{\underline{M}}] \text{ implies that } \bar{h} \cdot y \leq \bar{k},$$

and

(8) $$y \in [\ddot{L}] \text{ implies that } \bar{\bar{h}} \cdot y \leq \bar{\bar{k}}.$$

And, as shown in Appendix III, $\bar{k}$ and $\bar{\bar{k}}$ can be chosen to be positive numbers.

Let $\{y^1, y^2, \ldots, y^m\}$ be a cost-minimizing partition of y^I, and let $t^i y^i$, $\bar{t}^i y^i$, and $\underline{t}^i y^i$ be the points on the ray through y^i that are respectively in L, $\bar{M}$, and $\underline{M}$. By (1), either

(9) $$t^i = \bar{t}^i \geq 2\underline{t}^i$$

or

(10) $$t^i = 2\underline{t}^i \geq \bar{t}^i.$$

If (9) holds, then we now show that it must be true that $t^i \geq 1$. Otherwise, with $t^i < 1$, we would have $y^i > t^i y^i = \overline{t}^i y^i$, that is, y^i lying beyond the M locus. Then, by Assumption 1,

$$(11) \qquad\qquad C(y^i) > C(ty^i)/t \qquad \text{for } \underline{t}^i \leq t \leq \overline{t}^i.$$

We shall demonstrate that, in this case, y^i cannot be part of the cost-minimizing configuration because the output y^i could be produced more cheaply by $\lfloor 1/\underline{t}^i \rfloor$ firms, each producing $y^i/\lfloor 1/\underline{t}^i \rfloor$. To see this, note first that $t^i < 1$ and (9) together imply that $1/\underline{t}^i \geq 2$, so that $\lfloor 1/\underline{t}^i \rfloor \geq 2$. Second, $1/\underline{t}^i \geq \lfloor 1/\underline{t}^i \rfloor$, so that $\underline{t}^i \leq 1/\lfloor 1/\underline{t}^i \rfloor$. Finally, $1/\underline{t}^i < \lfloor 1/\underline{t}^i \rfloor + 1$ and $\lfloor 1/\underline{t}^i \rfloor \geq 2$ imply $(1/\underline{t}^i)/\lfloor 1/\underline{t}^i \rfloor < 2$, so that $1/\lfloor 1/\underline{t}^i \rfloor < 2\underline{t}^i \leq \overline{t}^i$. Consequently, (11) gives us $C(y^i) > \lfloor 1/\underline{t}^i \rfloor C(y^i/\lfloor 1/\underline{t}^i \rfloor)$, which is the contradiction needed to show that (9) implies $t^i \geq 1$.

If (10) holds, then $t^i \geq 1$. Otherwise, with $t^i < 1$, we would have $y^i > t^i y^i = 2\underline{t} y^i \geq \overline{t} y^i$, that is, y^i lying beyond the M locus and beyond twice minimum efficient scale. Then, $y^i/2 > \underline{t} y^i$, so that $C(y^i) > C(y^i/2)/(\frac{1}{2}) = 2C(y^i/2)$. This contradicts the role of y^i in the cost-minimizing configuration.

We have established that, whether (9) or (10) holds, $t^i \geq 1$. Since $t^i y^i \in L \subseteq [\ddot{L}]$, (8) implies that $\overline{\overline{h}} \cdot t^i y^i \leq \overline{k}$. Then, $t^i \geq 1$ and $0 < \overline{\overline{h}} \cdot y^i$ imply that $\overline{\overline{h}} \cdot y^i \leq \overline{k}$. Summing over the m firms gives $\overline{\overline{h}} \cdot y^I \leq m\overline{k}$. Then $m \geq \overline{\overline{h}} \cdot y^I/\overline{k} = 1/\overline{t}$, by (6). This establishes (2).

To prove (3), note that, by (1), $y \in L$ and $\underline{t} y \in M$ imply that $y \geq 2\underline{t} y$. Because $\overline{t} y^I \in [\ddot{M}]$, there exist $x^j \in \underline{M}$ such that $\overline{t} y^I = \sum \lambda_j x^j$, $\lambda_j > 0$, and $\sum \lambda_j = 1$. Let $a_j x^j \in L$. Then $a_j \geq 2$. By (8), $\overline{\overline{h}} \cdot a_j x^j \leq \overline{k}$, so that $\overline{\overline{h}} \cdot x^j \leq \overline{k}/a_j \leq \overline{k}/2$. Hence, $\sum \lambda_j \overline{\overline{h}} \cdot x_j \leq \sum \lambda_j \overline{k}/2$, or $\overline{\overline{h}} \cdot \overline{t} y^I \leq \overline{k}/2$. Then by (6) $\overline{t}\overline{k}/\overline{t} \leq \overline{k}/2$, and (3) follows.

Finally, to establish (4), note that we have already shown $\underline{t}^i > \frac{1}{2}$, except where (9) holds. In that case, we have proved that $t^i \geq 1$, so that y^i lies within the M locus, which extends past twice minimum efficient scale. If $\underline{t}^i \leq \frac{1}{2}$, then, we now show, the firm producing y^i can be replaced in the configuration, without an increase in total industry cost, by $\lfloor 1/\underline{t}^i \rfloor$ firms which each produce the output vector $y^i/\lfloor 1/\underline{t}^i \rfloor$, which is less than twice minimum efficient scale. With $\underline{t}^i \leq \frac{1}{2}$ and (9), the argument given earlier proves that $\underline{t}^i \leq 1/\lfloor 1/\underline{t}^i \rfloor < 2\underline{t}^i \leq \overline{t}^i$. Hence, $y^i/\lfloor 1/\underline{t}^i \rfloor \in M$, so that $C(y^i) = \lfloor 1/\underline{t}^i \rfloor C(y^i/\lfloor 1/\underline{t}^i \rfloor)$. Moreover, $1/\lfloor 1/\underline{t}^i \rfloor < 2\underline{t}^i$ means that the new firms each produce at points strictly less than twice minimum efficient scale. Consequently, we have constructed a new cost-minimizing configuration for which $\underline{t}^j > \frac{1}{2}$, for all firms j.

Then, $\underline{t}^j y^j \in M \subseteq [\ddot{M}]$ and (7) imply that $\overline{h} \cdot \underline{t}^j y^j \leq \overline{k}$. Because $\underline{t}^j > \frac{1}{2}$, $\overline{h} \cdot y^j < 2\overline{k}$. Summing over the m firms and using (6) gives $\overline{k}/\overline{t} < 2m\overline{k}$, from which (4) follows.   Q.E.D.

The next result generalizes Proposition 5G1. (See Section 5G for explanations of the notation.)

Proposition 2   *For a given industry output vector y^I, let*

$$\bar{t}^S \equiv \max\{t\,|\,ty^I_S \in [\ddot{M}_S]\}$$
$$\bar{\bar{t}}^S \equiv \max\{t\,|\,ty^I_S \in [\ddot{L}_S]\}.$$

Then, Assumption 1 implies that

(12) $$m_S(y^I) \geq 1/\bar{\bar{t}}^S$$

Also, there exists an industry cost-minimizing configuration with m_S firms producing positive quantities of some of the goods in the subset S such that

(13) $$m_S > 1/2\bar{t}^S \quad \text{and} \quad m_S \geq \lceil 1/2\bar{t}^S \rceil.$$

(14) $$1/2\bar{t}^S \geq 1/\bar{\bar{t}}^S.$$

   Proof   The proof is the same as that of Proposition 1, except that the constructions are carried out on the projections of the sets in the subspace corresponding to the subset of products S, and the summations over firms are limited to those which produce positive quantities of some of the goods in S.   Q.E.D.

Corollary   *Assumption 1 implies that*

(15) $$m_i(y^I) \geq y^I_i/y^L_i,$$

where $$y^L_i \equiv \max\{y_i\,|\,y \in L\}.$$

Also, there exists an industry cost-minimizing configuration with m_i firms producing positive quantities of good i such that

(16) $$m_i > y^I_i/2y^M_i,$$

where $y^M_i \equiv \max\{y_i\,|\,y \in M\}$.

(17) $$1/2y^M_i \geq 1/y^L_i.$$

6

Input-Price Changes, Cost Functions, and Efficient Industry Structure[1]

In the preceding chapters, and in most of the discussion in the remainder of the book, input prices are taken to be fixed, thus permitting production costs to be described as a function of the output vector alone. Yet, as recent experience makes dramatically clear, interest rates, fuel costs, wages rates, and other input prices *do* change. Such changes can have profound consequences for the shape of the cost hypersurface and, consequently, for both cost-minimizing and equilibrium industry structure. For example, it seems plausible that by changing an industry that is labor-intensive to one that uses relatively large amounts of capital equipment, a large change in the relative prices of labor and capital may move the minimum points of firms' ray average cost curves from the vicinity of the origin to a locus much further out. A likely result is a significant reduction in the number of firms that minimizes industry cost. Thereby, it may, for example, transform the industry from one that is naturally competitive into a natural oligopoly.

Casual observation seems to confirm such connections. The typical industry structure in a less developed country (LDC) seems to be characterized predominantly by enterprises far smaller than those often found in industrialized countries. India and Nigeria have no Mitsubishi, I.G. Farben,

---

[1] This chapter is based largely on an as-yet unpublished work by Herman Quirmbach.

or IBM.[2] Intuition suggests that this is so because relatively low wages make for relatively high labor–capital ratios in these countries and, hence, for a locus of minimum ray average costs that lies relatively close to the origin. Thus, it is natural to surmise that a sharp fall in the relative price of capital will lead to the employment of larger quantities of capital and to increased lumpiness, scale economies, and oligopoly, whereas the opposite price change, a sharp reduction in relative real wages, will make for smallness of firms and rapid exhaustion of scale economies. We shall return to this issue after a theoretical analysis of the effects of factor price changes on some of our more important cost concepts.

6A. The Single-Product Case

We begin our analysis by investigating the effect of a change in the price of an input on the cost function relating to single-output production. We examine the consequences of the price change in terms of its implications for subadditivity and economies of scale.

A natural monopoly was defined in Chapter 2 as an industry in which the firm's cost function is subadditive over the relevant range of outputs. Thus, at a given output level y an industry is a natural monopoly if

(6A1) $$C(y) < \sum_{j=1}^{m} C(y^j)$$

for any $m \geq 2$, and any positive $y^1, \ldots, y^m$ such that

$$\sum_{j=1}^{m} y^j = y.$$

Let

$$D(y^1, \ldots, y^m) \equiv \sum_{j=1}^{m} C(y^j) - C(y).$$

---

[2] In practice, the explanation of smallness of firms in an LDC may in fact reside in special local circumstances and institutional arrangements. In India, for example, encouragement of small enterprise and discouragement of bigness are both central tenets of government policy. Virtually any legal firm of substantial size requires both a license to operate and a license for its output size. There are many lines of economic activity in which large or even medium-sized firms are simply prohibited. In other cases, a firm will be fined heavily for producing an output larger than its license authorizes. To this is added a variety of cultural and historical influences which are strong "impediments" to large enterprise that are unrelated to either absolute or relative input prices.

We may interpret D as a measure of the absolute degree of subadditivity of single-firm production as against the specified multifirm allocation,

$$(y^1, \ldots, y^m), \qquad \sum_{j=1}^{m} y^j = y, \qquad m \geq 2.$$

An increase in the price of input i will increase the absolute degree of subadditivity of the cost function if it increases the cost of multifirm production of y (for any m-firm division, with $m \geq 2$) more rapidly than it increases the cost of single firm production of that output. Mathematically, the degree of subadditivity of the cost function will increase if

(6A2)
$$\frac{\partial C(y, w)}{\partial w_i} < \sum_{j=1}^{m} \frac{\partial C(y^j, w)}{\partial w_i}$$

with the same m and y^j as before. But, by Shephard's lemma,[3] (6A2) is equivalent to

(6A3)
$$X_i(y, w) < \sum_{j=1}^{m} X_i(y^j, w)$$

where $X_i(y, w)$ is the derived demand for input i at factor prices w and output y.

But comparison of the preceding inequality with (6A1) tells us immediately that the former, (6A3), is tantamount to the assertion that the demand for input i is subadditive. That is, it tells us that less of input i will be used if y is produced by a single firm than if it is produced by several smaller enterprises. Thus, we have

Proposition 6A1 *The degree of subadditivity of the cost function will be increased by an increase in w_i if, at the output in question and at the original input prices, the demand for input i is itself subadditive.*

Note that this holds even if the cost function is not subadditive: Whatever the original shape of $C(y, w)$, an increase in w_i in this case warps $C(y, w)$ to make it resemble more closely the shape of $X_i(y, w)$. The intuitive explanation is that as input i grows more expensive, savings in the use of that input become more important in determining the behavior of overall costs. We will see that a similar observation applies to our other measures of cost behavior as well.

---

[3] See Shephard (1970).

Next, we turn to the single-product measure[4] of scale economies, S, defined in Chapter 2 as the ratio of average to marginal costs:

(6A4)
$$S \equiv \frac{AC(y)}{MC(y)} = \frac{C(y)}{yMC(y)} = \frac{1}{e_{TC}}.$$

As one can readily see from (6A4), S is the reciprocal of the elasticity of total costs with respect to output. We use e_{TC} to represent this elasticity so that $e_{TC} \equiv 1/S$, as indicated in (6A4). A similar elasticity with respect to output can be calculated for the derived demand for input i:

(6A5)
$$e_{iy} \equiv \frac{y}{X_i} \frac{\partial X_i(y, w)}{\partial y}.$$

Using these definitions, we can determine the effect of an increase in w_i on S by calculating the elasticity of S with respect to w_i by direct differentiation of (6A4). Using Shephard's lemma:

(6A6)
$$\frac{\partial S}{\partial w_i} = \frac{CX_i}{y^2 MC^2} \left(\frac{1}{S} - e_{iy} \right).$$

Multiplying through by w_i/S, and using (6A4) we obtain

(6A7)
$$\left(\frac{w_i}{S} \right) \frac{\partial S}{\partial w_i} = \eta_i(1 - Se_{iy}) = \eta_i \left(1 - \frac{e_{iy}}{e_{TC}} \right),$$

where $\eta_i = w_i X_i / C$ is the share of total costs attributable to purchases of factor i.

Thus, $\partial S/\partial w_i$ will be positive, zero, or negative as e_{iy}/e_{TC} is less than, equal to, or greater than unity, respectively. This immediately yields

Proposition 6A2   *A rise in the price of input i in single-good production will increase the degree of scale economies if and only if*

$$e_{iy} \equiv (\partial X_i/\partial y)(y/X_i) < (\partial C/\partial y)(y/C) \equiv e_{TC}$$

(with scale economies decreased by a rise in w_i if the preceding inequality is reversed). That is, a rise in the price of input i will reduce e_{TC} if and only if

---

[4] For an excellent analysis of the relation of input prices to scale economies and optimal number of firms in the single-product industries see Bassett and Borcherding (1970). The analyses of Silberberg (1974) and Hanoch (1975) are also concerned with the effects of factor prices on the shape and position of the average cost curve.

e_{iy}, *the output elasticity of input i, is less than* e_{TC}, *the output elasticity of the firm's cost function. Thus, here again, a rise in* w_i *changes the shape of the cost function to resemble more closely the shape of the demand function for input i.*

An immediate and pertinent implication of this result is that an increase in the price of a fixed factor must increase the degree of scale economies. This follows because the use of such a factor does not change with the size of output, as long as any production is undertaken, so that the output elasticity of demand for such an input is necessarily zero (and, of course, $e_{TC} > 0$).

The obverse is also true: Where fixed factors are present, proportionate increases in the prices of all variable factors must decrease the degree of scale economies. This follows from the preceding observation because S is homogeneous of degree zero in factor prices, so that if it increases with the prices of all fixed inputs, it must decrease with proportionate increases in the prices of all variable factors.

To interpret Proposition 6A2 in a deeper manner, it is useful to recognize that the difference between the output elasticity of total cost and that of input i is equal to the difference between the output elasticity of average cost and that of expenditure on input i, averaged over output. That is,

$$e_{TC} - e_{iy} = \frac{\partial(C/y)}{\partial y}\frac{y}{(C/y)} - \frac{\partial(w_iX_i/y)}{\partial y}\frac{y}{(w_iX_i/y)}.$$

Thus, the effect of an input price change on the degree of scale economies can be assessed by comparing the average cost curve with the curve giving the average expenditure on the input.

For example, Figure 6A1 displays these curves for the cost function $C(y, w, r) = rF + wA + wy^2 + ry$, where r is the rental rate on capital K and w is the wage rate for labor L. As shown, the average cost and average labor utilization curves are both U-shaped with minima at y_M and y^1, respectively. Between y^1 and y_M the output elasticities of $C(y)/y$ and Lw/y are respectively negative and positive, so that their difference is negative, and Proposition 6A2 implies that a rise in the wage must decrease the degree of scale economies. This conclusion also holds at larger output levels; but, for y sufficiently small, the elasticity of Lw/y is negative and larger in absolute value than the elasticity of $C(y)/y$ and the conclusion is reversed.

In this example, the U shape of the average labor utilization curve is caused by the positive amount of labor, A, that is part of the fixed start-up cost of production. In other examples, with substitution occurring between capital and labor, such a shape can result from increased use of labor-saving equipment as output rises, and then, with even greater output expansion, from supervision and coordination problems that cause average labor utilization to turn upward.

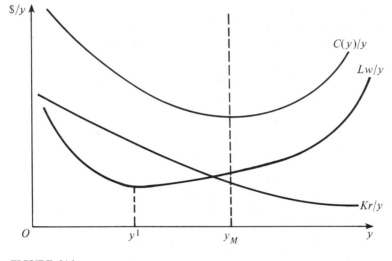

FIGURE 6A1

Equation (6A7) also yields additional information. It reveals that the magnitude of the percentage change in S resulting from a percentage change in w_i, whatever its sign, is larger the larger the share of expenditures on input i as a fraction of total costs. This is as one would expect, since the capacity of a change in w_i to influence the degree of scale economies (or any other cost property) will be negligible if factor i accounts for only a tiny portion of total costs. On the other hand, changes in the price of a factor on which expenditure looms large relative to total cost can be expected to have a large effect, and in the limit, as $\eta_i \rightarrow 1$, they tell the whole story. The case of homothetic production is also instructive, for then $e_{iy} = e_{TC}$, $\forall i$, and from (6A7) we see that in this case S is independent of factor prices. For an input utilized in a fixed proportion to output, as is true of some constituent materials, $e_{iy} = 1$ so that increases in its price increase the degree of scale economies where $S < 1$, decrease them where $S > 1$, and leave values of $S = 1$ unchanged. Examination of the factor share relationship yields another interesting result:

Proposition 6A3 $\partial S/\partial w_i \gtreqless 0$ *if and only if* $\partial \eta_i/\partial y \lesseqgtr 0$, *where* $\partial \eta_i/\partial y$ *is the effect of an increase in output on the proportion of expenditure on i to total cost.*

Proof

$$\frac{\partial \eta_i}{\partial y} = \frac{w_i}{C^2}\left[C\frac{\partial X_i}{\partial y} - X_i MC\right] = -\frac{\eta_i}{y}\left(\frac{1}{S} - e_{iy}\right) = -\frac{w_i}{yS^2}\frac{\partial S}{\partial w_i}. \quad \text{Q.E.D.}$$

Note that this result generalizes our preceding observation on homothetic production, because there cost shares are independent of output level.

6B. Multiproduct Cost Attributes

We turn next to formal analysis of the effects of a change in an input price on the most noteworthy attributes of multiproduct cost functions. We consider subadditivity, scale economies (including those for an individual product or for a limited set of products), economies of scope, and trans-ray convexity. The subadditivity analysis is literally the same as that in the single-product case, except that now we deal with an output *vector y* which is to be subdivided into a set of output vectors y^j such that $\sum y^j = y$. Thus (6A3) again constitutes the condition under which a rise in w_i will increase the absolute degree of subadditivity. Since, however, it is difficult to judge the degree of subadditivity of the multiproduct input demand function, the usefulness of this criterion is rather limited.

Economies of Scale

In Chapter 3, the multiproduct degree of scale economies was shown to characterize the local degree of homogeneity of the technology set and, when sufficient regularity conditions are satisfied, can be expressed as

$$(6B1) \qquad S_N(y, w) = \frac{C(y, w)}{\sum\limits_{j=1}^{n} y_j C_j(y, w)} = \frac{C}{y \cdot \nabla C}.$$

Differentiation with respect to w_i and use of Shephard's lemma yields

$$(6B2) \qquad \frac{\partial S_N}{\partial w_i} = \frac{X_i}{y \cdot \nabla C}\left(1 - S_N \sum\limits_{j=1}^{n} e_{ij}\right),$$

where $e_{ij} \equiv (y_j/X_i)(\partial X_i/\partial y_j)$ is the elasticity of the demand for factor i with respect to output j. Multiplying (6B2) by w_i/S_N and rearranging, we obtain

$$(6B3) \qquad \frac{w_i}{S_N}\frac{\partial S_N}{\partial w_i} = \eta_i\left(1 - S_N \sum\limits_{j=1}^{n} e_{ij}\right).$$

Equation (6B3) bears a strong resemblance to equation (6A7), especially when it is recalled that S_N can be interpreted as the elasticity of total cost with respect to movements along the ray through the point in question. It is also easily shown that the sum of the output elasticities of X_i is, analogously, precisely equal to the *ray* elasticity of demand for factor i. Once again, note that the relative share of total cost accounted for by factor i is an important determinant of the quantitative effect a change in its price will have. Thus, the interpretation of (6B3) is precisely the same as that of (6A7), as long as it is ray output quantities that are kept in mind.

Product-Specific Economies of Scale

The degree of scale economies specific to the product set $T \subseteq N$ was defined in Chapter 4 as

(6B4) $$S_T(y, w) = \frac{C(y, w) - C(y_{N-T}, w)}{\sum\limits_{j \in T} y_j C_j(y, w)} = \frac{IC_T(y, w)}{y_T \cdot \nabla C(y, w)}$$

Differentiating with respect to w_i and using Shephard's lemma, one obtains

(6B5) $$\frac{\partial S_T}{\partial w_i} = \frac{(y_T \cdot \nabla C)[X_i(y, w) - X_i(y_{N-T}, w)] - IC_T \sum\limits_{j \in T} y_j \partial X_i / \partial y_j}{(y_T \cdot \nabla C)^2}$$

This can be compared more readily with the preceding results by defining the concept of the *incremental demand* for factor i and its output elasticity in the natural manner. That is,

(6B6) $$IX_i^T(y, w) \equiv X_i(y, w) - X_i(y_{N-T}, w),$$

(6B7) $$e_{ij}^T = \frac{\partial IX_i^T}{\partial y_j} (y_j / IX_i^T).$$

Now, multiplying (6B5) by w_i / S_T and using (6B6) and (6B7), we obtain

(6B8) $$\frac{w_i}{S_T} \frac{\partial S_T}{\partial w_i} = \eta_i^T \left(1 - S_T \sum\limits_{j \in T} e_{ij}^T\right),$$

where $\eta_i^T \equiv w_i IX_i^T / IC_T$ is the relative share of *incremental* expenditures on factor i as a proportion of the total incremental cost of the product set T. The correspondence with the discussion of the effect upon overall scale economies is complete, as long as outputs and inputs are measured as increments over their quantities in stand-alone production of y_{N-T}.

Economies of Scope

By our definition in Chapter 4, if $P = \{T_1, \ldots, T_s\}$ is a partition of the product set N, then economies of scope are said to hold at output vector y with respect to this partition if

$$(6B9) \qquad C(y) < \sum_{k=1}^{s} C(y_{T_k})$$

where $\sum_{k=1}^{s} y_{T_k} = y$ and the components of y_{T_k} are zero for all products not in T_k. Thus, since economies of scope are related to the general subadditivity concept, an analog of (6A3) holds. Specifically, the absolute degree of economies of scope (defined analogously to the absolute degree of subadditivity) will be increased by a small rise in w_i if and only if

$$(6B10) \qquad X_i(y) < \sum_{k=1}^{s} X_i(y_{T_k}),$$

that is, if the demand for input i is characterized by economies of scope. This statement is, of course, valid whether or not the initial cost function is characterized by economies of scope. If economies of scope do hold for the cost function, then (6B9) can be written

$$(6B11) \qquad \sum_i w_i X_i(y) < \sum_i w_i \cdot \left[\sum_{k=1}^{s} X_i(y_{T_k}) \right].$$

Since this inequality holds for any particular partition of N, we know that (6B10) must hold for some input i.

The (relative) degree of economies of scope for a particular output and the partition of N into T and $N - T$ is defined as

$$(6B12) \qquad SC_T(y) \equiv \frac{C(y_T) + C(y_{N-T}) - C(y)}{C(y)}.$$

Analogously, we can define the relative degree of economies of scope in the demand for input i as

$$(6B13) \qquad SC_{T,i} = \frac{X_i(y_T) + X_i(y_{N-T}) - X_i(y)}{X_i(y)}.$$

Then, to derive our criterion we calculate

$$(6B14) \qquad \frac{w_i}{SC_T} \cdot \frac{\partial SC_T}{\partial w_i} = \frac{w_i X_i(y)}{C(y)} \cdot [SC_{T,i}/SC_T - 1].$$

The degree of economies of scope is the cost saving provided by simultaneous production of several goods as a percentage of their total production costs. If the percentage saving in the use of factor i achieved by producing the goods together is greater than that of overall costs, then a small increase in the price of input i increases the degree of economies of scope and thus makes production together relatively more attractive.

Thus, to understand the effects of input prices on economies of scope, it is necessary to consider the savings in the uses of various kinds of inputs that are made possible by common production. First, inputs i used in fixed proportions with individual outputs, such as one may expect of components of the products and some raw materials, will engender no savings in joint production, will have $SC_{T,i} = 0$, and will consequently yield $\partial SC_T/\partial w_i < 0$. Second, another extreme case (which was discussed in Chapter 4) is that of pure public inputs used in equal quantities, without any possibility of substitution, as part of the fixed cost of each separate production process. Then, $X_i(y) = X_i(y_T) = X_i(y_{N-T})$, $SC_{T,i} = 1$, and, since $SC_T < 1$, (6B14) implies that increases in such inputs' prices must increase the degree of economies of scope.

Other kinds of pure public inputs have effects whose directions are ambiguous. On the one hand, in joint production the same units of these inputs can be utilized simultaneously in both production processes. Thus, this sharing effect tends to yield savings in the use of public factors, i, tending to make $SC_{T,i}$ positive, as before. On the other hand, the sharing possibilities reduce the effective prices of the services of public inputs for their use in each production process. Consequently, to the extent that these services can be substituted for other inputs, the optimal utilization of public inputs will be larger in joint production than it is in stand-alone production. This substitution effect can outweigh the sharing effect to render $SC_{T,i} < 0$, with the implication of (6B14) that increases in the prices of such inputs will diminish the degree of economies of scope if that is initially positive.

It also follows from this argument that inputs that apparently cannot be shared physically among production processes may nevertheless be affected by joint production. There will be savings in the quantities of inputs substitutable for the services of shared factors in joint production, while the use of inputs complementary with the services of public factors will tend to be expanded in joint production. Thus, increases in the prices of the former may raise SC_T, while increases in the prices of the latter will diminish the degree of economies of scope.

Trans-Ray Convexity

The last multiproduct measure we discuss is trans-ray convexity. The effect of a small change in input price upon trans-ray convexity is easily seen. For any y_a and y_b lying in the trans-ray cross section in question, trans-ray convexity requires

(6B15) $C(ky_a + (1 - k)y_b) \le kC(y_a) + (1 - k)C(y_b)$

for $0 < k < 1$. Let DT represent the difference between the right and left side of (6B15), and DT the absolute degree of trans-ray convexity. By Shephard's lemma we conclude immediately that $\partial DT/\partial w_i$ will be positive only if a relationship directly analogous to (6B15) holds for the input demands $X_i(y)$ derived from the cost function in the inequality. Thus, a small increase in the price of input i will increase the absolute degree of trans-ray convexity (whether or not it is positive) if and only if the derived demand for input i is itself trans-ray convex.

Trans-ray convexity of the derived demand for input i can be interpreted graphically in terms of the shape of the input demand surface, $X_i(y)$, over output space. Figure 6B1 depicts this surface for a capital input, and shows

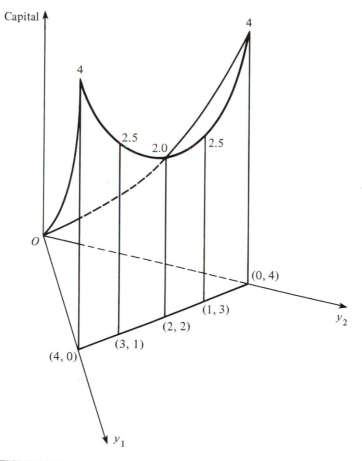

FIGURE 6B1

it to be trans-ray convex over the trans-ray hyperplane $\{y \geq 0 \,|\, y_1 + y_2 = 4\}$. Here, the trans-ray convexity arises for two reasons. First, movements from the edges toward the middle of the hyperplane reduce the output of each individual good, and, in the case depicted, contractions in output reduce capital utilization more than proportionately. Second, movements to the middle of the hyperplane decrease the specialization of production and increase the opportunity for sharing of capital facilities, thereby permitting a reduction in capital purchases. Our general interpretations of trans-ray convexity, given in Chapter 4, are pertinent here, as are our remarks on shared inputs earlier in this section.

6C. Input Prices and Efficient Industry Structure

The main objective of our analysis of the effects of changes in factor prices on cost properties is to determine the consequences for the typical size and number of firms in the efficient industry configuration for the production of a given output vector. Here we must use the bounds upon the cost-minimizing number of firms derived in Chapter 5.

In the single product case, let us assume, as in Chapter 5, that average costs are U-shaped, attaining a unique minimum at output y_M. Analogously, for the multiproduct case, we assume that the ray average cost curves are U-shaped, with the points that minimize the average costs along each ray constituting the M locus. The bounds upon the cost-minimizing number of firms were derived in Chapter 5 directly from the location of y_M or the M locus. Changes in input prices therefore affect the bounds on the cost-minimizing number of firms through their effects on y_M or on the M locus.

In the single-product case, the AC-minimizing output $y_M(w)$ is implicitly defined by the identity

$$(6C1) \qquad\qquad S(y_M, w) - 1 = 0$$

since at the minimum point of the AC curve returns to scale are locally constant. Hence, the effect of a change in the price of factor i can be evaluated using standard comparative statics analysis:

$$(6C2) \qquad\qquad \partial y_M / \partial w_i = -\partial S / \partial w_i / \partial S / \partial y.$$

From the assumption that the average (or ray average) cost curve reaches a well-behaved global minimum at y_M, we know that S decreases from a value greater than one to less than one as y passes through y_M, so that the denominator of (6C2) can be taken to be negative. Thus,

$$(6C3) \qquad\qquad \text{sign}(\partial y_M / \partial w_i) = \text{sign}(\partial S / \partial w_i).$$

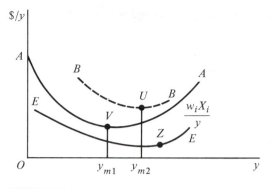

FIGURE 6C1

Using (6A6) and (6C1) we have

$$\text{(6C4)} \qquad \text{sign}(\partial y_M / \partial w_i) = \text{sign}(1 - e_{iy}).$$

Therefore an increase in the price of any factor which is *not* a *superior* input (i.e., $e_{iy} \leq 1$) will increase the size of the output at which average costs are minimized. This situation is depicted graphically in Figure 6C1, where a rise in the price of i shifts the AC curve from AA to BB, moving the minimum point rightward from V to U. While this is a well-known result (see Bassett and Borcherding, 1970), somewhat more insight into this process can be obtained by plotting $w_i X_i(y, w)/y$, the expenditure on factor i per unit of output. A useful way of interpreting condition (6C4), in keeping with our earlier discussion, is that the average cost curve will change to resemble more closely the shape of the average expenditure curve of the factor whose price has risen. For inputs whose utilization increases less than proportionally with output, the minimum point of that curve, Z, occurs at a *larger* output level than that which minimizes average cost. Consequently, increases in the prices of such inputs cause the minimum point of the AC curve to shift out from a point such as V on AA to U on BB. Note that, since costs are an increasing function of each factor price, the AC curve must shift upward as well. (Of course, as drawn, the move from V to U requires a nonlocal change in w_i, which will also shift EE. This effect will be examined in Section 6D.)

Finally, as one would expect intuitively, only changes in *relative* prices can affect the position of y_M. That is, if all input prices rise in proportion, the average cost curve merely shifts upward vertically by that same proportion. In other words, the function $y_M(w)$ is homogeneous of degree zero. To see this, we substitute (6A7) and (6C1) into (6C2), and sum over all inputs, obtaining

$$\text{(6C5)} \qquad \sum_i w_i \frac{\partial y_M}{\partial w_i} = \left(\sum_i w_i X_i - y \sum_i w_i \frac{\partial X_i}{\partial y} \right) \bigg/ C \frac{\partial S}{\partial y} = (C - yMC) \bigg/ C \frac{\partial S}{\partial y} = 0.$$

It requires only minor modifications to extend this analysis to the multi-product case. Let $d(\bar{y})$ denote the distance from the origin to where ray average cost reaches its minimum along the ray characterized by $\bar{y}$. Then, one can show, using (6B1), (6B3), and the multiproduct analog of (6C2), that

$$\text{sign } \frac{\partial d(\bar{y})}{\partial w_i} = \text{sign}\left(1 - \sum_{j=1}^{n} e_{ij}\right).$$

Thus, as in the scalar case, along any ray the point on the M locus moves outward (inward) in response to an increase in the price of factor i if the utilization of that input increases less than (more than) proportionately with an expansion of output along that ray. If input i exhibits the same (local) property along all rays, then, by the analysis of Chapter 5 this will tend to increase (decrease) the average size of firm and to reduce (enlarge) the efficient number of firms for the production of any given output vector. For it will increase (decrease) both the upper and lower bounds upon the efficient number of firms for the production *of any given level of industry output*.

While the conclusion about the probable effects upon the average size of firm is the end of *that* story, a complication besets the observation about the efficient number of firms. For a *ceteris paribus* rise in w_i will raise total cost and, very likely, average and marginal costs as well. Hence, product prices are likely to change as a result and, with that, the vector of industry outputs will also change. In the single-product case, if a rise in w_i raises y_M and reduces industry output y^I, it will obviously tend to decrease the efficient number of firms which is approximated by y^I/y_M. However, if the rise in w_i reduces y_M as well as y^I, we obviously cannot tell what will happen to the number of firms in the industry without examining demand conditions along with the analysis of cost.

In the multiproduct case, there are additional complications. First, a rise in w_i may move some portions of the M locus closer to the origin and others further from it. It may, for example, affect only specialized firms (whose outputs are given by points on the axes) or only multiproduct firms, and so increase or decrease the concavity (convexity) of the locus.

Second, even if the entire M locus shifts outward or shifts inward, the changes in product prices induced by the rise in w_i may well change the output proportions in the industry output vector. Even if prices rise on the average, the quantity of some one (or more) of some firm's products may actually rise. The industry output ray will therefore change, and the corresponding new bounds on the cost-minimizing number of firms may bear little direct relation to the old bounds derived from a different ray. Thus, without a full analysis of both cost and demand conditions, we can apparently do no more than speculate roughly about the effects of an input price change on equilibrium industry structure that hold in general.

6D. Nonlocal Input Price Changes

The analysis of Sections 6B and 6C deals with small changes in input prices for which the rate of change of total cost with respect to an input's price is, by Shephard's lemma, equal to the quantity of the input demanded at the initial price. However, when an input price changes by a substantial amount, we can expect corresponding changes in the quantity of that input demanded. The input demand surface, toward which the cost surface is bent by a rise in that input's price, will therefore itself change shape. As the price of an input rises, the derived demand surface will fall (or at least not increase) because cost functions are concave in input prices. Since less of the input is then used at any given output vector, one might perhaps expect its influence on the shape of the cost function to be decreased rather than increased as the results of Sections 6B and 6C seem to indicate.

Because the cost function is the sum of the expenditures on the various inputs, $C(y) = \sum w_i X_i$, intuition suggests that the net effect of a rise in w_i and the consequent fall in X_i upon the influence of input i on the shape of the cost function depends upon the effect of the change in w_i upon the relative expenditures on the different inputs. For those cost measures involving unitless ratios—degrees of scope, scale, and product-specific scale economies—this conjecture turns out to be exactly correct. The reason this is so is that each of these measures of the shape of the cost function can be written as a convex combination of the corresponding measures for the demand surfaces for the individual inputs, with the weight for input i being its share of relevant total cost. Specifically, it is easy to show that

$$(6D1) \qquad e_{TC} = \sum_i \frac{w_i X_i}{\left(\sum_k w_k X_k \right)} e_{iy} = S^{-1}$$

since, by (6A5), the middle term equals

$$\sum_i \frac{w_i X_i}{\sum_k w_k X_k} \frac{y}{X_i} \frac{\partial X_i}{\partial y} = y \sum_i w_i \frac{\partial X_i}{\partial y} \Big/ C = \frac{yMC}{C} \equiv e_{TC}.$$

Similarly, for the multiproduct case we obtain

$$(6D2) \qquad e_{TC} = \sum_i \frac{w_i X_i}{\left(\sum_k w_k X_k \right)} \sum_{j=1}^{n} e_{ij} = S^{-1},$$

$$(6D3) \qquad SC_T = \sum_i \frac{w_i X_i(y)}{C(y)} SC_{T,i},$$

and

(6D4) $$e_{IC_T} = \sum \frac{w_i IX_i^T(y)}{IC_T(y)} \sum_{j \in T} e_{ij}^T = S_T^{-1}.$$

Thus, *ceteris paribus*, the influence of the shape of a given input's demand surface will increase or decrease as a change in input price increases or decreases its relative share of the relevant total cost. Of course, expenditures on a given input rise when its price increases if and only if the input's demand is inelastic. For a rise in its price to raise an input's *relative share* in total cost, its elasticity of demand must be inelastic *a fortiori* since

$$\frac{\partial(w_i X_i/C)}{\partial w_i} = \frac{X_i}{C} \cdot (1 + \varepsilon_{ii} - w_i X_i/C),$$

where ε_{ii} is own price elasticity of demand for input i.

6E. Toward Application

In order to move toward application of our results to the discussion of the difference in industry structures between industrialized and less developed economies, we employ a very simple stylized model. We assume that there are well-developed world markets for capital and all other nonlabor inputs, so that the prices of these factors are equal in both countries. But the wage rate is taken to be much lower in the LDC. In addition, we assume that technology is such that *over all of the relevant range* an expansion in output leads to an expansion in labor input that is less than proportionate (i.e., labor is *not* a superior input). Finally, we must assume that the *output* of the industry in question is a nontraded good, such as construction or consumer services, because under our other postulates, the industry in the industrialized country would be driven out of world markets by its high wage rates.

Figure 6E1 depicts this situation. Because wages w^I are higher in the industrialized country, its average cost curve, $AC(y, w^I)$, lies everywhere above that of the LDC, $AC(y, w^L)$. If the demand for labor is price-elastic, the reverse relationship will hold between the labor expenditure curves, $w^L X(y, w^L)$ and $w^I X(y, w^I)$. The assumption that labor demand increases less than proportionately with o tput over the relevant ranges causes the average labor expenditure curves to attain their minima at larger outputs than is true for the average cost curves. This is so both in the industrialized country

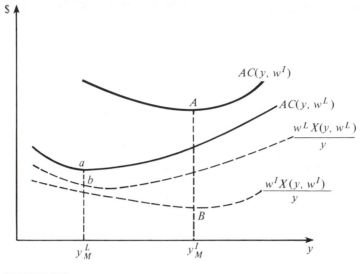

FIGURE 6E1

and the LDC, as well as in any country with intermediate wage rates. As was brought out in Sections 6C and 6D, this nonsuperiority of labor input is also responsible for the fact that $y_M^I \gg y_M^L$. Finally, the assumption that the demand for labor is (price) elastic means that the share of total cost comprised of payments to labor will be smaller in the industrialized country than in the LDC: that is, $by_M^L/aby_M^L > By_M^I/AB y_M^I$ in Figure 6E1.

The economic relationships behind these formal conclusions are straightforward. In either country, and at any relative wage rate, larger long-run outputs by an enterprise make it efficient for it to invest in larger amounts of capital using plants that require less labor per unit of output. As a consequence of this substitution of capital for labor, at any relative factor prices, the output elasticity of labor utilization is less than one. The overall technological process, including choice of plant and choice of labor–output ratio, is subject eventually to diminishing returns, so that the average cost and average labor utilization curves do eventually turn upward.

Now, as wages rise relative to capital rental rates, the financial savings from the substitution of capital for labor rise, so that efficient enterprises will be motivated to invest in large labor-saving plants at smaller output levels than they would otherwise, and to do so more persistently as the labor savings per unit of capital reach diminishing returns. Consequently, where wage rates are high, economies of scale persist well past the point at which they are exhausted in regions where low wages prevail. Even so—and despite the fact that the higher wage rates induce a smaller *share* of spending on labor—where wages are lower, total costs and average costs for any given output are also lower. And, in this scenario, industry is less concentrated as well.[5]

6F. Summary Remarks

This chapter is, in a sense, a digression. Throughout the rest of the book input prices are, for simplicity, held constant. But here we have paused to examine the consequences of changes in input prices for industry structure. We have provided some rough conclusions about the implications for the average size of firm in an efficient industry structure, and a bit about the consequences for the cost-minimizing number of firms. Since, as we see later, in a contestable market any sustainable industry structure must be efficient, the conclusions about efficient industry structure to which we have just referred also have some behavioral counterparts. In this chapter we have also sought to suggest, by means of our LDC parable, how profoundly industry structure can be affected by larger changes in industry prices. But in doing so we have also, perhaps, offered some further insights into the character of our formal analysis.

---

[5] Lest this be misread as an argument for low wages, we ask the reader to recognize that the authors are themselves wage earners and that we are concerned primarily with relative factor prices. Perhaps we should have cast our argument in terms of low energy prices—to which the theory applies equally well.

7

Natural Monopoly

Sufficient Conditions
for Subadditivity

In this chapter we examine economic conditions that are sufficient to ensure that an industry is a natural monopoly. In technical terms, we investigate sets of sufficient conditions for the cost function to be subadditive, so that production by a single firm is more efficient than any other industry structure.

7A. Subadditivity and Equilibrium in Contestable Markets

Our quest for the results discussed in this chapter began when we found that there was a somewhat surprising confusion in the literature over the precise meaning of the widely used term "natural monopoly." Before attempting to master the complexities besetting the definition, characterization, and evaluation of equilibrium in a monopoly industry, a thorough grasp of the supply or cost side of the analysis is essential. In particular, it is necessary to grapple with the concept of the subadditivity of the costs of a multiproduct firm. This investigation relies heavily upon the cost concepts examined in Chapters 3 and 4.

In order to provide motivation for our discussion of the *technological* determinants of subadditivity, we remind the reader of a result described briefly in Chapter 2 (which will be discussed fully in Chapter 8): For single-firm production to constitute an equilibrium in a perfectly contestable market, it is *necessary* that the firm be a natural monopoly at its output quantity.

Moreover, as shown in Chapter 11, for any firm to be in equilibrium in contestable markets, it is necessary that it be a natural monopoly with respect to its own output level. And this holds whether the industry is a natural monopoly, is perfectly competitive, or is an oligopoly. Therefore, the presence or absence of subadditivity in the relevant regions of the *firm's* cost function has profound implications for the existence and nature of equilibrium in any type of contestable market, competitive, monopolistic, or oligopolistic. While the importance of subadditivity is most obvious in the monopoly case, the technical characteristics of the cost function that lead to subadditivity implicitly play an important role, usually an unrecognized one, in all equilibrium analysis. The connections between these cost characteristics and subadditivity are analyzed in detail in the later sections of this chapter.

7B. Remarks on the Nature of Subadditivity

Before proceeding to this analysis, it is useful to discuss the concept of subadditivity itself. Intuitively, it is a simple and appealing concept. A cost function is subadditive for a particular output vector y when y can be produced more cheaply by a single firm than by *any* combination of smaller firms. That, surely, is what anyone has in mind, at least implicitly, when speaking of a monopoly's being "natural," and that is what economists were undoubtedly groping for when they (as it turns out, mistakenly) identified natural monopoly with economies of scale.[1]

Unfortunately, the intuitive appeal of the subadditivity concept is counterbalanced by its analytic elusiveness. We often cannot recognize whether a particular function is or is not subadditive simply by looking at its mathematical expression or its graph. Moreover, there apparently exist no straightforward mechanical criteria that permit us to test whether or not a particular function is subadditive, short of checking the applicability of the definition itself. That is, there exist no conditions necessary *and* sufficient for subadditivity that are analytically simpler than the definition. Instead, we will provide some conditions that are sufficient and others that are necessary for subadditivity. These conditions will also yield some intuitive grasp of the more recondite properties of the concept of subadditivity.

We state as a proposition the unfortunate attribute of subadditivity that unavoidably complicates empirical tests to determine whether or not

---

[1] See Sharkey (1982, Chap. 2) for a thorough discussion of the historical development of the applications and definitions of the term "natural monopoly."

it is satisfied:

Proposition 7B1 *Subadditivity is a global, not a local concept. Specifically:*
(i) To determine that a cost function $C(\cdot)$ is subadditive at some output vector
y, it is necessary *to have information (explicit or implicit) about the value of*
$C(\cdot)$ for every possible output vector smaller than y, that is, for all $y^* \neq y$,
$y^* \leq y$; (ii) information about the cost function throughout the region just
specified is sufficient *to determine whether the cost function is subadditive at y.*

Proof of Necessity Suppose the contrary, that there exists a $y^* \leq y$,
$y^* \neq y$, such that the value of $C(y^*)$ is unknown. Then, whatever the values
of $C(y)$ and $C(y - y^*)$, by hypothesis we do not know whether or not $C(y^*) \geq$
$C(y) - C(y - y^*)$. But if that inequality were violated, then

$$C(y^*) + C(y - y^*) < C(y)$$

and $C(\cdot)$ would not be subadditive at y. That is, complete ignorance of
$C(y^*)$ for any single output vector y^* in the region $y^* \leq y$ prevents the
conclusion that $C(\cdot)$ is subadditive at y.
Proof of Sufficiency If y is subdivided in any way, into $y^1, \ldots, y^m$
with $y^i \geq 0$, $y^i \neq 0$, $m \geq 2$, and $\sum_{i=1}^{m} y^i = y$, then clearly $y^i \leq y$ and $y^i \neq y$.
Hence knowledge of $C(y^i)$ for all such y^i permits a direct test of whether
or not $\sum_{i=1}^{m} C(y^i) > C(y)$.   Q.E.D.
This proposition reveals that, unlike the property of scale economies
at y, subadditivity at y cannot be conclusively assessed from data about
costs only in the vicinity of y. The cost surface must be scrutinized not merely
in the neighborhood of that point, but also all the way to the axes and the
origin. This is a very demanding task for empirical work because it is likely
to require data well outside the range of available observations. If in recent
decades a firm has produced no quantities varying by more than, say, 25
percent from its current output, then there will be no statistical data (except
perhaps for some obsolete figures) that can indicate the shape of the cost
function anywhere near the origin or the axes. Thus, the data requirements
of a statistical test of subadditivity can be very severe indeed.

Yet this problem is unavoidable, for reasons that are readily seen intuitively. We cannot know whether an industry is a natural monopoly for the
production of the output vector y if we have no information ruling out the
possibility that many small firms (or several intermediate-sized firms or
some combination of the two) can produce y more cheaply than can a single
producer. Thus, to prove subadditivity, we must have information on the
costs of *every* potential small or intermediate producer; and that is why
we must know the cost function for a firm for *every* $y^* \leq y$.

We can see now why analysts have attempted to use the analytically and
statistically tractable concepts of scale economies as a surrogate test of

natural monopoly. Unfortunately, as will be shown here, such traditional tests simply cannot do the job.

However, there are legitimate ways to go about testing for subadditivity in practice. Specifically, what can be done empirically is to test the various sets of necessary and sufficient conditions for subadditivity that are examined in the bulk of this chapter. Rejection of necessary conditions permits the inference that subadditivity does not hold. Failure to reject sufficient conditions is evidence of subadditivity. While we may not be able to recognize directly whether or not a cost function is subadditive, we may easily, for example, be able to test whether its ray average costs decrease throughout the relevant region, and that, as will be shown, is one of the recurrent attributes in sets of sufficient conditions for subadditivity. For these reasons, among others, we will examine the issue of sufficient conditions for subadditivity in considerable detail.[2]

7C. Some Improper Tests of Natural Monopoly

Our quest for a set of readily interpretable conditions sufficient to guarantee that a multiproduct cost function is characterized by subadditivity must begin with an intuitively appealing candidate, the concept of scale economies, which is widely associated with the idea of natural monopoly.[3] Our first result is a negative one.

Proposition 7C1 *Multiproduct economies of scale are neither* necessary *nor* sufficient *for subadditivity.*

Proof (i) Lack of *necessity* is an immediate corollary of Proposition 2A1, which provides an example of a subadditive scalar-output cost function with decreasing returns to scale over a portion of the relevant range. (ii) *Insufficiency* is demonstrated by the following example:

$$(7C1) \quad C(y_1, y_2) = y_1^a + y_1^k y_2^k + y_2^a; \quad 0 < a < 1, \quad 0 < k < \tfrac{1}{2}.$$

This function exhibits globally increasing returns to scale, as is confirmed by our Definition 3B1, by which

$$(7C2) \quad S_N \equiv \frac{C}{y_1 C_1 + y_2 C_2} = \frac{y_1^a + y_1^k y_2^k + y_2^a}{a y_1^a + 2k y_1^k y_2^k + a y_2^a} > 1.$$

---

[2] Of course, this by itself does not completely circumvent the difficulty caused by absence of cost observations for outputs outside the range of relevant experience. There, engineering estimates and other substitutes for direct observation must be found. We return to the issue of empirical testing of subadditivity in Chapter 15.

[3] Kahn (1970, p. 123) and Mansfield (1975, p. 267) are just two examples.

However, (7C1) is not subadditive everywhere, because

(7C3) $$C(1,0) + C(0,1) = 2 < 3 = C(1,1).$$ Q.E.D.

Since the cost function in (7C1) is globally strictly concave, we can also state

Proposition 7C2   *Global strict concavity of the cost function is not sufficient for cost subadditivity.*

The reader should not be surprised at these results. Indeed, given the crucial role of various forms of cost complementarity and economies of joint production, it is to be expected that economies of scale cannot tell the entire subadditivity story in the multiproduct case. This suspicion is increased by inspection of equation (7C3), which reveals that the cost function in (7C1) exhibits *diseconomies* of scope. Thus, since the production of each individual good adds to the costliness of production of the other, rather than complementing it (i.e., since $C_{12} > 0$), it is not surprising that subadditivity fails. Or, looked at in another way, subadditivity (natural monopoly) is violated when cost can be reduced either by breaking the firm into a set of smaller firms, each with the same output proportions as the large enterprise, or, instead, by replacing the large firm by a set of more specialized firms, with each of the latter perhaps producing only one item in the initial bundle of outputs. Now, economies of *scale* guarantee that money cannot be saved in the first of these ways—by the introduction of smaller firms with proportionate reductions in all outputs. But scale economies do not preclude economies of specialization, and so they cannot preclude failure of subadditivity through absence of complementarity.

Somewhat less obvious, however, is the fact that even the intuitively appealing combination of economies of scale and economies of scope is inadequate to guarantee subadditivity.[4]

Proposition 7C3   *Economies of scale* and *economies of scope do* not *imply subadditivity.*

Proof   Consider the cost function

(7C4) $$C(y_1, y_2) = 10v + 6(x - v) + z + \varepsilon, \qquad C(0,0) = 0,$$

where $x = \max(y_1, y_2)$, $v = \min(y_1, y_2)$, $z = \min(v, x - v)$, and ε is an arbitrarily small positive number. This function would be linearly homogeneous except for the fact that the presence of the fixed cost $\varepsilon > 0$ introduces

---

[4] Sharkey (1982, Chap. 4) provides an alternative demonstration of this result, as well as a discussion of some conditions which *are* sufficient for subadditivity.

global economies of scale.[5] Clearly, for the case of stand-alone production we obtain $C(y_i) = 6y_i + \varepsilon$ so that

(7C5) $C(y_1, 0) + C(0, y_2) = 6y_1 + \varepsilon + 6y_2 + \varepsilon = 6(x + v) + 2\varepsilon.$

Subtraction of (7C4) yields

(7C6) $C(y_1, 0) + C(0, y_2) - C(y_1, y_2) = 2v - z + \varepsilon > 0,$

thereby showing that the function exhibits economies of scope. However, without loss of generality assume that $y_2 > y_1$ and consider the division of outputs (y_1, y_2) into $(y_1, y_1) + (0, y_2 - y_1)$. This division must decrease costs and violate subadditivity for ε sufficiently small. This follows because

(7C7) $C(y_1, y_1) + C(0, y_2 - y_1) = 10y_1 + 6y_2 - 6y_1 + 2\varepsilon$

$$= 10v + 6(x - v) + 2\varepsilon$$

Subtraction of this from (7C4) yields

(7C8) $C(y_1, y_2) - C(y_1, y_1) - C(0, y_2 - y_1) = z - \varepsilon = \min(y_1, y_2 - y_1) - \varepsilon.$

Since ε can be chosen arbitrarily small without violating the properties of global economies of scale and scope, the expression in (7C8) can be rendered positive for any $y_2 > y_1$. Thus the cost function is not subadditive.   Q.E.D.

7D. Sufficient Conditions for Subadditivity without Strong Complementarity

Having established that the demonstration of subadditivity is no simple matter, we now provide a collection of conditions that *are* sufficient for subadditivity. The reader should be forewarned, however, that all of these conditions are excessively strong in that they contain many elements that are clearly *not* necessary for subadditivity. In fact, of the cost concepts presented so far, only the condition of economies of scope is necessary for subadditivity since, as shown in Section 4B, it is required for subadditivity over orthogonal output vectors and therefore is a prerequisite for overall subadditivity.

It is useful, first, to define and investigate another necessary condition for global subadditivity, one that may be considered to constitute a sort of partial subadditivity:

---

[5] An economic model yielding this cost function is presented in Chapter 9.

Definition 7D1: Ray Subadditivity   A cost function $C(\cdot)$ is said to be strictly ray-subadditive at y if, for any set of two or more positive numbers v_i that sum to one, $\sum C(v_i y) > C(y \sum v_i) = C(y)$.

By focusing upon subadditivity along a ray, or indeed along *all* rays, we reduce the concept to the one-dimensional phenomenon discussed in Chapter 2. We can therefore state

Proposition 7D1   (*Strictly*) *Decreasing ray average costs up to y imply ray subadditivity at y.*

Proof   Consider $m \geq 2$ output vectors along the same ray: $v_1 y, \ldots, v_m y$, with $1 > v_i > 0$ and $\sum_{i=1}^{m} v_i = 1$. Decreasing ray average costs ensure

(7D1) $$C(y) < C(v_i y)/v_i \qquad i = 1, \ldots, m.$$

or

(7D2) $$v_i C(y) < C(v_i y) \qquad i = 1, \ldots, m.$$

Summation of the expressions (7D2) for all i then yields the desired condition

(7D3) $$C(y) \equiv \sum_{i=1}^{m} v_i C(y) < \sum_{i=1}^{m} C(v_i y). \qquad \text{Q.E.D.}$$

Thus we confirm, as already suggested, that *DRAC* (decreasing ray average costs) ensures that it is not possible to divide a firm's output vector *proportionally* among several firms without increasing total cost. But, as the previous examples confirm, this is not the only threat to subadditivity: Other partitions must also be assessed. Nevertheless, this sufficient condition for subadditivity along rays in output space is important for our subsequent analysis. The preceding result immediately permits us, with the aid of results from Chapter 4, to describe the true relationship of economies of scale to subadditivity:

Proposition 7D2   *Increasing returns to scale up to y imply decreasing RAC up to y, and, hence, ray subadditivity at y.*[6]

(The converse of Proposition 7D2 does *not* hold, as demonstrated by the scalar-output example employed in the proof of Proposition 2A1.)

---

[6] The proof is essentially the same as that in the scalar-output case (Proposition 2A1).

As already noted, economies of scale and economies of scope together do not suffice for subadditivity. Since, speaking loosely, they represent rather weak measures of the savings effected by size and combination of outputs, respectively, it is obviously necessary to strengthen one or the other to obtain conditions sufficient for subadditivity. As shown in Chapter 4, the "primitive" concept of economies of scope is, in the analytic sense, a rather weak manifestation of the general phenomenon of cost complementarity. Most of our sufficient conditions for subadditivity therefore substitute an analytically stronger form of cost complementarity, retaining the assumption of economies of scale. Our next result, however, uses the opposite approach, combining economies of scope with a stronger substitute for economies of scale.

Proposition 7D3 *Decreasing average incremental costs of each product, up to y, and economies of scope at y imply subadditivity at y.*

The key to the argument behind this result lies in the fact that $DAIC$ (decreasing average incremental cost) in a product line, unlike decreasing RAC, implies that *that product line* must be monopolized if industry cost is to be minimized. For $DAIC_i$ implies that any way of subdividing the output of good i, holding all other outputs constant, must increase the total cost of the industry.

To analyze the role of $DAIC_i$, consider any output vector y, and divide it into two batches, y^a and y^b, each with a positive quantity of good i. Let y_i^a be the output of good i in batch y^a and let y_{N-i}^a be the vector containing the quantities of all other outputs in $y^a = y_i^a + y_{N-i}^a$. Then, as a step toward Proposition 7D3, we show (in the Appendix)

Lemma 7D1 *If $DAIC$ holds for product i, then either*

$$(7D4) \qquad C((y_i^a + y_i^b) + y_{N-i}^a) + C(0 + y_{N-i}^b) < C(y_i^a + y_{N-i}^a)$$
$$+ C(y_i^b + y_{N-i}^b) \equiv C(y^a) + C(y^b)$$

or

$$C((y_i^a + y_i^b) + y_{N-i}^b) + C(0 + y_{N-i}^a) < C(y^a) + C(y^b).$$

Intuitively, this asserts that in an industry in which the average incremental cost of good i is declining, it will save money, *ceteris paribus*, for all of good i to be produced by one of the firms rather than having the *same total quantity of i divided* among several firms. This is a surprising result, for one might have supposed that cost complementarities could require the dispersal of the production of the good to obtain the greatest industry-wide savings from economies of scope.

Now, to complete the proof of Proposition 7D3, we use the Lemma sequentially until an orthogonal partition is achieved; that is, until the two firms are producing totally nonoverlapping vectors of quantities $y^{a'}$ and $y^{b'}$, saving money for the industry at each step in the specialization process. Thus we have, from the Lemma,

(7D5) $$C(y^{a'}) + C(y^{b'}) < C(y^a) + C(y^b),$$

where

(7D6) $$y^{a'} + y^{b'} = y^a + y^b = y \quad \text{and} \quad y^{a'} \cdot y^{b'} = 0.$$

The Proposition now follows via (weak) economies of scope, which gives us $C(y^{a'}) + C(y^{b'}) \geq C(y^{a'} + y^{b'})$, so that by (7D5) and (7D6) we have the subadditivity result

$$C(y^a) + C(y^b) > C(y^{a'} + y^{b'}) = C(y^a + y^b). \qquad \text{Q.E.D.}$$

In Chapter 11, Lemma 7D1 will be shown to have other important implications for the analysis of more complex market forms. Its general usefulness derives from the fact that it establishes that savings are obtainable from single-firm production of a good even when $DAIC$ pertains to that good only. Thus, in our exploration of the territory between pure competition and pure monopoly, we are able to begin the characterization of market equilibrium in this area by noting that products for which $DAIC$ holds must be produced by just a single firm, which may also actively compete with other firms in the markets for the other goods produced by the industry.

7E. Conditions Sufficient for Subadditivity Using Stronger Complementarity Requirements

Our remaining propositions on subadditivity rely upon sets of premises that combine economies of scale and forms of cost complementarity somewhat stronger than economies of scope. Since the crucial ingredient in all these recipes is the ability to "support" a cross section of the cost hypersurface at the output vector in question by an appropriately defined hyperplane, it is not surprising that convexity properties of the cost function play a key role. Although we later present a sharper result, we choose as our first set of convexity conditions one that lends itself to a proof of sufficiency for subadditivity which is constructive and readily interpretable economically. We also find this result particularly useful throughout our subsequent discussions because of the relation of its assumptions to those employed in much of our analysis of the issue of sustainability.

Proposition 7E1 *Two conditions together sufficient to guarantee strict subadditivity of costs at an output vector y^0 are (1) (nonstrict) trans-ray convexity along any one hyperplane $H = \{y \,|\, w \cdot y = w_0, w > 0\}$ through y^0 and (2) strictly declining ray average costs up to hyperplane H.*

The argument is essentially straightforward, though it requires a bit of algebraic manipulation. In the text we discuss it only in heuristic terms, relegating the formal proof to the Appendix.

By Proposition 7D1 we know that declining ray average costs imply that costs are ray-subadditive at y^0. Therefore, we need only consider an arbitrarily chosen pair of output vectors on different rays, $y^1 \neq 0$, $y^2 \neq 0$, which satisfy

$$y^1 + y^2 = y^0$$

(Figure 7E1). We now proceed by locating any hyperplane H through y^0 along which trans-ray convexity holds. Then, y^1 is expanded proportionately to $v_1 y^1$, the point on H which intersects the ray through y^1; and, similarly, the outputs of y^2 are expanded proportionately to $v_2 y^2$ on H. Moreover, since all three points lie on H, y^0 can be expressed as a weighted average of $v_1 y^1$ and $v_2 y^2$. Thus, we proceed from y^1 and y^2 to y^0 in two steps: proportionate expansion of y^1 and y^2, then averaging of the two expanded vectors. The point of the proof is that some cost savings result from each of these two steps, so that the cost of y^0 must be smaller than the total costs of producing y^1 and y^2 separately. By declining ray average costs, the first of these steps saves money because total costs increase less than proportionately as y^1 and

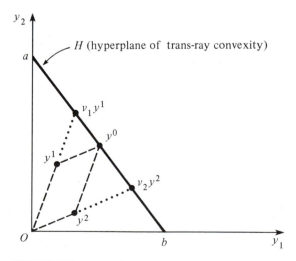

FIGURE 7E1

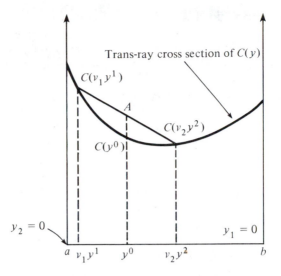

Trans-ray cross section of $C(y)$

FIGURE 7E2

y^2 are increased proportionately to H. Then, by trans-ray convexity, the second step reduces (or does not increase) cost because the averaging of $v_1 y^1$ and $v_2 y^2$ yields a cost of y^0 which is less than or equal to the average of the costs of $v_1 y^1$ and $v_2 y^2$ [Figure 7E2 is the standard convexity diagram for our trans-ray cross section, showing why $C(y^0)$ must be no greater than A, the weighted average of $C(v_1 y^1)$ and $C(v_2 y^2)$]. Thus, in each of our two steps going from points y^1 and y^2 to y^0, some savings are effected, and that is why $C(y^0) < C(y^1) + C(y^2)$ as subadditivity requires.[7]

Conditions for subadditivity that rely without modification upon convexity arguments have a serious and, as we see next, an unnecessary shortcoming: They require much more continuity than the analyst will want to count upon (for convex functions must be continuous "almost everywhere," except over a set of measure zero). Several times our discussion has reminded the reader that the cost functions we study may entail jump discontinuities

---

[7] By reducing the range over which convexity is required, Proposition 7E1 extends a well-known mathematical result, attributable to Rosenbaum (1950) which asserts

Proposition 7E2 *If C is (nonstrictly) convex and strictly subhomogeneous (DRAC), then C is strictly subadditive.*

Proof (Sharkey and Telser, 1978)

$$C(y^1 + y^2) = C[2(y^1 + y^2)/2] < 2C\left(\frac{y^1 + y^2}{2}\right) \le C(y^1) + C(y^2). \quad \text{Q.E.D.}$$

at the axes and the origin. Thus, while we assume, as a property of long-run cost functions, $C(0) = 0$, we must be prepared to deal with cases where $\lim_{y \to 0} C(y) = F > 0$, that is, where nonnegligible fixed costs are involved in the production process. While we have not yet dealt formally with this point, it should be obvious that the existence of such fixed costs for each firm only strengthens the subadditivity results described, because single-firm production permits the economy to escape some such fixed outlay(s). However, when fixed costs are incurred by the introduction of additional product lines (i.e., when there are product-specific fixed costs) the arguments used so far encounter a serious problem. This is illustrated in Figures 7E3 and 7E4. The first of these shows a three-dimensional cost surface $Okbcdef$ with a trans-ray cross section above line segment ag. Figure 7E4 again depicts this trans-ray cross section of the cost surface. Although cde, the portion of the cross section that is inside the axes, is convex, the product-specific fixed costs result in discontinuous declines, indicated by the intervals bc and fe, where the production of an individual product line is discontinued. Such cost functions obviously cannot be trans-ray convex.

However, a review of the diagrammatic argument accompanying the proof of Proposition 7E1 reveals that its logic does not require global trans-ray convexity. Rather, the argument relies upon the fact that, given any subdivision of an output vector $y^0 = y^1 + y^2$, it is possible to express y^0 as a convex combination of vectors constituting proportional expansions $v_1 y^1$ and $v_2 y^2$ of the outputs in y^1 and y^2, where $v_1 y^1$ and $v_2 y^2$ both lie on the trans-ray cross section in question, so that $C(y^0)$ is no greater than a weighted average of those costs.

This same technique can be employed using weaker postulates. Rather than assuming trans-ray convexity, we merely postulate the existence of a hyperplane Ohh (Figures 7E3 and 7E4) through $C(y^0)$ (point d in the figures) in cost space, that hyperplane having above y^0 a trans-ray cross section hh which nowhere lies above the corresponding cross section $bcdef$ of the cost surface.

That is, we now assume there exists a trans-ray direction in output space, given by ag, through point y^0, over which the cost surface can be *supported* at y^0 relative to that direction. (See the formal definition of trans-ray supportability in Chapter 4.)

Proposition 7E3 $C(\cdot)$ *is subadditive at* y^0 *if it is trans-ray-supportable there in any direction* $w > 0$ *and if DRAC holds for all y with* $w \cdot y \leq w \cdot y^0$.

While it is concise, the mathematical proof in the Appendix does not indicate the similarities between the argument and that underlying Proposition 7E1. Given any division of y^0: $y^1 + y^2 = y^0$, $y^1 \not\equiv 0 \not\equiv y^2$, the existence of a support line in Figures 7E3 and 7E4 permits one to construct proportionally expanded output vectors such as $v_1 y^1$ and $v_2 y^2$, $v_i > 1$, which lie on

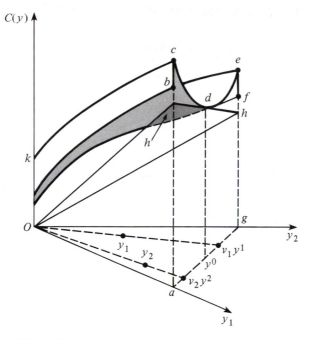

FIGURE 7E3

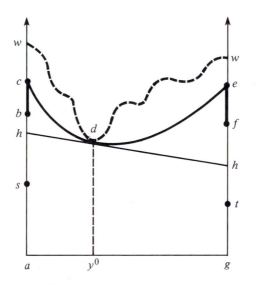

FIGURE 7E4

the desired hyperplane ag such that $\hat{y}_1 \equiv v_1 y^1 = (1/k)y^1$, $\hat{y}_2 \equiv v_2 y^2 = [1/(1-k)]y^2$, with $0 < k < 1$. Thus, as in the proof of Proposition 7E1, $y^0 = k\hat{y}^1 + (1-k)\hat{y}^2$. Then the argument proceeds just as before. Because the cost surface is *supported* at y^0 by hh, the convex combination $kC(\hat{y}^1) + (1-k)C(\hat{y}^2) \geq C(y^0)$. Strict subadditivity then follows from declining ray average costs: $C(y^1) = C(kv_1 y^1) > kC(v_1 y^1)$ and $C(y^2) = C[(1-k)v_2 y^2] > (1-k)C(v_2 y^2)$.

Note that this line of proof does not require that cost functions be nicely behaved over the hyperplane in question. As the dashed curve ww in Figure 7E4 indicates, the cross section of the cost surface may wiggle as much as one wishes, and need not even be continuous. This is what permits application of the proposition to cases involving product-specific fixed costs.

We can see now why convexity assumptions, while excessively strong, have served our purposes. As is well known, convexity guarantees that a supporting hyperplane exists for the cost surface at the points of interest. In particular, because trans-ray convexity implies trans-ray supportability (Proposition 4F2), Proposition 7E1 can properly be considered a corollary to Proposition 7E3. We have offered Proposition 7E1 first largely for expository reasons. We now also have as an immediate corollary of Proposition 7E3:

Corollary *Declining ray average costs up to the hyperplane $\{y \geq 0 \mid y \cdot \nabla C(y^0) = y^0 \cdot \nabla C(y^0)\}$ and quasiconvexity of the cost function at y^0 imply subadditivity at y^0.*

(By Proposition 4F1 and its Corollary we know that quasiconvexity implies trans-ray supportability.)

However, there are limits to the power of the trans-ray supportability postulate. Suppose, in Figure 7E4, that the existence of product-specific fixed costs, that is, fixed costs that are avoidable if the good is not produced, yields costs at the axes represented by the points s and t. In that case, it is clear that no point in the *interior* of the cost cross section cde can be supported. That is, any tangent hh to a point on cde must lie above s or t or both, so that some portion of the cross section of the cost surface must be below hh. However, there is a way out of this difficulty. Intuition suggests that the presence of fixed costs, product-specific or otherwise, must be favorable to subadditivity, not inimical to it. This is indeed true, and we turn to a reformulation of the expression for the cost function which makes this clear and precise.

Without loss of generality, any cost function can be broken down into its "fixed" and variable parts:

(7E1) $$C(y) = F(S) + c(y),$$

where $S = \{i \in N \,|\, y_i > 0\}$. The *total* value of the "fixed" costs $F(S)$ thus depends upon the precise set of goods of which *strictly* positive quantities are produced. The product-specific fixed costs associated with any product line $V \subseteq S$ are then given by $F(S) - F(S - V)$. This formulation permits the incremental fixed costs associated with the introduction of a particular product line to depend upon the set of goods already in production, as we often expect to be true in reality.

Thus, for example, it seems plausible that the fixed costs of production of *blue* widgets will be quite different when red widgets are already being produced and when they are not. Thus (7E1) includes, *but is not limited to,* the case in which $F(S) = \sum_{i \in S} F(i)$; that is, where each product i has its own given fixed cost whose magnitude does not vary with the other components of the firm's output vector.

In Chapter 4, we labeled cost functions of this type *transylvanian* if $c(y)$, the variable portion, is strictly subadditive. (The term "transylvanian" refers to the batlike shape that characterizes such cost surfaces in the three-dimensional case, as shown in Figure 7E3.) We offer for transylvanian functions a subadditivity result which does not require any stronger assumptions about their trans-ray behavior (e.g., convexity or supportability) other than those *necessary* to guarantee subadditivity of $c(y)$.

Proposition 7E4   *A transylvanian cost function is strictly subadditive at y^0 if $c(\cdot)$ is strictly subadditive there and if $F(S \bigcup T) \le F(S) + F(T)$, $\forall S,\, T \subset N$; that is, if the fixed cost of simultaneous introduction of S and T is no greater than the sum of the fixed costs of the two product sets, each introduced by itself.*

Proof   Consider any two nonzero vectors, y^a and y^b, that sum to y^0.

(7E2) $C(y^a) + C(y^b) = F(A) + F(B) + c(y^a) + c(y^b),$

where $A = \{i \in N \,|\, y_i^a > 0\}$ and $B = \{i \in N \,|\, y_i^b > 0\}$. Using the (weak) subadditivity of $F(\cdot)$, we have

(7E3) $C(y^a) + C(y^b) \ge F(S) + c(y^a) + c(y^b)$

where $N \supseteq S = \{i \in N \,|\, y_i^a + y_i^b \equiv y_i^0 > 0\}$. The result then follows by the *strict* subadditivity of $c(y)$.   Q.E.D.

Corollary   *A transylvanian cost function is strictly subadditive at y^0 if $c(y)$ is (nonstrictly) subadditive and if $F(S \bigcup T) < F(S) + F(T)$, $\forall S,\, T \subset N$.*

For, as the proof of Proposition 7E4 indicates, strict inequality is required for only one or the other of the two steps in the argument. Thus, it is convenient to be able to deal with the case in which $F(\cdot)$ is *strictly* subadditive, while $c(\cdot)$ is only weakly subadditive. This is helpful, for example, for consideration of the *generalized affine* cost function: $C(y) = F(S) + \sum_{i \in S} c_i y_i$, that is, the cost function whose variable portion is of the simple linear and homogeneous form indicated in the preceding expression.

Our final sufficient condition for subadditivity brings us a step closer to the analysis of market equilibrium under monopoly, the main subject of Chapter 8. First we must describe a cost concept introduced by Sharkey and Telser (1978):

Definition 7E1: Supportability   A cost function $C(y)$ is *supportable* at an output vector y^0 if there exists a vector $h(y^0) > 0$ such that $y^0 \cdot h(y^0) = C(y^0)$ and $y \cdot h(y^0) < C(y)$ for all $0 \le y \le y^0$, $y \ne y^0$.

This means that the cost function is supportable if it is possible to construct *some* hyperplane through the origin and $C(y^0)$ that lies everywhere below the cost surface for output vectors no greater than y^0. Thus, in Figure 7E5,

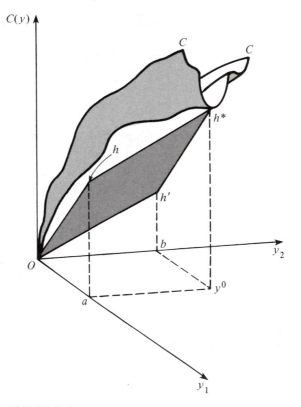

FIGURE 7E5

the cost surface is OCC: It is supported at y^0 (point h^*) by hyperplane Ohh^*h', which lies below OCC at every other point in the region Oay^0b which contains all points $y \leq y^0$.

Proposition 7E5 (*Sharkey-Telser*)   *If $C(\cdot)$ is supportable at y^0, then $C(\cdot)$ is subadditive there.*

    Proof   Consider any nonzero pair of output vectors, y^a and y^b, such that $y^a + y^b = y^0$. Then, by supportability,

(7E4) $$C(y^a) > y^a \cdot h(y^0)$$

and

(7E5) $$C(y^b) > y^b \cdot h(y^0).$$

Adding (7E4) and (7E5), we have

(7E6) $$C(y^a) + C(y^b) > (y^a + y^b) \cdot h(y^0) = y^0 \cdot h(y^0) = C(y^0).$$

    Proposition 7E5 immediately gives us access to an additional set of sufficient conditions for subadditivity, for anything which suffices for supportability now ensures subadditivity as well. The interested reader is urged to consult Sharkey and Telser (1978).

7F. Toward the Study of Monopoly Equilibrium

    Proposition 7E5 moves us toward the subject matter of Chapter 8 because it begins to cross the border between technological relationships and market equilibrium conditions. The $h(y^0)$ can be interpreted as a vector of prices; indeed Sharkey and Telser refer to it in this way. Then $y \cdot h(y^0)$ becomes the total revenue generated by output y at prices $p_i^* = h_i(y^0)$. The condition of supportability therefore implies that there exists a price vector which (i) just permits production of y^0 to cover its total cost $(h(y^0) \cdot y^0 = C(y^0))$ and which (ii) for any smaller output vector y yields a total revenue strictly lower than the cost of producing y $(h(y^0) \cdot y < C(y); y \leq y^0, y \neq y^0)$. This suggests that market equilibrium under a monopoly whose cost function is supportable possesses a weak stability property: For any *fixed* output vector y^0 there always exists a set of prices for the monopoly's outputs which just cover the monopolist's cost of producing y^0, but which preclude profitable entry at any *smaller* scale.

    However, a major gap still separates the idea of supportability from a genuine market relationship. Since nothing about demand conditions has been postulated, there is no guarantee that the set of price vectors which can support $C(\cdot)$ at y^0 intersects the set of price vectors which would, via the market demand mapping, induce consumers to purchase y^0. That is, it is very possible that, at prices $h_1(y^0), \ldots, h_n(y^0)$, consumers would choose to demand some vector of outputs other than y^0. Moreover, even if the demand conditions did satisfy $y^0 = D(h(y^0))$, the stability property of these supporting prices would still suffer from a significant limitation. While they preclude profitable entry by firms producing outputs *no greater than* those of the incumbent, they need not be inconsistent with profitable entry by a firm that produces more than does the incumbent of at least one good.

Chapter 8 begins our investigation of market equilibrium under monopoly. We discuss the issues that have just been raised and a variety of other matters, including the role of the invisible hand in the operation of monopoly. Throughout the discussion, the tools of analysis of subadditivity that have been provided in this chapter play a crucial role, for we find that if a monopoly operates in a contestable market it will be able to retain its market position only if no combination of smaller firms can produce the monopolist's output more cheaply, that is, unless the monopolist's cost function is subadditive at that output.

APPENDIX

Proof of Lemma 7D1

Without loss of generality, assume that the average incremental cost of shifting the production of y_i^b from firm B to firm A is no greater than the cost of shifting production of y_i^a in the opposite direction, that is,

$$(1) \qquad \frac{C(y_i^a + y_i^b + y_{N-i}^a) - C(y^a)}{y_i^b} \leq \frac{C(y_i^a + y_i^b + y_{N-i}^b) - C(y^b)}{y_i^a}.$$

From the $DAIC$ assumption we have

$$(2) \qquad \frac{C(y_i^a + y_i^b + y_{N-i}^b) - C(0 + y_{N-i}^b)}{y_i^a + y_i^b} < \frac{C(y^b) - C(0 + y_{N-i}^b)}{y_i^b}.$$

Cross-multiplying and adding and subtracting $y_i^b C(y^b)$ on the left-hand side yields

$$\frac{C(y_i^a + y_i^b + y_{N-i}^b) - C(y^b)}{y_i^a} < \frac{C(y^b) - C(0 + y_{N-i}^b)}{y_i^b}.$$

Along with (1) this implies

$$C(y_i^a + y_i^b + y_{N-i}^a) < C(y^a) + C(y^b) - C(0 + y_{N-i}^b),$$

thus completing the proof of the Lemma.   Q.E.D.

Proof of Proposition 7E1

Let $v_1 y^1$ and $v_2 y^2$ lie on H, a hyperplane of trans-ray convexity. Since by definition all points on the hyperplane satisfy its equation $w \cdot y = w_0$, we have immediately $w \cdot v_1 y^1 = w \cdot v_2 y^2 = w \cdot y^0 = w_0$ or $1/v_1 + 1/v_2 = wy^1/w_0 + wy^2/w_0 = wy^0/w_0 = 1$

$$(3) \qquad v_i = w_0/w \cdot y^i > 1 \qquad i = 1, 2.$$

Writing $k \equiv 1/v_1$,

(4) $1/v_1 \equiv k < 1, \qquad 1/v_2 = 1 - k < 1.$

By *DRAC*, we therefore obtain

(5) $C(y^1)/k > C(y^1/k) = C(v_1 y^1)$

and

(6)

$$C(y^2)/(1 - k) > C(y^2/(1 - k)) = C(v_2 y^2).$$

Now our proof follows from the (nonstrict) convexity of the cost function over H. As we know, because of the convexity assumption, the cost of a linear combination of output vectors in H is less than or equal to the same linear combination of their costs. Since by the definition of k in (4), $v_1 k = 1$ and $v_2(1 - k) = 1$, we have immediately $y^0 = y^1 + y^2$, $y^0 = kv_1 y^1 + (1 - k)v_2 y^2$. Thus, by the convexity assumption,

(7) $C(y^0) \le kC(v_1 y^1) + (1 - k)C(v_2 y^2).$

Finally, (5)–(7) yield our result

(8) $C(y^1) + C(y^2) > kC(v_1 y^1) + (1 - k)C(v_2 y^2) \ge C(y^0).$ Q.E.D.

Proof of Proposition 7E3

Let $y^1 + y^2 = y^0$, with $y^1 \ne 0 \ne y^2$ and $0 < wy^i < wy^0$ so that $y^i(wy^0/wy^i)$ is well defined and lies on the trans-ray hyperplane. Letting the vector v contain the coefficients of the supporting trans-ray hyperplane then

$$C\left[\frac{wy^0}{wy^i} y^i\right] \ge v\left(\frac{wy^0}{wy^i} y^i\right) + v_0$$

and

(9) $\dfrac{wy^i}{wy^0} C\left[\dfrac{wy^0}{wy^i} y^i\right] \ge vy^i + \dfrac{wy^i}{wy^0} v_0.$

Since $wy^0/wy^i > 1$, by *DRAC*

(10) $C(y^i) > \dfrac{wy^i}{wy^0} C\left(\dfrac{wy^0}{wy^i} y^i\right).$

Putting (9) and (10) together gives

$$C(y^1) > v \cdot y^1 + \frac{wy^1}{wy^0} v_0$$

$$C(y^2) > v \cdot y^2 + \frac{wy^2}{wy^0} v_0.$$

Then adding these

$$C(y^1) + C(y^2) > v(y^1 + y^2) + \frac{w(y^1 + y^2)}{wy^0} v_0 = vy^0 + v_0 = C(y^0). \quad \text{Q.E.D.}$$

8

Monopoly Equilibrium

The preceding chapters have laid the groundwork for the analysis of general multiproduct industry structures and prices consistent with equilibrium in contestable markets. The subject of this chapter, industry equilibrium with production by only a single firm, provided the original impetus for our more general study of equilibrium in contestable markets. The extension of this analysis to nonmonopoly markets is possible because of the way in which we happened to formulate the problem in the monopoly case. For, as discussed in Chapter 9, our model employs assumptions about entry conditions and equilibrium concepts that mesh precisely with the classical precepts of the theory of perfect competition. Moreover, as presented in Chapter 11, these same concepts can be used to describe market equilibrium in any type of industry from which entry barriers are absent. Finally, the analysis ultimately led us to the formulation of a theory of entry barriers largely independent of the characteristics of *post-entry* equilibrium and the controversial behavioral assumptions traditionally needed to define it. This theory, described in Chapter 10, implies that in the absence of such entry barriers, markets are contestable.

But this chapter presents the material from which the analysis began. Its focus is the availability or unavailability to natural monopolists of a vector of sustainable prices—fixed prices capable of deterring entry without the threat of any price changes in reaction to an entrant's choice of strategy. In other words, we investigate whether there exist any prices which, if chosen by a natural monopolist, will permit the monopoly firm to operate profitably in contestable markets and yet discourage any and all entrants who offer no innovations.

As was noted in Chapter 1, this question has a natural interpretation as the problem faced by a regulated monopolist whose prices, once approved, are fixed by the regulator. It is also the problem faced by a natural monopolist in markets where entry is perfectly reversible and frictionless. However, the introduction of the *entry cost function*, which will be discussed in greater detail in Chapter 10, makes the analysis applicable to a much wider class of markets.

This chapter is organized about a set of conditions—some necessary, some sufficient—for the existence of sustainable prices in an industry populated only by a monopoly. Put in this way, it would seem primarily to describe formal exercises designed to provide theoretical underpinnings to our equilibrium concept. But the analysis is far more than this. It can be interpreted, rather, as an exercise in welfare theory and a step toward reformulation of public policy in relation to monopoly.

The necessary conditions for sustainability are important for this purpose because they show that, in the absence of barriers to entry, a monopolist cannot hope to find prices that protect him from entry if his industry is not truly a natural monopoly—if he earns "excessive" profits, operates inefficiently, or forgoes opportunities for cost-saving innovation. Similarly, there are conditions under which a monopolist's good behavior, in the form of prices compatible with Ramsey optimality, is rewarded automatically by making him invulnerable to entry. In other words, the chapter shows that even where monopoly is natural and entry never actually occurs, much can be accomplished for the public welfare and efficiency in resource allocation by freedom of entry and exit, together with freedom of the monopolist to choose his prices and other features of his operations.

8A. Sustainability: Formal Definitions and Intuitive Discussion

Our formal definition of a sustainable monopoly equilibrium is identical in spirit with that of our notion of equilibrium in contestable markets, from which it differs only slightly in the details of its notation (see Definitions 1E1 and 2B2).

Definition 8A1: Sustainability of Monopoly   The announced prices of a monopolist, p^m, are sustainable if the monopolist is financially viable at these prices and if no potential entrant can find a marketing plan whose expected economic profits cover the costs of entry $E(y^e)$. More formally,[1] using the superscripts "e" and "m" to refer, respectively, to the potential entrant and the incumbent monopolist, the price vector p^m is sustainable if and only if $\pi^m \equiv p^m Q(p^m) - C(Q(p^m)) \geq 0$ and $p^e y^e - C(y^e) \leq E(y^e)$, for all p^e, y^e with $p_S^e \leq p_S^m$ and $y^e \leq Q(p_S^e, p_{(S)}^m)$; where $S = \{i \in N \mid y_i^e > 0\}$, N is the relevant set of products for the industry, S is any subset of N, $Q(p)$ is the vector demand function, and $C(y)$ is, as usual, the cost function.

Thus, a marketing plan of a potential entrant consists of a (sub)set of the relevant products $S \subseteq N$ which, including the costs of entry, can be marketed by the entrant at prices no higher than those of the incumbent. The entrant can offer to sell *any* quantities of its products no greater than the amounts demanded at the prevailing prices constituted by the effective price vector[2] $(p_S^e, p_{(S)}^m)$. These are the entrant's prices for the products in S, and the incumbent's prices for the products, offered only by him, in the set $N - S \equiv (S)$.

We will also find it useful to distinguish between the sustainability of a particular price vector p^m and the sustainability of the monopoly itself. Having already defined the former, we can now offer

Definition 8A2 A monopoly is sustainable if and only if there exists for that firm at least one sustainable price vector.

---

[1] Here, as before, x_S is a vector with components equal to x_i for $i \in S$ and 0 for $i \in (S) \equiv N - S$. Similarly $x_{(S)}$ is the vector with components equal to x_i for $i \in (S) \equiv N - S$, and Q^S denotes the vector Q^i for $i \in S$. Juxtaposition of two vectors, such as $p^e Q$, denotes their dot product. Where no ambiguity results, vectors of zeros will be suppressed. However, where it is necessary to merge the price vector for the products supplied at lowest prices by the entrants with those supplied most cheaply by the incumbent (former) monopolist to give us the vector of prices for the industry, p_S denotes a *truncated* vector referring to the prices for $i \in S$. Thus, the vector $(p_S^e, p_{(S)}^m)$ denotes the relevant argument of the market demand functions under entry.

[2] When the demands for some of the products are complementary, the concept of *effective prices* must be used with care. By the effective price for product i we mean the lower of the two price offers: that of the incumbent and that of the entrant. Interpreted literally, this formulation would suggest the possibility that a potential entrant who wished to enter the industry by selling electric stoves could quote a ludicrously low price for electricity, choose to sell none, and yet reap the benefits of a greatly stimulated demand for stoves. To preclude such absurd possibilities, one could stipulate that whenever an entrant refuses to serve the entire market at his quoted prices, the demand curves of all complementary products must be evaluated at the monopolist's quoted prices. However, even this may require some complicated adjustment. For if the entrant does choose to sell some electricity, the effects on demand will depend on the identity of the *particular consumers* who happened to obtain service at the low price (i.e., on their position on the aggregate demand curve). This issue can be handled with more precision, but only at the cost of additional special assumptions and much more technical apparatus. In any event, our analysis deals primarily with substitute products. The case of complementary demands is considered further in Appendix III.

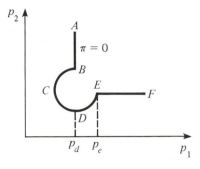

FIGURE 8A1

Before proceeding to our technical discussion of the sustainability of monopoly, it is helpful to illustrate the concept using some simple diagrams depicting the price vectors and iso-profit loci in a two-product industry—the price-space diagram. Later in the chapter we provide a second type of graph with outputs on the axes (the output-space diagram), which is also very helpful in explaining sustainability theory. To proceed, we first need

Definition 8A3: Undominated Prices   A price vector p which yields a given quantity of profit $\pi(p)$ is said to be undominated if there does not exist any *other* price vector $p^* \leq p$ such that $\pi(p^*) > \pi(p)$, that is, any vector of prices no higher than p which yields profits greater than p's.

Thus, suppose, in Figure 8A1, that the entire locus of price vectors yielding zero profit is[3] AF. Then only the negatively (nonpositively) sloped region CD is undominated. Thus, for example, price vector B is dominated by price vector C, since at the latter *both* prices are lower than they are at the former, and yet profit is equal at the two points.

The discussion that follows deals primarily with the undominated price vectors of the incumbent, and it assumes, at first, that potential entrants are willing to provide whatever quantities of the goods they offer that are demanded at the prices they quote. We refer to such an entrant as a *full supplier*. It is also assumed that the industry is a natural monopoly in the sense of subadditivity of costs; that is, it is cheaper for any given output vector to be produced by a single firm than by any combination of two or more firms.

We can now turn to the representation of sustainability in price space. Figure 8A2 represents an industry producing two substitute products. The

---

[3] The apparently peculiar shape of the locus as depicted is in fact the form it can be expected to assume in practice. For an explanation see Baumol, Fischer, and ten Raa (1979). In brief, for $p_1 > p_d$ the price of item 1 exceeds its profit-maximizing level, while beyond p_e the commodity is completely priced out of the market.

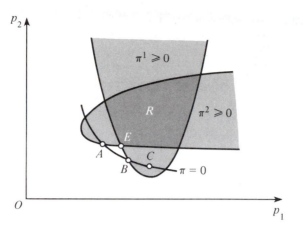

FIGURE 8A2

curve $\pi = 0$ is the locus of *undominated* p_1, p_2 pairs which just permit a firm marketing both products as a monopoly to break even. The set $\pi^1 \geq 0$ represents all those pairs of prices which permit a firm marketing only good 1 to make nonnegative profits. That is, for any given value of p_2 charged by some other firm supplying good 2, this set includes the values of p_1 which offer nonnegative profits to a firm supplying *only* good 1. The $\pi^2 \geq 0$ set is defined analogously.

The intersection, R, of these two nonnegative profit sets consists of all price pairs at which the two single-product firms can *simultaneously* avoid losses. At point E both such firms earn exactly zero profits. It follows from the hypothesis that the industry is a natural monopoly that the $\pi = 0$ locus must pass to the "southwest" of R.[4] That is, it tells us that a multiproduct

---

[4] This is easy to see, since, by definition

$$\pi(p_1, p_2) = p_1 Q^1(p_1, p_2) + p_2 Q^2(p_1, p_2) - C(Q^1, Q^2),$$

$$\pi^1 = p_1 Q^1(p_1, p_2) - C(Q^1, 0),$$

and

$$\pi^2 = p_2 Q^2(p_1, p_2) - C(0, Q^2).$$

Using the subadditivity (natural monopoly) relationship $C(Q^1, Q^2) < C(Q^1, 0) + C(0, Q^2)$, we have

$$\pi - \pi^1 - \pi^2 = C(Q^1, 0) + C(0, Q^2) - C(Q^1, Q^2) > 0.$$

Thus, over R, where $\pi^1 \geq 0$ and $\pi^2 \geq 0$, it follows that $\pi > 0$. By the assumed continuity of π between R and some price pair close enough to the origin to ensure that $\pi < 0$, the undominated section of the $\pi = 0$ locus must pass between R and the origin. Of course, continuity of π follows from an assumption of continuity of the cost and demand functions.

firm can break even by marketing the pair of outputs at prices lower than those which independent producers of the items would have to charge. This is all the natural monopoly condition tells us. For example, it does *not* rule out the possibility that the $\pi = 0$ locus intersects the $\pi^1 \geq 0$ set *below R*.

Thus, it is easy to see that an undominated zero-profit point such as C cannot be sustainable, since a firm marketing only good 1 can charge a price slightly less than the monopolist's and earn positive profits. In the case depicted, no such possibility arises if the incumbent chooses any price vector on the arc between A and B on its zero-profit locus. As Figure 8A2 is drawn, no portion of the $\pi^1 \geq 0$ or $\pi^2 \geq 0$ sets lies between arc AB and the origin (though we see next that this need not always be true). It follows in the case depicted that, if the incumbent's price vector is given by a point on the arc between A and B, *any* entrant's prices at or below the incumbent's must yield losses to the entrant. Therefore, any price vector on $\pi = 0$ between A and B is sustainable for the incumbent.

Figure 8A3, however, depicts a case in which no sustainable prices exist. Once again, the $\pi = 0$ locus lies between the origin and R, the set of price vectors profitable to two specializing firms. But now, some points for which $\pi^2 \geq 0$ lie below any point on the undominated portion of the incumbent's zero-profit locus.

The preceding argument indicates why zero-profit points to the right of B cannot be sustainable. But, now, while points such as F between A and B are still not profitable for either potential entrant at (or near) the quoted prices of the incumbent, a firm marketing only good 2 can expect to earn positive profits by cutting p_2 *drastically* to a level near G. Since we have ruled out all the undominated zero-profit points, Figure 8A3 depicts a situation in which

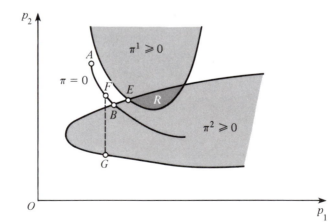

FIGURE 8A3

no sustainable prices exist. Hence, the natural monopoly here is *unsustainable*.

The preceding examples have illustrated the potential vulnerability of a natural monopolist to full-supplier entry—to the appearance of entrants prepared to meet all of the market's demands for their subset of products at the prices they offer. The natural monopoly may also be vulnerable to entry by *partial suppliers* who elect not to serve the entire market demand. This can easily be seen in the single-product case depicted in Figure 8A4. Here p^m is the only market clearing price at which the incumbent monopoly earns zero economic profit. At that price, y^m units of output are demanded. While the average cost curve is drawn so that costs are subadditive throughout the relevant range, average cost is not monotonically decreasing. Then, an entrant can charge a price slightly less than p^m and obtain positive profits if he supplies, for example, only ye units of output. Thus, price p^m, the only market clearing price which yields zero profits, is clearly not sustainable.

The analysis up to this point has served to demonstrate that the sustainability of natural monopoly is not a trivial issue. In Section 8C, by establishing a series of necessary conditions, we begin the task of sorting out the important forces working for and against sustainability. Later in the chapter we turn to conditions *sufficient* for the existence of sustainability. Because no analytically useful conditions that are both necessary and sufficient for sustainability are known, it is appropriate to proceed in this way, dealing with the two sides of the matter in turn. But first we digress briefly to indicate the relationship of our sustainability concept to the game-theoretic approach, which naturally recommends itself for analysis of our subject.

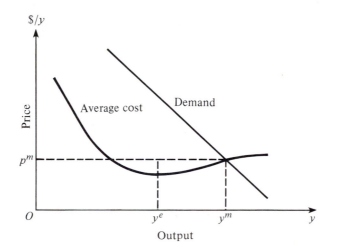

FIGURE 8A4

8B. Digression: Sustainability and Game Theory

As has already been indicated, we have constructed our notion of equilibrium in monopoly markets, that is, sustainability, to be in keeping with that employed in our analysis of contestable markets in general and perfectly competitive markets in particular. This was done to focus attention on the interactions of firms and consumers in traditional economic markets. Firms quote prices and consumers act upon them. Potential entrants can only hope to lure customers from the monopolist if they match (or *slightly* undercut) his price. The only "contracts" considered between firms and consumers involve simple, linear prices. The only latitude in this respect which we have permitted potential entrants is the option to sell *less* than the maximal quantity demanded of them.

While we believe that this view is entirely in keeping with the goals of our analysis, it should be pointed out that the objective circumstances of the monopoly case permit an alternative approach, which, on its own merits, is equally rich analytically; this alternative is cooperative game theory and its equilibrium concept, the core. The multiproduct natural monopoly situation which we analyze can also be interpreted as the problem of finding an equilibrium (general or partial) in a production economy with increasing returns. Interest in this problem was generated by H. E. Scarf (unpublished), and the conditions under which the core exists have been explored by a number of authors and remain a subject of current research.[5]

We have no quarrel with this analysis, and our intuition suggests that there may well exist a more fundamental and unifying approach which encompasses both views. However, until such a synthesis emerges, we must, regretfully, conclude that a full discussion of the points of contact and divergence between our analysis and that of the game theorists lies outside the scope of this volume. The main reason we avoid the game-theoretic approach is that when the players in the game are taken to be final consumers—as is usually (and appropriately) done—the primary engine of analysis, that is, the characteristic function of the game that describes the net benefits any coalition (subset) of consumers can obtain by acting on its own, tends to leave completely unspecified the market mechanism through which these payoffs are obtained. In particular, the core (natural monopoly equilibrium) may not exist because of a "blocking coalition" that can be formed only through a series of complex contracts (which vary from one person or group to another) between the consumers and the productive agent (i.e., the entrant). It appears to be quite difficult to build into this analysis our market-oriented constraint permitting only simple, nondis-

---

[5] A (partial) list of relevant papers includes Champsaur (1975); Littlechild (1975); Quinzii (1980); Scarf and Hansen (1973); Sharkey (1979, 1981); Sorenson, Tschirhart, and Winston (1978). See also Chapters 5 and 6 in Sharkey (1982).

criminating prices to be used in attacking the monopolist. Yet this restriction is fundamental for the analysis.

8C. Necessary Conditions for Sustainable Prices

This section reports several necessary conditions specifying the forms of cost advantage of the monopolist that are required for any sustainable price vectors to exist. It also reports several conditions relating to the behavior of a monopolist that are necessary for it to be compatible with sustainability.

Perhaps the most fundamental structural condition is cost subadditivity.

Proposition 8C1: Subadditivity *In the absence of entry costs, no sustainable prices exist unless the cost function for the monopolist is (at least weakly) subadditive at* y^m*; that is, unless* $C(y^m) \leq \sum_{i=1}^{k} C(y^i)$ *if* $\sum_{i=1}^{k} y^i = y^m$.

This result states that sustainable pricing is impossible for a monopolist if the industry is not a natural monopoly. Thus, the monopoly whose existence imposes a cost penalty upon society cannot find prices that will protect it from entry. To prove this, let p^m be any vector of prices at which the monopoly covers its costs; that is,

(8C1) $$p^m \cdot y^m \geq C(y^m).$$

The failure of subadditivity at the output vector y^m requires the existence of a set of output vectors y^i such that

(8C2) $$\sum y^i = y^m$$

and

$$\sum C(y^i) < C(y^m)$$

so that multifirm production is cheaper. Using (8C1) and (8C2), we have $\sum p^m \cdot y^i = p^m \cdot y^m \geq C(y^m) > \sum C(y^i)$, or $\sum (p^m \cdot y^i - C(y^i)) > 0$. It follows immediately that $p^m \cdot y^k - C(y^k) > 0$ for some k. Thus, a competing firm can offer y^k at prices no greater than p^m and earn a profit, so the prices p^m cannot be sustainable.   Q.E.D.

We come next to

Proposition 8C2: Efficiency *In the absence of entry barriers, sustainable prices exist for a monopoly only if it produces its output vector* y^m *at the lowest cost available to the industry.*

Proof   Assume there exist sustainable prices p^m. Let $C^I(y)$ be the industry's minimum cost function (see Chapter 5). Since $C^I(y^m)$ is, by definition, the industry's least cost of producing y^m, if the monopolist is not producing at least cost, it must be true that $p^m \cdot y^m - C^I(y^m) > 0$ because sustainable prices must permit the monopolist to cover its actual costs. Then, by continuity, there must be lower prices p^e at which one or more efficient competitors can make a profit and, therefore, enter successfully.

Q.E.D.

Proposition 8C3: Least Ray Average Cost *In the absence of entry costs, sustainable prices exist for a monopoly only if its output vector is such that any proportionate reduction in all outputs yields no more than a proportionate reduction in total cost. That is*

(8C3) $(1/\lambda)C(\lambda y^m) \geq C(y^m)$ for $0 < \lambda \leq 1$.

Proof An entrant can attempt to match p^m and sell λy^m. Sustainability implies

(8C4) $0 \geq p^m \lambda y^m - C(\lambda y^m)$.

Because the monopoly incurs no loss, $p^m \cdot y^m \geq C(y^m)$, so that $p^m \lambda y^m \geq \lambda C(y^m)$. Together with (8C4), this yields (8C3).   Q.E.D.

Violation of this requirement was illustrated in Figure 8A4. There, the output y^m is unsustainable because at $y^e < y^m$ average cost is lower than it is at y^m. A competitor can profitably serve a segment of the market selling y^e at a price between the average costs at y^e and y^m. Condition (8C3) rules out this possibility for the composite commodity $y^e = \lambda y^m$. We also have the following corollary:

Proposition 8C4 *For sustainable prices to exist for y^m, the production techniques used by the monopoly must yield nondecreasing returns to scale at y^m.*

Proof It is shown in Chapter 3 that with decreasing returns to scale at y^m there exists a $\lambda < 1$ with $(1/\lambda)C(\lambda y^m) < C(y^m)$; that is, ray average cost must be increasing, thus contradicting the necessary condition (8C3).   Q.E.D.

In addition, we can now state the result, attributable to Faulhaber (1975b), which indicates that the converse of Proposition 8C1 is false.

Proposition 8C5 *Strict subadditivity is not sufficient for sustainability.*

Proof Figure 8A4 depicts an output level at which costs are subadditive but returns to scale are decreasing. By Proposition 8C4, it is unsustainable.

Q.E.D.

Having examined several necessary conditions related to the monopolist's costs, we turn next to some pricing rules that must be satisfied for the prices in question to be sustainable. The first of these conditions places a ceiling on the monopolist's profits; that is, it specifies the largest earnings the monopolist can obtain without rendering himself vulnerable to entry. For this purpose we must first offer our definition of the concept of barriers to entry, which is explained fully in Chapter 10.

Definition 8C1: Entry Barrier An industry is characterized by entry barriers if, in effect, there is an entry cost[6] $E(y^e)$, which must be paid by any firm new to the industry but which does not have to be paid by the incumbent firm, that is, the monopolist (or former monopolist).[7]

Then we can treat the monopolist's cost advantage—the fact that he does not have to pay $E(y^m)$—as a *rent* yielded by his special position. We may now state

Proposition 8C6: Zero Profit after Rent *Sustainable prices can yield to the monopolist profits no higher than the value of the entry barrier, $E(y^m)$. That is, if $E(y^m)$ is considered as a rent and is included in the monopolist's costs, sustainability requires that the monopolist's total revenue not exceed those total costs.*

Proof Suppose the contrary. If $p^m \cdot y^m > C(y^m) + E(y^m)$, then, by continuity, a competitor can enter profitably at prices lower than p^m. Q.E.D.

Proposition 8C7: Prices Not below Marginal Cost *If a monopolist is a profit maximizer and therefore earns his full entry rent $E(y^m)$, or if there are no entry barriers, then for prices to be sustainable they cannot be below the corresponding marginal costs.*[8]

---

[6] For generality, these entry costs are taken to be function of the entrant's output y^e, although it makes no difference for the discussion if these costs are independent of y^e. Throughout the chapter it is convenient to assume that in the neighborhood of the monopolist's output y^m the entry costs are constant. This can be given simple interpretations. For example, suppose the entry costs are of the zero-one variety, attributable to the fact that the incumbent uses legal steps to discourage entrants but mounts an effort for this purpose only when the prospective entrant is sufficiently large. Then for sufficiently great y^e we have $\partial E(y)/\partial y = 0$.

[7] Note that while this definition differs substantially from many of those used in the literature [see, e.g., Bain (1956)], it is the same as our interpretation of Stigler's (1968) definition.

[8] *Proof* Let $C^*(y)$ represent full cost, including entry cost (rent) $E(y)$ and other costs $C(y)$. Now, one of the output plans open to an entrant is production and attempted sale at price vector p^m of the output vector

$$y^e = (y_1^m, \ldots, y_{i-1}^m, y_i^m - \Delta, y_{i+1}^m, \ldots, y_n^m) \equiv y^m - \Delta^i.$$

Sustainability requires that this entry plan not yield a profit; that is,

$$p^m \cdot (y^m - \Delta^i) \le C^*(y^m - \Delta^i).$$

But, by the assumption that the monopolist earns his full rent, $C^*(y^m) = p^m y^m$, it is then necessary for sustainability that $p_i^m \Delta > C^*(y^m) - C^*(y^m - \Delta^i)$, or

$$p_i^m > \frac{C^*(y^m) - C^*(y^m - \Delta^i)}{\Delta}$$

Taking the limit as $\Delta \to 0$, we obtain $p_i^m \ge \partial C^*(y^m)/\partial y_i = \partial C(y^m)/\partial y_i$, where the equality follows from the assumed constancy of $E(y)$ near y^m. Q.E.D.

The reason for this result is not difficult to explain. If the monopoly earns as much in revenues as it would cost the entrant to replicate his output, and if the price p_i^m of one of the monopolist's products is less than its marginal cost, then an entrant replicating the monopolist's output vector for $j \neq i$ and reducing the output of product i by a small amount can earn positive profits. For elimination of a unit of output of a good priced below marginal cost reduces revenues less than it reduces costs.

This result shows that, if a monopolist who is precluded from responsive limit pricing sets any of his prices below marginal cost, then he must either be vulnerable to entry (the price vector p^m must be unsustainable) or he must sacrifice some of his potential profit $E(y^m)$.

Next, we come to two pricing requirements which are particularly significant for analytic purposes and which we will use repeatedly.

Proposition 8C8: Nondomination *In the absence of entry costs, a sustainable price vector p^m must be undominated.*

Proof If there exists a $p^e \leq p^m$, $p^e \neq p^m$ with $\pi(p^e) > \pi(p^m)$, then an entrant can operate successfully at prices p^e. Q.E.D.

Proposition 8C9: Preclusion of Cross Subsidy *In the absence of entry costs, there must be no cross subsidy among any of the products or the subsets of products of the incumbent. Specifically, if S is any proper subset of the set N of outputs of the monopolist, then it is necessary for sustainability that*

(8C5) $$\sum_{i \in S} p_i^m y_i^m \leq C(y_S^m) \quad \text{where } y^m = Q(p^m).$$

That is, there can be no subset of products S, which contribute a surplus of revenues over the cost of producing them alone.

This condition follows immediately from the definition of sustainability. It is obviously of interest for regulatory and antitrust policy, for it shows that, even in the presence of monopoly, market forces may be able to prevent cross subsidy and the unfairness in competition it may imply—a subject that has been of considerable concern in discussions of public policy. Proposition 8C9 is directly related to an issue discussed by Faulhaber (1975b) in his analysis of cross subsidization. He had the insight that monopoly production of y^m at a cost of $C(y^m)$ naturally gives rise to an n-player cooperative "cost game" (one "player" associated with each product). The basic idea is quite intuitive. Let the charges paid by each user group (its imputation) be given by E_i; $i = 1, \ldots, n$. For total revenues to cover total costs it is necessary that

(8C6) $$\sum_{i=1}^{n} E_i \geq C(y^m).$$

One is not free to select just any vector of imputations that satisfy (8C6), since if any user group or (proper) subset of user groups is charged more than its stand-alone cost (i.e., the cost of serving only itself), the group will then have the incentive to secede from the cooperative arrangement (i.e., to block the grand coalition). Thus it is required that

(8C7) $$\sum_{i \in S} E_i \leq C(y_S^m), \qquad \forall S \subset N.$$

It should be easy to see that (8C6) and (8C7) can be interpreted as core conditions of a cooperative game. Thus, Proposition 8C9 can be interpreted as the requirement that the imputation $E \equiv (p_1^m y_1^m, \ldots, p_n^m y_n^m)$ lie in the core of the cost game. For this to be true, the cost game must have a nonempty core. Faulhaber (1975b) has demonstrated that this will not be so in general. However, if, as is frequently assumed,[9] weak cost complementarities are present—that is, if production of more of one good does not increase the marginal cost of producing a different good (Definition 4B3)—then the cost game will have a nonempty core.

Proposition 8C10   *If* $C(\cdot)$ *exhibits weak cost complementarities, then the cost game has a nonempty core.*

(The proof is given in Appendix I.)
      However, cost complementarities do not imply that market revenues are in this nonempty core, which is the requirement imposed by Proposition 8C9. In the two-dimensional case[10] depicted in Figures 8A2 and 8A3, this requirement is violated by all those undominated zero-profit price vectors which lie inside either the $\pi^1 \geq 0$ or $\pi^2 \geq 0$ sets. Any price vector p on a negatively sloping segment of the $\pi = 0$ locus is undominated because there is no other vector $p' \leq p$ which yields a return higher than total cost (including entry cost). If p lies inside $\pi^1 \geq 0$, for example, it means that an entrant can produce product 1 alone, in the same quantity as the monopolist, and sell it at the same price p_1^m, earning a positive profit. Thus, such monopoly prices and quantities violate (8C5), result in cross subsidy, and are not sustainable. As drawn, Figures 8A2 and 8A3 both show some price vectors that are free of cross subsidy. However, such price vectors do not necessarily exist.[11]

---

[9] See, for example, Sakai (1974).

[10] Note that every two-player game with a superadditive characteristic function (i.e., subadditive cost function) has a nonempty core.

[11] Formally, we have a natural correspondence from price space into price space. For every price vector p there is the demand vector $Q(p)$, the cost $C(Q(p))$, the nonempty convex set of imputations E in the core of the cost game, and finally the associated set of pseudoprice unit imputations $(E_1/Q^1, E_2/Q^2, \ldots, E_n/Q^n)$. If this correspondence has a fixed point, p^m, then $(p_1^m Q^1(p^m), \ldots, p_n^m Q^n(p^m))$ will be in the core of the game defined by $C(Q(p^m))$. Sharkey (1981) has shown that such a fixed point does exist for the case of independent demands which yield bounded revenue functions for each product, combined with a *generalized affine cost function* (discussed in Chapter 7), $C(y) = F(S) + \sum_{i=1}^{n} c_i y_i$ in which $-F(S)$ is convex in the game-theoretic sense.

Consider this very simple example:

$$C(y_1, y_2) = 200, y_1 > 0, y_2 > 0; \qquad C(0,0) = 0$$
$$C(0, y_2) = 90, y_2 > 0; \qquad C(y_1, 0) = 130, y_1 > 0$$
$$Q^1(p_1, p_2) = 20 - 2p_1 + p_2; \qquad Q^2(p_1, p_2) = 20 - 2p_2 + p_1.$$

The only price vector that can cover costs, with some positive quantity demanded, is $p = (10, 10)$. $Q(10, 10) = (10, 10)$, and the core of the cost game at $y = (10, 10)$ is the set of vectors (E_1, E_2) satisfying

$$130 \geq E_1, \qquad 90 \geq E_2, \qquad E_1 + E_2 \geq 200.$$

Market revenues, however, are not in the core. While $p_1 Q^1(p) = 100 \leq 130$, $p_2 Q^2(p) = 100 > 90$. Thus, the consumers of good 2 have a strong incentive to patronize a competitive entrant offering only good 2, and so the (cost-advantageous) natural monopoly is not sustainable. Figure 8C1 illustrates this situation, with the parameters adjusted slightly so that the undominated region of the $\pi = 0$ locus is more than a single point, and all of it lies inside the region $\pi^2 \geq 0$.

We conclude our discussion of necessary conditions for sustainability with a result attributable to Sharkey and Telser (1978).

Proposition 8C11   *A price vector p^m can be sustainable in the absence of entry barriers only if the cost function is (at least weakly) supportable at $y^m = Q(p^m)$.*

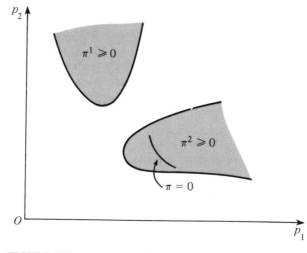

FIGURE 8C1

Proof Sustainability requires that $p^m y^m = C(y^m)$ and that $p^m \cdot y \le C(y)$, for all $y \le y^m = Q(p^m)$. Thus, the revenue hyperplane $z = p^m \cdot y$ supports the cost function at y^m over the region relevant for the Sharkey-Telser definition.

Q.E.D.

8D. Some Sufficient Conditions for Sustainability

We have now described a rather extensive set of *necessary* conditions for sustainability, most of which can be considered attributes of economic efficiency or to serve the public interest in other ways. We turn next to some sufficient conditions. However, it is first necessary to distinguish formally two types of entry behavior which may beset the monopolist.

Definition 8D1: Full-Supplier Entrant An entrant to an industry is said to be a full supplier *vis-à-vis* a (former) monopolist if it undertakes to supply a proper subset S of the monopolist's set of outputs N, charging for each product $i \in S$ a price $p_i^e \le p_i^m$, that is, a price no higher than the monopolist's, and offers to sell whatever quantity of each product in S is demanded by the market at the relevant prices. The relevant prices are those set by the entrant, p_S^e, for the goods in S and those set by the monopolist for the remaining goods in N, p_{N-S}^m.

Definition 8D2: Partial-Supplier Entrant An entrant to an industry is said to be a partial supplier if it undertakes to supply any of the goods it produces in positive amounts at a quantity below that demanded by the market at the relevant prices. The relevant prices are those described in Definition 8D1.

We can now begin our consideration of sufficient conditions for sustainability.

Proposition 8D1: Sustainability against Entry by Partial Suppliers *If the goods* 1, 2, . . . , *n are weak gross substitutes* (WGS)[12] *and the cost function exhibits decreasing average incremental costs* (DAIC),[13] *then for a monopoly to be sustainable at prices* p^m *against entry by partial suppliers, it is sufficient that it be sustainable at* p^m *against entry by full suppliers.*[14]

---

[12] Alternatively, it can be shown that the WGS hypothesis can be omitted if DAIC is strengthened to require costs to be globally concave. We prefer the version in the text because we find WGS and DAIC, unlike cost concavity, to be readily interpretable in economic terms.

[13] See Chapter 4 for the definition and a discussion of this concept.

[14] This result is proved in Appendix I.

Proposition 8D1 is useful primarily because, for the case of weak gross substitutes and declining average incremental cost, it tells us that we need only consider full-supplier entry in an analysis of sustainability. For in that case, if no full-supplier can enter profitably, the potential entrant cannot hope to succeed by attempting only partial supply of the quantities demanded by the market. Why is this so? It is natural to think that when an entering firm is given the option of serving only part of the market if it desires to do so, then this option can only benefit that firm: for the firm can then decide whether to serve all or part of the market on the basis of the relative profitability of the two options, a decision which, in turn, depends on the shapes of the cost and demand functions. However, $DAIC$ guarantees that a restriction in any of its outputs below the amount demanded by the market will never benefit the entrant, because reduced output must increase average incremental cost without increasing the price obtained for the product, since prices are bounded from above by p^m. It also follows from the $DAIC$ and WGS premises that it does not pay the entrant to undercut any of the monopolist's prices p_i^m for any product i for which it supplies less than total market demand. Obviously, such a cut will bring the entrant no direct additions to profit through increased sales of the product, since it has already decided to sell less than the market will take at price p_i^m. And the price cut will not add indirectly to the entrant's profits via effects on the sales of any other good j, since the (weak) substitutability of i and j means that a reduction in the price of i will reduce (or at least not increase) the demand for j.

Confining ourselves now to the case of entry by full suppliers, we have

Proposition 8D2: Independent Demands[15]   *With all demands independent and with weak cost complementarities, p^m is sustainable against entry by full suppliers if p^m is undominated and if market revenues $p_i^m Q^i(p^m)$ involve no cross subsidies [i.e., (8C5) is satisfied].*

Although the multidimensional discussion is somewhat hard to follow intuitively, the argument is fairly clear when only two products are involved. Figure 8D1 is our familiar price diagram for the two-good industry, with demands independent. The $\pi^1, \pi^2 \geq 0$ sets are unbounded vertical or horizontal rectangles because, with independence of demands, the set of profitable prices for a firm marketing only good 1 is unaffected by the price of good 2, and conversely. As before, because of cost subadditivity, the undominated $\pi = 0$ locus lies southwest of R, the intersection of the two nonnegative profit sets.

---

[15] This result is also proved in Appendix I. As indicated in Footnote 11, Sharkey (1981) obtains a similar result without *assuming* the absence of cross subsidies, but using stronger cost assumptions.

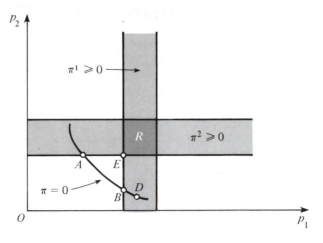

FIGURE 8D1

Points such as D involve cross subsidies and are therefore not sustainable. It can be shown that weak cost complementarity and demand independence together rule out the possibility that the undominated portion of the $\pi = 0$ locus extends either to the right of $\pi^1 \geq 0$ or above $\pi^2 \geq 0$. Thus, all undominated points which are subsidy-free lie between A and B. The rectangular shape of the $\pi^i \geq 0$ sets establishes the proposition, since it is impossible for an entrant to reduce p_i and thereby reach the $\pi^i \geq 0$ set from any point between A and B.

Another glance at Figure 8A3 reveals why the theorem fails when the goods are substitutes. In this case, the $\pi^2 = 0$ boundary need no longer be horizontal throughout and can assume the shape depicted. Then, any point, such as F, between A and B involves no cross subsidy, but an entrant marketing only good 2 can hope to earn nonnegative profits by reducing p_2 far enough below the monopoly price to attract a large volume of demand from substitute product 1. Thus Figure 8A3 contains no sustainable price vectors.

The two-dimensional case yields a simple analytic condition sufficient for sustainability, which brings out the underlying structural differences between the sustainable region bounded by A and B in Figure 8A2 and the analogous, but unsustainable, region in Figure 8A3.

Proposition 8D3: The Two-Good Case *Where there are only two substitute goods, a subsidy-free and undominated price pair (p_1^m, p_2^m) is sustainable against entry by full suppliers if*

$$(8D1) \quad \frac{\partial Q^2}{\partial p_2}(C_2(Q^1, Q^2) - C_2(0, Q^2)) + \frac{\partial Q^1}{\partial p_2}(1 - S_1(Q^1, Q^2))C_1(Q^1, Q^2) \geq 0$$

and

(8D2) $\dfrac{\partial Q^1}{\partial p_1} C_1(Q^1, Q^2) - C_1(Q^1, 0) + \dfrac{\partial Q^2}{\partial p_1} (1 - S_2(Q^1, Q^2)) C_2(Q^1, Q^2) \geq 0.$

(This result is proved in Appendix I.)

Requirement (8D1) and the symmetric condition (8D2) are readily interpretable economically despite the complexity of their appearance. The first term in (8D1) is the product of the effect of a change in good 2's own price upon its demand and any reduction in its marginal cost resulting from the presence of good 1. Thus, this term, and the analogous term in (8D2), is positive where there are cost complementarities. The second term multiplies the cross partial of demand (which is positive for substitutes) by 1 minus the degree of product-specific scale economies, which is negative so long as *DAIC* holds.

Expressions (8D1) and (8D2) make it clear then, that with given demand sensitivities to price, cost complementarities favor sustainability, while declining average incremental cost makes it more difficult for the monopolist to deter entry. Put another way, strongly increasing returns to the scale of a particular product line may permit an entrant to forgo cost complementarities and to succeed in specialized entry, offering a single product at a lower price and a larger quantity than the monopolist can. The monopolist, constrained to market all goods at prices at or below p^m, is unable to match this lower price if the average incremental costs of other products are increased too rapidly by the loss of demand caused by the concomitant substitution effects.

In addition, the second terms in (8D1) and (8D2) suggest that, with a given cost structure, strong substitution effects on the demand side may make it difficult for the monopolist marketing several goods to achieve the low average costs available to an entrant, who can specialize in a chosen product line. On the other hand, weakness in the substitution effects may permit the monopolist's cost complementarities to outweigh the advantages of a competitor's specialization, because there is little competition among the monopoly's own products impeding it from enjoyment of the full scale economies derivable from expansion of a particular output. In the case of independent demands (zero substitution effects), cost complementarities suffice for satisfaction of (8D1) and (8D2). For then, this effect cannot be offset by *DAIC* [the otherwise negative second term in (8D1) and (8D2)], and so Proposition 8D2 tells the entire story.

8E. The Main Weak Invisible Hand Theorem

We come now to the basic proposition relating the concept of sustainability to the standard welfare analysis of pricing and resource allocation. This is undoubtedly the most surprising of our sustainability conditions.

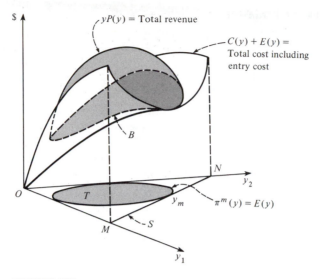

$yP(y)$ = Total revenue

$C(y) + E(y) =$ Total cost including entry cost

FIGURE 8E1

Proposition 8E1: Sustainability of Ramsey Prices *Under a set of assumptions given below, which include a cost function exhibiting both economies of scale and trans-ray convexity, Ramsey-optimal price-output vectors are sufficient to guarantee sustainability.*[16]

In other words, if the monopoly selects prices that satisfy the conditions for Pareto optimality under a profit constraint, then its profits can equal E, the maximal amount permitted by barriers to entry, and furthermore its virtue will be rewarded by protection from the threat of entry! The Ramsey-optimal prices will deter entry of competing firms attempting to market goods identical to those produced by the monopolist, and will also deter entry of firms proposing to market other goods in N, which may be very close substitutes for the monopolist's.

Since the proof of this result is rather lengthy and technical, we begin by discussing it heuristically, leaving a more rigorous analysis until later. Our initial discussion is formulated diagrammatically, but this time our graphs are drawn in output space rather than price space. In Appendix II we show that the results hold for the entire natural monopoly product set N, for corner as well as interior solutions, and for somewhat weaker assumptions for the cost function.

Figure 8E1 displays the total cost surface $C(y) + E(y)$. The particular shape shown for this function satisfies the assumptions of declining ray

---

[16] But see Faulhaber (1975b) for an example involving a cost function which does not exhibit the characteristics we will assume, and in which Ramsey pricing is not sustainable.

average costs and trans-ray convexity (complementarity) that were defined in Chapters 3 and 4, respectively. Figure 8E1 also shows a portion of the total revenue surface, the shaded dome $yP(y) \equiv \sum y_i P^i(y)$, given by the market's inverse-demand function. The heavy closed curve B is the intersection of the total cost surface with the total revenue surface. Because of its shape, we refer to it as a (floating) hyperbagel.[17] The projection of the curve B onto the floor of the diagram gives the boundary of the set of outputs T for which the profit of the monopolist, $\pi^m(y)$, is greater than or equal to the entry costs $E(y)$. We assume that T is convex. Further, we shall suppose that region T contains every output vector from which the entrant can hope to earn a profit. This assumption is plausible[18] since we may well expect that the revenue an entrant can earn from the production of any output vector will not exceed the revenue the monopoly could have earned from the sale of the same vector.[19]

We now consider the pricing decision facing a monopolist seeking a set of prices sustainable against entry. The monopolist announces a profitable set of fixed prices, $h = (h_1, \ldots, h_n)$, for his n outputs, and he offers to sell as much of his products to his customers as they desire to buy at these prices. Since a potential entrant must plan to set prices at or below the monopolist's if he is to be able to sell any of the corresponding products, the revenues he anticipates, R^e, are represented by points on or below a hyperplane H, satisfying $R = \sum h_i y_i$. That is,

(8E1) $$R^e(y^e) \leq \sum h_i y_i^e.$$

While $\sum h_i y_i^e$ has the form of a revenue function, it does not represent market revenue because the prices h_i are held fixed, while quantities vary freely without regard to demand conditions. Thus, we shall refer to H as a pseudo-revenue hyperplane.

The position of the hyperplane H with respect to the augmented cost function, $C^*(y) = C(y) + E(y)$, is critical for sustainability. First, suppose

---

[17] It may not be apparent at first glance that the somewhat twisted B locus really does have the shape of what George Borts has described, felicitously, as "that delectable desiccated dough-nut." But, viewed from above, a (slightly deformed) bagel it surely is. Alas, as any *maven* will confirm, bagels are not what they used to be.

[18] The assumption is plausible, but it is possible to devise exceptions stemming from demand complementarity between the outputs of the entrant and those of the (former) monopolist. Thus, consider our earlier example of an entrant who (profitably) produces electric stoves for which the former monopolist supplies the electricity. If the monopolist were to produce the entrant's output vector composed of stoves but no electricity, he could hardly expect to make any profit, because no one would buy the stoves if no electricity were available. We are able, in Appendix III, to expand the region of profitable entry to correspond to some greater range of demand complementarity.

[19] That is, if the entrant's revenue is $R^e(y^e) \leq R^m(y^e)$, then if $\pi^m(y^e) = R^m(y^e) - C(y^e) < E(y^e)$, the entrant's net profit from y^e must be negative since $\pi^e(y^e) = R^e(y^e) - C(y^e) - E(y^e)$.

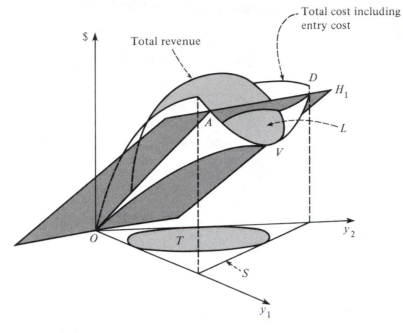

FIGURE 8E2

the monopolist charges prices $p_i^m = h_i$ and that at those prices consumers buy the vector y^m. Then the prices $p_i^m = h_i$ will be sustainable if H lies on or above C^* at y^m, so that the monopolist can cover his costs, and if C^* lies above H at every other point in the zone of potential profits. For then, at every possible vector of entrant outputs y^e, we have $R^e(y^e) \le \sum h_i y_i^e < C^*(y^e)$, so that entry is impossible without loss.

On the other hand, as in Figure 8E2, the pseudorevenue hyperplane H_1 may cut through the total cost surface above T. Then, for at least some demand–price relationships, it seems clear that the corresponding prices may not be sustainable because a competitor might enter, offer some output combination in T at lower prices (for example, point L), and earn more than the cost of production and entry.

To avoid this problem, if H is not initially below C^* throughout region T, the monopolist can reduce all the prices proportionately so that the hyperplane H swings downward toward the floor of the diagram, until it reaches the lowest hyperplane H_2 in Figure 8E3 that still has (at least) one point in common with the hyperbagel B, which we can take to be a point of tangency between H and B.[20] Let the quantities associated with this point

---

[20] If the functions are smooth and if $y^m > 0$, then H and B must be tangent.

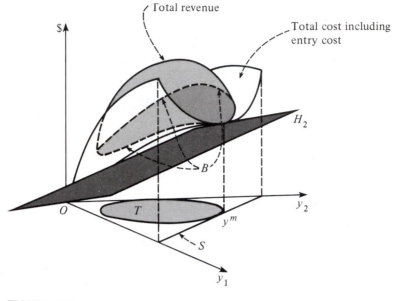

FIGURE 8E3

be y^m, and the prices defining H_2 be h^m. If these prices happen to satisfy the inverse demand function $h_i^m = P^i(y^m)$, then H_2 is the pseudorevenue hyperplane associated with y^m and pseudorevenue and market revenue coincide at y^m. By construction, H_2 lies below the total cost surface (when it has the shape pictured)[21] over T (except at y^m), and so the sustainability of h^m is assured.

This completes our informal discussion of the geometry of sustainability. Since we have indicated that, with the cost surface depicted, tangency between C^* and H at market prices is sufficient for sustainability, it remains for us to provide a heuristic proof that an interior Ramsey optimum corresponds to this sustainable tangency point.

After this argument we return briefly to our two cost attributes (transray convexity and declining ray average costs) to show explicitly why, when they are satisfied, tangency of H and C^* yields sustainable prices.

For diagrammatic simplicity, we continue to restrict our attention to the two-goods case. Tangency between H and B above y^* requires $dH = dC + dE$ at y^* locally along the hypercurve $\pi(y) = E(y)$. Since in the neighborhood of an optimal output vector y^*, we have assumed that the costs of entry are equal to the constant E, we have $dE = 0$. The requirement $dH = dC$ then

---

[21] The relevance of the shape of C will be discussed explicitly in Section 8F.

yields

(8E2) $$dH = \sum P^i(y^*)\,dy_i = \sum C_i(y^*)\,dy_i = dC$$

while the requirement that dy satisfy $\pi(y) = E(y)$ gives us

(8E3) $$d\pi = \sum \{MR_i(y^*) - C_i(y^*)\}\,dy_i = dE = 0$$

where MR_i denotes marginal revenue of output i, $\partial R(y)/\partial y_i$. Solving (8E2) and (8E3) for the two products $i = 1, 2$, we have

(8E4) $$\frac{MR_1(y^*) - C_1(y^*)}{MR_2(y^*) - C_2(y^*)} = -\frac{dy_2}{dy_1} = \frac{P^1(y^*) - C_1(y^*)}{P^2(y^*) - C_2(y^*)}$$

which is the Ramsey-optimality condition for any two products whose outputs are positive.[22]

Conversely, suppose $\pi(y^*) = E$, $\partial E/\partial y_i$ at $y^* = 0$, and that the Ramsey conditions (8E4) hold. $\pi = E$ implies that H and $C(y) + E$ coincide at y^*, and (8E4) implies that (8E2) holds for dy satisfying (8E3). Thus, the Ramsey conditions (8E4) imply that a meeting point of H and B must, under our cost assumptions, be a point of tangency between them.

As a final part of our proof that the Ramsey solution is sustainable, we must show that, with costs having the properties we have assumed, the pseudorevenue hyperplane is not only a local support for the cost surface at the Ramsey point y^*, as was just demonstrated, but, as was just suggested, that it supports costs throughout the potentially profitable region T so that the entrant must lose money anywhere in that region. We can see immediately that this follows from the assumed curvature properties of the cost function (trans-ray convexity and declining ray average cost) for, as a result, at y^* the total cost surface "bends away" from H in every direction over T (Figure 8E4). Thus, at every point in T, an entrant's total revenue [which, we know from (8E1), cannot be above the height of H at that point] must be insufficient to cover his operating and entry costs. Hence, the Ramsey prices will be sustainable, as our sufficient condition asserts.

To show explicitly the role of our cost assumptions, we now state the geometric argument a bit more carefully. Let S be a hyperplane tangent to T at the monopolist's output y^m on the floor of the diagram (Figure 8E3). Then the region S^- between S and the origin will contain T. Consider any

---

[22] If there are locally increasing returns to scale, the Ramsey rule (8E4) implies that marginal profit yields are negative; that is, the firm will produce more than it would if it were a profit-maximizing monopoly immune from entry. More than that, (8E4) may even involve negative marginal revenues at the solution point. Thus, Ramsey theory can account for those observed cases in which prices appear to leave the firm in the inelastic portions of its demand curves.

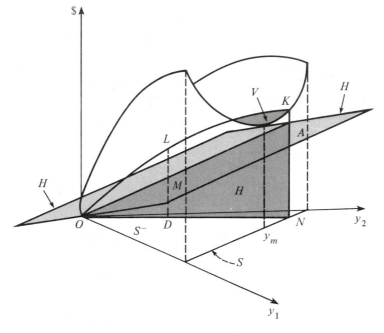

FIGURE 8E4

point D in T, and, hence in S^- (Figure 8E4). With the pseudorevenue hyperplane satisfying the Ramsey tangency condition, we want to show that *any*
output D in T must lose money for an entrant because L lies above M,
where L and M are, respectively, the corresponding points on the cost surface
and pseudorevenue hyperplane H. For the cost surface, draw the cross
section $ONKL$ through point L above ray ODN. Then we know that if the
cost surface and hyperplane H are tangent at V, at N the cost point K must
(by trans-ray convexity) lie on or above point A on hyperplane H. By strictly
decreasing ray average cost, point L must lie above the line segment connecting the origin and K, and this in turn must lie on or above the line
segment OA containing M. Hence, L must lie above M, and so any point
D in S^- must be unprofitable to the entrant, which is what Proposition 8E1
asserts.

Thus, we have provided a geometric argument showing that Ramsey-
optimal pricing offers a reward to the monopolist seeking anticipatory
protection from entry. We have, then, a new "invisible hand" notion, applicable outside the well-understood world of perfect competition. Our result
is a consequence of the same essential fact that underlies the welfare optimality of perfect competition. With prices held fixed at their market levels,
the derivatives of profit with respect to quantities are proportional to the
corresponding derivatives of consumers' plus producer's surplus. At the

given prices, changes in profit can therefore be used as a local approximation to increments in welfare.

Under perfect competition, producers take prices as parametric because they are so small relative to their markets that they are aware of no influence over prices. They set quantities to maximize profit, given the market prices. Welfare is, fortuitously, also maximized (if functions have the right shapes) because of the essential fact that, with parametric prices, profit approximates welfare locally.

In our model, the monopolist knows (Proposition 8C6) that his profits are limited to E. He also knows that potentially viable entry threats are limited to T, and that the profit of an entrant will be no greater than pseudo-profit (pseudorevenues less costs, including the cost of entry) calculated at the monopoly market prices, held fixed. Hence, if he chooses an output vector at which profit is equal to the entry cost (so that pseudoprofit is zero) and which at his fixed prices happens to maximize pseudoprofit over T, then pseudoprofit must be less than zero everywhere else over T. That is, profits of all potential entrants must be negative, and the monopolist's unique market position is guaranteed to be sustainable.

The monopolist who seeks stationary prices that can protect him from entry, like the perfect competitor, has an incentive to choose outputs that maximize profit calculated at those parametrically fixed market prices. But by doing so, the firm inadvertently maximizes net social welfare. Thus, the same invisible hand that guarantees welfare-optimal pricing under perfect competition, may guide the farsighted monopolist, seeking protection from entry, to the Ramsey welfare optimum.

Even under the conditions postulated, Ramsey pricing is sufficient but not necessary for sustainability. Consequently, there may exist sustainable price vectors, perhaps even many of them, that do not satisfy the Ramsey conditions. If so, one may well ask whether there is any reason to expect the firm to prefer the Ramsey solution: that is, whether it has any motivation, under those circumstances, to behave in a manner that promotes the social welfare. We have no unambiguous answer. However, we do have one source of comfort on this score, for, as will be shown next, at any non-Ramsey prices h (i.e., prices at which H does not support B) the monopolist cannot be sure of being able to prevent entry without knowing the shape of his demand and cost functions at outputs that may be far from his current output y^m, and hence far from his range of experience. It should be noted, however, that even sustainability of Ramsey pricing does depend on global properties of the *cost* function, notably, declining ray average cost throughout the region S^- in Figure 8E4. However, to assess this property, qualitative information can suffice, while to be sure of the sustainability of other price vectors, one needs global and quantitative information about costs *and demands*.

The argument is simple. By the assumption that H does not support B, there must be some points y^e in T at which $\sum h_i y_i^e > C(y^e) + E$, as at point L in Figure 8E2. But, as we have argued in (8E1), the entrant's revenue

$R^e(y^e) \leq \sum h_i y_i^e$, where the difference $\sum h_i y_i^e - R^e(y^e)$ depends upon the magnitude of the inverse demand function at y^e. Hence, at y^e, $R^e(y^e) \lessgtr C(y^e) + E$ and we cannot tell, without knowledge of the demand and cost functions at every point y^e in T, whether at the prices in question there will or will not exist points y^e in T at which successful entry is possible.

To discuss this result in a more concrete manner, we now present a simple peak-load pricing example, and show through it that non-Ramsey price vectors may or may not be sustainable. Let the cost relationship be

$$C(y_1, y_2) = 1 + \max(y_1, y_2) + y_1 + y_2 \qquad \text{for } \max(y_1, y_2) > 0$$

which is shaped like a V in cross section.[23] In the cost function, capital cost is $1 + \max(y_1, y_2)$ and the marginal operating expense of each service is constant and equal to unity. We suppose that there are no entry costs, and that the inverse demand functions, which involve independent demands for the two products or services, are

$$P^1(y) = 6 - y_1 \quad \text{and} \quad P^2(y) = 4 - y_2.$$

We now pick two candidate non-Ramsey price vectors and show that one is sustainable while the other is not. For our sustainable non-Ramsey solution we set price just above marginal cost (by the small amount $\delta > 0$) in the off-peak period 2, so that $p_2^m = 1 + \delta$. We then solve the $\pi = 0$ equation for the lowest price in market 1 which just permits the firm to break even.

We first show that these are not Ramsey prices. To see this, note that if we had set $p_2^m = 1$, we would have had $y_2^m = 3$, so that total revenue from that service covers only its variable cost, \$3. Therefore, by continuity, with $p_2^m = 1 + \delta$, the total revenue from market 2 must be \$3 + ε where we can make ε as small as we like by selecting δ sufficiently small. This means that if the firm is to cover its total cost, market 1 must cover its own variable cost, $1 \cdot y_1^m$ plus the capacity cost minus ε. This is clearly not possible if its price is set near its marginal cost. Therefore, these are not Ramsey prices, since $p_1^m > MC_1$, $p_2^m \approx MC_2$. They are, however, sustainable prices. This is so since for any price at or below marginal cost, the entrant will get no contribution to capital cost from his off-peak sales. Hence, he must obtain all capital cost from peak market 1. Since p_1^m is the lowest price that permits the firm to break even in market 1, any $p_1^e \leq p_1^m$ will yield a negative profit if the entrant serves all demand. But because of the fixed component of costs, revenues will be lower than costs for smaller outputs as well. Thus, these non-Ramsey prices of the monopolist are sustainable.

---

[23] This is a version of the cost function usually found in peak-load pricing models [see, e.g., Bailey and White (1974)]. The function is not differentiable everywhere. However, calculations using this function are much simpler to follow than those using a smooth approximation. Note that the function does exhibit trans-ray convexity and declining ray average costs, as our cost assumptions require.

For our non-Ramsey prices that are unsustainable, we set off-peak price so as to equate marginal revenue and marginal cost. Then, $p_2^m = 2.5$, $y_2^m = 1.5$. Solving the $\pi = 0$ equation, peak price and output turn out to be $p_1^m = 4 - \sqrt{21}/2$, $y_1^m = 2 + \sqrt{21}/2$. These are not Ramsey prices because $MR_2 = MC_2$, while $MR_1 < MC_1$. The prices are, however, *not* sustainable, since the entrant can cover his costs at the price output vectors $p_1^e = 1.5$, $y_1^e = 2$, $p_2^e = 2$, $y_2^e = 2$. That is, $\pi^e = (1.5)2 + (2)2 - 1 - 4 - 2 = 0$, where the entrant serves only part of the peak demand. Thus, with the same cost and demand conditions, some candidate non-Ramsey prices are sustainable, while others are not. But this clearly depends on the shapes and positions of those demand functions.

8F. Implications of Unsustainability

The analysis presented in this chapter has demonstrated that the sustainability or unsustainability of a monopolist's price vector, or indeed of the monopoly itself, is not a trivial matter. On the one hand, we have presented several examples in which some or all of the feasible price vectors available to the monopolist are unsustainable in the absence of entry barriers. However, we have also shown that while the pressure of potential entry can, in some circumstances, lead to instability in the market place, it is also a potent force extending the benefits of Adam Smith's "invisible hand" to monopoly markets. In some cases it may even induce the natural monopolist to follow the socially optimal rules of Ramsey pricing! Clearly, our analysis has important implications both for regulatory policy and for public policy toward monopoly in general. These issues are examined in Chapter 12. But we must immediately urge the reader to bear in mind that unsustainability does *not per se* constitute a case for governmental restrictions on entry. In this section, we describe some theoretical results that may rule out some apparently attractive and potentially popular policies toward monopoly unsustainability, primarily to indicate that, in such cases, hard policy choices may be inevitable: Easy answers just may not work. We begin by examining the sustainability of regulated oligopoly.

When the regulator of a natural monopoly encounters a firm eager to enter the industry at current prices, and no sustainable prices exist, it is tempting to suppose the problem will solve itself once the entrant is taken into the regulatory fold. Unfortunately, the analysis reveals that this is only wishful thinking. If a natural monopoly is unsustainable, then any multifirm equilibrium is also likely to be unsustainable.[24]

Let us turn to the details of this proposition. We define an oligopolistic market structure to be a collection of firms k, each selling a different set

---

[24] Sustainability is, by definition, sufficient for the existence of equilibrium in contestable markets. Here, we see that it is necessary as well.

of goods T_k, and each earning zero profit π^k at a single, market-wide, vector of prices p^o (the superscript "o" indicating that they apply to the oligopoly). Each good in N must be offered by one of the firms. Formally,

$$\pi^k(p^o) = \sum_{i \in T_k} p_i^o Q^i(p^o) - C(Q^{T_k}(p^o)) = 0,$$

(8F1)

$$k \geq 2, \qquad T_j \cap T_k = \varnothing, \qquad \bigcup_k T_k = N.$$

In general, there need not exist a feasible oligopolistic division of the market for the goods in N. In Figures 8A2, 8A3, and 8D1, the points E correspond to price vectors p^o which satisfy (8F1) with $T_1 = \{1\}$ and $T_2 = \{2\}$. In Figure 8C1 no such market division is possible.

In Figure 8A3, as in Figures 8A2 and 8D1, the prices at E fulfill (8F1), but here there are problems. At E, the firm producing good 2 can drop its price and earn greater profits. However, when p_2 is lower than it is at E, the firm producing good 1 cannot survive. Moreover, freedom of entry renders E unsustainable in Figure 8A3 because a new competitor can offer good 2 at a lower price. For this same reason, in Figure 8A3 no price point is sustainable for an oligopoly.

This example clearly suggests that the definition of the sustainability of monopoly (Definition 8A2) can be extended to oligopolistic market structures. However, in the case where cost subadditivity holds, there is an inherent difficulty. Given the cost advantages of natural monopoly, all the oligopolistic firms satisfying (8F1) can merge and, because of the resulting cost reduction, earn positive profits at the prices p^o. This new monopoly then can, by continuity, reduce all its prices and break even. Hence, under our subadditivity hypothesis, no regulated oligopoly can be sustainable against total merger, or against an entrant who proposes to monopolize the market.

We proceed here by excluding this case from our definition of sustainable oligopoly.[25] This exclusion does not deprive the definition of substance since by that definition an oligopoly may or may not be sustainable (compare Figures 8A2 and 8D1 with 8A3 and 8C1), and it only strengthens the consequent lessons for policy. We now state formally

Definition 8F1: Sustainability of Oligopoly   A regulated oligopoly characterized by the product sets T_k and by the price vector p^o which together satisfy (8F1) is sustainable if and only if

(8F2) $$\pi_S^e(p_S^e, y_S^e, p_{(S)}^o) = p_S^e y_S^e - C(y_S^e) \leq 0$$

---

[25] However, we do not exclude this case from our definition of sustainable oligopolies given in Chapter 11, because there we focus on natural oligopolies rather than on the natural monopoly industries that are the subject of this chapter.

for all $S \subset N$ (strict subset), with

$$p_S^e \leq p_S^o, \ y_S^e \leq Q^S(p_S^e, p_{(S)}^o),$$

except

$$y_S^e = Q^S(p^o) \qquad \text{for } S = \bigcup_i T_{k_i}.$$

This form of sustainable oligopoly need not be invulnerable to an entrant offering the complete product set N. In addition, a trivial merger of some of the firms, which affects neither prices nor costs, is not considered a violation of sustainability.[26] However, a cost-saving merger must permit positive profits at reduced prices and must, therefore, violate (8F2). As defined, sustainability also requires that the oligopoly market structure be invulnerable to any other full-supplier entrant who is not simply duplicating the prices and output of some of the existing firms. Moreover, the market structure must be invulnerable to entry by a partial-supplier marketing a proper subset of the goods in N. The following result is proved in Appendix I.

Proposition 8F1   *Given subadditivity of cost in the production of the goods in N, where these goods are weak gross substitutes for one another, if monopoly over N is not sustainable against entry by a full supplier, then any oligopoly that supplies N must also be unsustainable.*

We can now state

Proposition 8F2   *Given subadditivity of costs, weak gross substitute demands, and decreasing average incremental costs for the goods in N, if monopoly over N is not sustainable, then any oligopoly over N must also be unsustainable.*

Proof   By Proposition 8D2, under these hypotheses, if a monopoly is unsustainable against partial-supplier entry, then it must also be unsustainable against full-supplier entry. Hence, Proposition 8F1 now applies.

Q.E.D.

The implications should now be clear. If the natural monopoly is unsustainable, then entry restrictions may be required for the entire product set to be offered, regardless of the market structure that emerges or is established by a regulator.

Suppose that a natural monopoly industry with a product set N of weak gross substitutes is unsustainable. What are some policy options? First, free

---

[26] This stipulation permits sustainability if costs are exactly additive over some of the products. It can be interpreted as assuming that the cost of merger is positive.

entry can be maintained in the expectation that a sustainable natural monopoly equilibrium will result, albeit at the loss of some of the products in N.
Second, natural monopoly production of N can be protected via entry restrictions or, more generally, by the establishment of institutions that raise
entry costs sufficiently. Third, regulated oligopoly production of N can be
maintained via restriction of entry and prohibition of merger. Within the
theoretical framework of this chapter, the second option is superior to the
third, because of the cost savings offered by monopoly. Nevertheless, it can
be argued that competition can yield substantial counterbalancing advantages such as correction of regulatory defects and stimulation of innovation.

The evaluation of the first policy option depends on the social desirability
of the product set N.[27] The second policy option preserves the production
of the full product set by the natural monopoly. The disadvantages of this
second option increase with the severity of the entry restrictions needed to
carry it out—a subject on which more will be said in Chapter 12.

8G. Subsidized Marginal Cost Prices Do Not Ensure Sustainability

The earlier focus of the sustainability literature upon the regulated firm
may have led to the impression that unsustainability is something that
plagues only privately owned utilities, which are forced to cover their costs.
This would suggest that public enterprises, which set their prices equal to the
corresponding marginal costs, with the shortfall made up by the treasury,
are immune from this malady. Unfortunately, this is simply not true, as
shown by the following simple example.

Let the cost function be given by

$$(8G1) \qquad C(y_1, y_2) = \begin{cases} 15 + y_1^2 & y_1 > 0, y_2 = 0 \\ 20 + y_2 & y_1 = 0, y_2 > 0 \\ 25 + y_1^2 + y_2 & (y_1, y_2) > 0 \end{cases}$$

and let $Q^1(p_1) = 12 - p_1$. A firm employing marginal cost pricing will then
set $p_1 = 8 = C_1 = 2y_1$ and sell four units of y_1 and set $p_2 = 1$ regardless of
demand conditions. Such a firm requires a subsidy of 9 units per period. It
is easy to verify that the cost function is subadditive and exhibits both
economies of scale and economies of scope in the relevant range. However, at
these *first best* prices $(p_1, p_2) = (8, 1)$, a firm specializing in product 1 can

---

[27] If it is not N that is optimal, but instead some sustainable subset, then, of course, there
is no problem. However, it is easy to construct examples in which natural monopoly over a set
of socially essential (infinitely desirable) products is unsustainable.

enter and make a profit of 1 unit without any subsidy by charging a price (slightly) less than 8 and capturing the entire market for good 1.

This example reveals that public ownership and subsidy is not a guranteed cure for the unsustainability of natural monopoly in perfectly contestable markets. As long as the economy possesses a private sector searching for profitable opportunities, complete freedom of entry may be incompatible with the requirements of an *optimum optimorum*.

8H. Conclusions: The Virtues of Entry

It is time to pause and take stock. In Chapter 12 we delve more thoroughly into the policy issues raised by our analysis of the different market forms. But our view of natural monopoly easily lends itself to a misunderstanding that is best cleared up at once.

Threat of entry has played the central role in our analysis of market equilibrium. It has been shown not to be devoid of blemishes. Indeed, our results that emphasize that subadditivity does not imply sustainability have been misinterpreted as rationalizations of restrictions upon freedom of entry. The argument is simple: In an industry characterized by subadditive costs but no sustainable prices, freedom of entry threatens the viability of the (natural) monopoly. If entry succeeds because the incumbent has no prices available that are capable of preventing it, then the public will bear the resulting increase in costs. It will be served by a multiplicity of firms whose production is *in toto* more expensive, rather than by a single firm whose relatively low costs reflect subadditivity. Therefore, to prevent such waste, it would appear appropriate for the policy maker to remove the threat by prohibiting entry.

The argument is not entirely incorrect, but it is highly misleading. First, it applies with full force only to perfectly contestable markets. Outside such markets, where entry costs are present, the nonsustainability of natural monopoly means that a monopoly cannot find prices that permit it to earn rent equal to the entry costs and prevent wasteful entry simultaneously. But that does not preclude the possibility that a natural monopoly can find prices that yield a slightly lower rent and leave it invulnerable to entry. In such a case, despite the unsustainability, the threat of entry may still force the monopoly to behave better than it otherwise would have.

Figure 8H1 depicts such an example. Here, AC^m and AC^e are the average cost curves of the incumbent monopoly and potential entrants, respectively, with the latter higher because of entry costs. The market clearing price that enables the incumbent to earn the full entry cost as rent is p^m. However, as the figure is drawn, p^m is unsustainable. This unsustainability need not cause the industry to be unstable, and it need not lead to wasteful entry. Rather, it may induce the monopoly to prevent entry by the adoption of p^m_*, a price

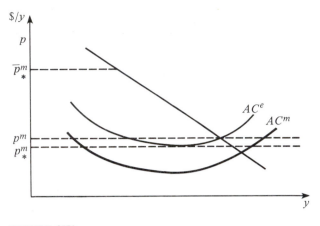

FIGURE 8H1

lower than p^m, which earns less rent for the incumbent. And, without any threat of entry, if freedom of entry were curtailed, an unconstrained monopolist would have the incentive to set a high price, such as $\bar{p}^m$ which maximizes profits under the circumstances in the figure. Thus, in this unsustainability case, restrictions upon freedom of entry are clearly socially injurious.

But there are more important reasons for denying emphatically that our analysis is intended to support restrictions on freedom of entry—a conclusion that is virtually the reverse of the truth. Our analysis, if anything, should lend itself to interpretation as a powerful argument for freedom of entry. The theme of the chapter has been that freedom of entry, indeed the mere threat of incursions by entrants into the market, may effectively discipline the monopolist, *even if entry is never successful.* It can force the monopolist to curb his avarice and forgo profits he might otherwise have enjoyed. Indeed, in the absence of entry barriers, in perfectly contestable markets, it can force him to accept earnings no higher than those available under perfect competition. Potential competition can also force the monopolist to produce with maximal efficiency, and to hunt down and utilize fully every opportunity for innovation. Perhaps most surprising of all, it can induce the monopolist to institute those (Ramsey) prices which welfare theory has shown to be requisites of Pareto optimality under a profit constraint. In short, the threat of entry can force virtuous behavior upon the monopolist, for if he behaves badly his monopoly becomes vulnerable. In our analysis, it is freedom of entry alone that is capable of accomplishing all these things.

Consequently, we feel that cases in which entry may lead to some social inefficiency should not be taken to tilt the scales against the presumption that freedom of entry is socially beneficial.

Yet, it is true that our analysis shows the possibility of grounds for occasional and limited restrictions upon freedom of entry. Where markets are perfectly contestable and monopoly is natural but unsustainable, limitations upon entry may be needed to ensure that the socially optimal set of products can be produced in the most efficient manner. This theoretical finding must be interpreted as no more than a first step in the process of identifying the economic characteristics of the special situations that may warrant concern over the effects of free entry.

Our analysis in this chapter has indicated that there are two basic sets of circumstances in which unsustainability may be troublesome.[28] The first is characterized by natural monopoly with decreasing returns to scale at its market-clearing output levels (see Figure 8A4 and Proposition 8C3). Here, in perfectly contestable markets, there are incentives for profitable entry by partial suppliers who destabilize the industry and raise total production costs. While such circumstances may conceivably warrant control of entry, it may be argued that they are unlikely to occur. If (ray) average cost curves have substantial flat bottoms (see Chapters 2 and 11 for descriptions of the single and multiproduct cases, respectively), as some empirical evidence suggests, then the combination of natural monopoly and decreasing returns to scale must be considered unlikely. Nevertheless, it may be prudent to view with suspicion the entry of partial-suppliers into a perfectly contestable industry. However, where significant sunk costs and other entry barriers are present, socially beneficial entrants may well have the incentive to serve only part of the demand for their products, in order to conserve on irreversible investment put at risk.

The second unsustainability scenario found by our analysis (Proposition 8D3 and Figures 8A3 and 8C1) entails goods with relatively strong product-specific returns to scale that are linked by relatively weak cost complementarities and relatively strong substitution effects in demand. Apparently, the primary social danger here is that contestable markets will fail to provide some goods whose total (welfare) value to consumers exceeds the required incremental production cost. One possible remedy for this type of market failure is a policy that condones or even encourages allocation mechanisms that are more effective in raising revenues than are nondiscriminatory simple prices. Multipart prices, quantity discounts, and group-differentiated prices are all examples of mechanisms that may permit suppliers' revenues to correspond more closely to consumers' total benefits than simple prices do, and that may, therefore, encourage the supply of socially beneficial goods that the market might otherwise fail to provide.

The other conceivable remedy that must be considered is restriction of the competitive entry that here renders unstable production by a natural monopoly of the optimal product set. However, it seems implausible that even the best-intentioned authorities can, except in very special cases, restrict

---

[28] Chapter 14 identifies some other significant unsustainable cases involving intertemporal effects, and, of course, those need not exhaust all the possibilities.

whatever entry would genuinely be socially injurious without, simultaneously, impeding beneficial and innovative entry and the potential entry needed to discipline effectively the performance of incumbents.

Moreover, our theoretical evaluation of the scenario changes drastically with a modification that is only moderate in magnitude and is therefore likely to be difficult to distinguish in practice. If the cost complementarities among the products in question are *strong*, relative to the product-specific returns to scale, then, as discussed in Chapter 4, trans-ray convexity (or trans-ray supportability) of costs can be expected to hold. But in that case, the scenario beset by unsustainability is replaced by one consistent with sustainability.

And, here, the weak invisible hand theorem tells us that the socially (Ramsey) optimal set of products is sustainable and that their Ramsey-optimal prices are sustainable. Not only are restrictions on entry unnecessary here, they are distinctly inimical to the public interest. It is the existence of potential entrants that constrains the incumbent monopolist to operate efficiently and to hold his profit down to the minimum level. It is the threat of unrestricted entry that can lead the incumbent to serve the public interest by choosing to operate at the Ramsey optimum, led to this only by pursuit of self-interest.

APPENDIX I

Proof of Proposition 8C10

The following argument demonstrates that if $C(\cdot)$ exhibits weak cost complementarities, then the cost game has a nonempty core.

Proof As stated above, the vector $(E_1, E_2, \ldots E_n)$ is in the core of the cost game defined for a fixed y if $\sum_{i=1}^{n} E_i \geq C(y)$ and if $\sum_{i \in S} E_i \leq C(y_S)$, $\forall S \subset N$. A fundamental theorem by Shapley (1971) on "convex games" implies that there is such an imputation E if

$$(1) \qquad C(y_T + y_R) - C(y_T) \leq C(y_S + y_R) - C(y_S) \qquad \forall T, R, S \subseteq N$$

with $S \subseteq T$ and $T \cap R = S \cap R = \varnothing$ (the null set).

We show that this condition follows directly from $C_{ij} \leq 0$:

$$C(y_T + y_R) - C(y_T) = \int_{\Gamma} \sum_{i \in R} C_i(y_T + x_R) \, dx_i,$$

$$C(y_S + y_R) - C(y_S) = \int_{\Gamma} \sum_{i \in R} C_i(y_S + x_R) \, dx_i,$$

where Γ is a smooth monotonic arc from the origin to y_R. These are line integrals along the common path Γ:

$$[C(y_T + y_R) - C(y_T)] - [C(y_S + y_R) - C(y_S)]$$

$$= \int_{\Gamma} \sum_{i \in R} [C_i(y_S + y_{T-S} + x_R) - C_i(y_S + x_R)] \, dx_i$$

$$= \int_{\Gamma} \sum_{i \in R} \left[\int_{\Lambda} \sum_{j \in T-S} C_{ij}(y_S + z_{T-S} + x_R) \, dz_j \right] dx_i \leq 0,$$

where Λ is a smooth monotonic arc from the origin to y_{T-S}. Thus, (1) is satisfied and the core of the cost game is nonempty. Q.E.D.

This result can be extended directly to the case of the "transylvanian" cost function discussed in Chapter 7; that is, $C(y) = F(S) + c(y)$ where

$S = \{i \in N \,|\, y_i > 0\}$. The argument goes through exactly as before under the assumptions that $c_{ij} \le 0$ and F satisfies the analog of (1).

Proof of Proposition 8D1

We demonstrate that the necessary conditions for profit maximization for a partial-supplier entrant imply that his (maximized) profits must be less than those of a hypothetical full-supplier entrant (which are negative by hypothesis).

Assuming that $S \subset N$ is the entrant's profit optimal product set, conditions for maximal profits are obtained from the program

$$\max_{p_S^e, y_S^e} \pi^e = p_S^e y_S^e - C(y_S^e)$$

subject to $p_S^m - p_S^e \ge 0$ and $Q^S(p_S^e, p_{(S)}^m) - y_S^e \ge 0$. The Lagrangian expression is

$$\mathscr{L} = p_S^e y_S^e - C(y_S^e) + \sum_{i \in S} \lambda_i \{Q^i(p_S^e, p_{(S)}^m) - y_i^e\} + \sum_{i \in S} \mu_i(p_i^m - p_i^e).$$

Kuhn-Tucker necessary conditions for an interior maximum ($p_S^e, y_S^e > 0$) are given ($\forall i \in S$) by

(2)
$$\frac{\partial \mathscr{L}}{\partial p_i^e} = y_i^e + \sum_{j \in S} \lambda_j \frac{\partial Q^j}{\partial p_i} - \mu_i = 0$$

(3)
$$\frac{\partial \mathscr{L}}{\partial y_i^e} = p_i^e - C_i(y_S^e) - \lambda_i = 0$$

(4)
$$\frac{\partial \mathscr{L}}{\partial \lambda_i} = Q^i - y_i^e \ge 0 \qquad \lambda_i \ge 0 \qquad \lambda_i(Q^i - y_i^e) = 0$$

(5)
$$\frac{\partial \mathscr{L}}{\partial \mu_i} = p_i^m - p_i^e \ge 0 \qquad \mu_i \ge 0 \qquad \mu_i(p_i^m - p_i^e) = 0.$$

Let T be the set of products for which the entrant finds it optimal *not* to meet the market demand. That is, $Q^i > y_i^e$ for $i \in T$. ($T \ne \varnothing$ because the entrant is assumed to be a partial supplier.) From (4) we know

(6)
$$\lambda_i = 0, \qquad \forall i \in T.$$

Substituting this into (2), we obtain

(7)
$$\mu_i = y_i^e + \sum_{j \in S - T} \lambda_j \frac{\partial Q^j}{\partial p_i} > 0 \qquad \forall i \in T.$$

Since $y_i^e > 0$, expression (7) is positive by WGS ($\partial Q^j/\partial p_i \geq 0$, $i \neq j$). Thus, by (5)

(8) $$p_i^m = p_i^e \qquad \forall i \in T.$$

Combining (3), (6), and (8), we have

(9) $$p_i^e = C_i(y_S^e) = p_i^m, \qquad \forall i \in T.$$

Expression (9) establishes the intuitive result that if sale of a quantity smaller than that permitted by the demand curve is optimal, it is optimal to take the monopolist's price as given, and adjust output so that marginal cost is equated to that price. Using the results thus far, we can express the partial supplier's (maximized) profits as

(10)
$$\pi^e = \sum_{i \in S-T} p_i^e Q^i [p_{S-T}^e, p_{(S-T)}^m]$$
$$+ \sum_{i \in T} y_i^e C_i(y_T^e + Q^{S-T}) - C(y_T^e + Q^{S-T}).$$

We now add and subtract $C(Q^{S-T})$ on the right-hand side of (10). Upon rearrangement, we obtain,

(11)
$$\pi^e = \left\{ \sum_{i \in S-T} p_i^e Q^i [p_{S-T}^e, p_{(S-T)}^m] - C(Q^{S-T}) \right\}$$
$$+ \left\{ \sum_{i \in T} y_i^e C_i(y_T^e + Q^{S-T}) + C(Q^{S-T}) - C(y_T^e + Q^{S-T}) \right\}.$$

The first bracketed term in (11) is precisely the anticipated profit of a full-supplier entrant offering the product set $S - T$ at prices p_{S-T}^e. By hypothesis, this term must be nonpositive (since p^m is assumed to be sustainable against full-supplier entry). The second bracketed term is nonpositive by $DAIC$. Thus the (maximized) anticipated profits of any partial-supplier entrant must be nonpositive.   Q.E.D.

Proof of Proposition 8D2

Proof   Suppose the contrary, that there is $S \subseteq N$ and $p_S^e \leq p_S^m$ with

(12) $$p_S^e Q^S(p_S^e, p_{(S)}^m) > C(Q^S(p_S^e, p_{(S)}^m)).$$

(8C5) requires $p^m Q(p^m) \geq C(Q(p^m))$ and $p_S^m Q^S(p^m) \leq C(Q^S(p^m))$. Subtraction gives

(13) $$p_{(S)}^m Q^{(S)}(p^m) \geq C(Q^S(p^m) + Q^{(S)}(p^m)) - C(Q^S(p^m)).$$

Because demands are independent and $p_S^e \leq p_S^m$, $Q^S(p^m) \leq Q^S(p_S^e, p_{(S)}^m)$. Then, by the same reasoning that established Proposition 8C10, the weak cost complementarities hypothesis implies

$$C(Q^S(p^m) + Q^{(S)}(p^m)) - C(Q^S(p^m))$$

(14)
$$\geq C(Q^S(p_S^e, p_{(S)}^m) + Q^{(S)}(p^m)) - C(Q^S(p_S^e, p_{(S)}^m)).$$

Also, $Q^{(S)}(p^m) = Q^{(S)}(p_S^e, p_{(S)}^m)$, by independent demands. Hence, (13) and (14) together give

$$p_{(S)}^m Q^{(S)}(p_S^e, p_{(S)}^m) \geq C(Q^S(p_S^e, p_{(S)}^m) + Q^{(S)}(p_S^e, p_{(S)}^m)) - C(Q^S(p_S^e, p_{(S)}^m)).$$

Adding this to (12) yields

$$p_S^e Q^S(p_S^e, p_{(S)}^m) + p_{(S)}^m Q^{(S)}(p_S^e, p_{(S)}^m) > C(Q(p_S^e, p_{(S)}^m)),$$

or $\pi(p_S^e, p_{(S)}^m) \geq 0$. Thus p^m is dominated, which contradicts the hypothesis.
                                                                                              Q.E.D.

It can also be shown that if demands are independent or complementary, if $C_{ij} \leq 0$ and if $C_{ii} \leq 0$, then the necessary conditions for sustainability given in Proposition 8C7, 8C8, and 8C9 suffice for the sustainability of p^m against market entry. Of course, $C_{ii} \leq 0$ is implied by the concave costs assumption used in Footnote 12 to extend Proposition 8D1 to the case of demand complementarities.

Proof of Proposition 8D3

We obtain our result by assuming that (8C5) is satisfied at the undominated price vector p^m. We then take a full supplier to have entered successfully and derive an economically interpretable consequence whose negation is thus sufficient for the sustainability of p^m. Specifically, suppose a full supplier can offer good 2 profitably at some $p_2^e < p_2^m$. Then

(15) $$p_2^e Q^2(p_1^m, p_2^e) - C[0, Q^2(p_1^m, p_2^e)] > 0.$$

The former monopolist's prices p^m are assumed to be undominated, so, in particular, monopoly profits, π^m, at (p_1^m, p_2^e) must be nonpositive:

(16) $$0 \geq \pi^m(p_1^m, p_2^e) = p_1^m Q^1(p_1^m, p_2^e) + p_2^e Q^2(p_1^m, p_2^e)$$ (16)

$$- C[Q^1(p_1^m, p_2^e), Q^2(p_1^m, p_2^e)]$$

$$= (p_2^e Q^2(p_1^m, p_2^e) - C[0, Q^2(p_1^m, p_2^e)]) + (p_1^m Q^1(p_1^m, p_2^e)$$

$$- C[Q(p_1^m, p_2^e)] + C[0, Q^2(p_1^m, p_2^e)]).$$

Equation (16) has expressed $\pi^m(p_1^m, p_2^e)$ as the sum of two terms, the first of which is (15), the potential entrant's profit. If the second of these terms were positive or zero, then the positivity of (15) would contradict the undomination of p^m. It would then follow that no full supplier offering good 2 could do so profitably. This sufficient condition for sustainability against this sort of entrant, then, is

$$p_1^m Q^1(p_1^m, p_2^e) - C[Q(p_1^m, p_2^e)] + C[0, Q^2(p_1^m, p_2^e)] \geq 0$$

or

(17)
$$\frac{C[Q(p_1^m, p_2^e)] - C[0, Q^2(p_1^m, p_2^e)]}{Q^1(p_1^m, p_2^e)} \leq p_1^m.$$

The condition in (17) can helpfully be related to (8C5), which requires that

$$p_2^m Q^2(p_1^m, p_2^m) \leq C[0, Q^2(p_1^m, p_2^m)].$$

Together with nonnegativity of the monopolist's profits at p^m, this implies that

$$p_1^m Q^1(p_1^m, p_2^m) \geq C[Q(p^m)] - C[0, Q^2(p_1^m, p_2^m)],$$

or

(18)
$$\frac{C[Q(p^m)] - C[0, Q^2(p_1^m, p_2^m)]}{Q^1(p_1^m, p_2^m)} \leq p_1^m.$$

The left-hand sides of both (17) and (18) can be viewed as values of this function of p_2:

(19)
$$\frac{C[Q(p_1^m, p_2)] - C[0, Q^2(p_1^m, p_2)]}{Q^1(p_1^m, p_2)}.$$

The inequality in (18) and $p_2^e < p_2^m$ together ensure (17) if the function in (19) is nondecreasing in p_2, or if its partial derivative is nonnegative. This is equivalent to

(8D1)
$$\frac{\partial Q^2}{\partial p_2} (C_2[Q^1, Q^2] - C_2[0, Q^2])$$

$$+ \frac{\partial Q^1}{\partial p_2} \left(C_1[Q^1, Q^2] - \left(\frac{C[Q^1, Q^2] - C[0, Q^2]}{Q^1} \right) \right) \geq 0.$$

The analogous chain of reasoning establishes that (8D2) is sufficient for sustainability against full suppliers who would enter by offering good 1.

Q.E.D.

Proof of Proposition 8F1

We demonstrate the contrapositive. Suppose an oligopoly consistent with (8F1) supplying N is sustainable at prices p^o which satisfy (8F2). We construct a price vector p^m at which the monopoly is sustainable against entry by full suppliers.

First, note that $\pi(p^o) = p^o Q(p^o) - C(Q(p^o)) > 0$, since

$$\pi(p^o) - \sum_k \pi^k(p^o) = \sum_k C(Q^{T_k}(p^o)) - C(Q(p^o)),$$

and the right-hand side is positive because $Q(p^o) = \sum_k Q^{T_k}(p^o)$. By continuity of π, there is a compact set of vectors p, bounded strictly above by p^o and below by 0, at which $\pi = 0$. Among these choose an undominated $p^m < p^o$. This can be done, for example, by minimizing p_i over the set of p just mentioned. A continuous function, p_i, achieves its minimum over a compact set.

Because p^m is undominated, the monopoly is invulnerable to full-supplier entry by a competitor offering all goods in N. Suppose a full-supplier entrant can earn nonnegative profit by offering the goods in $S \subset N$ at prices $p_S^e \leq p_S^m$. Then

(20) $$p_S^e Q^S(p_S^e, p_{(S)}^m) \geq C(Q^S(p_S^e, p_{(S)}^m)).$$

But, since $p^m < p^o$, $p_S^e < p_S^o$. Further, by the weak gross substitute demands and $p_{(S)}^m < p_{(S)}^o$, $Q^S(p_S^e, p_{(S)}^m) \leq Q^S(p_S^e p_{(S)}^o)$. Thus, the successful competitor of the monopolist can also succeed on the same terms as an entrant (possibly a partial supplier) into the regulated oligopoly. That is, setting $y_S^e = Q^S(p_S^e, p_{(S)}^m)$, (20) contradicts (8F2). Hence, sustainability of the oligopoly implies sustainability of the monopolist against entry by full suppliers.   Q.E.D.

APPENDIX II

A More General Proof of Proposition 8E1:
The Weak Invisible Hand Theorem

Before proving the weak invisible hand theorem, we first present the Ramsey rule for welfare optimal pricing under a profit constraint.[1] This rule is pertinent to a firm which, because of scale economies, would suffer losses if it were to set the prices of its products equal to the corresponding marginal costs. The rule is usually derived under the assumption that the optimal product set is known; however, here we seek both prices *and outputs* that are welfare-optimal, subject to the constraint that the firm earns profits π equal to the maximal economic profit E permitted by barriers to entry.

We take as our welfare measure the sum of consumers' plus producers' surplus. The Lagrangian expression is

$$\mathcal{L} = S + \pi + \lambda(\pi - E)$$
$$= S[P(y)] + (1 + \lambda)[y \cdot P(y) - C(y)] - \lambda E,$$

where $P(y)$ is the vector of inverse demand functions and $S(\cdot)$ is consumers' surplus as a function of output prices. Differentiation yields the necessary conditions

$$\frac{\partial \mathcal{L}}{\partial y_i} = -\sum_j y_j \frac{\partial P^j}{\partial y_i} + (1 + \lambda)\left(P^i - C_i + \sum_j y_j \frac{\partial P^j}{\partial y_i}\right) \le 0$$

$$y_i \ge 0, \qquad y_i \frac{\partial \mathcal{L}}{\partial y_i} = 0, \qquad \lambda \ge 0, \qquad \pi = E.$$

Letting $R(y) = P(y) \cdot y$ denote the total revenue function and R_i its partial

---

[1] We adopt this name since the rule was first derived by Ramsey (1927) in a taxation context. For other important contributions on the subject see Boiteux (1971) and Diamond and Mirrlees (1971a,b). For a history and some related forms of the first-order conditions, see Baumol and Bradford (1970).

derivatives, these conditions can be rewritten as

$$P^i - C_i = -\lambda(R_i - C_i) \qquad \text{for } y_i^* > 0,$$

$$P^i - C_i \leq -\lambda(R_i - C_i) \qquad \text{for } y_i^* = 0,$$

$$\lambda \geq 0, \qquad \pi = E.$$

All optimal price-output vectors satisfy these *Ramsey conditions* by the Kuhn-Tucker theorem.

In the case of independent demands these conditions lead to the familiar inverse elasticity rule popularized by Baumol and Bradford (1970). Then, $R_i = P^i + y_i \partial P^i / \partial y_i = P^i(1 - 1/e_i)$, where e_i is the price elasticity of demand for product i. Substitution yields $(P^i - C_i)/P^i = \lambda/(1 + \lambda)e_i$ for all $y_i > 0$. Thus, for outputs produced, the percentage deviation of price from marginal cost is inversely proportional to the elasticity of demand.

Our analysis of the sustainability of the Ramsey price-output vector in the text was partially heuristic and unnecessarily restrictive in several respects. The geometric argument assumed that there were only two products, and that positive quantities of *each* were sold by the monopolist. Moreover, the cost function exhibited trans-ray convexity and declining ray average costs, and entry was limited to the potentially profitable region T. We now provide a proof which relaxes each of these assumptions significantly. We begin by providing a complete list of our assumptions and commenting on their features:

A1 (Assumption 1) (a) *Potential entrants only consider marketing goods in the natural monopoly product set N. The potential entrants and the monopolist have acces to the same production techniques.* (b) *An entrant incurs entry costs $E(y)$ which attain a maximum value E at "large" output vectors; in particular, $E(y) = E$ for all welfare-optimal (monopoly) output vectors which satisfy the profit constraint $\pi^m(y) \geq \pi^0$, for $0 \leq \pi^0 \leq E$. The profit ceiling E is less than the unconstrained maximum monopoly profit; i.e., $E < \max\{\pi^m(y)|y \geq 0\}$.*

A2 (Assumption 2) (a) *There is a one-to-one relationship between market prices and demand vectors.* (b) *Market prices depend on the industry output vector, and not on the number of firms that produce it.* (c) *The cost and inverse demand functions are differentiable over the region $Y = \{y \geq 0 | \pi^m(y) \geq E\}$.*

A3 (Assumption 3) (a) *There exists a Ramsey optimum y^* at which (1) is satisfied with $\pi^m(y^*) = E$.*[2] (b) *y^* lies in $Y^* = \{y | \pi^m(y) = E, \pi_i^m(y) < 0, \forall i\}$,*

---

[2] Alternatively, assume that the set $Y = \{y | \pi^m(y) \geq E\}$ is compact and satisfies a Kuhn-Tucker constraint qualification [see Diamond and Mirrlees (1971a,b) for a discussion of these properties in a similar context]. Thus a y^* exists which maximizes welfare over Y, and the Kuhn-Tucker conditions of this program (1) are satisfied at y^*.

i.e., the outer "northeast" boundary of Y where increases in output decrease profits.

We show below that when demands are weak gross substitutes and "normal" in a sense defined there, an optimal Ramsey solution must indeed lie in Y^*. Furthermore, it can be shown that no Ramsey solution outside Y^* is sustainable. Thus, only Ramsey points in Y^* need be considered.

A4 (Assumption 4)   The net profit function $\pi^m(y) - E(y)$ is strictly quasi-concave over the potentially profitable set, $T = \{y \geq 0 \mid \pi^m(y) - E(y) \geq 0,$ $y \neq 0\}$. More broadly, T is strictly supported at the y^* given by A3 by the tangent hyperplane $S = \{y \geq 0 \mid -\sum \pi_i^m(y^*)y_i = -\sum \pi_i^m(y^*)y_i^*\}$.

A2–A4 imply that the region which is relevant for our Ramsey optimization is Y, the set where profit equals or exceeds the largest entry cost E. The set T, to which entry threats are limited, does not coincide with Y since, for some y, $E(y) < E$.

A5 (Assumption 5)   The set T [or, given A3(b) and A4, more broadly, the set $S^- = \{y \geq 0 \mid -\sum \pi_i^m(y^*)y_i < -\sum \pi_i^m(y^*)y_i^*, \ y \neq 0\}$] contains every non-trivial output vector from which an entrant can hope to earn a nonnegative net profit.

We show below that the set T does contain all possible entrant vectors if demands are weak gross substitutes and "normal." On the other hand, as we have mentioned, limiting entry threats to T is restrictive if there are complementarities in demand. The advantage of using the larger set S^- (the region bounded by the origin and the hyperplane S, whose slopes match those of the outer boundary of T at y^*) is that it permits us to deal with a very broad range of complementaries in demand. Appendix III provides assumptions that render S^- the relevant set of entry threats.

A6 (Assumption 6)   (a) The total cost function $C(y) + E(y)$ has decreasing ray average costs over $S^- \cup S$. (b) $C(y) + E(y)$ is trans-ray convex on S.

The cost concepts of A6 can be replaced by the somewhat weaker A6$'$.[3]

A6$'$ (Assumption 6$'$)   The total cost function $C(y) + E(y)$ is strictly supported at y^* by the pseudorevenue hyperplane H above S^-, i.e., $C(y) + E(y) > P(y^*)y$ for $y \in S^-$ and $C(y^*) + E = P(y^*)y^*$.

The virtue of this substitution is that there may be costs which rise sharply with the introduction of a new product. In terms of Figure 8E4 this

---

[3] See Telser (1978), whose "kind characteristic function" has analogous properties.

means that the cross section of the cost surface above S may first increase and then turn downward as it approaches the axes, thus violating trans-ray convexity. Nevertheless, H may still serve as a support for the cost function above S^- and A6' permits us to deal with this case.

The cost concepts described in Chapters 3 and 4 can be used to provide an intuitive set of sufficient conditions for A6' to hold.

Lemma 1   *Trans-ray supportability of total costs, at y^*, along the trans-ray hyperplane S, and declining RAC are together sufficient for A6'. The proof is essentially the same as that given below for Lemma 3.*

We now proceed with the proof of Proposition 8E1. First, we establish the connection between Ramsey optimality and the points in Y^* at which the pseudorevenue hyperplane supports the total cost surface above S. The argument is a straightforward application of Kuhn-Tucker analysis.

Lemma 2   *Let $y^* \in Y^*$, and with S defined by A4, let the total cost function, $C(y) + E(y)$, be locally (trans-ray) convex over S at y^*. Then, y^* satisfies the Ramsey conditions if and only if the pseudorevenue hyperplane defined by $P(y^*)$ locally supports the total cost surface on S at y^*; i.e., if and only if*

$$(1) \qquad P(y^*)y \le C(y) + E(y) \qquad \text{for } y \in S \cap \mathcal{N}(y^*)$$

where $\mathcal{N}(y^*)$ is some neighborhood of y^* and $P(y^*)y^* = C(y^*) + E$.

Proof   The relations (1) are tautologically equivalent to y^* being a local maximum of the net pseudoprofit function $P(y^*)y - C(y) - E(y)$ over S. Forming the corresponding Lagrangian

$$\mathcal{L} = P(y^*)y - C(y) - E(y) + \mu\left[\sum \pi_i^m(y^*)y_i - \sum \pi_i^m(y^*)y_i^*\right],$$

we obtain the Kuhn-Tucker conditions necessary and sufficient[4] for y^* to be this local maximum of net pseudoprofit over S and to satisfy (1). Since we are in the region over which entry cost $E(y)$ is at a maximal plateau [A1(b)], its derivatives $E_i(y^*)$ must all be zero. Therefore the Kuhn-Tucker conditions are

$$P^i(y^*) - C_i(y^*) + \mu\pi_i^m(y^*) = 0, \qquad \text{for } y_i^* > 0$$

$$P^i(y^*) - C_i(y^*) + \mu\pi_i^m(y^*) \le 0, \qquad \text{for } y_i^* = 0 \text{ and } y^* \in S.$$

---

[4] Note that this is true since total costs are assumed convex locally, so that net pseudoprofit is concave locally at y^*. Further, S is a convex set defined by linear constraints.

Since $\pi_i^m = MR_i - C_i$ and, by construction, $\pi(y^*) = E$ and $y^* \in S$, these conditions are equivalent to those for Ramsey optimality.

We next show that cost assumption A6 implies the support assumption A6'.

Lemma 3   *Given the entry condition of A1(b) and A3, then A6 implies A6'.*

Proof   A1(b) and A3 provide the context needed for Lemma 2 to apply. A6(b) implies that $C(y) + E(y)$ is locally convex over S at y^*. Then, with y^* satisfying the Ramsey conditions, Lemma 2 asserts that the pseudorevenue hyperplane supports $C(y) + E(y)$ locally over S at y^*. But with the trans-ray convexity of A6(b), a local support is a global support. Hence, $P(y^*)y \leq C(y) + E(y)$, for $y \in S$.

Now, let $y \in S^-$ and define $\hat{y}$ to be its ray extension which lies in S. That is, $\gamma y = \hat{y} \in S$. Clearly, by the definitions of S and S^-, $\gamma > 1$,

$$P(y^*)y - C(y) - E(y) = \frac{1}{\gamma}\left[P(y^*)\hat{y} - C(\hat{y}) - E(\hat{y})\right]$$

$$+ \left[\frac{C(\hat{y}) + E(\hat{y})}{\gamma} - C\left(\frac{\hat{y}}{\gamma}\right) - E\left(\frac{\hat{y}}{\gamma}\right)\right] < 0.$$

To see that this expression is indeed negative, note that the first term in square brackets is nonpositive since $\hat{y} \in S$. The second term is negative by declining ray average costs, A6(a). Thus, A6' is established.

Having shown the relationship between the Ramsey optimality conditions and the condition that the pseudorevenue hyperplane acts as a support for the total cost surface, we finally give our more general proof of Proposition 8E1:

Given A1–A5, either A6 or A6' implies that the Ramsey optimum y^* is sustainable.

Proof   Lemma 3 allows us to proceed using A6'. Thus $P(y^*) \cdot y - C(y) < E(y)$ for $y \in S^-$. By A5, all nontrivial entry threats lie in S^-. But the anticipated profits of an entrant, $p^e y^e - C(y^e)$, are not greater than $p^m y^e - C(y^e) = P(y^*)y^e - C(y^e)$, since $p^e \leq p^m$. Thus $p^e y^e - C(y^e) < E(y^e)$ over S^-, no entrant can anticipate covering his entry costs, and the monopoly is sustainable at y^*, $P(y^*)$.

Alternative Set of Assumptions for the Weak Invisible Hand Theorem

We prove that Assumptions A2(a), A3(b), and A5 could be dispensed with, if we were to adopt the premise that the goods in N are weak gross substitutes for one another and that demands are "normal," as defined in A8.

A7 (Assumption 7)   The goods in N are weak gross substitutes. That is, a rise in prices of goods in $N - \{i\}$ will never reduce the demand for good i:

$$p' \geq p^2, p_i^1 = p_i^2 \rightarrow Q^i(p^1) \geq Q^i(p^2).$$

A8(a) [Assumption 8(a)]   (See Sandberg, 1974.) Demands for the goods in N are normal, i.e.,

$$p^1 \neq p^2 \rightarrow [Q^i(p^1) - Q^i(p^2)] \cdot (p_i^1 - p_i^2) < 0 \qquad \text{for some } i.$$

This latter premise asserts that whenever one set of prices is replace by another, there is at least one good whose demand moves in the opposite direction from its price.[5] If the price of only one product is changed, the response in the quantity demanded will be *normal* (its demand curve will have a negative slope). Moreover, no matter what the two sets of prices, the premise precludes demand interdependences among the various goods so strong as to cause *all* quantities to respond perversely to the price changes.

It is easy to see that with such demands, there is a unique price vector which calls forth each vector of quantities.[6]

Lemma 4   If demands for goods are normal $[A8(a)]$, then there is a one-to-one relationship between market prices and demand vectors $[A2(a)]$.

Proof   Suppose the contrary; i.e., $p^1 \neq p^2$ and $Q(p^1) = Q(p^2)$. This immediately contradicts A8(a).

Sandberg has proved that A2(c), A7, and A8(a) together imply that all prices are nonincreasing with quantities. If greater amounts of one or more goods are to be sold, then no prices may rise, and at least one must fall.

Lemma 5   (See Sandberg, 1974.) A2(c), A7, and A8(a) together imply that $\partial P^i(y)/\partial y_j \leq 0, \forall i, j$.

Proof   (See Sandberg, 1974.) A self-contained proof, provided to us by Thijs ten Raa, is given here. Consider two vectors of outputs, y^1 and y^2. Without loss of generality, we show that when the demand of only one good (good j) increases, then no price increases. Assume

(2) $y_j^1 < y_j^2$ and $y_i^1 = y_i^2, \qquad i \neq j.$

Define $K = \{k | p_k^1 \geq p_k^2\}$ and $L = \{l | p_l^1 < p_l^2\}$. We will show that L is the null set. Define p^3 as follows. For $k \in K, p_k^3 = p_k^1$ and for $l \in L, p_l^3 = p_l^2$. Thus, each

---

[5] It is well known [see Samuelson (1947) for example] that integrable compensated demands have this property for all but price changes that are proportional across the board.
[6] Of course, in the background are fixed outside prices and incomes.

component of p^3 is the larger of the corresponding components of p^1 and p^2. Since $p_l^3 = p_l^2$ for $l \in L$ and $p_k^3 \geq p_k^2$ for $k \in K$, A7 (weak gross substitutes) implies that

$$(3) \qquad\qquad\qquad y_l^3 \geq y_l^2, \qquad \forall l \in L.$$

Suppose L is not empty. Then $p^1 \neq p^3$ and $p^1 \leq p^3$, so A8(a) (normal demands) implies that there is an m with $(p_m^1 - p_m^3)(y_m^1 - y_m^3) < 0$. Since $p_k^1 = p_k^3$ for $k \in K$, m must be in L, with $y_m^1 > y_m^3$. Then, using (3) $y_m^1 > y_m^2$. But this contradicts (2) so it is proved that L is empty, and the result is established. If, moreover, A2(c) is assumed, then $\partial P^i(y)/\partial y_j \leq 0$, $\forall i, j$.

We need to strengthen this result slightly by assuming that each product's demand curve is downward sloping with no critical points.

A8(b) [Assumption 8(b)]   *At any y, for each i, $\partial P^i(y)/\partial y_i < 0$.*

We also need this technical condition:

A9 (Assumption 9)   *The Ramsey program of choosing y to maximize welfare over Y admits strict complementary slackness in the Kuhn-Tucker conditions. That is, the necessary conditions can be replaced by the slightly stronger conditions*

$$(4) \qquad\qquad P^i(y^*) - MC_i = -\lambda(MR_i - MC_i) \qquad \text{for } y_i^* > 0$$
$$P^i(y^*) - MC_i < -\lambda(MR_i - MC_i) \qquad \text{for } y_i^* = 0.$$

with $\pi = E$ and $\lambda \geq 0$.

With these new assumptions, we can establish A3(b).

Lemma 6   *Under A2(c), A7, A8, and A9, if y^* is the Ramsey optimum given by A3(a), then $\pi_i(y^*) < 0$.*

Proof   By definition

$$\pi_i(y^*) = MR_i - MC_i = P^i(y^*) - MC_i(y^*) + \sum_j y_j^* \frac{\partial P^j(y^*)}{\partial y_i}.$$

Together with the conditions (4), this yields

$$(5) \qquad\qquad \pi_i(y^*) = \frac{1}{1+\lambda} \sum_j y_j^* \frac{\partial P^j(y^*)}{\partial y_i} \qquad \text{for } y_i^* > 0$$

$$(6) \qquad\qquad \pi_i(y^*) < \frac{1}{1+\lambda} \sum_j y_j^* \frac{\partial P^j(y^*)}{\partial y_i} \qquad \text{for } y_i^* = 0.$$

Lemma 5 gives us $\partial P^j(y^*)/\partial y_i \leq 0$. Then, (6) immediately yields, recalling that $\lambda \geq 0$, $\pi_i(y^*) < 0$ for $y_i^* = 0$. With $y_k^* > 0$, $y_k^*(\partial P^k(y^*)/\partial y_k) < 0$ by A8(b), and then (5) implies that $\pi_k^*(y^*) < 0$.

It can be seen that this result is surprisingly sensitive to the assumptions. In particular, it is easy to construct well-behaved counterexamples with two complementary goods.[7]

What is most interesting for our present purposes is that Assumption A5 can be shown to be unnecessary for demands which are normal and weak gross substitutes. To show this we first prove

Lemma 7   *A7, A8(a) and continuity of demands imply*

$$(7) \qquad \pi^e \equiv p_A^e y_A^e - C(y_A^e) \leq \sum_{i \in A} P^i(y_A^e)y_i - C(y_A^e) \equiv \pi^m(y_A^e).$$

Here (7) asserts, in effect, that π^e, the entrant's profit from the production of y_A^e in the presence of the former monopolist, will never exceed $\pi^m(y_A^e)$, the amount a monopolist could earn by producing the same quantities.

Proof   Referring to Definition 8A1 (of sustainability),

$$\pi^e = p_A^e y_A^e - C(y_A^e), \qquad \text{where } p_A^e \leq p_A^m \text{ and } y_A^e \leq Q_A(p_A^e, p_{N-A}^m).$$

From the definitions,

$$(8) \qquad \pi^m(y_A^e) - \pi^e = \sum_{i \in A} y_i^e [P^i(y_A^e) - p_i^e].$$

Note that, by construction, $y_A^e \leq Q_A(p_A^e, p_{N-A}^m)$. Then, under A7, A8(a), and continuity, Sandberg's theorem (the differentiable version is here in Lemma 5) yields

$$P(y_A^e) \geq P[Q(p_A^e, p_{N-A}^m)] = (p_A^e, p_{N-A}^m).$$

In particular, for $i \in A$, $P^i(y_A^e) \geq p_i^e$. Hence, the right-hand side of (8) is nonnegative, and (7) follows.

Corollary   *A2(c), A7, and A8(a) imply A5.*

Proof $\pi^e \geq E(y_A^e)$ implies $\pi^m(y_A^e) \geq E(y_A^e)$, by Lemma 7.

Thus, with demands that are normal and for goods that are all weak gross substitutes, an entrant can earn at least as much profit alone in the market

---

[7] R. D. Willig, W. J. Baumol, and D. E. Bradford are preparing a paper in which these matters are discussed.

as he can in the presence of the former monopolist. However, Lemma 7, like Lemma 6, is very sensitive to its assumptions. For example, in the case of two complementary goods, the single-product entrant may well be able to earn far more, when the former monopolist offers the other good, than he can earn in isolation. Fortunately, however, even in such cases, A5 can still hold. Where there are many goods, some substitutes and some complements, (8) will not hold for all entrant output vectors, while A5 may well remain plausible.

Finally, we have

Proposition   *Given A1, A2(b), A2(c), A3(a), A4, A7, A8, and A9, either A6 or A6′ implies that the Ramsey-optimal prices are sustainable.*

Proof   Lemmas 3, 4, 6, and 7 show that these assumptions together imply A1–A6. Thus, Proposition 8E1 applies.

APPENDIX III

Complementary Products

In the bulk of this chapter we assume that the set of outputs potentially profitable to entrants is a subset (not necessarily a proper subset) of the set of outputs potentially profitable to the monopolist (set T). That is to say, we assume that any output vector which may be financially viable for an entrant operating alongside the former monopolist would certainly be financially viable if it were produced by a monopoly having the industry entirely to itself. This assumption served to restrict the domain in output space with which the analysis had to be concerned.

While the assumption is plausible, it is not absolutely defensible except for the case in which there is no demand complementarity among the products of the industry (see the Corollary to Lemma 7 in Appendix II). Where there is complementarity between the outputs of the entrant and those of the (former) monopolist, the assumption can be violated. For example, consider an entrant who can profitably produce gas heaters for which the former monopolist supplies the gas. If the monopolist (as the only supplier) were to produce only the entrant's output vector composed of heaters but no gas, he could hardly expect to make any profit, because no one would buy the heaters if no gas were available.

We can, however, readily extend the analysis to cover a fairly broad range of complementarity. As we see, this extends the relevant region to one potentially much larger than T, the set of points potentially profitable to a monopolist. But the analysis nevertheless does place some limits upon the relevant region, by implicitly imposing some limits upon the permissible degree of complementarity.

To explain the assumption we use for this purpose, we have to define another concept—a *destructively parasitic entrant*. A destructive parasite is an entrant who cannot operate profitably without the continuation of the pre-entry mode of operation by the monopolist, but whose presence prevents the continuation of those operations of the monopoly. That is, as a consequence of destructive parasitic entry, either the monopolist must shut down and the entrant consequently is left with a loss $\pi(y^e) < 0$, or, for any new monopoly output vector $\hat{y}^m \neq 0$, which is economically viable for the incumbent in the presence of the entrant so that $P(\hat{y}^m + y^e) \cdot \hat{y}^m \geq C(\hat{y}^m)$, the entrant will earn a negative profit, $P(\hat{y}^m + y^e)y^e - C(y^e) < 0$. The new assumption described in this appendix rules out the possibility of entry by destructively parasitic[1] firms.

In our discussion the issue arises, for example, where an entrant produces a piece of equipment (say a gas heater stove) which is useless without some other output (gas) of the former monopolist, but where the terms of sale of that piece of equipment make the difference between profit and loss for the monopolist; that is, the gas supplier cannot break even without an exclusive market in heaters. It is the unlikeliness of such a situation that makes our assumption reasonable.

Assumption I: No Destructive-Parasitic Entry *Suppose* y^e *is the output vector of a profitable entry plan. Then, either there exists an alternative vector of quantities[2] for the former monopolist which makes both firms viable, or* y^e *is profitable in isolation; that is, there exists* $\hat{y}^m \geq 0$, *or* $\hat{y}^m = 0$, *with* $P(y^e + \hat{y}^m)y^e - C(y^e) - E(y^e) \geq 0$ *and* $P(y^e + \hat{y}^m)\hat{y}^m - C(\hat{y}^m) - E(\hat{y}^m) \geq 0$.

We show in Appendix II that Assumption I is redundant when the goods are weak gross substitutes and demands are "normal." However, Assumption I is more general, permitting considerable complementarity in demands, but excluding the most extreme cases. Let us see just how Assumption I extends the region relevant for our analysis. In Figure 1, let S be the supporting hyperplane of the strictly convex set $T = [y \mid \pi^m(y) - E(y) \geq 0]$ at the candidate sustainable point $y^* = Q(p^*)$. We show that Assumption I broadens the relevant region from T to the larger region S^-, which is the analytically convenient region lying between the origin and the supporting hyperplane S in the nonnegative orthant. The role of Assumption I is to limit the set of viable entry threats to S^-, as is needed for Assumption A5 in Appendix II.

Under Assumption I, for any output vector ye that is part of a profitable entry plan, there are only two possible cases: *Case i*: The entrant drives the monopolist out of business, but the entrant's outputs earn a nonnegative profit in the absence of the former monopolist. Then, since the entrant is simply a (new) monopolist, his output vector y^e can only be profitable if it is in T, a subset of S^-. *Case ii*: Instead, despite the presence of the new entrant, the monopolist can find a new quantity vector $\hat{y}^m$ that enables him *and* the entrant to earn nonnegative profits. Then, by subadditivity of costs, a single

---

[1] This colorful label is inspired by the behavior of such biological villains as the strangler fig—a destructively parasitic plant which commits both murder and suicide by strangling the host tree on which it depends for survival.

[2] That is, unlike the remainder of the chapter, here we inquire whether the former monopolist's operations could remain viable after entry if he were permitted a single quantity readjustment, should that prove necessary for the purpose.

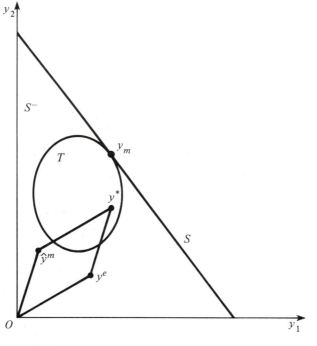

FIGURE 1

firm selling $y^* = y^e + \hat{y}^m$ can earn a positive profit at the same prices. There-fore, $\bar{y} \in T$, $y^e \leq \bar{y}$, and so $y^e \in S$ (Figure 1). This completes the proof, in heuristic terms. Now, more formally, we have

Proposition 1 *Given Assumptions A1, A2, A3, A4, and A6 or A6′ of Appendix II, Assumption I implies that the broad version of A5 is satisfied, that is, if y^e is the output vector of a profitable entry plan, then $y^e \in S^-$.*

Proof By definition, the hyperplane S supports the set T at y^*. By A3 and A1, $\pi_i^m(y^*) - E_i(y^*) = \pi_i^m(y^*) < 0$, $\forall i$. By A3, $\pi^m(y^*) - E(y^*) = 0$. Thus, for $y \leq y^*$ and $|y - y^*|$ sufficiently small, $\pi^m(y) - E(y) > 0$, and $y \in T$. Then, because T is strictly convex, by A4, it lies entirely (except for y^*) in S^-.

Suppose y^e is the output vector of a profitable entry plan. By Assumption I, there is a $\hat{y}^m$ with $P(y^e + \hat{y}^m)y^e - C(y^e) - E(y^e) \geq 0$ and $P(y^e + \hat{y}^m)\hat{y}^m - C(\hat{y}^m) - E(\hat{y}^m) \geq 0$. If $\hat{y}^m = 0$, then $y^e \in T - \{y^*\} \subseteq S^-$ and we are finished. Otherwise, add the inequalities and invoke the strict subadditivity of $C(\cdot) + E(\cdot)$ (which follows from A6 or A6′ by the results of Chapter 7) to obtain $P(y^e + \hat{y}^m)(y^e + \hat{y}^m) > C(y^e + \hat{y}^m) + E(y^e + \hat{y}^m)$. Thus, $\pi^m(y^e + \hat{y}^m) > 0$ and $y^e + \hat{y}^m \in T - \{y^*\} \subset S^-$. So, since $\pi_i^m(y^*) < 0$, $-\sum \pi_i^m(y^*)y_i^e < -\sum \pi_i^m(y^*)$ $(y_i^e + \hat{y}_i^m) < -\sum \pi_i^m(y^*)y_i^*$, and $y^e \in S^-$.   Q.E.D.

9

Equilibrium in the Multiproduct Competitive Industry

As we have already said, the theory of contestable markets generalizes the classical theory of perfect competition and permits the same analytic methods to be used to study other market structures as in the monopoly model of Chapter 8. The discussion of perfect competition in this chapter is designed to serve two purposes. The analysis is conducted along purely classical lines and can, in a limited sense, be considered an extension of the Vinerian tradition to the multi-output world.[1] Even in those terms, the number of difficulties encountered and insights obtained by going from one to several products is noteworthy. However, in assembling the apparatus necessary to carry out this extension we make a surprising discovery: The methodological approach which we design for the analysis of multiproduct industries that are perfectly competitive is equally applicable to any perfectly contestable market, regardless of industry structure. This parallel will be pursued further in Chapter 11, where it will offer some substantial new insights into the case of the industry with a small number of firms.

It should not be surprising that some of our theorems are almost obvious multiproduct analogs of the familiar single-product results. These show, for example, that even with many outputs the competitive firm must still produce

---

[1] Viner (1931) is the classic reference. See also Knight (1921) and Stigler and Boulding (1952) for other seminal contributions. Stigler (1957) puts the perfectly competitive model in historical perspective.

at minimum (ray) average cost, manifesting (locally) constant returns to scale and zero profits under marginal cost pricing. Still, these extensions are not self-evident, and proofs of the results cannot be avoided. However, we find that multi-output production also provides an abundance of new results about economies of scope, the coexistence of specialized and unspecialized firms, the relation between product-specific fixed costs and the set of items in an efficient firm's product line, and so forth. In addition, we derive conditions under which, in competitive equilibrium, all firms in an industry will produce precisely the same output vector, so that one can speak of a single representative firm. This contrasts with cases in which several different types of firms, each with its own output mix, are required for equilibrium. Finally, we present a rather general existence theorem for multiproduct competitive equilibrium.

9A. The M Locus, Industry Cost Minimization, and Multiproduct Competitive Equilibrium

To lay the groundwork for our analysis of multiproduct competitive industries, we begin by formulating a precise definition of the concept: a definition in keeping with the standard model of perfect competition in the scalar output case. We assume that prices are fixed outside the firm and that production techniques are freely available and can be represented by the cost function $C(\cdot)$. In addition, to limit the scope of our inquiry, we assume that the product set $N = \{1, 2, \ldots, n\}$ within which cost interactions may occur is given exogenously. Consider, then, the feasible industry configuration characterized by the prices $p \in R^n_{++}$ and by output vectors $y^1, \ldots, y^m$ of the m types of firms which produce some or all of the goods in N. Suppose that total industry supply $\sum_{i=1}^{m} a_i y^i$ is equal to market demand $Q(p)$, where $Q(\cdot)$ is a continuous vector demand function of the industry prices (with constant outside prices), and a_i is the number of firms of type i. As part of a feasible industry configuration, a firm of type i must earn no less than zero profit from its production and sale of y^i. Feasibility requires

(9A1) $p \cdot y^i - C(y^i) \geq 0 \qquad i = 1, \ldots, m.$

Definition 9A1: Multiproduct Competitive Equilibrium A feasible industry configuration yields a competitive equilibrium when there is freedom of entry if and only if (i) no firm in the market can increase its profit by changing output levels; (ii) no potential entrant expects to earn a positive profit by joining the industry; and (iii) in their evaluation of the profitability of output changes, both potential entrants and incumbents in the market are price takers.

Thus, assuming there are no cost complementarities between any products in N and any goods outside N, the industry position is a competitive equilibrium if, in addition to (9A1), we have

(9A2) $$p \cdot y - C(y) \leq 0 \qquad \text{for all } y \in R^n_+.$$

Together, these conditions have several familiar implications. First, each firm in the market must earn zero economic profit.

(9A3) $$p \cdot y^i - C(y^i) = 0 \qquad i = 1, \ldots, m.$$

Second, each firm that produces a positive amount of commodity j must equate its marginal cost to the market price. This follows from the formal Kuhn-Tucker conditions for the maximization of profit by a firm of type i, with respect to its output levels (which are constrained to be nonnegative), given the price vector p:

(9A4) $$\quad (p_j - C_j(y^i))y^i_j = 0, \ p_j - C_j(y^i) \leq 0, \ y^i_j \geq 0; \qquad \begin{array}{l} j = 1, \ldots, n. \\ i = 1, \ldots, m. \end{array}$$

We use these conditions to generalize to the multiproduct case a classic property of competitive industries:

Proposition 9A1 *In competitive equilibrium, all firms in the industry produce output vectors at which (i) there are locally constant returns to scale, that is, $S_N(y^i) = 1$; and (ii) ray average cost is at a global minimum. That is, all firms operate at points on the M locus.*[2]

Proposition 9A1 is merely the multiproduct analog of the classic result that, in equilibrium, a perfectly competitive firm earns zero economic profits by operating at the output y_m at which price = average cost = marginal cost.[3]

---

[2] *Proof* (i) Summation of (9A4) from $j = 1$ to n and substitution into (9A3) yields

$$\sum_j y^i_j C_j(y^i_j) - C(y^i) = \nabla C(y^i) \cdot y^i - C(y^i) = 0.$$

Using the definition of the degree of scale economies, $S_N(y^i)$, this can be rewritten as

$$C(y^i)\left(\frac{1}{S_N(y^i)} - 1\right) = 0.$$

Thus, $S_N(y^i) = 1$.
   (ii) Setting $y = ty^i$ in (9A2), and using (9A3), we have

$$C(ty^i) \geq p \cdot ty^i = tC(y^i), \qquad \forall t > 0 \text{ and } t \neq 1.$$

Thus,

$$\frac{C(y^i)}{\sum_j y^i_j} \leq \frac{C(ty^i)}{t \sum_j y^i_j} \qquad \forall t > 0, t \neq 1. \qquad\qquad \text{Q.E.D.}$$

[3] See, for example, Mansfield (1975, Chap. 8).

We now turn to the task of integrating fully the combinatorial analyses of Chapter 5 into our theory of multiproduct perfectly competitive markets. In doing so, we will find that the analysis of cost-minimizing structures permits us to characterize competitive equilibrium in an illuminating way.

We begin by rewriting, in a slightly modified fashion, the definition given in Chapter 5 of the industry cost function which results from the cost-minimizing distribution of a given industry output vector y^I among the firms of various types:

Definition 9A2: Industry (minimum) Cost Function   The industry cost function is given by

$$C^I(y) \equiv \min_{m,a_i,y^i} \sum_{i=1}^m a_i C(y^i),$$

where m and the a_i are positive integers, and $\sum_{i=1}^m a_i y^i = y$. Here, a_i is the number of firms (of type i) which produce output vector y^i, and m is the number of firm types of this sort.

The following propositions show the usefulness of this cost construct for our investigation of competitive equilibrium. The first of these concludes that competitive equilibrium can be characterized as a configuration of firms which produces industry output at minimum industry-wide cost [equation (9A5)], which earns total revenues equal to industry-wide costs [(9A6)], and which maximizes industry-wide profit with respect to industry-wide output, given the prices that clear the markets.

Proposition 9A2   *Consider the industry configuration comprised of sets of a_i firms, each producing outputs y^i, $i = 1, \ldots , m$, and prices p such that $Q(p) = \sum_{i=1}^m a_i y^i = y^I$. This is a competitive equilibrium in that (9A1) and (9A2) hold, if and only if* [4]

(9A5) $$C^I(y^I) = \sum_{i=1}^m a_i C(y^i)$$

---

[4] *Proof*   First we show that (9A5), (9A6), and (9A7) together imply (9A1) and (9A2). By definition, $C^I(y) \le C(y)$, so that $p \cdot y - C(y) \le p \cdot y - C^I(y)$, and hence (9A7) implies (9A2). Together, (9A5) and (9A6) imply $0 = \sum a_i(py^i - C(y^i))$. Then, (9A2) suffices for $p \cdot y^i - C(y^i) = 0$, and (9A1) follows.

To show that (9A1) and (9A2) imply (9A5), (9A6), and (9A7), first assume that (9A5) is violated. Then there exist $b_1, \ldots , b_k$ and $\hat{y}^1, \ldots , \hat{y}^k$ such that $\sum b_j \hat{y}^j = y^I$ and $\sum b_j C(\hat{y}^j) < \sum a_i C(y^i)$. Hence, $\sum b_j(p \cdot \hat{y}^j - C(\hat{y}^j)) > \sum a_i(p \cdot y^i - C(y^i)) \ge 0$, where the weak inequality follows from (9A1). But, then $p \cdot \hat{y}^j - C(\hat{y}^j) > 0$ for some j, in contradiction to (9A2). So, (9A5) does follow from (9A1) and (9A2).

Now, (9A1) and (9A2) imply $p \cdot y^i - C(y^i) = 0$. Then, in view of (9A5), $0 = \sum a_i(p \cdot y^i - C(y^i)) = p \cdot y^I - C^I(y^I)$, as required by (9A6).

Finally, suppose that (9A7) were violated, so that for some y, $p \cdot y - C^I(y) > 0$. Let $C^I(y) = \sum b_j C(\hat{y}^j)$, where $\sum b_j \hat{y}^j = y$. Then $0 < p \cdot y - C^I(y) = \sum b_j(p \cdot \hat{y}^j - C(\hat{y}^j))$, so that, for some j, $p \cdot \hat{y}^j - C(\hat{y}^j) > 0$, in contradiction to (9A2).   Q.E.D.

(9A6) $$p \cdot y^I = C^I(y^I)$$

and

(9A7) $$p \cdot y - C^I(y) \le 0 \qquad \forall y \in R_+^n.$$

Proposition 9A2 shows how competitive equilibrium can be characterized by means of the industry cost function. Together with Definition 9A2, (9A5) yields a fundamental efficiency property of competitive equilibrium.

Proposition 9A3   *In competitive equilibrium, the firms comprising the industry produce the total industry output at least industry cost. In particular, no firm's output can be produced by two or more firms at a smaller total cost.*

Thus, a competitive multiproduct industry with free entry *acts as if* it were operated by a planner who allocates the industry output optimally among its various "plants," sets the resulting social marginal costs equal to market clearing prices, and permits the "plants" to earn exactly zero profits in the process. Propositions 9A2 and 9A3 underscore the importance of the analysis of Chapter 5. The bounds calculated there emerge as powerful indicators of feasible market structure. For, since a competitive equilibrium *requires* that industry costs be minimized, a low upper bound upon the optimal number of firms producing the output vector y^I or a low upper bound on the optimal number of firms producing any *component* y_j^I makes it highly unlikely that a competitive market structure will emerge in the production of that output vector. This is because the "large-numbers" argument adduced to justify the assumption of price-taking competitive behavior may lose its plausibility when the equilibrium itself requires that only a "small" number of firms participate in it.

Further, the preceding results illustrate the crucial role played by the M locus in the structure of multiproduct competitive equilibrium. Indeed, Proposition 9A1 implies that (in the presence of the competitive behavioral hypothesis) all one needs to know about the cost surface is its properties over the M locus, since any competitive equilibrium consists *only* of firms operating in that region. The single-product example is instructive here. In that case, once we know that a free-entry competitive equilibrium exists,[5] we know immediately that the equilibrium price is equal to $AC(y_m)$ and that all firms produce quantities y_m that achieve this minimal average cost.

While, as usual, the problem is more complex in the multiproduct case, the logical structure of the analysis is the same. Now, however, knowledge of the demand surface is required to calculate the product mix and output

---

[5] In Section 9D we deal directly with the generic existence problems posed by models of this type, and propose a method to deal with them.

quantities selected by each individual firm, while in the scalar case the demand function is used only to determine total industry output.

Finally, Proposition 9A3 implies that there is a surprising similarity between industries that are perfectly contestable natural monopolies and industries that are perfectly competitive. In each case, any firm that participates actively in an equilibrium must have subadditive costs at its output levels. Consequently, the characterization of multiproduct competitive equilibria requires the very same trans-ray cost concepts that are essential in the analysis of multiproduct natural monopoly.

9B. The Role of Economies of Scope

Minimization of industry costs in competitive equilibrium is certainly part of the conventional wisdom on the subject of perfect competition.[6] However, this condition acquires new analytic force when it is related to multiproduct firms. We begin by employing Proposition 9A3 to establish the direct connection between competitive equilibrium and economies of scope, giving the first result of this chapter that clearly offers insights beyond those provided by single-product analysis. We show that in the choice of the set of goods to be produced, economies are associated with the presence of multiproduct firms under perfect competition, as they are under monopolistic conditions.

Proposition 9B1 *Multiproduct firms in competitive equilibrium must enjoy (at least weak) economies of scope over the set of products which they produce; that is economies of scope are* necessary *for the existence of multiproduct competitive firms.*

Proof Suppose y^i is an equilibrium output vector for some firm, with $V = \{j \in N \mid y^i_j > 0\}$. Suppose that there are diseconomies of scope at y^i, so that there is a set $T \subset V$ such that

$$C(y^i_T) + C(y^i_{V-T}) < C(y^i).$$

This contradicts Proposition 9A3; that is, it implies that the output of a firm in competitive equilibrium can be produced more cheaply by two or more firms that are more specialized than the one in question.   Q.E.D.

The reason for the result, which may at first be surprising, is obvious once the matter is considered. If there were universal diseconomies of scope, perfect specialization would always be most efficient, and competitive

---

[6] See, for example, Mansfield (1975, Chap. 15).

equilibrium would, therefore always involve only single-product firms. We can also reverse the argument, to obtain the following sufficiency result:

Proposition 9B2 *Strict economies of scope over the M locus implies that, if a competitive industry produces two or more outputs in equilibrium, then it must contain some multiproduct firms. That is, economies of scope are* sufficient *for the presence of multiproduct firms.*

Proof Suppose the contrary, that no firm produces more than one good. In particular, consider two firms producing outputs y^a and y^b where $y_i^a = 0$, $\forall i \neq 1$, and $y_i^b = 0$, $\forall i \neq 2$. Strict economies of scope imply that $C(y^a + y^b) < C(y^a) + C(y^b)$. But this means that industry cost is *not* minimized, thereby contradicting Proposition 9A3. Q.E.D.

Propositions 9B1 and 9B2 spell out the intimate connection between economies of scope and the emergence of multiproduct firms in competitive industries. However, these results do not exhaust the implications of the industry cost minimization theorem and must be interpreted carefully. We now employ that result to show that, while Proposition 9B2 requires the emergence of multiproduct firms, it does not mean that competitive equilibrium must involve multiproduct firms exclusively. Indeed, we shall see that cost minimization for the industry may require the presence of some single-product firms in equilibrium, even when the cost function exhibits global economies of scope.

We prove this with an example based on a modification of a simple Marshallian model of joint production. Suppose an industry has available three techniques for the production of meat (M) and fiber (F). The first technique, sheep raising, provides a unit of meat and a unit of fiber in fixed proportions, at a constant cost of \$10 per sheep. The other two techniques involve independent production of meat by, say, the raising of chickens, or of fiber by, say, the cultivation of flax, each at constant costs of \$6 per unit. However, we complicate the process by assuming that sheep and chicken raising interfere with the growing of flax and with one another, so that a firm engaging in two of these activities must "fence in" the activity *whose quantity is smallest*, at a constant cost of \$1 per unit.

For simplicity, this example is constructed to exhibit constant returns to scale so that the M locus covers the entire positive orthant. Consequently, the *scale* of operation of the firms in competitive equilibrium is indeterminate. But this permits us to focus more clearly upon the implications of industry cost minimization for the equilibrium mix of products in individual firms.

Figure 9B1 can help the reader to follow the derivation of the minimum cost function from the characteristics of the available techniques. The choice among the three production techniques and the choice of the processes to be fenced in give rise to five "activities" relevant for the analysis. To produce

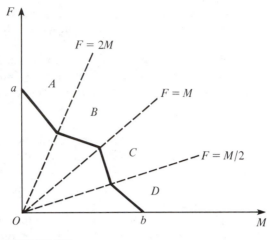

FIGURE 9B1

in region A where $F > 2M$, it is optimal to raise M sheep and $F - M$ flax plants and put a fence about the sheep, since they are smaller in number than the flax plants ($F - M > M$). In region B, where $M < F < 2M$, the same rules determine the number of flax plants and sheep, but it is best to fence in the plants since their number is smaller. Along the ray $F = M$, obviously only sheep are used. In regions C and D, the same reasoning as before yields the symmetric, but opposite production plans, this time using chicken production to supplement sheep raising.

Using this intuitive foundation, one can write the multiproduct *minimum cost function* as

(9B1) $$C(M, F) = 10y + 6(x - y) + z,$$

where $x = \max(M, F)$, $y = \min(M, F)$, and $z = \min(y, x - y)$. Here, y is the optimal number of sheep, $x - y$ is the optimal number of chickens or flax plants, and z is the efficient usage of fencing. The iso-cost curves have the shape indicated by the solid piecewise linear curve ab in Figure 9B1. We see next that this cost function exhibits economies of scope everywhere. Clearly, $C(0, F) + C(M, 0) = 6M + 6F = 6(x + y)$. Subtraction of equation (9B1) yields

$$C(0, F) + C(M, 0) - C(M, F) = 2y - z > 0 \qquad \text{for } F > 0 \text{ and } M > 0,$$

where the inequality follows from the fact that, by definition, $z \leq y$. This is, of course, the criterion of economies of scope.

Given this result, it might be expected that once demand conditions were specified, industry equilibrium would involve some number, say m, of identical firms, each producing its share of industry output $(M^I/m, F^I/m)$, in accord with the cost function given by (9B1). However, surprisingly enough, this is true only if demand conditions yield the special case in which $M^I = F^I$! To see this, suppose $F^I > M^I$. With prices given, a representative firm producing $(F^I/m, M^I/m)$ could then increase its profits by splitting into *two* firms, one producing $(M^I/m, M^I/m)$ by raising that many sheep, and the other producing $(F^I - M^I)/m$ units of fiber by growing that many flax plants. Because this sort of division avoids the need for fencing, it reduces total cost;

$$C(M^I/m, F^I/m) - [C(M^I/m, M^I/m) + C(0, \{F^I - M^I\}/m)]$$
$$= \min[M^I/m, (F^I - M^I)/m] > 0.$$

Thus, despite the presence of global economies of scope, in this case industry costs cannot be minimized without the presence of some specialized firms. Nonetheless, as shown by Proposition 9B2, equilibrium does require the presence of *some* multiproduct firms.

This section has emphasized the important roles of the industry cost minimization condition and of economies of scope in the determination of multiproduct competitive industry structure. Of course, it is possible that a firm will produce a multiplicity of outputs in competitive equilibrium even if its production processes are completely independent (i.e., $C(y^i) \equiv \sum_{j=1}^{n} C^j(y_j^i)$ and $SC_T = 0, \forall T \subset N$). In that case, however, one must search for some additional reasons explaining why the firm chooses to produce a variety of products.[7] Therefore, we shall proceed under the presumption that there are strict economies of scope for the product set N over the relevant region in output space—the M locus. However, as just shown, this does not mean that all firms will produce all of the industry's products, nor even that every firm will produce more than one of them.

9C. The Importance of Product-Specific Returns to Scale

The association of economies of scope and the existence of multiproduct firms in competitive equilibrium is established in Section 9B, while earlier in this chapter it is shown that all the firms involved must operate on the M locus and must, therefore, manifest locally constant returns to scale.

---

[7] One possible reason that springs to mind immediately is a desire to avoid risk by diversification when the firm faces prices and/or production uncertainty. Uncertainty lies outside the scope of our main analysis. However, if we consider risk to give rise to real costs, it becomes clear that risk reduction through costless diversification is an economy of scope that contradicts the independence assumption.

These results, together with the intimate quantitative connections established in Chapter 4 between the degree of overall scale economies, the degree
of *product-specific* scale economies, and the degree of economies of scope,
should lead one to expect that *new* conditions for competitive equilibrium
will emerge from the confluence of these relationships.

We examine first what determines whether a particular commodity will
be among the items supplied by a firm in equilibrium. We find that this is a
matter that cannot be settled by examining only the behavior of costs and
revenues in the neighborhood of the equilibrium point. Such local information simply is not enough to settle the issue.

This can be seen geometrically with the aid of Figures 9C1 and 9C2.
In Figure 9C1, surface $OCC'C''C'''$ is a transylvanian cost function of the
sort described in Chapters 4 and 7. It will be recalled that the wing OCC'
arises because a cost CC' must be incurred when the firm adds product 2
to its product line. Thus, as we move away from the y_1 axis, the cost surface
must rise by precisely the amount of this product-specific fixed cost. The
same explanation holds for the other vertical wing, $OC''C'''$, of the cost
surface. ORR' is the revenue hyperplane corresponding to the prices at
tangency point E. Figure 9C2, provided for clarity, shows the same relationship for the cross section above line segment AB in Figure 9C1.

Now, if point y^i below tangency point E is a candidate output for a firm
in competitive equilibrium, we know that the hyperplane ORR' defined
by market prices must reach the same "height" as the cost surface there

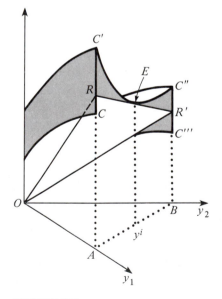

FIGURE 9C1

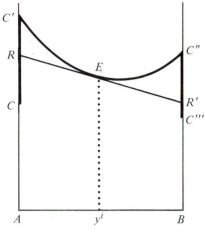

FIGURE 9C2

(for zero profit) and be tangent to it (for price-taking profit maximization). However, these local conditions are not enough. That hyperplane must also never lie (strictly) above the cost surface *anywhere* else. If it did, y^i would not be produced by an informed competitive firm facing the prices in question. It would recognize that it could obtain higher profits from a different vector of outputs.

In particular, for this purpose it is necessary to compare the heights of the cost surface and the market price hyperplane along, say, the y_1 axis. If the latter were to exceed the former at point A, then the firm would find it more profitable to dispense with the production of product 2 entirely. This problem is apt to be particularly serious when there are product-specific costs, as shown in the figures. If, as in the diagram, the cost surface is trans-ray convex in the interior of the orthant, one would not expect from information about the neighborhood of the tangency point that the surface would dip below the hyperplane. However, with product-specific setup costs, and the resulting discontinuity, the *limit point* C' of the cost surface on the y_1 axis will be above the revenue hyperplane, but the *actual* level of total cost C may not be.

In the diagram, the bottom edges of the wings of the cost surface are so low that this problem arises for any interior (two-product) tangency point. That is, there is no tangent to any interior point on the cost surface which simultaneously lies below both wings. Thus, it will never pay a competitive firm to produce both goods. This argument establishes

Proposition 9C1 *A necessary condition for a firm to produce y in competitive equilibrium is that the cost function be trans-ray supportable at y over any trans-ray hyperplane in output space.*

In particular, if the competitively produced output vector y involves positive outputs of more than one good, the economies of scope must sufficiently outweigh product-specific scale effects to yield the requisite trans-ray supportability.

Next, we show that the formula for the degree of scale economies with respect to a product subset T, in effect, constitutes a "benefit–cost ratio" for a firm contemplating abandonment of the "project" consisting of the production of the locally profit-maximizing quantities of the products in T. For this purpose, we use the result (9A4) that a firm in competitive equilibrium chooses its positive output quantities so that marginal costs equal market prices. Then, the cost to a firm producing y^i of dropping the products in T is the revenue forgone—the revenues that could have been derived from the sale of those products at prices equal to marginal costs: $\sum_{j \in T} y_j^i C_j(y^i)$. The benefit obtained by dropping the products in T is the production cost saved thereby: $C(y^i) - C(y_T^i) \equiv IC_T(y^i)$. The ratio between this benefit and cost is $IC_T(y^i)/y_T^i \nabla C(y^i) \equiv S_T(y^i)$, the degree of scale economies with respect to the product subset T. Hence, if $S_T(y^i) > 1$, the benefit that could be attained by abandoning the production of the goods in T would exceed the cost for a competitive firm, and y^i could not be part of a competitive equilibrium. These observations are summarized formally in the following result:

Proposition 9C2[8]   *If a firm in competitive equilibrium produces y^i with $y_j^i > 0$ for $j \in V \subseteq N$, then, for any nonempty proper subset $T \subset V$, $S_T \leq 1$ at y^i. Furthermore, if there are economies of scope over V at y^i, there must be product specific diseconomies of scale at y^i over either T or $V - T$.*

Thus, a multiproduct firm in competitive equilibrium must manifest locally constant returns to scale with respect to its entire product set, and decreasing returns to the scale of at least one, and perhaps every, proper subset of its products. This paradox is resolved by the presence of economies of scope. It may be recalled that equation (4B2) shows that, when there are economies of scope, the overall degree of scale economies must *exceed* the weighted average of the specific degrees of scale economies of all product lines. In particular, when $SC_T > 0$, $S_N = 1$ is therefore consistent with $S_T < 1$ and $S_{N-T} < 1$.

---

[8] *Proof*   Let T be any proper subset of $V \subseteq N$, with $y_j = 0$ for $j \in N - V$, and set $y = y_T^i$ in (9A2). Then, using (9A4), we have $y_T^i \cdot \nabla C(y^i) - C(y_T^i) \leq 0$. Subtracting this from $y^i \cdot \nabla C(y^i) - C(y^i) = 0$, the condition that the overall degree of scale economies be unity, yields $y_{V-T}^i \cdot \nabla C(y^i) \geq C(y^i) - C(y_T^i) = IC_{V-T}(y^i)$, so that $S_{V-T}(y^i) \equiv IC_{V-T}(y^i)/y_{V-T}^i \cdot \nabla C(y^i) \leq 1$. The obvious symmetric argument establishes $S_T(y^i) \leq 1$. We show, by contradiction, that at least one of these inequalities is strict. Suppose $S_T(y^i) = S_{V-T}(y^i) = 1$. Then, with economies of scope [i.e., $SC_T(y^i) > 0$] equation (4B2) applied to the product set V yields $S_V(y^i) > 1$, which contradicts Proposition 9A1.   Q.E.D.

Next, we offer an example demonstrating that the requirements of Proposition 9C2 do indeed impose effective constraints on the character of a competitive equilibrium above and beyond those stemming from the familiar necessary conditions (9A3) and (9A4).

$$(9C1) \qquad C(y_1, y_2) = \begin{cases} 25 + y_1^2 + y_2^2 & \text{for} \quad y_1 > 0, y_2 > 0 \\ 15 + y_1^2 & \text{for} \quad y_1 > 0, y_2 = 0 \\ 15 + y_2^2 & \text{for} \quad y_1 = 0, y_2 > 0. \end{cases}$$

This cost function embodies setup costs of 15 for each good if produced by itself. However, there are economies of scope in that the setup cost required to produce the two items together is 25, an amount less than the sum of the setup costs of specialized production. Note that this means that in multi-product operation, the incremental setup cost is equal to 10 for each good, since that is the increase in total setup cost resulting from the introduction of a second product. In this example, all ray average cost curves are U-shaped with minima that satisfy $y_1^2 + y_2^2 = 25$, for $(y_1, y_2) > (0, 0)$. However, in completely specialized production, $y_1 = \sqrt{15}$ minimizes average cost. This gives rise to the discontinuous M locus $ABDE$, including the points A and E on the axes in Figure 9C3, but excluding the points B and D.

We know that some firms must produce both goods in this case because of Proposition 9A3 and the economies of scope reflected in (9C1). Since

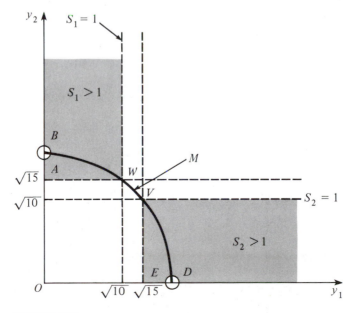

FIGURE 9C3

marginal costs are given by $C_i = 2y_i$ for $y_i > 0$, any competitive equilibrium involving production of both goods must have the property that $y_1/y_2 = p_1/p_2$, where (y_1, y_2) is the output vector of a multiproduct firm and (p_1, p_2) is the vector of market prices. Let the market demand functions be given by $Q_1(p_1) = k(9 - p_1)$ and $Q_2(p_2) = k(12 - p_2)$, where k is a large positive integer. In attempting to calculate the prices and quantities characterizing competitive equilibrium in this market, one's natural inclination is to look for a point on the M locus—a "representative firm"—satisfying the relative price condition, and to employ enough such representative firms to satisfy aggregate demand. We see, however, that in this case there is a fly in the ointment.

If we equate p_1/p_2 and y_1/y_2, with (y_1, y_2) on the M locus, we find that they are both equal to $\frac{3}{4}$. Since the output combination $(3, 4)$ lies on the M locus, the corresponding candidate competitive equilibrium involves k firms, each producing output $(y_1^e, y_2^e) = (3, 4)$, with candidate equilibrium prices given by $(p_1^e, p_2^e) = (6, 8)$. It is easy to see that conditions (9A3) and (9A4) are satisfied, since $p_1^e = C_1(3, 4) = 6$; $p_2^e = C_2(3, 4) = 8$; and $p_1^e y_1^e + p_2^e y_2^e = C(y_1^e, y_2^e) = 50$. Thus, each firm producing $(3, 4)$ earns zero profits at prices equal to marginal costs, and those prices clear the output markets when there are k active firms of this type.

Despite all these facts, however, this is not a competitive equilibrium because at this point there are product-specific economies of scale with respect to product 1:

$$S_1(3, 4) = \frac{C(3, 4) - C(0, 4)}{3C_1(3, 4)} = \frac{19}{18} > 1.$$

Therefore, by Proposition 9C1, this point *cannot* be a competitive equilibrium. This is easy to check, since a price-taking firm facing price vector $(p_1, p_2) = (6, 8)$ and the cost function given by (9C1) can produce $(0, 4)$ and earn a positive profit equal to 1.

Viewed another way, the requirements of Propositions 9C1 and 9C2, that the cost function be trans-ray supportable and that there be no product-specific increasing returns to scale, when added to the requirements of Proposition 9A2, can be interpreted to mean that all potential output vectors for competitive firms are restricted to a *(possibly proper) subset* of the M locus. In our example, it is easy to show that $S_i > 1$ for all $y_i < \sqrt{10}$, except for vectors on the y_i axis, where, as we have seen, $S_i > 1$ for $y_i < \sqrt{15}$. In Figure 9C3, part of the set of points for which $S_1 > 1$ is shaded vertically and part of the set of points for which $S_2 > 1$ is shaded horizontally. Thus, the subset of points on the M locus at which it is *possible* for firms in competitive equilibrium to produce are the points $A = (0, \sqrt{15})$, $E = (\sqrt{15}, 0)$ and the arc WV between $(\sqrt{10}, \sqrt{15})$ and $(\sqrt{15}, \sqrt{10})$, inclusive.

We have seen in this example that there does not exist any *single* representative firm type (any point on the M locus) which can be replicated to constitute a competitive equilibrium. Further calculations show that the only possible equilibrium in this example requires some firms to produce $(\sqrt{10}, \sqrt{15})$, where $S_1 = 1$, *and* other firms to specialize in product 2 and produce $y_2 = \sqrt{15}$. Both these points lie on the M locus and yield zero profits at the competitive prices $(2\sqrt{10}, 2\sqrt{15})$.

Our final example of this section shows that, when there are economies of scope, the requirements of Proposition 9C2 may preclude the existence of competitive equilibrium, even though all functions are continuous and ray average costs in multi-output production are U-shaped. Consider the cost function

$$C(y_1, y_2) = \left[y_1^{b_1} + y_2^{b_2} \right]^\rho, \qquad \text{with } b_1 > 0, b_2 > 0, \text{ and } 0 < \rho < 1.$$

This function (discussed in detail in Section 15G) exhibits cost complementarities, and hence economies of scope. For $b_1 = \frac{3}{2}$, $b_2 = 3.63$, and $\rho = \frac{1}{3}$, calculation confirms that everywhere in M there is increasing average incremental cost in the production of goods 1 and 2. Yet, there are significant scale economies (of degree 2.0) in the stand-alone production of good 1 (i.e., in its production in isolation). If, instead, $b_1 = 1$, $b_2 = 3$, and $\rho = \frac{1}{2}$, at every point in M there are increasing returns to the scale of good 1; that is, $S_1 \leq 1$ and $S_N = 1$ are inconsistent, and, by Proposition 9C2, competitive equilibrium does not exist.

Yet, in this case, every positive combination of outputs has a U-shaped ray average cost curve with a well-behaved minimum. The set M is characterized by $y_1 = y_2^3$ for $y_2 > 0$. For the combination of outputs $y_2/y_1 = k$, the point $y_1 = k^{-3/2}$, $y_2 = k^{-1/2}$ is in M. Thus, if the mixture of quantities demanded is weighted heavily toward good 2, k can be large and the scale of operation of the firm that minimizes ray average cost can be very small relative to market demand. Nevertheless, regardless of demand conditions, and despite the fact that ray average costs are U-shaped, the cost conditions do not permit any competitive equilibrium.

9D. The Existence and Structure of Multiproduct Competitive Equilibria

The preceding example indicates that there may be existence problems for multiproduct competitive equilibria. Indeed, as was discussed in Chapter 2, there is a serious existence problem that occurs quite generally. It stems from the twin requirements that industry costs must be minimized and that no profitable entry possibilities be generated. Generally, for single-product industries where average costs are U-shaped, this requires that the market demand curve intersect the *industry* average cost curve at an output which is an *integer* multiple of y_m, the quantity at which average costs are minimized. In Chapter 2 we indicated how assumption of "flat-bottomed" average cost curves can be used to eliminate this difficulty. We shall again consider that approach in Chapter 11. Here, we simply extend to the multiproduct world the classic technique employed in traditional scalar output discussions. The results derived here will play a central role in the analysis of the corresponding problem for equilibrium under oligopoly.

In the remainder of this section then, we see how the integer problem can be dealt with formally in the case of multiproduct perfect competition, and in the process we learn about the conditions determining equilibrium and the circumstances under which there can be a single representative firm for the industry. In brief, the existence issue which has just been described has not been considered a serious problem in the case of perfectly competitive industries, presumably because the scallops in the industry's average cost curve (see Figure 2C3) became shallower and shallower as industry output increases relative to y_m and, in the limit, disappear altogether. Theoretical analysis has, therefore, generally proceeded as if the industry's average and marginal cost curves were horizontal and, consequently, that its supply curve were in fact represented by the dashed horizontal line $p = AC(y_m)$ connecting the bottoms of the scallops.[9] Competitive equilibrium is then said to occur where this supply curve intersects the market demand curve. Equilibrium can in these circumstances be characterized by the equations

(9D1) $$p - MC(y_m) = 0$$

(9D2) $$py_m - C(y_m) = 0$$

(9D3) $$y^I \equiv my_m = Q(p)$$

which determine the values of the equilibrium price p, the output of the representative firm y_m, and the number of firms m. No one worries about whether or not m is an integer so long as the number of firms is "sufficiently large."[10]

In terms of the discussion in Section 9A, what this amounts to is the use of an approximate industry cost function obtained by dropping the restriction in Definition 9A2 that the a_i be integers. In the scalar output case, this alternative industry cost function $C^*(y)$ is given by

$$C^*(y^I) = (y^I/y_m) \cdot AC(y_m).$$

---

[9] See, for example, Silberberg (1974) and Hanoch (1975).
[10] As in Samuelson (1973, p. 484).

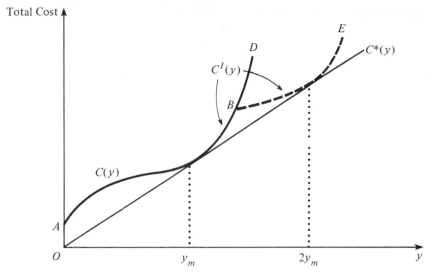

FIGURE 9D1

The procedure is illustrated in Figure 9D1. There we see the *firm's* total cost function $C(y)$ represented by curve ABD. From this we can derive the representation ABE of the industry's total cost function $C^I(y)$, where point B is that at which industry cost minimization requires the entry of a second firm. Clearly, the standard method of circumventing the integer problem consists of replacement of the ill-behaved function $C^I(y)$ by the linearly homogeneous approximation, $C^*(y) = (y/y_m)C(y_m) \leq C^I(y) \leq C(y)$, whose graph is a ray in Figure 9D1. Notice that determination of $C^*(y)$ requires information about the firm's cost function $C(y)$ only at y_m, the point constituting the M locus in this single-product case.

   This approximation procedure has in fact worked very well in the analysis of competitive industries offering only a single product, since under perfect competition the number of firms is expected to be large. The task that is consequently undertaken in the remainder of this section is to extend this way of dealing with the integer problem to multiproduct competitive industries.

   By analogy to the classic single-output case, we assume that ray average cost curves are roughly U-shaped, meaning that along each ray from the origin ray average cost attains a unique minimum, with $y \neq 0$, so that there exists a well-defined M locus. We begin by defining precisely the multiproduct version of $C^*(y)$, the approximation to the true industry cost curve that we have just discussed.

Definition 9D1   The *fractional-firm cost function* of the industry is defined by

(9D4) $$C^*(y^I) = \min_{m,a_i,y^i} \sum_{i=1}^{m} a_i C(y^i), \qquad \text{where} \sum_{i=1}^{m} a_i y^i = y^I.$$

Here, in contrast to Definition 9A2, only m, the number of *types* of firms, is required to be an integer, where firms are said to be of the same type if their output quantities all have the same proportions, that is, if their output vectors all lie on the same ray in output space. Since the a_i, which represents the "number" of firms of type i, is not required to be an integer, $C^*(\cdot)$, as before, is not precisely equal to $C^I(\cdot)$, the true industry cost function. However, $C^*(\cdot)$ is, once more, a powerful tool for the analysis of competitive equilibrium that permits us to circumvent the integer problem when the number of firms is large. We will provide a more intuitive explanation of this important concept with the aid of diagrams. First, however, it is necessary to establish

Proposition 9D1   *If all ray average cost curves are U-shaped, each of the m output vectors $y^1, \ldots, y^m$ that appear in the solution to the program given by (9D4) must lie on the M locus.*[11]

Proposition 9D1 and (9D4) establish that $C^*(\cdot)$, the fractional-firm cost function for the industry, is a positive linear combination of costs of output vectors on the M locus.

Here we may interpret a fractional firm in the following way: Let all firms of type k together produce the total output vector y^*, and let y_m^* be the point on the M locus that lies on the same ray as y^*. Then $k = y^*/y_m^*$ is the corresponding number of fractional firms. For example, if $y^* = 3.8$ times the magnitude of the corresponding output bundle that minimizes ray average cost, that is, if $y^* = 3.8 y_m^*$, then in our construction y^* is taken to be produced by 3.8 firms, each producing at minimum ray average cost.

The nature of $C^*(y)$ is indicated in Figures 9D2–9D4. Figure 9D2 shows a cross section along the ray OR. The curve $C(y)$ is the cost function for a firm, and $C^*(y)$ is simply the linear tangent to $C(y)$ at the point on $C(y)$ above the M locus. Because $C(y)$ is linearly homogeneous locally at the

---

[11] *Proof*   Suppose the contrary; that is, that the solution to (9D4) involves an output vector y^i not on the M locus. Then there exists a $t \neq 1$ such that $C(ty^i)/t < C(y^i)$. Replace y^i and a_i on the right-hand side of (9D4) by ty^i and a_i/t. (Note that y^I remains unchanged.) Then one can define

$$C^0(y^I) = \sum_{j \neq i} a_j C(y^j) + (a_i/t)C(ty^i) < \sum_{j \neq i} a_j C(y^j) + a_i C(y^i) = C^*(y^I),$$

which is a contradiction of the definition of $C^*(y^I)$.   Q.E.D.

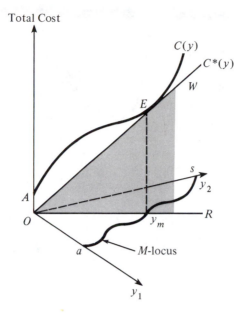

FIGURE 9D2

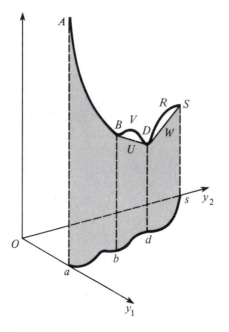

FIGURE 9D3

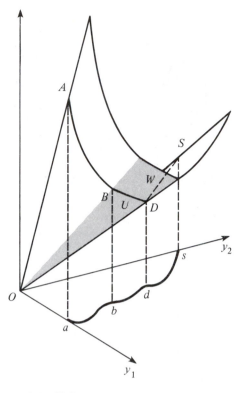

FIGURE 9D4

tangency point E, the linear cross section of $C^*(y)$ goes through the origin. This construction is obviously identical with that in the single-product case.

In Figure 9D3, to make the construction clearer we have chosen to deal with a cost surface whose trans-ray cross sections are very irregular. Actually, the figure shows the behavior of total cost above the M locus, *abds*, which happens to be represented by the curve *ABVDRS*. The curve *ABUDWS* is the highest convex locus which lies everywhere on or below *ABVDRS*, above the M locus. The curve *ABUDWS* may then be referred to as the boundary of the convex hull of the cost cross section above the M locus, and it can be seen that $C^*(\cdot)$ includes every point on this convex boundary.

Figure 9D4 shows the entire shape of $C^*(y)$. To draw it, we reproduce the convex hull boundary *ABUDWS* from Figure 9D3 in Figure 9D4. We then draw rays such as OA, OB, OD, and OS through every point of *ABUDWS*. The resulting figure, which is the lower boundary of a convex cone, is the graph of $C^*(y)$.

Notice how $C^*(\cdot)$ eliminates all trans-ray "wiggles," so that the resulting graph is trans-ray convex. In summary, two crucial properties of $C^*(y)$ have been observed in the diagrams: linearity along each ray (linear homogeneity) and convexity. These are established rigorously in the following propositions:

Proposition 9D2 $C^*(y)$ is a convex function of the output vector y.[12]

Proposition 9D3 $C^*(y)$ is linearly homogeneous in y. That is, $C^*(ty) = tC^*(y)$, $\forall t > 0$.[13]

Having now characterized C^*, we are ready to use it to study multiproduct competitive equilibrium, and the means of evading the integer problem.

Definition 9D2   A *fractional firm configuration of an industry* is a price vector p and a set of values $a_i \geq 0$, each indicating the number of firms producing y^i, $i = 1, \ldots, m$, where $Q(p) = \sum_{i=1}^{m} a_i y^i = y^I$.

Here, each a_i need not be an integer, so that fractions of firms are permitted. We are now able to establish, in perfect analogy to Proposition 9A2,

Proposition 9D4   *A fractional firm configuration of an industry is a* fractional firm competitive equilibrium, *in that (9A1) and (9A2) hold, if and only if*

(9D5) $$C^*(y^I) = \sum_{i=1}^{m} a_i C(y^i), \qquad \text{where } \sum_{i=1}^{m} a_i y^i = y^I,$$

(*that is, $y^1, \ldots, y^m$ and $a_1, \ldots, a_m$ minimize the (fractional firm) industry cost of the industry output which they comprise*),

(9D6) $$p \cdot y^I - C^*(y^I) = 0,$$

and

(9D7) $$p \cdot y - C^*(y) \leq 0, \qquad \forall y \in R^n_+.$$

The proof is omitted since it is the same as that of Proposition 9A2, except that the a_i need no longer be integers.

---

[12] *Proof*   Let $C^*(y') = \sum_{i=1}^{m'} a_i C(y^{i'})$ and $C^*(y'') = \sum_{j=1}^{m''} b_j C(y^{j''})$. Then $C^*(\lambda y' + (1 - \lambda) y'') \leq \sum_{i=1}^{m'} \lambda a_i C(y^{i'}) + \sum_{j=1}^{m''} (1 - \lambda) b_j C(y^{j''}) = \lambda C^*(y') + (1 - \lambda) C^*(y'')$.   Q.E.D.

[13] *Proof*   Let $C^*(y) = \sum_{i=1}^{m} a_i C(y^i)$. Clearly, $C^*(ty) \leq \sum_{i=1}^{m} a_i t C(y^i) = tC^*(y)$, $\forall t > 0$. Then $C^*[t^{-1}(ty)] \leq t^{-1} C^*(ty)$, so that $C^*(ty) \geq tC^*(y)$. Thus, $C^*(ty) = tC^*(y)$ for $t > 0$.   Q.E.D.

The important and useful implication of Proposition 9D4 is that, to circumvent the integer problem, one can characterize multiproduct fractional-firm competitive equilibria with the aid of the C^* function—the fractional-firm cost function of the industry. The industry's price and output vectors must satisfy relationships (9D5) and (9D6), which can be utilized to determine the equilibrium levels of these industry-wide magnitudes. Then, (9D4) indicates that the behavior of individual (fractional) firms can be determined from the industry configuration that minimizes the total cost of producing the equilibrium output of the industry.

As before, we find it necessary here to pay particular attention to the behavior of the costs over the M locus, which determine the shape of the C^* function. The special role of the C^* function is also indicated by the observation that the cost function of any firm in competitive equilibrium must be supported at its output vector y^i by the hyperplane defined by the industry price vector; that is, at no other output vector must the revenue hyperplane lie above the cost surface. Now, (9A5) and (9A6) indicate that under freedom of entry the industry cost function must also be supported at y^I by the revenue hyperplane generated by the equilibrium price vector. Since $C^I(y)$ is not convex when ray average costs are U-shaped, it is not generally possible to satisfy this necessary support property. This is the central analytic difficulty that is overcome with the aid of C^* since, as we have seen from Proposition 9D2, this function *is* convex and therefore everywhere supportable. For this reason, C^* is a more powerful concept than C^I for competitive equilibrium analysis.[14] There is, however, an important class of functions for which the two constructs are identical.

Proposition 9D5 $C^*(\cdot)$ *and* $C^I(\cdot)$ *are identical if* $C(\cdot)$ *exhibits constant returns to scale.*[15]

It is tempting to conclude also that when the firm's cost function is linearly homogeneous it, too, must be equal to $C^I(\cdot) = C^*(\cdot)$. However, this is not generally true because $C(\cdot)$ may well be linearly homogeneous without being convex, as we know C^* must be. An illustration is provided by the meat-fiber example in Section 9B. As one can see from the iso-cost curves in Figure 9B1, this cost function, while linearly homogeneous, is not convex.

---

[14] This approach contrasts sharply with recent approaches to this problem in a general equilibrium context. There, the integer problem is overcome by replicating demand and adjusting units to per capita terms. See Novshek (1979) and Sonnenschein (1980) and the references cited therein.

[15] *Proof*   In general, $C^*(y) \leq C^I(y) \leq C(y)$. With $C(ty) = tC(y)$ for $t > 0$, let $\sum_{i=1}^{m} a_i C(y^i) = C^*(y)$. Then, $C^I(y) \leq \sum_{i=1}^{m} C(a_i y^i) = \sum a_i C(y^i) = C^*(y)$. But $C^*(y) \leq C^I(y)$. Therefore, $C^*(y) = C^I(y)$. Q.E.D.

As is revealed in the example, industry costs need not be minimized by a single type of firm in such a case.

This discussion, together with Proposition 9D5, indicates that if the firm's cost function is linearly homogeneous, whether or not it is convex, the process of cost-minimizing aggregation must yield an industry cost function that is convex, as well as linearly homogeneous. The process is illustrated in Figure 9D5, which depicts the stylized "cross section" GH of the cost surface above the M locus. Suppose the industry output vector involves the output proportions represented by point D. Obviously, it can be produced by a set of firms each with those same output proportions. But instead, the industry's output can be produced more cheaply if some firms operate at point A and others operate at point B, where A and B are points at which C and C^* coincide. Then the corresponding average output and cost of the operations of such firms must be represented by some point such as E along the line segment AB. In effect, then, C^* "convexifies" C over the M locus and then extends it proportionally in to the origin and proportionally out to infinity. Proposition 9D5 reveals that where the firm's cost function is linearly homogeneous this is already done automatically by the aggregation of the firms' cost functions into the industry cost function. And even where the firms' cost functions are not linearly homogeneous, this will always be accomplished by the construction of C^*.

The preceding discussion of Figure 9D5 begins to suggest the circumstances under which equilibrium will, or will not, require different *types* of firms, that is, firms whose output vectors lie at different points on the M locus. We now define one additional cost concept, the homogenized cost function, $C^H(\cdot)$, which enables us to provide a complete characterization of

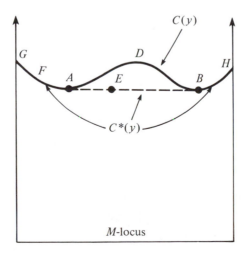

FIGURE 9D5

the set of points on the M locus that can be involved in a competitive equilibrium. $C^*(\cdot)$ is obtained graphically in Figures 9D3 and 9D4 by drawing a ray through every point of the boundary of the convex hull of the curve $ABVDRS$ on the cost surface above the M locus. Now we obtain $C^H(\cdot)$ in almost the same way, by drawing a ray through each point in $ABVDRS$, rather than the boundary of its convex hull. Thus, $C^H(\cdot)$ is the linear homogeneous extension of the cross section of C above the M locus. Unlike C^*, the new cost concept, C^H, does not "convexify" C. More formally.

Definition 9D3   The *homogenized cost function* is defined as $C^H(y) \equiv min_a \, aC(y/a)$.

     As in Figure 9D2, the homogenized cost function is defined for any point y by the lowest ray, OW, in cost and output space which anywhere touches the cost surface over the ray (OR) through y. Given our assumption in this section that the ray average cost curves are U-shaped, the point E at which OW touches the cost surface must lie above the M locus, so that $C^H(y) = aC(\bar{y})$, where $\bar{y} \in M$ and $y = a\bar{y}$. It follows that $C^H(\cdot)$ must be linearly homogeneous, since for any y on a ·given ray OR in output space, $\bar{y}$ is the same. Now, C and C^H coincide only at all points on the M locus. C and C^* may coincide at some points on the M locus and not at others. We see next that the relationships among these cost functions determines the output vectors that can be produced in equilibrium by a competitive firm.

Proposition 9D6   *Firms with an output vector $\hat{y}$ can participate in a fractional firm competitive equilibrium if and only if $C^*(\hat{y}) = C(\hat{y})$. Further, if and only if $C^*(\hat{y}) = C(\hat{y})$ can there be a fractional-firm competitive equilibrium in which every firm produces the output vector $\hat{y}$, that is, an equilibrium in which there exists a single representative firm.*

     The proof is provided in the Appendix, but an intuitive explanation, following along the lines of our previous discussion of Figure 9D5, is easily provided. In competitive equilibrium, a firm's cost function must be supported, at its output vector $\hat{y}$, by the revenue hyperplane defined by the industry price vector. That is, $C(\cdot)$ must coincide with the revenue hyperplane at $\hat{y}$, and $C(\cdot)$ must be at or above the revenue hyperplane everywhere else. This means that the cost function must be at least locally convex at $\hat{y}$. If it is not, then there do not exist *any* price vectors at which such a firm can be in equilibrium. In Figure 9D5, the output vector corresponding to the point D on the cost surface cannot be part of a fractional-firm equilibrium because the cost surface cannot be supported by *any* revenue hyperplane there. On the other hand, at points such as F, A, and B, the cost surface *can* be supported by different revenue hyperplanes (each corresponding to a different industry price vector); and each of these points corresponds, therefore, to a candidate output vector for a firm in a fractional-firm competitive equilibrium. Moreover, if the price vector yields a revenue hyperplane that supports the cost surface at just one of these points, then the corresponding output vector is the only one that can be produced by a firm in equilibrium.

Operationally, and perhaps more intuitively, once one characterizes a fractional-firm competitive equilibrium by means of (9D6) and (9D7), one then "shrinks" back the industry output y^I proportionately and moves its corresponding point on C^* [i.e., $C^*(y^I)$] along the ray through the point $(y^I, C^*(y^I))$ until one reaches the point on C^* that lies above the M locus. If it transpires that this point also lies on $C(y)$, then competitive equilibrium can be characterized by some (not necessarily integer) number of firms of that *one* type. In Figure 9D5, this is true, *with the appropriate equilibrium prices*, for any point on arcs GFA or BH. However, if this process of backward projection of $C^*(y^I)$ ends up at a point such as E, which is *not* on $C(y)$, then the equilibrium will require some (not necessarily integer) numbers of firms of both type A and of type B. Any output vector can be part of some fractional-firm competitive equilibrium if, at that point, the homogenized cost surface can be supported by some revenue hyperplane in some trans-ray cross section. No points along the arc ADB can be supported in this manner, and the associated output vectors cannot be included in any competitive equilibrium configuration. This is the import of Proposition 9D6.

We now provide, for the interested reader, additional conditions that are equivalent to those of Proposition 9D6. (The proofs are provided in the Appendix to this chapter).

Proposition 9D7   *Any one of the following is a necessary and sufficient condition for $C^*(\hat{y}) = C(\hat{y})$. Thus, all three conditions are equivalent:*

(i) $C^*(\hat{y}) = C^H(\hat{y})$ *and* $\hat{y} \in M$
(ii) $C^H(\cdot)$ *is subadditive at* $\hat{y}$ *and* $\hat{y} \in M$
(iii) $C^H(\cdot)$ *is trans-ray supportable at* $\hat{y}$ *and* $\hat{y} \in M$.

The possibility that an equilibrium may occur at a point such as E in Figure 9D5, requiring a multiplicity of types of firms, immediately raises a question about the complexity of the corresponding equilibrium configuration. Since each type of firm produces a different output vector, with m types of firms and n products, the number of output variables mn could be unmanageable for analysis if m were very large. Is there an upper bound on the number of different firm types (points on the M locus) that are required for a fractional-firm competitive equilibrium? The following result provides such an upper bound:

Proposition 9D8[16] *In an industry that produces n different goods, fractional-firm competitive equilibria never require firms of more than n different types.*

It is interesting to compare this result with our earlier examples of competitive equilibria that require more than one type of firm. Each of the examples involves two-product industries, and each equilibrium requires two types of firms. Thus, the upper bound given in Proposition 9D8 is reached in the examples.

The primary analytical contribution of Proposition 9D8 is that it permits one to write down a *given number* of simultaneous equations to describe multiproduct competitive equilibrium. In analogy to (9D1)–(9D3), these equations are

(9D8) $(p_j - C_j(y^i))y^i_j = 0,$ for $j = 1, \ldots, n$ and $i = 1, \ldots, m \leq n$

(9D9) $p \cdot y^i - C(y^i) = 0,$ for $i = 1, \ldots, m \leq n$

and

(9D10)
$$Q(p) - \sum_{i=1}^{m} a_i y^i = 0.$$

Since we are dealing with $m \leq n$ types of firms, each producing no more than n outputs, this is a system of, at most, $n(n + 2)$ equations in the same number of unknowns; y^i_j, p_j, and a_i. Such a system can be used to construct a model of the behavior of multiproduct perfectly competitive industries, in much the same way as was done with (9D1)–(9D3) in the scalar output case.[17]

The analysis of multiproduct competitive equilibria is simplified greatly where the following result applies:

Proposition 9D9 *All fractional-firm competitive equilibria permitted by a given cost function can (must) be composed of firms all offering the same output vector if $C^H(\cdot)$ is everywhere (strictly) trans-ray convex. This will be so if $C(\cdot)$ is (strictly) convex over all line segments connecting points on the M locus.*

---

[16] Varian (personal communication) suggested that one should seek a result of this kind. The proof of the proposition, relegated to the Appendix, hinges on the fact that any point on the boundary of the convex hull of an n-dimensional set can be expressed as a linear combination of n points on the frontier of that set.

[17] See Silberberg (1974) and the references cited therein.

The proof of this result, too, is left to the Appendix. But the intuitive explanation is clear. If $C(y)$ is convex in Figure 9D5, then $C^H(\cdot)$ must also be convex. If that is so, the C^* cost surface will contain no points like E which are not also on the C cost surface. If the convexity is strict, any given revenue hyperplane can support only *one* point on the C^* cost surface. And if fractional-firm competitive equilibria are each characterized by a unique output mix and type of firm, then comparative statics analysis of the system (9D8)–(9D10) is simplified considerably. With $m = 1$, that system is composed of only $2n + 1$ equilibrium conditions, instead of the $2n + n^2$ which are generally possible.

We conclude this chapter, appropriately enough, with a rather general theorem on the existence of fractional-firm competitive equilibria.

Proposition 9D10 *If all ray average cost curves are U-shaped, $C(y)$ is continuously differentiable on R^n_+, and if $Q(p)$ is continuous and nonzero on $P \equiv \{p \in R^n_+ \,|\, \exists y \in M \text{ such that } \nabla C(y) \geq p\}$,[18] then there exists a fractional firm competitive equilibrium.*

This, also, is proved in the Appendix.

9E. Concluding Comment:
Competitive Equilibrium and Sustainability

Multicommodity production gives rise to a series of issues that do not occur in the single-product case. When will the firm find it profitable to produce all of the goods offered by the industry? When will all firms find it advantageous to offer the same output proportions? When they do not, how many different output vectors can one expect firms to offer in competitive equilibrium? This chapter has considerably extended the standard analysis of perfect competition by providing at least formal answers to all of these questions and, in addition, by determining the relevant equilibrium conditions and criteria for existence of a multiproduct competitive equilibrium. All of this requires the use of one or more of our new concepts— economies of scope, product-specific scale economies, the M locus, trans-ray supportability, and so forth.

In dealing with these issues, we attend to a problem that also arises in the single-product case—the problem that occurs when industry output is not an integer multiple of the output which minimizes average cost for the

---

[18] This last condition merely means that quantities demanded are not all zero at any prices less than or equal to the marginal costs experienced by firms operating at points of minimum ray average cost.

firm. This problem and its multiproduct analog may, in principle, make the existence of an equilibrium very unlikely. However, when the number of firms grows sufficiently large, the influence of the integer problem can be taken to be negligible and can safely be assumed away. This difficulty returns to trouble us when we deal with oligopolies and their relatively small numbers of firms. In Chapter 11, we explore the issue further and offer some possible avenues of solution. There, as in Chapter 2, we are able to indicate more precisely what we mean by the statement that the integer problem becomes negligible when the number of firms grows sufficiently large.

There is one final issue that requires discussion here. The reader may well wonder why the term "sustainability"—the central issue in so much of the analysis of this book—has not even made an appearance in this chapter. The answer, in brief, is that perfect competition is the one market form for which there is no need to introduce this novel concept.

The classical notion of perfectly competitive equilibrium already incorporates potential entrants who evaluate the profitability of entry in terms of current market prices. They expect to be able to sell any quantities of output they wish without causing market prices to fall. An industry configuration can only be a competitive equilibrium if it permits no such production plan to appear profitable to any potential entrants.

Potential entrants who take current market prices to be fixed are more optimistic than the potential entrants incorporated in the definition of sustainability. In the latter, latent rivals only consider the prices of active firms to be fixed (temporarily) in making their entry decisions. They recognize, however, that they will have to accept lower prices if their attempted production increases the total quantities supplied by the industry. The distinction is that between the view that takes market prices to be fixed and the one that assumes this only for rivals' prices. And, clearly, it is the former that invites actual entry more readily.

Consequently, an industry configuration that entirely discourages the entry of potential rivals who take market prices as fixed must also completely discourage entry by those who just consider active rivals' prices to be constant; that is, it must automatically be sustainable. Thus, *an industry configuration that is in perfectly competitive equilibrium is sustainable* virtually by definition. That is why sustainability is a redundant concept for the analysis of perfectly competitive industries. And that also indicates how the concept of sustainability, rather than being absent from the analysis of this chapter, implicitly pervaded it, in the more conventional guise of the classical concept of perfectly competitive equilibrium.

The requirements of perfectly competitive equilibrium also appear to be stronger than those of sustainability in other ways. In addition to the preclusion of profitable entry possibilities, a competitive equilibrium industry configuration must be comprised of actively producing firms who choose their outputs so as to maximize profit at given and fixed market prices.

However, this is an illusory addition. As indicated in Section 9A, the requirements that competitive firms earn nonnegative profits and that no market-price-taking entrants can earn positive profits together imply that the actively producing firms must maximize their profits at market prices.

Thus, the only essential difference between the requirements of competitive equilibrium and sustainability lies in the price anticipations of the latent rivals whose entry must be discouraged. And, since the anticipations of potential rivals under perfect competition are more optimistic than those under perfect contestability, competitive equilibrium does impose stricter requirements upon an industry configuration than does sustainability in some circumstances, most notably natural monopoly. But, clearly, competitive equilibrium is generally an inappropriate (and infeasible) requirement for natural monopoly.

However, as was suggested in Chapter 2, and will be shown in Chapter 11, where technological and demand conditions require at least two firms to produce each good, sustainability alone requires perfectly competitive behavior by all firms that are producing actively. Hence, under these conditions, which are sure to be satisfied in industries for which the competitive model is thought to be roughly appropriate, among many others, sustainability suffices for the strong results usually imputed only to perfect competition. For these reasons, we assert that sustainability is an economically substantive generalization of classical competitive equilibrium, and, correspondingly, that the notion of a perfectly contestable market is an equally substantive generalization of the market that is perfectly competitive.

APPENDIX

Proof of Proposition 9D6

Since $C^*(\cdot)$ is a convex, linearly homogeneous, and increasing function, there is a $p \in \mathbf{R}^n_{++}$ such that $p \cdot \hat{y} = C^*(\hat{y})$ and $C^*(y) \geq p \cdot y$, $\forall y \in \mathbf{R}^n_+$. Suppose market demands were such that $Q(p) = t\hat{y}$. Then, since $p \cdot t\hat{y} = tC^*(\hat{y}) = C^*(t\hat{y})$, by Proposition 9D4 if $C^*(t\hat{y}) = \sum_{i=1}^m a_i C(y^i)$, the a_i firms, each producing y^i, constitute a fractional-firm competitive equilibrium. If $C^*(\hat{y}) = C(\hat{y})$, then $C^*(t\hat{y}) = tC^*(\hat{y}) = tC(\hat{y})$. Thus, t firms, each producing $\hat{y}$, can comprise a fractional-firm competitive equilibrium.

Suppose $C^*(\hat{y}) \neq C(\hat{y})$. Then, by definition, $C^*(\hat{y}) < C(\hat{y})$. Suppose a firm producing $\hat{y}$ does participate in a fractional-firm competitive equilibrium, with total industry output y^I. Then, by Proposition 9D4, $C^*(y^I) = \sum a_i C(y^i) + aC(\hat{y})$. Let $C^*(\hat{y}) = \sum b_j C(y^j)$. Then, since $C^*(\hat{y}) < C(\hat{y})$, $aC(\hat{y}) > aC^*(\hat{y}) = \sum ab_j C(y^j)$, and $C^*(y^I) > \sum a_i C(y^i) + \sum ab_j C(y^j)$. But $y^I = \sum a_i y^i + a\hat{y} = \sum a_i y^i + \sum ab_j y^j$, which provides a contradiction of the definition of $C^*(\hat{y}^I)$. Q.E.D.

Proof of Proposition 9D7

Note first that by Proposition 9D1, $C^*(\hat{y}) = C(\hat{y})$ implies that $\hat{y} \in M$. Further, $C^H(\hat{y}) = C(\hat{y})$ if and only if $\hat{y} \in M$. These observations immediately establish (i).

For (ii), let $C^*(\hat{y}) = \sum a_i C(y^i)$, where $\sum a_i y^i = \hat{y}$. By definition, $C^*(\hat{y}) \leq C^H(\hat{y})$. By Proposition 9D1, $y^i \in M$ so that $C(y^i) = C^H(y^i)$ and $a_i C(y^i) = C^H(a_i y^i)$. Then $C^*(\hat{y}) = \sum C^H(a_i y^i) \geq C^H(\hat{y})$ if C^H were subadditive at $\hat{y}$, and $C^*(\hat{y}) = C^H(\hat{y}) = C(\hat{y})$ would follow from this and from $\hat{y} \in M$. Suppose that C^H is not subadditive at $\hat{y}$ so that there are vectors y^1 and y^2 such that $y^1 + y^2 = \hat{y}$ and $C^H(y^1) + C^H(y^2) < C^H(\hat{y})$. Then $C^*(\hat{y}) \leq a_1 C(\bar{y}^1) + a_2 C(\bar{y}^2) < C^H(\hat{y})$, where $\bar{y}^i \in M$ and $a_i \bar{y}^i = y^i$, $i = 1, 2$. This establishes (ii). Part (iii) follows from (ii) and the following:

Lemma 1 *A linearly homogeneous function on $\mathbf{R}^n_+$ is subadditive at x^0 if and only if it is trans-ray supportable at x^0. Further, it is strictly subadditive at x^0 for nonproportional decompositions if and only if it is strictly trans-ray supportable at x^0.*

Proof Suppose f is trans-ray supportable at x^0; that is, there exist $w \in \mathbf{R}^n_{++}$, $v \in \mathbf{R}^n$, and $v_0 \in \mathbf{R}$ such that $f(x^0) = v \cdot x^0 + v_0$ and

(1) $$w \cdot x = w \cdot x^0 \Rightarrow f(x) \geq v \cdot x + v_0$$

Let $x^1 + x^2 = x^0$ with $x^i \geq 0$ and $x^i \neq 0$. Then $w \cdot x^i > 0$ and $w \cdot (w \cdot x^0 / w \cdot x^i) x^i = w \cdot x^0$ so that by (1) and linear homogeneity we have

(2) $$v \cdot \left(\frac{w \cdot x^0}{w \cdot x^i} \right) x^i + v_0 \leq f \left[\left(\frac{w \cdot x^0}{w \cdot x^i} \right) x^i \right] = \frac{w \cdot x^0}{w \cdot x^i} f(x^i).$$

Hence,

(3) $$f(x^i) \geq v \cdot x^i + \frac{w \cdot x^i}{w \cdot x^0} v_0,$$

and

(4) $$f(x^1) + f(x^2) \geq v \cdot x^0 + v_0 = f(x^0),$$

as is required for subadditivity.

If f were strictly trans-ray supportable, the inequality in (1) would be strict for $x \neq x^0$. Then, the inequalities in (2), (3), and (4) would also be strict unless $x^i = t_i x^0$, $t_i \in \mathbf{R}$; that is, unless x^1 and x^2 were a proportional decomposition of x^0.

Now, suppose f were not trans-ray supportable at x^0. Fix $w \in \mathbf{R}^n_{++}$, let $H = \{(x, z) \in \mathbf{R}^{n+1}_+ \mid w \cdot x = w \cdot x^0, z \geq f(x)\}$, and consider the convex hull of H, $[H]$. Then $(x^0, f(x^0)) \in H \subseteq [H]$. If it were a boundary point of $[H]$, there would be a separating hyperplane such that

(5) $$h \cdot x + h_0 z \geq h \cdot x^0 + h_0 f(x^0), \qquad \text{for all } (x, z) \in [H].$$

Then x^0 would minimize $h \cdot x + h_0 f(x)$ subject to $x \in R^n_+$ and $w \cdot x = w \cdot x^0$. By the Kuhn-Tucker theorem, there would hence be $\alpha \in \mathbf{R}$ and $\beta \in \mathbf{R}^n_+$ with

(6) $$h + h_0 \nabla f(x^0) - \alpha w - \beta = 0 \quad \text{and} \quad \beta \cdot x^0 = 0.$$

Together, (6) and (5) imply that $(\alpha w + \beta - h_0 \nabla f(x^0)) \cdot x + h_0 f(x) \geq (\alpha w + \beta - h_0 \nabla f(x^0)) \cdot x^0 + h_0 f(x^0)$ for $w \cdot x = w \cdot x^0$, since, then $(x, f(x)) \in H \subseteq [H]$. Because $\beta \cdot x^0 = 0$ and, by linear homogeneity $f(x^0) = \nabla f(x^0) \cdot x^0$,

$w \cdot x = w \cdot x^0$ implies

(7) $$h_0 f(x) \leq (\beta - h_0 \nabla f(x^0)) \cdot x.$$

By (5) and the definition of H, $h_0 > 0$. Consequently, division of (7) by h_0 shows that if $(x^0, f(x^0))$ were on the boundary of $[H]$, $f(x)$ would be trans-ray supportable at x^0. Hence, by hypothesis, $(x^0, f(x^0))$ is interior of $[H]$.

Then, there is a $z^0 < f(x^0)$ with $(x^0, z^0) \in [H]$. By Carathéodory's theorem there are points $(x^i, z^i) \in H$ and $\lambda_i \in [0, 1]$, $i = 1, \ldots, n + 2$, such that $x^0 = \sum \lambda_i x^i$ and $z^0 = \sum \lambda_i z^i$. By the definition of H, $z^i \geq f(x^i)$, and by linear homogeneity $\lambda_i f(x^i) = f(\lambda_i x^i)$, so that $f(x^0) > z^0 = \sum \lambda_i z^i \geq \sum \lambda_i f(x^i) = \sum f(\lambda_i x^i)$. Thus, since $x^0 = \sum \lambda_i x^i$, f is not subadditive at x^0, given that it is not trans-ray supportable there.

Similarly, if f were not *strictly* trans-ray supportable at x^0, $(x^0, f(x^0))$ could not be an extreme point of $[H]$. Then there would be members of $[H]$, $(y^0, f(y^0))$, and $(u^0, f(u^0))$, with $\gamma y^0 + (1 - \gamma) u^0 = x^0$ and $\gamma f(y^0) + (1 - \gamma) f(u^0) = f(x^0)$ for some $\gamma \in (0, 1)$. By Carathéodory's theorem there are members of H, (y^i, z^i), and $(u^i, \hat{z}^i)$, $\lambda_i \in [0, 1]$, and $\theta_i \in [0, 1]$, $i = 1, \ldots, n + 2$, with $\sum \lambda_i y^i = y^0$, $\sum \theta_i u^i = u^0$, $\sum \lambda_i z^i = f(y^0)$, and $\sum \theta_i \hat{z}^i = f(u^0)$. Then,

$$f(x^0) = \gamma \sum \lambda_i z^i + (1 - \gamma) \sum \theta_i \hat{z}^i \geq \gamma \sum \lambda_i f(y^i) + (1 - \gamma) \sum \theta_i f(u^i)$$
$$= \sum f(\gamma \lambda_i y^i) + \sum f((1 - \gamma) \theta_i u^i).$$

Since $x^0 = \sum \gamma \lambda_i y^i + \sum (1 - \gamma) \theta_i u^i$, f could not be strictly subadditive for nonproportional decompositions at x^0 if it were not strictly trans-ray supportable there. This completes the proof of the Lemma.

Since from (ii), C^H is subadditive at $\hat{y}$, by the Lemma it must be trans-ray supportable there.   Q.E.D.

Proof of Proposition 9D8

For any fixed $\hat{y} \in \mathbf{R}^n_+$, let $k > 0$ and $w \in \mathbf{R}^n_{++}$ be chosen such that $w \cdot \hat{y} = k$. Define $H = \{(x, z) \in \mathbf{R}^{n+1}_+ \mid w \cdot x = k, z \geq C^H(x)\}$ and $[H]$ to be the convex hull of H. These are n-dimensional sets. Let $C^*(\hat{y}) = \sum a_i C(y^i)$, $\sum a_i y^i = \hat{y}$, where, by Proposition 9D1, $y^i \in M$. Then

$$C^*(\hat{y}) = \sum a_i C^H(y^i) = \sum a_i [(w \cdot y^i)/k] C^H((k/w \cdot y^i) y^i).$$

Since $w \cdot [(k/w \cdot y^i) y^i] = k$, the point $[(k/w \cdot y^i) y^i, C^H(\cdot)]$ is an element of H. Since

$$\sum a_i(w \cdot y^i/k) = (w/k) \cdot \sum a_i y^i = (w/k) \cdot \hat{y} = 1,$$
$$(\hat{y}, C^*(\hat{y})) = [\sum a_i y^i, \sum a_i(w \cdot y^i/k) C^H((k/w \cdot y^i) y^i)]$$
$$= \sum [a_i(w \cdot y^i)/k][(k/(w \cdot y^i)) y, C^H((k/w \cdot y^i) y^i)] \in [H].$$

Then, by an extension of Carathéodory's theorem (Reay, 1965), there are points in H, (x^i, z^i), and $\lambda_i \in [0, 1]$, $i = 1, \ldots, n$, with $\hat{y} = \sum \lambda_i x^i$ and $C^*(\hat{y}) = \sum \lambda_i z^i$. From the definition of H, $z^i \geq C^H(y^i)$, so that $C^*(\hat{y}) \geq \sum \lambda_i C^H(x^i) = \sum b_i \lambda_i C(\bar{x}^i)$, where $\bar{x}^i \in M$ and $b_i \bar{x}^i = x^i$. But, by definition, $C^*(\hat{y}) \leq \sum b_i \lambda_i C(\bar{x}^i)$. Thus, firms with outputs $\bar{x}^i$, $i = 1, \ldots, n$, can realize $C^*(\hat{y})$ and can constitute a fractional-firm competitive equilibrium with industry output $\hat{y}$.   Q.E.D.

Proof of Proposition 9D9

Together, Propositions 9D6 and 9D7 show that any output vector $\hat{y}$ on the M locus can be the sole constituent of a fractional-firm competitive equilibrium if and only if $C^H(\cdot)$ is subadditive at $\hat{y}$. It is straightforward to show, also, that a fractional-firm competitive equilibrium with industry output $t\hat{y}$, for $\hat{y} \in M$ and $t > 0$, *must* be comprised of firms producing $\hat{y}$ if and only if $C^H(\cdot)$ is strictly subadditive at $\hat{y}$, with respect to nonproportional decompositions of $\hat{y}$. Moreover, it is a standard result that a linearly homogeneous function is subadditive if and only if it is convex. It is straightforward to extend this result to show that a linearly homogeneous function is strictly subadditive for nonproportional decompositions if and only if it is strictly trans-ray convex.

Then, it remains only to show that $C^H(\cdot)$ is (strictly) trans-ray convex if $C(\cdot)$ is (strictly) convex over all line segments connecting points on the M locus. For $\lambda \in (0, 1)$, and $w \cdot y^1 = w \cdot y^2$, $w > 0$,

$$(8) \qquad \lambda C^H(y^1) + (1 - \lambda)C^H(y^2) - C^H(\lambda y^1 + (1 - \lambda)y^2)$$
$$= \lambda t_1 C(\bar{y}^1) + (1 - \lambda)t_2 C(\bar{y}^2) - C^H(\lambda y^1 + (1 - \lambda)y^2),$$

where $\bar{y}^i \in M$ and $y^i = t_i \bar{y}^i$. By the assumed (strict) convexity of $C(\cdot)$ between $\bar{y}^1$ and $\bar{y}^2$,

$$\frac{\lambda t_1}{\lambda t_1 + (1 - \lambda)t_2} C(\bar{y}^1) + \frac{(1 - \lambda)t_2}{\lambda t_1 + (1 - \lambda)t_2} C(\bar{y}^2) \geq (>) C\left(\frac{\lambda t_1 \bar{y}^1 + (1 - \lambda)t_2 \bar{y}^2}{\lambda t_1 + (1 - \lambda)t_2}\right)$$
$$= C\left(\frac{\lambda y^1 + (1 - \lambda)y^2}{\lambda t_1 + (1 - \lambda)t_2}\right).$$

Thus,

$$\lambda t_1 C(\bar{y}^1) + (1 - \lambda)t_2 C(\bar{y}^2) \geq (>) (\lambda t_1 + (1 - \lambda)t_2)C\left(\frac{\lambda y^1 + (1 - \lambda)y^2}{\lambda t_1 + (1 - \lambda)t_2}\right)$$
$$\geq C^H(\lambda y^1 + (1 - \lambda)y^2),$$

by the definition of $C^H(\cdot)$. Consequently, the right-hand side of (8) is $\geq (>) 0$, and the result follows.   Q.E.D.

Proof of Proposition 9D10

In order to establish the required smoothness properties for C^*, we first prove the following:

Lemma 2   *Under the assumptions of Proposition 9D10, C^* is continuously differentiable for all $y \in \mathbf{R}^n_{++}$ and $\nabla C^*(y) \in P$.*

Proof   In view of Proposition 9D8,

$$C^*(y) = \min_{a_i, y^i \geq 0} \sum_{i=1}^{n} a_i C(y^i), \text{ subject to } \sum_{i=1}^{n} a_i y^i = y.$$

Thus, C^* is continuous on $\mathbf{R}^n_+$ by the Maximum theorem (see Berge, 1963). The Lagrangian of this program is

$$\mathscr{L} = \sum_{i=1}^{n} a_i C(y^i) + \sum_{j=1}^{n} \mu_j \left(y_j - \sum_{i=1}^{n} a_i y_j^i \right),$$

and the Kuhn-Tucker necessary conditions are

$$\frac{\partial \mathscr{L}}{\partial a_i} = C(y^i) - \sum_{j=1}^{n} \mu_j y_j^i \geq 0, \qquad a_i \left(\frac{\partial \mathscr{L}}{\partial a_i} \right) = 0, \qquad a_i \geq 0$$

$$\frac{\partial \mathscr{L}}{\partial y_j^i} = a_i \frac{\partial C(y^i)}{\partial y_j^i} - a_i \mu_j \geq 0, \qquad y_j^i \frac{\partial \mathscr{L}}{\partial y_j^i} = 0, \qquad y_j^i \geq 0$$

$$y_j = \sum_{i=1}^{n} a_i y_j^i.$$

By the generalized Envelope theorem, C^* is continuously partially differentiable, at y, with respect to y_j and $\partial C^*(y)/\partial y_j = \partial \mathscr{L}/\partial y_j$, if $\partial \mathscr{L}/\partial y_j$ is invariant over the set of solutions to the program.

The Kuhn-Tucker conditions show that, at a solution $a_1, \ldots, a_n$; $y^1, \ldots, y^n$, $\partial \mathscr{L}/\partial y_j = \partial C(y^h)/\partial y_j^h$ if $a_h y_j^h > 0$. And, with $y \in \mathbf{R}^n_{++}$, $a_h y_j^h > 0$ for some h. Suppose there were another solution, $b_1, \ldots, b_n$; $\tilde{y}^1, \ldots, \tilde{y}^n$ with $b_k y_j^k > 0$ and $\partial C(\tilde{y}^k)/\partial \tilde{y}_j^k \neq \partial C(y^h)/\partial y_j^h$. Then $C^*(y) = \sum_i a_i C(y^i) = \sum_i b_i C(\tilde{y}^i) = \frac{1}{2} \sum_i a_i C(y_i) + \frac{1}{2} \sum_i b_i C(\tilde{y}^i) > \frac{1}{2} \sum_{i \neq h} a_i C(y_i) + \frac{1}{2} \sum_{i \neq k} b_i C(\tilde{y}^i) + \frac{1}{2} a_h C(y^h + \delta e^j) + \frac{1}{2} b_k C(\tilde{y}^k - \delta e^j)$ for some $\delta \in \mathbf{R}$, where $e^j \in \mathbf{R}^n$ with $e_i^j = 0$ for $i \neq j$ and $e_j^j = 1$. But this would contradict the definition of C^*. So on $\mathbf{R}^n_{++}$, $\partial \mathscr{L}/\partial y_j$ is invariant over all solutions to the program and $\partial C^*(y)/\partial y_j = \partial C(y^h)/\partial y_j^h$ for some $a_h y_j^h > 0$. For this same y^h, the Kuhn-Tucker conditions show that $\mu_j = \partial C(y^h)/\partial y_j^h > 0$. Then, if $a_i > 0$, $\partial C(y^i)/\partial y_j^i \geq \mu_j = \partial C(y^h)/\partial y_j^h = \partial C^*(y)/\partial y_j$. Thus, $\nabla C^*(y) \leq \nabla C(y^i)$ if $a_i > 0$. Since $y^i \in M$ and since $a_i > 0$ for some i, $\nabla C^*(y) \in P$. This completes the proof of Lemma 2.

By Proposition 9D2 and Lemma 2, C^* is continuous and convex on $\mathbf{R}^n_+$ and is continuously differentiable on $\mathbf{R}^n_{++}$. Then, its gradient can be continuously extended to the function $p^s: \mathbf{R}^n_+ \to \mathbf{R}^n_+$, with $C^*(y) \geq C^*(y^0) + (y - y^0) \cdot p^s(y^0)$ for all $y^0 \in \mathbf{R}^n_+$ and $y \in \mathbf{R}^n_+$. Further, by the linear homogeneity of C^*, p^s is homogeneous of degree zero, and $C^*(y^0) = p^s(y^0) \cdot y^0$ for $y^0 \in \mathbf{R}^n_+$. Hence, $C^*(y) \geq p^s(y^0)y$.

Now let S^n be the unit simplex in $\mathbf{R}^n_+$, and consider $g: S^n \to S^n$ with $g_i(y) \equiv Q_i[p^s(y)]/\sum_{j=1}^n Q_j(p^s)$. The function g is well defined because $Q(p)$ is nonzero on P and $p^s(y) \in P$ by Lemma 2. Further, g is continuous because both Q and p^s are continuous functions. Hence, by Brower's fixed point theorem, there is a $\bar{y} \in S^n$ with

$$\bar{y}_i = \frac{Q_i(p^s(\bar{y}))}{\sum_j Q_j(p^s(\bar{y}))}$$

Then, by Proposition 9D4, $(\sum Q_j(p^s(\bar{y})))\bar{y} = y^I$ is the industry output in a fractional-firm competitive equilibrium because $p^s(y^I) = p^s(\bar{y})$, $Q[p^s(y^I)] = y^I$, $C^*(y^I) = p^s(y^I) \cdot y^I$ and $C^*(y) \geq p^s(y^I) \cdot y$ for all $y \in \mathbf{R}^n_+$.   Q.E.D.

10

Fixed Costs, Sunk Costs, Entry Barriers, Public Goods, and Sustainability of Monopoly

Costs lie at the center of the analysis of this book, and it is consequently important for us to disentangle a variety of cost concepts, such as fixed costs, sunk costs, and entry costs. This chapter attempts to explore these concepts and to determine their significance for our analysis. Specifically, it is shown that (i) fixed costs of sufficient magnitude ensure the presence of natural monopoly cost conditions and the existence of a vector of sustainable prices for the natural monopolist; (ii) nevertheless, costs that are truly fixed do not constitute barriers to entry; that is, they do not have the welfare consequences normally attributed to barriers to entry; (iii) indeed, in perfectly contestable markets large fixed costs are completely compatible with many, if not most, of the desirable attributes of competitive equilibrium; (iv) sunk costs do, however, constitute barriers to entry that lead to losses in efficiency and welfare; finally, (v) the resource allocation problems that stem from fixed costs are formally identical to those that accompany public goods—both their sources and their character are the same. Thus, the rationale for public supply of public goods may be no different from that pertaining to nationalization of natural monopolies.

10A. Definitions

Inevitably, terminology plays a crucial role in the discussion that follows; therefore, we must begin with a few definitions. As usual, taking "the long run" to refer to a period of time sufficient for all current commitments to be liquidated, it follows that in the long run all sunk costs are zero.

Long-run fixed costs are those costs that are not reduced, even in the long run, by decreases in output so long as production is not discontinued altogether. But they can be eliminated in the long run by total cessation of production. They comprise the height of the "jump discontinuity" near the origin in the graph of the long-run cost function.

Sunk costs, on the other hand, are costs that (in some short or intermediate run) cannot be eliminated, even by total cessation of production. As such, once committed, sunk costs are no longer a portion of the opportunity cost of production.

We can then provide the following formal definitions of the terms "fixed costs" and "sunk costs" as they are used here[1]:

Definition 10A1: Long-Run Fixed Cost   Long-run fixed cost is the magnitude $F(w)$ in the long-run total cost function

$$C_L(y, w) = \delta F(w) + V(y, w) \qquad \delta = \begin{cases} 0 & \text{if } y = 0 \\ 1 & \text{if } y > 0, \end{cases}$$

where

$$\lim_{y \to 0} V(y, w) = V(0, w) = 0,$$

$V(\cdot)$ is nondecreasing in all arguments, and y and w are, respectively, the vectors of output quantities and input prices.

Definition 10A2   Let $C(y, w, s)$ represent the short-run cost function, applicable to plans for the flow of production, that occurs s units of time (years) in the future. Then, $K(w, s)$ are the costs sunk for at least s years, if

$$C(y, w, s) = K(w, s) + G(y, w, s),$$
$$G(0, w, s) = 0.$$

---

[1] We are indebted to Charles Berry for this way of defining the concepts.

Here, since in the long run no costs are sunk,

$$\lim_{s \to \infty} K(w, s) = 0.$$

It should be emphasized that here fixed costs mean costs that are fixed in the long run as well as the short run. Thus, investments in large-scale plant and equipment do not generally qualify. For, as the textbook aphorism says, such costs do indeed become variable eventually. They are *sunk costs*, as distinguished from fixed costs that can only be diminished, even in the long run, by closing the enterprise down altogether.

To illustrate, contrast the operations of a railroad and an auto assembly line. If the price of petroleum were to rise to a point at which a market for only a few hundred cars a year remained, huge auto assembly lines would ultimately disappear. Cars might be hand-tooled, albeit at a rather high cost per vehicle, and *total* equipment cost would be reduced drastically. Thus, the equipment cost of an assembly line may be sunk, but it is not fixed.

On the other hand, if a railroad is to run between New York and Chicago, there is a minimum outlay on track and roadbed that must be incurred, even if the trains run virtually empty. The service can be discontinued altogether; but even in the longest of long runs, its roadbed cost cannot be reduced to a negligible level if the amount of service is to be positive.

Thus, sunk costs need not be fixed. And, even more important, fixed costs need not be sunk. To operate with current production techniques, a railroad requires at least a locomotive and one car the costs of which, while not normally done in the literature, must be included among its fixed costs. That is, with the *long-run* cost function written as $F + V(y)$ for $y > 0$, where $V(0) = 0$, at least the rental[2] cost of this locomotive and this car must be included in F. Yet, because they constitute capital on wheels, most of their cost can easily and quickly be recovered by rolling them to another market, should

---

[2] It should be noted that sunk costs are not eliminated by the availability of a rental market in which the equipment that gives rise to them can be obtained even on a very short lease. True, rental eliminates the financial commitment of the person who *uses* the equipment, but it does not eliminate the commitment of the person who owns it and rents it out. The point is that the cost that must be sunk to get into the business of renting out such equipment can then discourage entry and confer a corresponding degree of protection from the threat of entry upon the incumbent in the rental market. In other words, the availability of a rental market need not eliminate sunk costs; it can merely transfer them and their attendant problems from the user of the equipment to its owner.

However, a rental market can reduce or eliminate sunk costs if it augments the mobility and fungibility of durable inputs among alternative uses. A locomotive is not a sunk cost of rail service between New York and Chicago if it can be transferred, without loss, to the provision of rail services elsewhere. To the extent that they reduce the costs of such transfers, rental markets do reduce sunk costs.

the railroad's management decide (and be permitted) to close down the line in question. Thus, little or none of this portion of fixed cost is sunk, in contradistinction to the roadbed cost, which typically is sunk. Airlines and postal delivery are probably better examples, since they are industries with relatively low sunk costs along particular routes, and their fixed costs may considerably exceed their sunk costs.

The distinction between sunk and fixed cost that we emphasize here is not a mere terminological quibble. We see later that it makes a substantial difference for analysis and policy if the costs of the firms in an industry include the one rather than the other. For example, for reasons that will be discussed later, the advisability of deregulation of air transportation is increased by the smallness of the airlines' sunk costs relative to their fixed costs. Moreover, this judgment is not undermined by our conclusion that fixed costs of sufficient magnitude make it possible for incumbents to adopt prices that can prevent entry.

We also have

Definition 10A3: Entry Barrier   An entry barrier is anything that requires an expenditure by a new entrant into an industry, but imposes no equivalent cost upon an incumbent.

This definition requires some comment. It is essentially the same as Stigler's (1968, p. 67). However, it differs radically from the more traditional definitions which stem largely from the work of Bain (see, for example, Bain, 1956, Chap. 1). Most notably, this definition can, in some circumstances, exclude per unit cost advantages of large-scale operation (scale economies), which Bain took as a fundamental source of barriers to entry. The reason we adopt the definition we do can only be explained fully below, but the explanation is suggested by comparison with von Weizsäcker's (1980a,b) alternative approach to the matter. Somewhat inaccurately, but in essence correctly, we can say that von Weizsäcker defines an entry barrier as *any* (unspecified) advantage over an entrant that an incumbent firm enjoys *if that advantage produces a welfare loss.* That is, such an advantage is a barrier to entry if its consequences are undesirable, and it is not an entry barrier otherwise. While, unlike von Weizsäcker's, our definition seeks to specify operationally what types of impediments meet its criteria, we hope to show below that our criterion and his overlap in substance. That is, we argue that anything that is an entry barrier by our definition does reduce the sum of consumers' and producers' surplus, while phenomena such as fixed costs and scale economies need not do so.

In interpreting our definition of an entry barrier, a word must be said about cases in which an incumbent has costs lower than those of an entrant simply as a result of the superior efficiency of the former. In our analysis, such a cost difference should not be treated as an entry barrier, but as an efficiency rent which can properly be included among the incumbent's costs, thus formally eliminating this difference between the cost of the incumbent and the entrant. We can distinguish such efficiency rents from entry costs by means of a conceptual experiment. Assume that the entrant, with the skills and the set of inputs available to him, is magically substituted for the incumbent, having operated in the place of the incumbent for a period comparable to the incumbent's lifetime. In this hypothetical scenario, any excess of the entrant's total cost over the actual cost of the incumbent is an efficiency rent of the incumbent or of some of his inputs.

10B. Fixed Costs and Sustainability: Preliminary

To provide an intuitive indication of the way in which fixed costs contribute to sustainability, let us begin with an unsustainable case in which fixed costs are zero and in which there is only a single product. Suppose (Figure 10B1) that a monopolist produces output y^* and charges p^* yielding total revenue R^*, in accord with the industry's inverse demand function.

Next, draw in the cost curve $OCC^eC'R^*$, which involves no fixed cost since it approaches the origin smoothly as we move from right to left. We see at once that, as the figure is drawn, the projected entry quantity y^e will incur a loss for the entrant, since it will cost him C^e to provide, while his total

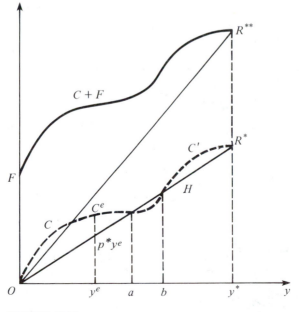

FIGURE 10B1

revenue will be no greater than R^e. However, as drawn, there *are* output quantities that may permit successful entry should the monopoly price remain fixed at p^*. For, since the cost curve dips below the pseudorevenue function between $y = a$ and $y = b$, it is clear that an entrant may be able to provide an output in this range and make a profit. Thus, we recall from Chapter 8 that a sufficient condition for sustainability of a price vector p^* is that the cost surface $C(y)$ nowhere equal or dip below the corresponding pseudo-revenue hyperplane $H(y)$ for any output in the relevant range; that is, $C(y) > H(y)$.

We return to our single-product diagram in a moment. First let us con-sider the multiproduct case. Figure 10B2 describes a case in which a sustain-able solution exists. Here ORS and BDE are, respectively, the pseudorevenue hyperplane and the cost hypersurface over the relevant range. The cost surface curls away from the pseudorevenue hyperplane in the direction of the origin [ray cross section $BC(y^I)$] and exhibits decreasing ray average cost. In addi-tion, cost curls away from the pseudorevenue hyperplane in the direction of the axes (the trans-ray cross section DE is convex). As a result, the point of tangency, $C(y^I)$, is the only point in the relevant range where cost does not exceed pseudorevenue. Since, as we know, the pseudorevenue hyperplane constitutes a ceiling upon the entrant's revenues, it follows that the entrant cannot break even at any point in the relevant set of output vectors, OVW, except if the entrant replaces the incumbent completely and provides his output vector y^I.

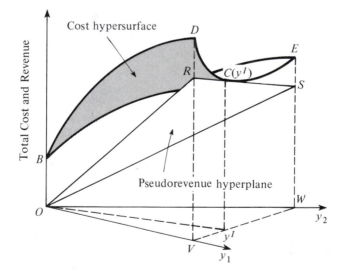

FIGURE 10B2

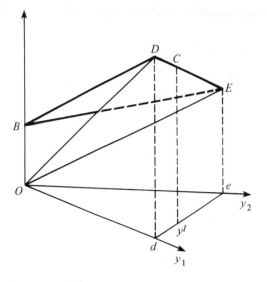

FIGURE 10B3

Next, suppose the *variable costs* of a particular firm do not exhibit the curvature of the cost surface of Figure 10B2, so that there exists no pseudo-revenue hyperplane through the origin whose height at the incumbent's output point is at least equal to that of his cost surface and which lies below the cost surface everywhere else in the relevant region. The greater the share of fixed costs (and other affine costs) in the total costs of the firm, and the smaller the share of (other) variable costs, the closer will be the approximation between the firm's total cost surface and a hyperplane whose value is positive at the origin in output space.[3]

Now the point is that any such hyperplane satisfies the two relevant "curvature" conditions: (strictly) declining average costs along any ray in output space (e.g., Oy^I in Figure 10B2) and (nonstrict) trans-ray convexity (e.g., cross section VW in Figure 10B2). Any hyperplane (such as BDE in Figure 10B3) with a positive value at the origin must satisfy both these conditions. First, every ray cross section (which is piecewise linear and

---

[3] That is, the cost function can be expressed as $kA(y) + (1 - k)V(y), 0 \le k \le 1$, where $A(y)$ is an affine cost function $F + \sum a_i y_i$, with y_i an element of vector y. Here, the weight k will approach unity as the share of fixed and other affine costs increases toward unity over the relevant region in output space. Indeed, as the proportion of fixed to all variable costs increases, the total cost hypersurface will approach a "horizontal" hyperplane, that is, one whose partial derivative with respect to any output quantity is zero. But the discussion holds for *any* hyperplane cost surface.

concave) clearly exhibits strictly declining ray average costs. Second, any trans-ray cross section, such as DE, must be linear and consequently non-strictly trans-ray convex. With both these properties satisfied, it is possible to find a pseudorevenue hyperplane (ODE in Figure 10B3) that is "tangent" to the cost surface BDE [above the boundary of the region where $R(y) > C(y)$] at any given point C. This pseudorevenue hyperplane will lie almost everywhere below the cost surface.[4] Thus, the incumbent's prices, which (by construction) are those that lead to the sale of output vector y^I, are sustainable since no output vector in the relevant region, Ode, can be supplied without financial loss by an entrant.

There is another illuminating way of looking geometrically at the relation between fixed costs and sustainability. Let us return to the single-product case of Figure 10B1, assuming for simplicity that demand is absolutely inelastic over the relevant range, so that equilibrium output will not change as fixed costs and equilibrium prices rise.[5]

In Figure 10B1 we have also drawn in a second cost curve, $C + F$, which is identical to $CC^eC'R^*$ except for the addition of a fixed cost F. There is a new pseudorevenue hyperplane OR^{**} which, because profit must be zero, once again meets the cost curve at the equilibrium output y^*. But, unlike the case without fixed costs, at every output $y < y^*$ total cost is now above the pseudorevenue figure. Thus, the addition of fixed costs F has given the monopolist the price p^{**} equal to the slope of OR^{**}, which is sustainable against entry.

The reason is that an increased fixed cost raises the average cost of smaller outputs relative to those of larger outputs. Thus, suppose we start off with two output levels $y_e < y^*$ such that their average variable costs satisfy $AVC_e < AVC^*$ without fixed costs. Any price that covers the average cost of y^* must permit viable entry by a firm proposing to produce y_e. But if there were an additional fixed cost, $(AVC^* - AVC_e)/(1/y^e - 1/y^*) > 0$, then $AC_e = AVC_e + F/y_e$ must exceed $AC^* = AVC^* + F/y^*$, since the fixed costs are spread over more units of output by the larger firm. Ultimately, with F sufficiently large, no entry output will be viable at a price that just covers

[4] In such a linear case there will be an $(n - 1)$-dimensional continuum of points at which cost and pseudorevenue are identical (DE in Figure 10B3). However, it must be remembered that the true market revenue function R is generally nonlinear. At the point C where the market permits the firm to cover its costs, R coincides with the pseudorevenue hyperplane. If R lies below pseudorevenue everywhere along DE except at point C, then there will be no point on DE other than C at which market expenditures can cover costs. Therefore, profitable entry cannot occur at any other point on DE, and as we have seen, it cannot occur at any point in output space closer to the origin.

[5] Of course, in general, as the share of fixed cost varies, equilibrium prices and quantities will change; and this, in turn, will affect entry opportunities. This, in essence, is the reason for the complexity of the mathematical arguments in the Appendix.

the cost of the incumbent's output y^*. By placing smaller firms at a relative disadvantage, fixed costs make sustainable prices possible for the incumbent.

10C. Fixed Costs, Affine Cost Functions, and Sustainability

Because demands are not usually perfectly inelastic, in our formal analysis we vary the share of fixed costs in a manner different from that in Figure 10B1. We hold the *total* cost of the equilibrium output constant and permit the share of fixed costs to change. Let $V(y)$ represent the variable costs of the firm, F its fixed costs, and $C(y)$ its total cost. Then, taking some output vector y^* as a reference point, in the limiting case in which all costs are fixed, we must have $F = C(y^*)$. With only some portion, k, of costs fixed, we must then have

(10C1) $$C(y) = kF + (1 - k)V(y);$$

that is

$$C(y) = kC(y^*) + (1 - k)V(y)$$

where $V(y^*) = C(y^*) = F$, and $0 \leq k \leq 1$. Thus, as k approaches 1, the cost function approximates a surface involving pure fixed costs, but with that surface always maintaining the same height at y^* no matter what the value of k.

We then prove in the Appendix, under a set of what may be considered plausible assumptions the most restrictive of which is that the goods are weak gross substitutes, that for any variable cost function $V(y)$ there exists a k^*, $0 \leq k^* < 1$, such that for any k, $k^* \leq k \leq 1$, sustainable prices exist. In other words, fixed costs of sufficient magnitude will, under the assumed conditions, always constitute an entry deterrent of sufficient magnitude to make sustainable prices possible.

The theorem proved in the Appendix is, in fact, more general. A cost function consisting exclusively of fixed cost is a special case of an *affine* cost function

$$A(y) = F + \sum a_i y_i \qquad \text{all } a_i \text{ constants, all } a_i \geq 0, F \geq 0,$$

which is characterized by its constant fixed cost and constant marginal costs. Then, with $V(y)$ any given function that is continuously differentiable with positive marginal costs in the relevant region, we may define a hybrid cost function to be one which satisfies

(10C2) $C(y) = kA(y) + (1 - k)V(y)$ for $y \geq 0$, and $C(0) = 0$.

It is shown in the Appendix that, with all goods weak gross substitutes[6]

Proposition 10C1   *For any affine cost function $A(y)$ with positive fixed costs, the enterprise is sustainable if its cost function resembles $A(y)$ sufficiently closely. That is, there exists a price vector p^m, with corresponding demanded quantities $Q(p^m)$, that is sustainable for the enterprise if its cost function has the hybrid form (10C2), for any variable cost function with $V(Q(p^m)) = A(Q(p^m))$, and with $k^* \leq k \leq 1$ for some $k^* < 1$. Further, any price vector p^m will be sustainable in this fashion if its revenues equal total cost, if each price in that vector exceeds the corresponding affine marginal cost and if the prices are strictly undominated (i.e., no lower price vector is financially feasible). Moreover, the prices that are Ramsey-optimal for the affine cost function will satisfy these conditions and will therefore be sustainable.*

In interpreting Proposition 10C1 and its assessment of the role of fixed costs in ensuring the existence of sustainable prices, it is important to bear in mind that the fixed costs in question are of the "all-or-none" variety. That is, they are incurred, in their entirety, when production of *any one* of the outputs in question is begun and do *not* increase as other product lines are added. Thus, moving k toward 1 simultaneously increases the degrees of overall economies of scale and economies of scope *without* any concomitant increase in *product-specific* returns to scale. This sheds additional light on the matter, since, as we learned in Chapter 8, the first two properties favor sustainability while the third is inimical to it. Here, only effects favorable to sustainability are present.

10D. Fixed Costs and Entry Barriers Distinguished

The preceding analysis has shown that fixed costs of sufficient magnitude can effectively prevent entry, which suggests that they might be classed as a type of entry barrier, as is sometimes done in the literature. However, it is possible to formulate at least three important distinctions between

---

[6] It is also assumed that the demand function is continuous and that the cost function is continuously differentiable, except at the origin; that, when prices change, demand for at least one good will change in the opposite direction from its price; and that, if outputs were expanded indefinitely in any direction, they would reach a point at which market revenues could not cover costs.

fixed costs and entry barriers:

1. The most important distinction is that fixed costs, unlike entry barriers, do not lead to industry performance inconsistent with optimality. Proposition 10C1 tells us that a preponderance of fixed costs enables an incumbent to deter entry if its prices yield zero profits and, in particular, if its prices are welfare-optimal, given the constraint of financial viability. Obversely, we learned in Chapter 8 that entry costs may permit an incumbent to earn more than normal profit. Moreover, we show in this section and in Section 10G that entry barriers permit incumbents to cause substantial welfare losses.

2. The cost figure corresponding to an entry barrier need not be fixed, but may vary with the magnitudes of the entering firm's outputs. This is why we use the notation $E(y)$ for these costs. For example, in most regulated industries there is a substantial entry barrier that takes the form of legal costs and delays to obtain the regulatory agency's permission to enter. But the magnitude of this cost undoubtedly rises substantially with the strength of the opposition mounted by firms whose sales are likely to be affected. Since their opposition will typically be stronger, the larger the scale of entry proposed, entry costs can be expected to rise with y^e.

3. As the preceding point suggests, the magnitude of the entry cost which must be incurred by a particular entrant can be affected by deliberate acts of incumbent firms, such as legal countermoves, capacity expansion, or advertising.[7] And (as will be discussed in Section 10G) incumbents may well be tempted to try these in order to raise their profits.

Thus, we conclude:

Proposition 10D1 *Fixed costs are not, and do not raise, entry barriers. Fixed costs, unlike entry barriers, do not prevent optimal welfare performance by the industry.*

The source of the difference is easy to identify. Fixed costs affect incumbent firm and entrant alike. They offer an advantage to the incumbent only to the extent that his output is greater, and this permits him to spread his costs more thinly than the entrant can. Entry costs, on the other hand, are *defined* here as costs that fall upon the entrant but *not* upon the incumbent at equal outputs. For example, if the incumbent were established before regulation of the industry began, but future entrants must incur heavy legal and delay expenses before they can start business, then these costs do constitute an entry barrier in the sense defined. It is clear why they then permit the incumbent to enjoy a positive rent.

---

[7] See the deep discussions of entry deterrence models in Dixit (1980), Eaton and Lipsey (1980, 1981), Gilbert and Newbery (1979), Hay (1976), Salop (1979b), Schmalensee (1978), Spence (1977), Stiglitz (1981), and von Weizsäcker (1980a,b).

To see just how an entry barrier $E(y)$ affects welfare it is helpful to proceed by analogy. Imagine a regulatory agency that initially keeps the economic profits of a natural monopoly under its jurisdiction exactly equal to zero, requires all prices to be set at the corresponding Ramsey levels p_0, and prohibits all entry. Suppose, now, that the regulatory agency changes its rules and permits the monopoly to earn positive economic profits equal to $E(y) \leq \pi_{g\,max}$, where $\pi_{g\,max}$ is the highest profit the monopoly could earn, given its cost and demand relationships, if it were immune both from regulation and entry threats. Then there will be a clear reduction in welfare even if the monopoly were now to adopt the welfare-maximizing prices under this eased profit constraint, that is, the Ramsey prices p_E under the constraint $\pi \geq E(y)$. That welfare loss is roughly equal to the reduction in consumers' surplus resulting from the change in prices from p_0 to p_E [i.e., from the change in prices needed to raise profits from zero to $E(y)$], minus the $E(y)$ itself, which represents a rise in producer's surplus.

It should be clear that the rise in regulatory profit ceiling in the preceding parable leads precisely to the same sort of change in behavior as a rise in entry costs from zero to $E(y)$. This entry cost can render the incumbent immune from entry so long as it restricts its profits to $\pi \leq E(y)$ and adopts the corresponding sustainable vector of prices. Thus, the entry barrier will lead to just the same sort of loss in welfare as a regulatory decision to permit supernormal profit $\pi = E(y)$.

10E. Sunk Costs as Entry Barriers

Sunk costs to some degree share with entry barriers the ability to impede the establishment of new firms.[8] The need to sink money into a new enterprise, whether into physical capital, advertising, or anything else, imposes a difference between the incremental cost and the *incremental risk* that are faced by an entrant and an incumbent. The latter's funds are already committed and are already exposed to whatever perils participation in the industry entails. On the other hand, a new firm must take the corresponding amount of liquid capital and turn it into a frozen asset if it enters the business. Thus, the incremental cost, as seen by a potential entrant, includes the full amount of the sunk costs, which is a bygone to the incumbent. Where the excess of prospective revenues over variable costs may prove, in part because of the actions of rivals, to be insufficient to cover sunk costs, this can constitute a very substantial difference. This risk of losing unrecoverable entry costs, as perceived by a potential entrant, can be increased by the threat (or the imagined threat) of retaliatory strategic or tactical responses of the incumbent.

Entry can be expected to be profitable only if the profits expected in the event of success outweigh the unrecoverable entry costs that will be lost in

---

[8] See Caves and Porter (1977) for lucid views on this and related matters.

the case of failure. And the likelihood of low revenues and failure is apt to be higher for potential entrants who know that they face active rivalry by the incumbent than it is for the incumbent who was initially without rivals (and may well have expected to remain so). The additional expected revenue that a potential entrant requires as compensation for the excess of its incremental cost and incremental risk over those of the incumbent becomes an entry cost as defined here and permits the incumbent to earn corresponding profit (rent). Thus, we have

Proposition 10E1 *The need to sink costs can be a barrier to entry.*

Some final points may be offered on the subject of sunk costs: Their role as barriers to entry depends on the risk to which they subject the entrant.[9] Futures contracts that assure the entrant that his sunk outlays will not be wasted can reduce that problem, so that *at the time the contracts are being negotiated* they can decrease the size of the entry barrier which the sunk cost would otherwise entail.[10] But in doing so they substitute a new and more formidable entry barrier during the life of the contract, by precluding any further potential entrants from the business that the contract has tied up. The greater the length of time for which the contract holds, the lower will be the risk of the initial entrant when he sinks his cost, so the lower will be the magnitude of entry barriers at the time of contract negotiation. But the longer the life of the contract, the longer the later entry barriers will endure. Calculation of the life of a futures contract optimal for consumers and for society must involve a balancing of these two costs, as well as any other associated social costs such as those resulting from stultified innovation.

Finally, it should be emphasized that the need to sink substantial costs upon entry is not an entry barrier only in itself. Rather, sunk costs permit a

---

[9] However, it may be noted as an aside that, if used in a sufficiently aggressive manner by a potential entrant, sunk costs can, curiously, become a means to overcome other barriers to entry. The entrant who deliberately incurs substantial sunk costs, perhaps substituting them for variable costs to a greater degree than would otherwise be most profitable, may thereby make it far more difficult for the incumbent to dislodge him. The entrant, in effect, chooses to burn his bridges so that he is left with far less to lose by remaining in the field. Thereafter, the incumbent will have to realize that the entrant will *not* leave the market unless revenues are driven below his (reduced) total variable costs. Whether a rational entrant will or will not decide to burn his bridges in this way is, of course, not a fortuitous matter, but depends on the perceived trade-off between sunk and variable costs (which, in turn, depends on the nature of the available productive techniques) and, perhaps, on the entrant's subjective estimate of the way in which the incumbent is likely to react. If the bridge-burning strategy is not optimal for the potential entrant, sunk costs remain a barrier to entry. Otherwise, they can no longer be interpreted that way.

[10] In the extreme case, competition for such futures contracts entails bidding for long-term and exclusive natural monopoly franchises, as suggested by Demsetz (1968). However, see Williamson (1976) on the problems inherent in such an arrangement, and how they are best alleviated by well-functioning secondary markets. In our terminology, such secondary markets reduce the extent to which fixed costs are sunk.

wide variety of other influences to affect and increase the entry barriers. In particular, for a given positive value of sunk cost, the associated barrier will be higher the greater the economies of scale, as Bain suggested. It may be observed that virtually the entire literature on entry deterrence and entry barriers has implicitly assumed the presence of substantial sunk costs in its analysis.[11] In contrast, our purpose here is to distinguish sunk from fixed costs, and to bring out their vastly different implications.

10F. Contestable Markets with Fixed Costs

We conclude that sunk costs, unlike fixed costs, can constitute a barrier to entry. In particular, we argue now that fixed costs need not have any detrimental welfare consequences, *unless they also happen to be sunk.* In an industry whose firms use only capital on wheels (or wings), some or all of that capital may be fixed, but it is not sunk. This means that in the absence of other entry barriers, natural or artificial, an incumbent, even if he can threaten retaliation after entry, dare not offer profit-making opportunities to potential entrants because an entering firm can hit and run, gathering in the available profits and departing when the going gets rough. Such a situation fits our definition of a contestable market, that is, a market vulnerable to costlessly reversible entry, even when it is currently occupied by an oligopoly or a monopoly. The contestable market is a generalization of the case of pure competition, and it offers many of the same benefits. Even if it is run by a monopoly, a contestable market will yield only zero profits[12] and offer inducements for the adoption of Ramsey-optimal prices; in addition, it will enforce efficiency of production, the adoption of new improved techniques (as they become available), and avoidance of cross subsidy in pricing.[13]

This resolves the apparent contradiction between our conclusion that fixed costs of sufficient magnitude permit the incumbent to adopt entry-preventing prices (sustainable prices) and the preceding assertion that, in themselves, they constitute no barrier to entry. The availability of sustainable prices *does* permit the incumbent to preclude entry. But he can do so *only* by offering the public the very same benefits that actual competition would otherwise have brought with it. With entry barriers, supernormal profits,

---

[11] The recent works of Eaton and Lipsey (1980, 1981) constitute an important exception. They too recognize that the durability of capital, which affects its "sunkenness," can be responsible for barriers to entry and that this concept is quite distinct from the "natural monopoly" issue.

[12] If the monopoly also sells some of its products in markets that are not contestable, then the firm clearly need not earn zero profits on its operations overall. However, it will earn only "normal" incremental profits on its incremental investment used to produce the outputs sold in its contestable markets.

[13] See Chapters 8 and 12.

inefficiencies, cross subsidies, and nonoptimal prices all become possible. But in a contestable market, which is perfectly consistent with the presence of fixed costs that are not sunk, matters change drastically, and government intervention can contribute far less, if anything, to the general welfare.

Proposition 10F1   *In contestable markets, fixed costs do not impede most of the efficiency properties normally associated with pure competition.*

10G. Erection of Entry Barriers and Their Social Costs

Incumbent firms can generally increase the barriers to entry into their industry and they have a motive for doing so. But they can accomplish this only at a cost, either in the form of a direct outlay or of an opportunity cost. For example, consider the incumbent firm's advertising budget, a. Let $E(a)$ represent the cost of entry, a_k represent what would be the profit-maximizing advertising budget for the incumbent if E were constant, and a_e represent the most profitable advertising budget that takes account of dE/da, the effect of increased advertising on entry cost. Then $a_e - a_k$ may be taken as the incremental cost of the incumbent's optimal advertising-entry barrier.

Letting $G(E)$ be the cost to an incumbent of erecting an entry barrier equal to E, the incumbent's maximal net profit will be given by

$$\pi_n = \max\{E - G(E)\,|\,E \le \pi_{g\,\max}\}$$

where $\pi_{g\,\max}$ is the maximum gross profit permitted by the industry's revenue and cost functions in the absence of any threat of entry. Let E^* be the unique value of E which maximizes $E - G(E)$. Then if $E^* < \pi_{g\,\max}$, E^* will be the net profit-maximizing height of entry barrier for the incumbent. In many cases we may expect $E^* > E_0$, where E_0 is the autonomous entry barrier that would characterize the industry in the absence of any barrier-erection activity by the incumbent.

Now we have seen that entry barriers may permit inefficiency in the firm's (and the industry's) operations, and they can undermine the other advantages offered by a contestable market. Thus, while the rise in the value of E from E_0 to E^* serves the purposes of the firm, it imposes a social cost. We can say somewhat more about that social cost.

It will be recalled that sustainable price vectors need not all be Ramsey-optimal. However, we show in Chapter 8 that, where both types of sustainable prices are available, there are influences that may well drive the incumbent to select the Ramsey prices, which can be done with no sacrifice in profit. Accordingly, let us assume that the incumbent is a natural monopolist who does select a Ramsey price vector as a set of prices that is sustainable against entry. As the magnitude of the entry barrier increases, so long as $E \le \pi_{g\,\max}$, the incumbent's total gross profit will increase equally. This means that the Ramsey prices will change correspondingly. As E rises higher and higher, the Ramsey solution will represent the maximization of producer's and consumers' surplus under an increasingly tighter profit constraint. Thus, total surplus must be a monotonically nonincreasing function of E and, presumably, will generally decrease with E. It follows from the argument that since total surplus may decrease and certainly does not increase with E, while the producer's surplus, as given by gross profit, must rise with E so long as $E \leq E^*$, consumers' surplus must, *a fortiori*, fall with E over the same range of values. Thus, unless $E^* \leq E_0$, the resulting change in the Ramsey solution will result in a clear loss in consumers' surplus and, very likely, a loss in the sum of consumers' plus producer's surplus. The expenditures on entry-barrier erection may be an additional source of social loss or gain.[14]

We can deal more explicitly with the relation between the magnitude of E and the behavior of consumers' surplus, on the assumption that the incumbent selects the Ramsey prices. Write $CS(p)$ for consumers' surplus expressed as a function of the price vector p. Then the Ramsey prices are those which

$$\text{maximize } CS(p) + \pi_g(p)$$

subject to the requirement that profit net of rent be nonnegative; that is,

$$\pi_g(p) - E \geq 0.$$

By the envelope theorem this requires

$$d[CS(p^*(E)) + \pi_g(p^*(E))]/dE = -\lambda$$

where λ is the Lagrange multiplier and $p^*(E)$ is the optimal price vector when entry cost is E.

If the constraint is not binding, $\lambda = 0$ and p^* is composed of the "first-best" prices, each price equal to its corresponding marginal cost. Of course, this case is ruled out if $E \geq 0$, returns to scale are increasing, and no demand is perfectly inelastic.[15] Then $\lambda > 0$, $d[CS + \pi_g]/dE < 0$, and $\pi_g(p^*(E)) = E$. Consequently, $d\pi_g/dE = 1$, and so

$$dCS/dE = -\lambda - 1.$$

---

[14] In the studies of entry deterrence cited earlier, such barrier-raising activities as advertising, capacity investment, research and development have been shown to be capable of producing various ancillary effects on the incumbent's incentives with respect to price and output decisions.

[15] See Chapter 3.

This relationship has the following general characteristics: For E sufficiently small (perhaps negative), the p^* are marginal cost prices whose values do not change with E. When E reaches the value at which marginal cost prices exactly satisfy the profit constraint, we obtain $\lambda = 0$, so that $dCS/dE = -1$ for small increases in E from that level. The curve depicting the relation between CS and E will be decreasing. It will not necessarily be concave at intermediate values of E. However, as E approaches $\pi_{g\,max}$, it is easy to see that $dCS/dE \to -\infty$. This follows because at the prices that yield $\pi_{g\,max}$ we must have $\partial \pi_g/\partial p_i = 0$ for all i. But, the first-order Ramsey conditions are, as usual,

$$-y_i + (\lambda + 1)\,\partial \pi_g/\partial p_i = 0.$$

Thus, at such a value of E, if some output quantities y_i remain bounded above zero, then $(\lambda + 1) = -dCS/dE$ must be infinite.

At an intermediate value of E the slope of the $CS(E)$ curve can be evaluated most easily in the case in which the cross elasticities of demand are zero so that the inverse elasticity form of the Ramsey rule applies. Then

$$\left(\frac{p_i - mc_i}{p_i}\right) e_i = \alpha = \lambda/(\lambda + 1) < 1$$

where e_i is the own-price elasticity of demand and mc_i the marginal production cost of good i. Combining this with our expression for the slope of the CS curve we obtain[16]

$$dCS(p^*(E))/dE = -(\lambda + 1) = -1/(1 - \alpha).$$

This is larger in absolute value, the larger is α and the greater the price–cost margins and the elasticities of demand at the relevant[17] $p^*(E)$.

Perhaps the most significant implication of the preceding calculations for the industry structures under consideration is the conclusion that damage to consumers must ultimately accelerate as E increases, and its rate of increase must ultimately approach infinity. This implies that society derives some protection from the fact that the erection of entry barriers is costly to the incumbent, for this cost will discourage the imposition of high barriers and may therefore tend to keep the value of E below the range in which it

---

[16] Empirical estimates of the values of α are available for several industries (see Willig and Bailey, 1979) and these numbers may be suggestive. They range between $\alpha = 0.1$ and $\alpha = 0.2$ This implies that a $\Delta E = \$1$ will cause a welfare loss of about 10–25 cents and if profits rise equally with E, a loss to consumers amounting to $\$1.10$ to $\$1.25$.

[17] One must be careful in the formulation of such a statement, since as E varies, price–cost margins and the elasticities will also vary. The value of α, of course, gives us information strictly applicable only to the point at which it is observed.

becomes most damaging. It also suggests that it may be good strategy for the antitrust agencies to concentrate their resources on the toppling of the most serious entry barriers rather than diffusing those resources in simultaneous attacks on a larger number of barriers of smaller average magnitude.

10H. Toward a General Model of Entry Barriers

Our goal in this section is to construct a general model of incumbent–entrant interaction that makes precise the critical role of sunk costs in the process of entry deterrence when the incumbent's response to the establishment of a new competitor is not prespecified. We show that the absence of sunk costs is necessary for the beneficial results of contestability when incumbents are not restricted in their responses to entry. The model of this section is general in the sense that it enables us to establish the validity of the preceding propositions relating sunk costs, entry costs, and welfare *without* any restrictions on the nature of post-entry market equilibrium, that is, on the nature of the incumbent's countermoves after entry occurs. As has already been observed, such restrictions play an important role in earlier entry deterrence models, notably in those constructed by Bain (1956), Spence (1977), Salop (1979b), and Dixit (1980).

We do not seek to provide an exhaustive analysis of our model. Clearly, additional results could be derived from it if we were to assume more about the nature of the post-entry equilibrium. However, our more limited objective here is to provide a formal characterization of the difference between the cases in which sunk costs are present and those in which they are absent, and to derive an entry cost function from the magnitude of those sunk costs.

Since entry must be viewed as an intertemporal process, our general model is given at least a modicum of dynamic structure. Specifically, we divide time into three periods: (1) the Past, which lasts until until time zero; (2) a "disequilibrium" period of length τ; and (3) the Future, which starts at time τ, the beginning of period 1. The disequilibrium period represents an interval during which the incumbent, for lack of information, because of regulation, or for other reasons, is unable to adjust his posted prices in response to any entry that happens to occur.

At time zero the incumbent is taken to have K_i^0 units of capital and an associated cost function $V^i(y_i, K_i^0)$, where y_i is the output flow. (Here, "i" refers to the incumbent.) This gives the sum of the production costs that are fully variable during the disequilibrium period, expressed in instantaneous terms.

At time zero, the potential entrant is assumed to have available a production process represented by its variable cost function $V^e(y_e, K_e^0)$, where y_e is the output flow, and "e" refers to the potential entrant. It can buy capital at the price β_e^0 per unit. Thus, in common parlance, an entrant coming into the market with K_e^0 units of capital is saddled with $\beta_e^0 K_e^0$ dollars in capitalization.[18] However, this investment may only be partially sunk, since we assume that at the end of the disequilibrium period (the beginning of period 1) the entrant may, if he chooses, "scrap" his capital for a salvage price α_e^1 per unit. (The incumbent is assumed to have a similar disinvestment option.) In the absence of inflation, the assumption $0 \leq \alpha_e^1 \leq \beta_e^0$ is clearly plausible. If $\alpha_e^1 = 0$, all capital costs are sunk, whereas if $\alpha_e^1 = \beta_e^0$ no costs are sunk since all investments can be reversed fully at time 1. (We assume for simplicity, that *physical* deterioration of capital during the disequilibrium period is negligible.) The magnitude of the capitalization chosen by an entrant can be determined endogenously from the properties of V^e and the proportion of this outlay $(1 - \alpha_e^1)/\beta_e^0$, which is sunk.

While we adopt no restrictions on detailed characteristics of the post-entry equilibrium, it may plausibly be assumed that the present values (discounted to the beginning of period 1) of future profits of the incumbent and actual entrant are functions of the *state variables* of the system at the time period 1 begins. Thus, future profits for the entrant and the incumbent can be expressed as $\pi_e^f(K_e^0, K_i^0)$ and $\pi_i^f(K_e^0, K_i^0)$, respectively. While the precise forms of these functions depend upon the behavioral assumptions of the appropriate model of future market rivalry, generally applicable lower bounds are yielded by the fact that either firm always has the option of selling its plant for its salvage value at the beginning of period 1. Thus, we have

(10H1) $$\pi_e^f(K_e^0, K_i^0) \geq \alpha_e^1 K_e^0$$

(10H2) $$\pi_i^f(K_e^0, K_i^0) \geq \alpha_i^1 K_i^0.$$

We are now in a position to analyze the entry decision at time zero. Following our earlier discussion, an entry plan is defined as a price vector $p_e^0 \leq p_i^0$ for the products offered by the entrant and an instantaneous flow of outputs $y_e^0 \leq Q(p_e^0)$, where Q is the instantaneous market demand function. Entry will occur if and only if the potential entrant concludes that the total profit π_e^T yielded by the best of these plans is positive. That is, entry will occur if and only if the sum of entry period and future profits is greater than zero. Formally,

(10H3) $$\pi_e^T \equiv \max_{p_e^0, y_e^0, K_e^0} \{ \gamma_\tau [p_e^0 \cdot y_e^0 - V^e(y_e^0, K_e^0)] - \beta_e^0 K_e^0$$
$$+ \pi_e^f(K_e^0, K_i^0)e^{-r\tau} \} > 0,$$

---

[18] Obviously, such capitalization may or may not be a fixed cost in the long run. It is fixed if it is the minimum total cost of doing business at the smallest nonzero output level.

where $\gamma_\tau = \int_0^\tau e^{-rt} \, dt = (1 - e^{-r\tau})/r$, and where r is the discount rate. However, using the lower bound (10H1), we have immediately,

$$(10H4) \qquad \pi_e^T \geq \max_{p_e^0, y_e^0, K_e^0} \gamma_\tau [p_e^0 \cdot y_e^0 - V^e(y_e^0, K_e^0) - \rho_e^0 K_e^0],$$

where ρ_e^0, the effective rental cost of capital to the potential entrant, is given by

$$(10H5) \qquad \rho_e^0 = [(\beta_e^0 - \alpha_e^1 e^{-r\tau})/\gamma_\tau].$$

This enables us to rewrite (10H4), for any given planned prices and outputs, in terms of an instantaneous cost function that is fully minimized for the given output vector.

$$(10H6) \qquad \pi_e^T \geq \gamma_\tau [p_e^0 \cdot y_e^0 - C^e(y_e^0, \rho_e^0)]$$

Here

$$C^e(y, \rho) \equiv \min_K [V^e(y, K) + \rho K].$$

Equation (10H6) is readily interpreted. It tells us that the potential total profit is bounded from *below* by what the potential entrant can earn during the disequilibrium period alone by disposing of all assets at the end of that period. If that value is positive, entry will occur. For the interesting case in which the production techniques are freely available to everyone, that is, where $V^i(\cdot) = V^e(\cdot)$, $C^e(\cdot, \cdot) = C^i(\cdot, \cdot) \equiv C(\cdot, \cdot)$, (10H6) and (10H3) together have the usual sustainability interpretation that entry will occur if the revenues resulting from an entry plan exceed the costs as calculated from the incumbent's and entrant's common cost function. That is, a sufficient condition for entry to occur is

$$p_e^0 \cdot y_e^0 - C(y_e^0, \rho_e^0) > 0.$$

Therefore, a *necessary* condition for the incumbent's monopoly to be sustainable (i.e., for *no* entry to occur) is

$$(10H7) \qquad p_e^0 \cdot y_e^0 - C(y_e^0, \rho_e^0) \leq 0,$$

for *all* feasible entry plans (p_e^0, y_e^0).

If all firms face the same factor prices ($\beta_e^0 = \beta_i^0 = \beta$), the only difference between (10H7) and the usual definition of sustainability is that the effective rental rate of capital for the entrant as given by (10H5) may exceed that of the incumbent, which equals $r\beta$, the interest per unit of invested capital. Any such difference in rental rates must be attributed to the possibility that the entrant may find himself forced to depreciate his capital fully during the disequilibrium period. That is, with $\beta_e^0 = \beta_i^0 = \beta$,

$$\rho_e^0 - \beta r = \frac{re^{-r\tau}(\beta - \alpha_e^1)}{1 - e^{-r\tau}} = \frac{r(\beta - \alpha_e^1)}{e^{r\tau} - 1}$$

which is positive as long as $\beta > \alpha_e^1$, that is, as long as any costs are sunk. This possibility arises because the next period's response of the incumbent is not preset and this exposes the entrant to the risk that he will be driven out of business in the immediate future, that is, at the beginning of period 1.

Despite this possibility, however, one of our key observations is

Proposition 10H1 *If all the entrant's capital equipment can be resold without loss* ($\alpha_e^1 = \beta$) *so that no costs are sunk, then by equation (10H5)* $\rho_e^0 = r\beta$; *and so if all order costs are otherwise equal, the potential entrant is faced with no cost disadvantage vis-à-vis the incumbent. That is, no barrier to entry exists, regardless of the (optimal) capitalization* $\beta_e^0 K_e^0$ *that the entrant chooses to incur. And this is true even if the technology entails substantial fixed costs, in that production may require* $\beta_e^0 K_e^0 \geq F > 0$.

Thus, this model provides the instrument for a formal proof of the necessity of sunk costs for the presence of entry barriers under the conditions postulated. Similarly, when $\alpha_e^1 < \beta$, so that some costs are effectively sunk, $\rho_e^0 > r\beta$ and the incumbent effectively enjoys a cost advantage over a potential entrant who may be led to scrap his plant during period 1. Thus, with the aid of our model, we have formally derived Proposition 10E1, the sufficiency of sunk costs, under the conditions postulated for the presence of entry barriers.

Our model also permits an analytic derivation of an entry cost function $E(y)$, used in earlier discussions. Consider a single-product monopoly whose operations just make impossible the violation of the necessary condition, (10H7), for sustainability. In particular

(10H8) $$p_i^0 = C(Q(p_i^0), \rho_e^0)/Q(p_i^0).$$

Figure 10H1 depicts, in instantaneous terms, the profits thereby earned by the incumbent because of the fact that his effective rental rate for capital, $r\beta$, is less than that of the entrant, ρ_e^0. The instantaneous rate of excess profit is the shaded area.

To consider a simple example, suppose that $C(y, \rho) = cy + \tilde{F}\rho$, for $y > 0$. Then price p_i^0 just meets the necessary condition for sustainability if

$$(p_i^0 - c)Q(p_i^0) = \tilde{F}\rho_e^0 = r\tilde{F}(\beta + (\beta - \alpha_e^1)/(e^{r\tau} - 1)).$$

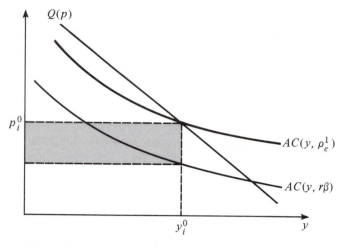

FIGURE 10H1

The associated instantaneous rate of excess profit is

(10H9) $$(p_i^0 - c)Q(p_i^0) - r\beta\tilde{F} = r\tilde{F}(\beta - \alpha_e^1)/(e^{r\tau} - 1).$$

If p_i^0 were stationary, as would be the case if the incumbent had to guard against possible entry at each instant of time in a stationary environment, then it is possible to calculate, by integrating (10H9), this upper bound on the discounted present value of the incumbent's excess profit:

(10H10) $$\tilde{F}(\beta - \alpha_e^1)/(e^{r\tau} - 1).$$

In this example, (10H10) can be interpreted as the entry cost function that faces an entrant who expects to operate only until period 1 begins. Here, this is an upper bound on the true entry cost function, because entrants may expect to engage in activities more profitable then salvage during period 1. Also, (10H10) only provides an upper bound on the incumbent's profit because the full (unimpeded) monopoly profit may be less. Nevertheless, we can conclude

Proposition 10H2 *In the absence of other cost disadvantages to the entrant, entry costs are zero if either (a) there are no sunk costs ($\beta = \alpha_e^1$), or (b) the monopolist is* never *permitted to respond, that is, $\tau \to \infty$.*

The logic of Condition (b) becomes intuitively clear when we recall that the difference between the (effective) instantaneous cost curve of the incumbent and that of the potential entrant is the result of the entrant's inability to foresee with confidence the market conditions he will face once the incumbent has had a chance to respond. It is for this reason that the entrant must, in this model (implicitly), use a rental rate higher than the incumbent's, $\rho_e^0 > r\beta$, for his capital expenditures. Consequently, if the incumbent is never permitted to respond, as may perhaps be true in some regulatory situations, this risk premium effectively disappears. Of course, if physical deterioration of capital were included in the model, then the condition $\tau \to \infty$ could be replaced by τ greater than or equal to the physical life of the entrant's capital.

Our argument requires only one additional step toward generality. The entry cost function in (10H10) was taken to be independent of the size of the output envisioned by the potential entrant. This was a consequence of the special form of $C(\cdot, \cdot)$, which was taken to permit no choice in the magnitude of capital costs. But our discussion remains valid when, instead of (10H10), we employ the relationship

(10H11) $$E(y) \leq [(\beta - \alpha_e^1)/(e^{rt} - 1)]K^*(y, \rho_e^0),$$

where K^* denotes the optimal size of capital expenditure, given $V(\cdot)$, y, and ρ_e^0. In particular, the sufficient conditions for nonpositive entry costs of Proposition 10H2 still remain valid under this more general formulation.

To sum up briefly, this section provides a rather general model of the entry process which enables us to *derive* an economically significant entry cost function, *and* which yields the zero-profit, price-taking characteristics of the equilibrium conditions employed throughout our analysis as a readily interpretable limiting case.

Most important (combining Propositions 10F1 and 10H2), it has led us to one of the basic conclusions of this volume:

Proposition 10H3 *Where no other entry barriers and frictions are present, either the absence of sunk costs or the prevention of post-entry responses by incumbents are (individually) sufficient for markets to be contestable.*

Thus, among the key instruments of public policy that emerge from these discussions are (a) measures that help to reduce the share of sunk costs in an industry by making it easier to transfer or sell assets, and (b) measures that restrict *ex post* responses to entry by incumbents. More will be said about these when we return in later chapters to more extended discussions of policy.

10I. Fixed Costs and Public Goods

Pure public goods can be interpreted as a limiting case of a fixed cost relationship, although this does not generally seem to be recognized. A pure public good is usually defined to satisfy what has been called the *supply-jointness* or *nondepletability* condition: That is, at least over a substantial range, once the good is supplied to anyone, the cost of making its benefits available to additional persons is zero (or relatively small). The difficulty for resource allocation posed by such supply-jointness is that the bulk of total cost is fixed, so that if the good is supplied at a price equal to its marginal cost, then its sale will incur a loss. But that is precisely the problem that plagues the optimal allocation of *any* good whose production involves economies of scale. Thus, public goods are simply a class of goods whose production has a large fixed cost component and for which there is a comparatively low marginal cost (in the sense of the cost of serving another customer rather than the cost of providing another unit of physical product).

It may also be noted that public goods and any other goods that incur large fixed costs have the same sort of Ramsey-optimal prices. For it is easy to show that if *all* prices are really to be Pareto-optimal subject to a zero-loss budget constraint, then the prices of public goods *must not be* zero as commonly asserted, but must follow exactly the same pattern and obey the same rules as do those of ordinary fixed-cost products (see Baumol and Ordover, 1977). To summarize:

Proposition 10I1   *There is no essential difference between the case of fixed costs and the case of public goods.*

One may well wonder, consequently, why there seem to be so many products with heavy fixed costs in private industry. After all, we tend to think of public goods as goods whose production should not be, and generally is not, left to the private sector. Yet, it would seem that the automobile industry, aircraft production, steel manufacturing, and many other industries do have large fixed costs, which would appear to make them equivalent to public-good producers. The confusion here is that the pertinent costs in these examples are largely sunk, not fixed, and it is this that distinguishes them from the public-goods case. This is yet another example of the significance of the distinction between fixed and sunk costs.

All of this suggests a somewhat novel policy conclusion. Suppose there were a public good with exclusion from its consumption possible and easy, and suppose its production involved little sunk cost.[19] Then the supply of the item might well be entrusted to private enterprise, for a contestable market should be just as effective in protecting the public interest in the production of this good as it is in the supply of an "ordinary" item involving fixed costs.[20]

---

[19] Police protection of property, postal delivery, broadcast programs, and aerial traffic watch may be examples of public goods with low sunk costs and excludability.

[20] Demsetz (1970) has also concluded that private producers may be able to provide excludable public goods efficiently. His argument rests on competition among active suppliers of substitute public goods. The critical difference between our position and his is that we limit our conclusion to contestable markets.

10J. Recapitulation

This chapter shows that sunk costs constitute entry barriers which, as is well known, can give rise to monopoly profits, resource misallocation, and inefficiencies. On the other hand, fixed costs (which are shown to be analytically indistinguishable from the costs incurred by public goods) do not constitute barriers to entry and do not entail the misallocation problems to which entry barriers lead. It is true that, as shown here, fixed costs of sufficient magnitude do guarantee that an industry will be a natural monopoly in the sense of subadditivity of costs. In addition, such a monopoly will have available to it a vector of equilibrium prices (prices sustainable against entry). But where fixed costs are not sunk and there are no artificial entry barriers, the market will be contestable. Profits will consequently be driven toward zero, efficiency will be enforced, cross subsidy will be ruled out, and prices can protect the firm from entry. In sum, the analysis confirms that fixed costs pave the way for monopoly; but for the monopoly which operates in a contestable market the analysis shows also that potential entry can control undesirable behavior.

APPENDIX

The Basic Theorem on Fixed Costs

The theorem stated and proved below requires that the market demand function facing the enterprise satisfy the following conditions:

(i) The market demand function $Q(p)$ is differentiable, with $\partial Q^i(p)/\partial p_i < 0$.

(ii) Demands are weak gross substitutes; that is, $Q^i(p') \geq Q^i(p)$ if $p'_i = p_i$ and $p'_j \geq p_j$ for $j \neq i$.

(iii) Demands are Sandberg-normal; that is $[Q^i(p) - Q^i(p')][p_i - p'_i] < 0$ for some i whenever $p \neq p'$. It follows that $Q(\cdot)$ has an inverse $P(\cdot)$.

(iv) Revenues from all goods fall to zero as quantities demanded expand indefinitely in any fixed proportions; that is, $\lim_{t \to \infty} ty \cdot P(ty) = 0$.

Theorem *Assume that conditions (i)–(iv) hold, and let $A(y)$ be any affine cost function with positive fixed costs: $A(y) \equiv F + a \cdot y$, $F > 0$, $a \geq 0$. Then the enterprise is sustainable if its cost function resembles $A(y)$ sufficiently closely. That is, there exists a price vector p^m such that for any continuously differentiable variable cost function $V(y)$, with $V(Q(p^m)) = A(Q(p^m))$ and with positive marginal costs, there is a k^*, $0 < k^* < 1$, such that the enterprise is sustainable at prices p^m if its cost function has the form*

$$(1) \qquad C(y) = kA(y) + (1 - k)V(y), \qquad \text{for } 1 \geq k \geq k^*.$$

Further, any price vector p^m has this property if it fulfills the following conditions:

$$(v) \qquad p^m y^m = A(y^m), \qquad \text{where } y^m = Q(p^m)$$

$$(vi) \qquad p^m > a$$

$$(vii) \qquad pQ(p) < A(Q(p)) \qquad \text{for } p \leq p^m$$

$$(viii) \qquad \partial \pi_A(p^m)/\partial p_i > 0, \qquad \text{for all } i, \text{ where } \pi_A(p) \equiv pQ(p) - A(Q(p)).$$

Finally, the prices that would be Ramsey-optimal for an enterprise with the cost function $A(y)$ fulfill conditions (v)–$(viii)$ above. Thus, such Ramsey prices are sustainable for an enterprise with a cost function that resembles $A(y)$ sufficiently closely.

Proof Before proceeding to the proof that conditions (i)–(viii) imply that p^m is a sustainable price vector for an enterprise with the cost function given in (1), we establish this preliminary result:

Lemma   *Given conditions (i)–(iv), the Ramsey prices for an enterprise with the cost function $A(y)$ satisfy conditions (v)–$(viii)$.*

Proof of Lemma By definition, these Ramsey prices p^* maximize $W(p)$, subject to $\pi_A(p) \geq 0$, where $W(p)$ is a welfare function with the property that its gradient is positively proportional to that of the sum of consumers' and producers' surplus: $\partial W/\partial p_i = \gamma[-Q^i + \partial \pi_A/\partial p_i]$, $\gamma > 0$. Such a maximum exists because in view of (iv), $\{p \geq 0 \,|\, \pi_A(p) \geq 0\}$ is compact. Then p^* satisfies the Kuhn-Tucker necessary conditions,

$$(2) \qquad\qquad 0 = \partial W/\partial p_i + \lambda \partial \pi_A/\partial p_i,$$

for all i, $\lambda\pi_A = 0$, $\lambda \geq 0$, and $\pi_A \geq 0$. It follows immediately that, at p^*,

$$(3) \qquad\qquad \partial \pi_A/\partial p_i = \gamma Q^i/(\lambda + \gamma) > 0,$$

for all i, so that (viii) is satisfied. Differentiation of π_A yields these rearrangements of the system of equations in (3):

$$(4) \qquad\qquad \left[\frac{\partial Q^i}{\partial p_j}\right][p_i^* - a_i] = \frac{-\lambda}{(\lambda + \gamma)}[Q^i]$$

$$(5) \qquad\qquad [p_i^* - a_i] = \frac{-\lambda}{(\lambda + \gamma)}\left[\frac{\partial Q^i}{\partial p_j}\right]^{-1}[Q^i].$$

Here, the heavy brackets denote matrices and column vectors. The inverse of the Jacobian of the mapping from prices to quantities that appears in (5) exists because of (i) and (ii). Further, in Chapter 8 it is shown to be a nonpositive matrix with negative diagonal elements under conditions (i)–(iii). Consequently, if λ were positive, (5) would imply that p^* satisfied (vi) and, moreover, (2) would imply that p^* satisfied (v). If, instead, λ were zero, then $p^* = a$ by (5). However, $\pi_A(a) = -F < 0$, so that such p^* would not satisfy the nonnegative profit constraint, and $\lambda = 0$ and a nonbinding profit constraint are thereby ruled out. Finally, if (vii) were not satisfied by p^*, with $\pi_A(p^*) = 0$, there would exist lower prices for which π_A were nonnegative. The existence of such prices would contradict the Ramsey optimality of p^*.

Q.E.D.

We first sketch the proof of the result that (i)–(viii) imply that p^m is sustainable with the cost function given by (1) and then proceed to fill in the necessary details. A potential entrant's marketing plan consists of prices p^e and quantities y^e such that $p^e \leq p^m$, $y^e \neq 0$, $y^e \neq y^m$, and $y^e \leq Q(p^e)$. Let us call the set of such marketing plans $\tilde{M}$. With restrictions on k^*, we are able to narrow $\tilde{M}$ to a compact set M which is bounded away from $y^e = 0$ and $y^e = y^m$, by eliminating some market plans that are assured to be unprofitable.

In view of (1), an entrant's profit can be expressed as a convex combination of the profits that would be earned by entrants with cost functions $A(y)$ and $V(y)$. That is,

$$(6) \qquad p^e y^e - C(y^e) = k[p^e y^e - A(y^e)] + (1-k)[p^e y^e - V(y^e)].$$

Let

$$\pi^A = \max_M \left[p^e y^e - A(y^e) \right]$$

and

$$\pi^V = \max_M \left[p^e y^e - V(y^e) \right].$$

These maxima exist because M is compact and because the functions are continuous over M. Then, it follows that

$$(7) \qquad \max_M \left[p^e y^e - C(y^e) \right] \leq k\pi^A + (1-k)\pi^V.$$

π^V may be positive, but it will be shown below that $\pi^A < 0$. If $\pi^V < 0$ were satisfied, it would follow from (7) and $\pi^A < 0$ that $p^e y^e - C^k(y^e) < 0$ over M, so that the enterprise would be sustainable without further restriction of k^*. If, instead, $\pi^V > 0$, set

$$(8) \qquad k_1^* = \pi^V / (\pi^V - \pi^A).$$

With $\pi^V > 0$ and $\pi^A < 0$, it follows that $0 < k_1^* < 1$. Then, $1 \geq k > k_1^*$ implies that $k\pi^A + (1-k)\pi^V < k_1^* \pi^A + (1-k_1^*)\pi^V = 0$. Consequently, from (7), $p^e y^e - C(y^e) < 0$ over M for all such k, and thus the enterprise would be proved to be sustainable at prices p^m for some value of $k^* < 1$.

To fill in the details of this summary of the proof, it is necessary to show that the set $\tilde{M}$ can be narrowed to M and that $\pi^A < 0$. $\tilde{M} \equiv \{p^e, y^e \,|\, 0 \leq p^e \leq p^m, y^e \geq 0, y^e \neq y^m$ and $y^e \leq Q(p^e)\}$. $\tilde{M}$ is to be narrowed in several ways.

First, we find a bounded set in which all potentially profitable entry plans must lie. Lemma 4 of Appendix III to Chapter 8 shows that under conditions (i)–(iii), if $(p^e, y^e) \in \tilde{M}$ and $p^e y^e - A(y^e) \geq 0$, then

$$y^e \in Y^A = \{y \,|\, P(y) \cdot y - A(y) \geq 0\}.$$

Similarly, $(p^e, y^e) \in \tilde{M}$ and $p^e y^e - V(y^e) \geq 0$ imply

$$y^e \in Y^V = \{y \,|\, P(y) \cdot y - V(y) \geq 0\}.$$

If $p^e y^e - C(y^e) \geq 0$, for $0 \leq k \leq 1$, then either $p^e y^e - A(y^e) \geq 0$ or $p^e y^e - V(y^e) \geq 0$ or both, must hold. Thus, if $p^e y^e - C(y^e) \geq 0$ for $(p^e, y^e) \in \tilde{M}$, it must follow that $y^e \in Y^A \cup Y^V$. And, by (iv), this is a compact set.

Second, note that over $\tilde{M}$, $p^e y^e - C(y^e) \leq p^m y^e - C(y^e) \leq p^m y^e - kF$. Hence, a necessary condition for $p^e y^e - C(y^e) \geq 0$ is

$$p^m \cdot y^e \geq kF \geq k^*F \geq 0.01F > 0,$$

for $k^* \geq 0.01$. Therefore, restricting y^e in this way causes no loss of generality. This restriction, made possible by the positive fixed cost, permits M to be bounded away from plans with $y^e = 0$.

Third, we construct an open neighborhood N of (p^m, y^m) which can be eliminated from consideration because all marketing plans in it would yield an entrant negative profit. Choose k^* close enough to 1 so that at p^m, for all i, and for $k^* \leq k \leq 1, 0 < \partial[p \cdot Q(p) - (kA(Q(p)) + (1 - k)V(Q(p)))]/\partial p_i = k\partial \pi_A/\partial p_i + (1 - k)\partial[p \cdot Q(p) - V(Q(p))]/\partial p_i$. This can be accomplished by some value, $k_2^* < 1$, because of (viii). Then, for $k \geq k_2^*$ at p^m, $\partial[pQ(p) - C(Q(p))]/\partial p_i > 0$. Consequently, there is some open neighborhood of p^m in which $p \leq p^m$ implies

(9) $$pQ(p) - C[Q(p)] < p^m Q(p^m) - C[Q(p^m)] = 0.$$

Denote by $\tilde{N}_y$ the image of this neighborhood under the demand function.

At $Q(p^m)$, $\partial C/\partial y_i = ka_i + (1 - k)\partial V/\partial y_i$, so that, because of (vi), there is a $k_3^* < 1$ such that $\nabla C(y^m) < p^m$ for $k \geq k_3^*$. Then, since $C(y)$ is continuously differentiable around y^m, (p^m, y^m) is in the interior of $\{p, y \,|\, \nabla C(y) < p\}$. Consequently, and in view of the continuous invertability of $Q(\cdot)$, there is an open rectangular neighborhood of y^m, $N_y \subset \tilde{N}_y$, which is bounded away from $y = 0$ and which has the property that

(10) $$N \equiv P[N_y] \times N_y \subset \{p, y \,|\, \nabla C(y) < p\}.$$

We now show that $p^e y^e - C(y^e) < 0$ for all $(p^e, y^e) \in N \cap \tilde{M}$. Suppose, otherwise, that there is a $(p^e, y^e) \in N \cap \tilde{M}$ with $p^e y^e - C(y^e) \geq 0$. If $y^e = Q(p^e)$, then $p^e \neq p^m$, and this contradicts (9). Therefore, $y^e \leq Q(p^e)$.

$$(11) \qquad p^e Q(p^e) - C(Q(p^e)) = p^e y^e - C(y^e) + \oint_\Gamma (p^e - \nabla C(t)) \cdot dt$$

$$\geq \oint_\Gamma (p^e - \nabla C(t)) \cdot dt,$$

where Γ is a monotone nondecreasing path from y^e to $Q(p^e)$. By the construction of N in (10), $(p^e, y^e) \in N$ implies that $(p^e, Q(p^e)) \in N$. Also by (10), the rectangular construction of N_y, and the definition of Γ as nondecreasing, we have $\{p^e\} \times \{y \mid y \text{ on } \Gamma\} \subset N$. Then, by (10), the integrand in (11) is positive and, since Γ is everywhere nondecreasing and nontrivial, the integral is positive and $p^e Q(p^e) - C(Q(p^e)) > 0$. This contradicts (v) if $p^e = p^m$ and it contradicts (9) if $p^e \neq p^m$. Consequently, we have established that for $k \geq k_2^*$ and $k \geq k_3^*$,

$$p^e y^e - C(y^e) < 0 \qquad \text{for } (p^e, y^e) \in N \cap \tilde{M},$$

so that N can be eliminated from consideration without loss of generality. At last we are ready to define M:

$$(12) \quad M = \{p^e, y^e \mid (p^e, y^e) \in \tilde{M} - N, \ y^e \in y^A \cup y^V, \quad \text{and} \quad p^m \cdot y^e \geq 0.01 F\}.$$

The preceding arguments have proved that $p^e \cdot y^e - C(y^e) < 0$ for $(p^e, y^e) \in \tilde{M} - M$, so that we may restrict our attention to M. It is now a straightforward matter to show that M is compact from (12), the definition of $\tilde{M}$, and the fact that N is open and $Y^A \cup Y^V$ is compact. As a result, both π^A and π^V, defined earlier, do exist.

Now, to prove that $\pi^A < 0$, it suffices to show that $p^e y^e - A(y^e) < 0$ for any $(p^e, y^e) \in M$. Suppose instead that $p^e y^e - a \cdot y^e - F \geq 0$ for $(p^e, y^e) \in M$. By (12), $y^e \leq Q(p^e)$. Define $\hat{p}^e$:

$$\hat{p}_i^e = p_i^e \ \text{ if } \ p_i^e > a_i \quad \text{and} \quad \hat{p}_i^e = p_i^m \ \text{ if } \ p_i^e \leq a_i.$$

Note that $\hat{p}^e \leqq p^m$. We now show that $\hat{p}^e \cdot Q(\hat{p}^e) - A(Q(\hat{p}^e)) > 0$, which is a contradiction of (v) or (vii).

Let $I = \{i \mid p_i^e > a_i\}$ and $J = \{i \mid p_i^e \leq a_i\}$. Then

$$(13) \quad [\hat{p}^e \cdot Q(\hat{p}^e) - A(Q(\hat{p}^e))] = [p^e \cdot y^e - A(y^e)] + [(p_J^m - a_J) \cdot Q_J(\hat{p}^e)]$$

$$+ [(p_I^e - a_I) \cdot (Q_I(\hat{p}^e) - y_I^e)] - [(p_J^e - a_J) y_J^e].$$

The first bracketed term on the right-hand side of (13) is nonnegative by hypothesis. Provisionally, let us assume that J is nonempty. Then, the second term is positive. The third term is nonnegative because $p_I^e > a_I$ and because

$$(14) \qquad\qquad y_I^e \leqq Q_I(p^e) \leqq Q_I(\hat{p}^e).$$

Here, the first inequality follows from $(p^e, y^e) \in \tilde{M}$, and the second from the assumption that the products are weak gross substitutes. The fourth bracketed term on the right-hand side of (13) is nonpositive because $p_J^e - a_J \leq 0$. Hence, given the provisional assumption that J is nonempty, the right-hand side of (13) is positive, which contradicts (vii).

Then, suppose J is empty, so that the second and fourth terms on the right-hand side of (13) are zero. Here $\hat{p}^e = p^e$, and, by (12), $y^e \leq Q(p^e)$. Thus, the third term on right-hand side of (13) is positive, and either (v) or (vii) is again contradicted. Thus, it has been demonstrated that $\pi^4 < 0$.

Hence, as indicated earlier, there is a $k_1^* < 1$ such that the enterprise with cost function given by (1), for $k \geq k_1^*$, is sustainable against entry plans in the set M. Finally, with $k^* = \max\{0.01, k_1^*, k_2^*, k_3^*\}$, and $1 \geq k \geq k^*$, the enterprise is sustainable at p^m against all feasible entry plans.   Q.E.D.

11

Sustainable Industry Configurations

General Industry Structures in Contestable Markets

Although sustainability is defined in Chapter 1 to apply to any industry configuration, our discussion in the intervening chapters focuses primarily upon the sustainability of the two polar cases of industry structure—natural monopoly and perfect competition. Chapter 5 does treat industry structure generally, but it does so in terms of the requirements of industry-wide efficiency, without reference to the behavioral and equilibrium issues to which sustainability refers. Thus, the time has come, in this chapter, to analyze the implications of sustainability for general multiproduct industry structures and, thereby, to explore what properties industry configurations must possess in order for them to satisfy the requirements of equilibrium in contestable markets.

We deal in this chapter with such familiar market forms as oligopoly and Chamberlinian monopolistic competition. We also encounter firms which are monopolies in the markets for some of their products but which sell other products in markets that are oligopolistic or even competitive. This is a case for which the formulation of rational regulatory policy has proved very troublesome.

In this chapter we also take a substantial step toward "solution of the oligopoly problem," that is, toward construction of a general theoretical analysis of oligopoly.

The chapter begins by investigating the requirements of sustainability in multifirm industry configurations. We demonstrate that to be in equilibrium in contestable markets, a configuration must minimize the industry's production costs of supplying the industry's outputs. We show that sustainable price and output behavior are very different in the case in which efficiency requires a commodity to be produced by a single firm and the case in which the good must be produced by at least two firms. In particular, we establish the surprising result that in the latter case all firms must price at marginal cost so that sustainable behavior even under duopoly must, in this sense, be the same as that of a perfect competitor.

Next we return to the special problems of existence of sustainable solutions under oligopoly described in Chapter 2. Where, as we just said, firms must price at marginal cost, financial feasibility requires that the outputs of firms be consistent with local constant returns in production. It is shown that a generalization of the concept of the flat-bottomed AC curve to the multiproduct case removes much of the difficulty of reconciliation of this requirement with the fortuitous position of the industry's demand hypersurface.

Next we explore the cases of monopolistic competition and of firms that sell some products in monopolistic markets and others in nonmonopolistic markets. We consider the consistency of Ramsey optimality with sustainability in such cases. We show that under oligopoly this problem is much more difficult than it is under perfect competition or even under monopoly. That is because under oligopoly it may be efficient for some firms to earn higher profits than others do, even when the industry as a whole earns zero profits. Indeed, if *each* firm is required to earn zero profit, we show the surprising result that (viable-firm) Ramsey optimality may be inconsistent with minimization of industry cost. Consequently, since sustainability requires cost minimization, the two must be incompatible in such cases.

We see that viable-firm Ramsey optimality generally requires some firms to place themselves or their patrons at a disadvantage in order to effect implicit interfirm transfers of net revenues. We argue that, as a result, optimality of this sort is inconsistent with the operation of decentralized markets under simple price systems. However, we conjecture that autarkic Ramsey optima, in which each firm maximizes only its own contribution to social real income, are consistent with equilibrium in contestable markets.

11A. Feasible and Sustainable Industry Configurations

In Chapters 2 and 5, we define and discuss the concept of a *feasible and efficient* industry configuration. Here, we begin with industry configurations that are only required to be feasible.

Definition 11A1 A *feasible industry configuration* over the set of potential products $N = \{1, \ldots, n\}$ is composed of m firms producing output vectors $y^1, \ldots, y^m \in R^n_+$, at prices $p \in R^n_{++}$, such that $\sum_{i=1}^m y^i = Q(p)$, where $Q(\cdot)$ is the vector-valued market demand function, and such that $p \cdot y^i - C(y^i) \geq 0$, for $i = 1, \ldots, m$.

Thus, the industry configuration is taken to be comprised of m firms, where m can be any positive integer, so that the industry structure is monopoly if $m = 1$, competitive if m is sufficiently large, or an oligopoly for intermediate values of m. The term "feasibility" refers to the requirements that each of the firms involved select a nonnegative output vector that permits coverage of its production costs $C(\cdot)$ at the market prices p, and that the sum of the outputs of the m firms satisfy market demands at those prices. It is especially important for this discussion that N be defined as the set of products *potentially* producible by the industry, rather than as the set of products it actually produces. Thus, in any particular industry configuration, the quantity of industry output of some of the products in N may be zero. In such cases, we define the price associated with an unproduced good to be the smallest one that chokes the market demand for it to zero. Thus, the equality of market demand and supply for unproduced goods is formally preserved.

For a feasible industry configuration to afford no opportunities for profitably entry in contestable markets, it is necessary that it be sustainable. We refer to such an arrangement as a sustainable and feasible industry configuration, or more briefly, as a *sustainable configuration*.

Definition 11A2   A feasible industry configuration over N, with prices p and firms' outputs $y^1, \ldots, y^m$, is *sustainable* if $p^e \cdot y^e \leq C(y^e)$ for all $p^e \in R^n_{++}$, $y^e \in R^n_+$, $p^e \leq p$, and $y^e \leq Q(p^e)$.

Here, entry plans may involve positive outputs of products in the set N that are not offered by any of the firms in the industry configuration. To allow for all such possibilities that are capable of disrupting the original industry configuration, N should be defined to be sufficiently large to include all products with a significant substitution or complementarity relationship (in terms of demand) with the goods that actually are offered in the configuration.

Clearly, Definition 11A2 is the natural extension of the definition of a sustainable monopoly to configurations with unspecified numbers of firms. It is evident immediately, as discussed in Section 9E, that competitive equilibria are sustainable configurations. Moreover, the explicit dynamic model of potential entry that is constructed in Chapter 10 for the case of an incumbent monopolist can, with trivial modifications, be adapted to any feasible configuration of incumbent firms. The same analysis applied to this modified model would then show once again that in the absence of entry barriers, sustainability of the industry configuration is a necessary condition for equilibrium. Thus, the notion of sustainable and feasible industry configurations constitutes a single unifying framework for the study of industry structure in contestable markets, for any equilibrium industry structure in a contestable market must be a sustainable configuration.

The bulk of this chapter is concerned with the salient features of such configurations and with examples of types of such structures—some familiar, and some not generally discussed in the literature. It is reassuring to find that some of the novel configurations appear to approximate important cases in reality, and that they promise to lead to rich theoretical analysis (whose beginnings only are undertaken here).

11B. Attributes of Sustainable Configurations

The attributes of firms in sustainable configurations include several that coincide with necessary conditions for the sustainability of a monopoly firm. Because these have already been discussed, and because the proofs are essentially the same, they are simply listed here.

Proposition 11B1 *In a sustainable configuration each firm must* (i) *operate efficiently,* (ii) *earn zero economic profit,* (iii) *avoid cross subsidies,* (iv) *select prices for each of its products that are at least as large as marginal costs, and* (v) *select an output vector at which ray average cost is no larger than it is at any proportionately smaller output vector.*

Next we offer a powerful result that generalizes to industry structures with more than one firm the proposition that a sustainable monopoly must be a natural monopoly:

Proposition 11B2 *A sustainable configuration must minimize the total cost to the industry of producing the total industry output. That is, no different number, size distribution, or output vectors of the industry's firms can provide the industry's output at a lower total cost than that incurred by the firms of a sustainable configuration.*

Proof   The proof is by contradiction. Suppose there were another group of firms' output levels, $\hat{y}^1, \ldots, \hat{y}^k$, that could produce the same total output, $\sum_{i=1}^{m} y^i$, as that offered by a given sustainable configuration at lower cost; so that $\sum_{j=1}^{k} \hat{y}^j = \sum_{i=1}^{m} y^i$ and $\sum_{j=1}^{k} C(\hat{y}^j) < \sum_{i=1}^{m} C(y^i)$. Then at the original prices p the profits of the new group would, in total, be positive; that is, $\sum_{j=1}^{k} (\hat{y}^j \cdot p - C(\hat{y}^j)) = \sum_j (\hat{y}^j \cdot p) - \sum_j C(\hat{y}^j) = p \cdot \sum_j \hat{y}^j - \sum_j C(\hat{y}^j) = p \cdot \sum_i y^i - \sum_j C(\hat{y}^j) > p \cdot \sum_i y^i - \sum_i C(y^i) = \sum_i (p \cdot y^i - C(y^i)) \geq 0$. Consequently, to some firm j in the new group, the old prices would have yielded a positive profit, $\hat{y}^j \cdot p - C(\hat{y}^j) > 0$. Hence, $p^e = p$ and $y^e = \hat{y}^j \leq \sum_i y^i = Q(p)$ would have constituted a profitable entry plan that rendered the given configuration unsustainable.   Q.E.D.

Proposition 11B2 shows the power of frictionless and unimpeded entry to discipline an industry into industry-wide efficiency. Thus, the invisible hand is revealed to possess a remarkable amount of strength even outside the realm of perfect competition. Not the least noteworthy characteristic of this power is its ability to overcome the computational difficulties which, as we saw in Chapter 5, beset the very determination of the efficient industry structure, especially where economies of scope require the presence of multiproduct firms. Thus, once again, the market mechanism proves itself to be a remarkable analog computer, capable of determining optimal solutions which our most sophisticated techniques cannot easily find.

Students of industry structure have sometimes sought a different mechanism making for the minimization of industry costs. They have argued that markets provide incentive for industry-wide efficiency through the profit offered by any cost-saving merger not prohibited by government restrictions or any cost-saving act of fission (unless deterred by the prospect of loss of market power).

However, we now show that, in a multiproduct setting, these incentives are not generally sufficient to bring about cost minimization for the industry. For example, suppose that there are cost complementarities between goods 1 and 2, decreasing average incremental cost in the production of good 1, and that two firms respectively produce (y_1, y_2) and (y'_1, y'_2), $y_2 > y'_2$. It follows that $C(y_1 + y'_1, y_2) + C(0, y'_2) < C(y_1, y_2) + C(y'_1, y'_2)$. Yet, it may nevertheless be true that the average incremental costs of good 2 increase with such rapidity that $C(y_1 + y'_1, y_2 + y'_2) > C(y_1, y_2) + C(y'_1, y'_2)$, so that the firms cannot reduce their total costs by merging. How, then, can the market bring about the agglomeration of the production of good 1 that reduces the industry's cost? In general, a firm will not be in a position to purchase just the other's good 1 production facilities because, as a result of the common plant that provides the cost complementarities, these are not separable from the facilities that produce good 2. One firm can bribe the other to refrain from producing good 1, but the receipt of such a payment would be a powerful incentive for the entry of (many) other firms whose primary business would be the threat of producing good 1.

However, free entry in contestable markets does provide the required mechanism. As the proof of Proposition 11B2 shows, any market prices that enable the firms producing (y_1, y_2) and (y'_1, y'_2) to cover their costs must also provide revenues that exceed the costs of an entrant's production of $(y_1 + y'_1, y_2)$, or an entrant's production of $(0, y'_2)$, or both. Consequently, the two firms with outputs (y_1, y_2) and (y'_1, y'_2) cannot both participate in any industry equilibrium that is open to potential entrants who have access to the same productive techniques and who assess the profitability of entry in terms of the current prices of incumbent firms. In this way, free entry in contestable markets enforces industry-wide efficiency even where profit-seeking incumbent firms cannot achieve it among themselves.

An immediate corollary of Proposition 11B2 is

Proposition 11B3 (i) *In a sustainable configuration each firm must select an output vector at which its cost function has the property of (possibly weak) subadditivity.* (ii) *Any sustainable configuration must be a feasible and efficient* *configuration (as defined in Chapter 5), but not conversely. (iii) If several firms in a sustainable configuration all produce positive amounts of a given good, their marginal costs for that good must all be equal.*

Proposition 11B3(i), the requirement of (weak) subadditivity, follows because, if it did not hold for some firm's output vector, industry costs could be reduced by subdividing the firm's output among some set of smaller enterprises, thus violating the efficiency requirement of the sustainable configuration.

Similarly, 11B3(iii), the equality of marginal costs, must hold because, otherwise, if two firms, A and B, produced the same product i with A's marginal costs higher than B's, then industry costs could be reduced by a small reduction in A's output of i and an equal offsetting increase in B's production of that good.

Proposition 11B3(i) reports explicitly a property that was foreshadowed in Chapter 7: the fact that any firm in equilibrium in a contestable market must operate at a point of subadditivity. In particular, such a firm must exhibit (possibly weak) economies of scope among the various goods it produces. Conversely, economies of scope between two goods produced in contestable markets implies that they cannot each be produced by different specialized firms—industry costs would then be reduced if the two were instead produced jointly by a multiproduct enterprise. And the theory of subadditivity is now shown to be applicable to the costs of each firm in any contestable industry, whether the industry is a natural monopoly, competitive, or oligopolistic.

Proposition 11B3(ii) provides the crucial link between the general theory of industry structure in contestable markets and the normative theory of cost-minimizing industry structure that comprised Chapter 5. Since any sustainable configuration must be feasible and efficient, the necessary conditions for the latter, derived in Chapter 5, all automatically pertain to the former. In particular, the bounds on the cost-minimizing number of firms that are active in the industry and that actively produce each good substantively delimit the structure of an industry in a contestable market. This way of looking at the matter proves to enhance our analytic power considerably.

Proposition 11B3(iii) is a reminder that if minimization of industry costs involves production of a given good by more than one firm, it is necessary that the marginal costs of production of that good be equal for all of them. It will be shown, soon, that in contestable markets this equalization of firms' marginal costs is achieved via the price of the product in question.

Our next result follows immediately from Propositions 11B2, 11B1, and from Lemma 7D1:

Proposition 11B4   *In a contestable industry, any product with average incremental costs that decline throughout the relevant range must be produced by* *only a single firm (if it is produced at all). Further, such a product must be priced above marginal cost, except in the knife-edge case in which the degree of product-specific scale economies is unity.*

Regardless of any other features of the productive techniques, and regardless of the number of firms in the industry, with decreasing average incremental cost ($DAIC$) holding globally for some product, Lemma 7D1 tells us that it cannot be produced by more than one firm for industry costs to be minimized. Thus, free entry will enforce monopoly production of any such good in a contestable industry that offers it at all. Of course, several suppliers may offer some other products of a firm that has a monopoly in the production of a good with $DAIC$. However, any product characterized by global $DAIC$ must be supplied by only one firm.

Such a product's average incremental cost must exceed its marginal cost, by Definition 4A5 and Proposition 4A1, except in the knife-edge case in which the degree of product-specific scale economies is precisely unity. Then, since Proposition 11B1(iii) tells us that price must (weakly) exceed average incremental cost in a sustainable configuration, it follows that the price must generally exceed the marginal cost of a product with global $DAIC$.

Next, we come to what is probably the most surprising result of this chapter. Where, in the absence of global $DAIC$, a product is offered by several firms, free entry not only enforces efficiency, but it also dictates that price actually be set *equal* to marginal cost.

Proposition 11B5 *If two or more firms produce a given good in a sustainable configuration, then they must all select outputs at which their marginal costs of producing it are equal to one another and to the good's market price.*[1]

---

[1] *Proof* Suppose $y_j^k < \sum_{h=1}^m y_j^h = Q_j(p)$ and $p_j > C_j(y^k)$. Consider this following function of the scalar t:

$$\psi(t) = p \cdot (y^k + t\,\Delta^j) - C(y^k + t\,\Delta^j),$$

where Δ^j is the vector with zeros for each component except for the jth, which is unity. $\psi(t)$ is the profit earned by an entry plan replicating all of the activities of firm k with the exception of an increase by amount t in the output of good j. Evaluating it at $t = 0$, $\psi(0) = p \cdot y^k - C(y^k) = 0$, since by Proposition 11B1, a firm in a sustainable configuration must earn exactly zero profit. Differentiation of $\psi(t)$ yields $\psi'(t) = p_j - C_j(y^k + t\Delta^j)$, and, at $t = 0$, $\psi'(0) = p_j - C_j(y^k) > 0$, which is positive by hypothesis.

Thus, the profit earned by the entry plan increases from zero as t is increased from 0. Hence, there exists some $\bar{t} > 0$ such that $\psi(t) > 0$ for $0 < t < \bar{t}$. Moreover, the entry plan is feasible for $0 < t \le Q^j(p) - y_j^k$, so that the entrant's output of good j, $y_j^k + t$, remains no greater than the amount demanded by consumers $Q^j(p)$. Consequently, for $0 < t \le \min(\bar{t}, Q^j(p) - y_j^k)$, the entry plan is both feasible and profitable, which contradicts the hypothesis that the industry configuration is sustainable. Our result follows: p_j must be equal to $C_j(y^k)$ for firm k's output vector to be part of a sustainable configuration in which firm k is not the sole producer of good j. Q.E.D.

Intuitively, suppose firm k is one of two or more firms producing good j in a sustainable configuration. We know from Proposition 11B1 that its marginal cost of producing j cannot exceed the price of j. Thus, to prove that the marginal cost is equal to price, it only remains to show that if marginal cost were below price, a profitable entry plan would exist. Such a profitable plan is, indeed, always possible in this case. To see this, let the price of good j which is produced by several firms exceed its marginal cost for firm k, one of its producers. Then, to earn a profit, an entrant need merely replicate firm k's output of all other goods and exceed slightly k's output of good j. This must yield a profit higher than the zero earnings of firm k because the additional output of good j will yield more incremental revenue than it costs to produce. Such an entry plan is necessarily feasible here because the production of good j by at least one firm other than k assures the entrant that it is possible to market more of good j than firm k alone does, without exceeding the total quantity demanded by consumers. Note that such an entry plan need not have been profitable if firm k had instead been the sole producer of the good in the configuration, and therefore itself met the entire market demand. For in that case, any entry plan involving an increase in the output of good j must cope with the corresponding cut in the price of good j required to induce consumers to purchase the added output. This unavoidable price cut may well cause the net profit yielded by an increase in the output of a good with price greater than marginal cost to be negative. Consequently, as we have already seen in the case of a natural monopoly, for goods produced by a single firm in a contestable industry, prices certainly can, and often will, be above the corresponding marginal costs. Our new observation here is that any firm that is *not* the sole supplier of a good will find any price that is above marginal cost to be unsustainable.

The implications of this result are surprisingly strong. The discipline of sustainability in contestable markets forces firms to adopt prices just equal to marginal costs, provided only that they are not monopolists of the products in question. Conventional wisdom implies that, generally, only perfect competition involving a multitude of firms, each small in its output markets, can be relied upon to lead to marginal-cost pricing. Here, with the focus on potential competition by prospective entrants rather than on rivalry among incumbent firms, we have just shown that marginal-cost pricing is a requirement of equilibrium in contestable markets, even those containing as few as two active producers of each product. The conventional view holds that the enforcement mechanism of full competitive equilibrium requires the smallness of each active firm in its product market, in addition to freedom of entry. We see that the smallness requirement can be dispensed with, almost entirely, with exclusive reliance on the freedom of entry that characterizes contestable markets.

We also have, from Propositions 11B1(ii) and 11B5,

Proposition 11B6   *If each good offered by an industry is produced by at least two firms, then every firm's output vector in a sustainable configuration must occur at a point at which its production exhibits locally constant returns to scale.*

The proof follows immediately from the requirement that all prices equal marginal costs; for then, if returns to scale were not locally constant, the firm's total profits would be unequal to zero.

11C. Contestable Oligopolies with Competitive Properties

All of this leads us to inquire whether perfectly competitive behavior can be expected to characterize a particular oligopolistic industry operating in contestable markets. We see now that the answer rests on the answers to two basic questions: (i) Do technological and demand conditions ensure that each product will be produced by more than one firm? (ii) If so, does the requisite sustainable configuration exist? We take up the first of these basic questions in this section and defer the second to Section 11D.

The conceptual techniques already described in this book help us to determine whether two or more producers for each good are required by technological and demand conditions in a contestable industry. Proposition 11B4 provides the first test: A good with $DAIC$ from a zero output to the maximal output that is relevant cannot be produced by more than one firm in a sustainable configuration, and will generally be priced above marginal cost. Thus, perfectly competitive behavior cannot be expected in the markets for such goods. To give substance to this principle, it is necessary to indicate explicitly what is meant by the "relevant" output levels. These include all outputs that can possibly be produced by a sustainable configuration. Since such a configuration must also be feasible and efficient, the set of relevant industry outputs studied in Chapter 5 contains the set that is pertinent here and, in fact, more restrictions are now germane. Specifically, no firm in a sustainable configuration can produce an output at which price falls below either marginal cost or average incremental cost. Thus, to investigate whether a good must be produced by a monopoly in a sustainable configuration, it suffices, using Proposition 11B4, to check for $DAIC$ up to the output at which the market demand curve crosses the AIC curve.

As illustrated in Figure 11C1, the AIC curve may be U-shaped, but may nevertheless exhibit $DAIC$ in the relevant range. It must be noted, though, that the shape and location of an AIC curve will generally depend on the quantities of other goods produced by the firm. Thus, the relationship shown in Figure 11C1 must be robust under alterations in the firm's other output

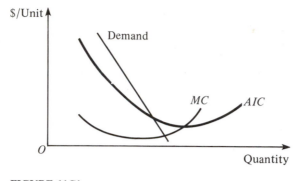

FIGURE 11C1

levels before one can infer from Proposition 11B4 that a good must not be produced by a multiplicity of firms in a sustainable configuration.

On the other hand, the Corollary to Proposition 5G1 provides a sufficient condition for two or more firms to be required to produce a given good in a sustainable configuration. This must be true if all relevant industry output vectors involve at least twice the maximal output of the good in question at any point on the M locus. Since in this case efficiency requires that at least two firms actually produce the good, any sustainable configuration must entail marginal-cost pricing of the good by all firms producing it.

Finally, if the methods discussed in Chapter 9 indicate that there is a competitive equilibrium in which at least two firms produce each good offered, then that equilibrium must be a sustainable configuration. Although under conventional reasoning rivalry between two firms in each product line is generally not sufficiently vigorous to ensure an outcome with the properties of perfect competition, in contestable markets such an outcome is in fact ensured by the discipline of potential entry.

Our assumptions here do not preclude the possibility that the sustainable configurations will entail a large number of active firms, but in that case the conclusion that behavior must be competitive comes as no surprise. The surprising aspect of the result is that it holds for true natural oligopolies—where market demands and productive techniques require that only a handful of firms produce each good. Here (as discussed later), sustainable configurations may well exist, and all of them must involve competitive behavior and an optimal allocation of resources that is not merely second-best.

The unexpected character of this result is brought out more clearly when it is contrasted with those of the received theory of oligopoly. Conventional analysis usually assumes some pattern of interaction among firms (e.g., quantity-Cournot behavior, cartellization, limit pricing) and analyzes the corresponding equilibrium, taking the number of active firms to be fixed. The typical model is completed with the aid of an assumption relating to freedom of entry. This usually requires that when the number of firms is in equilibrium, each incumbent firm must earn nonnegative profit, while if one additional firm were present, the profits of some enterprises (at least those of the last firm to enter) would be negative. It is concluded throughout the literature that "the oligopoly problem" remains unsolved, because the different postulates about patterns of interfirm rivalry yield different "free-entry" equilibria with substantially different properties. And no one seems able to determine with any degree of persuasiveness which behavioral patterns are most common, or are theoretically most compelling.

In contrast, we take the position that the preceding analysis "solves" the oligopoly problem, under one special set of well-defined and observable circumstances. The behavior we impute to potential entrants does not stem from arbitrary assumptions. Rather, this behavior is a rational response to the structural conditions that characterize contestable markets. By focusing on the behavior of potential entrants in contestable markets, we have made it unnecessary to take an *a priori* position on the most likely pattern of rivalry and interaction among incumbent firms. *Yet we have been able to derive clear-cut results about their equilibrium behavior patterns and the resulting prices and outputs.*

11D. Existence of Sustainable Configurations with Competitive Properties

Let us now return to the question of the existence of sustainable configurations with competitive properties. Proposition 11B6 establishes that in such configurations, all firms must produce with locally constant returns to scale, that is, where RAC curves are (locally) horizontal. As we show in the discussion of the scalar output case in Chapter 2, this requirement, when combined with the classical assumption of U-shaped (ray) average cost curves, seems to make the existence of a sustainable industry configuration extremely unlikely. Essentially, existence would then require an apparently fortuitous intersection of the multi-output demand surface with the gradient of the cost function at an industry output vector that can be produced via *integer* multiples of (at most n) points on the M locus.[2]

---

[2] Even when the integer problem is most serious, say when cost minimization calls for only two firms, entry costs of sufficient magnitude can, of course, prevent the establishment of new firms and render a two-firm configuration sustainable. One can then investigate, in a particular case, the minimum entry cost necessary to render a configuration sustainable. Ten Raa (1980) has carried out an illuminating study of this issue. He has pointed out that the magnitude of this minimum entry cost necessary to ensure sustainability can be used as a measure of the "degree of unsustainability." He has also confirmed rigorously the intuitive conclusion that the degree of unsustainability caused by the pressure of the integer problem decreases monotonically toward zero as the cost-minimizing number of firms increases, and has provided explicit expressions for the minimum entry cost necessary for sustainability.

Chapter 9 deals with this difficulty by constructing the well-behaved fractional-firm industry cost function $C^*(\cdot)$ and by appealing to the traditional "large-numbers" argument as justification for an analysis of multi-product competitive equilibrium using that approximation to the true industry cost function. Chapter 2, however, treats the integer problem in single-product industries with small numbers of firms by means of the assumption of "flat-bottomed" average cost curves. In this section, we extend that approach to multiproduct industries, using the analytic apparatus provided in Chapter 9 to begin rigorous theoretical analysis of the existence of oligopoly equilibria in contestable markets.

We start by expanding the notion of flat-bottomed AC curves (as depicted in Figure 2C4) to *ray* average cost curves:

Definition 11D1   A ray average cost curve is *flat-bottomed* if there is a point y_m on the ray and a positive number k such that RAC is decreasing up to y_m, is constant between y_m and $(1+k)y_m$, and is increasing at ty_m for $t > (1+k)$.

Here, y_m is the *minimum* efficient scale of operation along the pertinent ray, rather than the only efficient scale, as in the case of U-shaped RAC curves. When RAC curves are flat-bottomed, we redefine the M locus to be the set of all output vectors at which there are locally constant returns to scale ($S_N = 1$). Thus, as shown in Figure 11D1, the M locus extends from y_m to $(1+k)y_m$ along each ray. Throughout this portion of any given ray, RAC is constant, $C(\cdot)$ is linearly homogeneous, and marginal costs are constant.

With flat-bottomed RAC curves, the fractional-firm industry configurations studied in Section 9D do not necessarily require fictitious fractions of

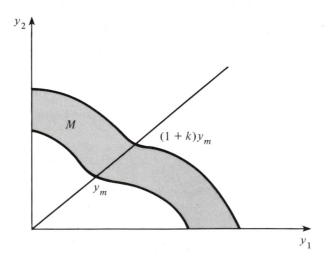

FIGURE 11D1

firms as part of their mechanism. For example, if $k = 1$ and if a fractional-firm configuration includes 3.8 firms, each producing y_m, then these can be replaced without increase in cost and without decrease in profit either by two firms each producing $(1.9)y_m$, or by three firms, each producing $(3.8/3)y_m$. Replacements of this kind can transform a fractional-firm configuration into an equivalent one with an integer number of firms, and a fractional-firm competitive equilibrium into an equivalent and ordinary competitive equilibrium. The reasoning is precisely that described in Chapter 2, only here it is applied along multiproduct output rays rather than to single-output production.

This multiproduct extension of the assumption that the RAC curves are flat-bottomed (that is, the total-cost curves are locally linearly homogeneous over some interval along each ray) runs into one complication. The firms that comprise the industry do not, in general, employ the same output proportions, and those proportions may change as the industry's output expands. Thus, suppose increased industry output were to enhance the advantages of specialization. Proportional expansion of each firm's output vector might yield constant returns for each firm by itself but economies of scale for the industry overall. That is, the relevant RAC curve for the industry may not be flat-bottomed at the output in question. As we will see, this problem can be dealt with by using our handy complementarity assumption of trans-ray convexity (supportability), in essence, to rule out such savings from increased specialization. But problems such as this unavoidably complicate the discussion in the multiproduct case.

We begin our analysis of this case with the following general result (which is perfectly analogous to the pertinent single-product theorem):

Proposition 11D1 *Suppose that RAC curves are all flat-bottomed. Without loss of generality, any fractional-firm industry configuration can be normalized so that all participating fractional firms produce at the minimum efficient scale along their rays. Then, a normalized fractional-firm industry configuration is equivalent to a regular industry configuration, and a normalized fractional-firm competitive equilibrium is equivalent to a normal competitive equilibrium, if the fractional numbers of firms of each type involved are all at least equal to the smallest integer greater than or equal to $1/k$, denoted by $\lceil 1/k \rceil$.*[3]

---

[3] *Proof* Let h be the number of fractional firms, where $y^* = hy_m^*$, y^* is the vector of total outputs of firms of the type in question, and y_m^* is the point on the same ray as y^* which is nearest the origin on the flat-bottomed portion of the firm's RAC curve. We can be sure that the integer number of firms, $\lfloor h \rfloor$ (which is the largest integer no larger than h), can produce y^* in total, with each firm producing an equal output on the flat-bottomed portion of the RAC curve, if $h/\lfloor h \rfloor \leq 1 + k$. We now show that a sufficient condition for this inequality is $h \geq \lceil 1/k \rceil$: If $h \geq \lceil 1/k \rceil$, then $\lfloor h \rfloor \geq 1/k$, $1/\lfloor h \rfloor \leq k$, and $(\lfloor h \rfloor + 1)/\lfloor h \rfloor \leq 1 + k$. Since $h < \lfloor h \rfloor + 1$, the result follows. Q.E.D.

This proposition can be utilized in conjunction with Proposition 9D10, our existence result for fractional-firm competitive equilibria, to establish sufficient conditions for the existence of sustainable configurations with competitive behavior. The remaining issue is whether the normalized fractional-firm competitive equilibria (whose existence is ensured by Proposition 9D10) entail (fractional) numbers of firms of each type that are at least $\lceil 1/k \rceil$. If they do, then they are equivalent to competitive equilibrium configurations each comprised of an integer number of firms producing somewhere on the flat portions of their RAC curves. Such configurations are best interpreted as true competitive equilibria if they include large numbers of firms. However, where such configurations involve small numbers of firms, they can only be interpreted as sustainable oligopolies with competitive properties.

We can offer two methods of ensuring that $\lceil 1/k \rceil$ is exceeded by each of the numbers of firms of each type in a normalized fractional-firm competitive equilibrium. The first method, germane only to the analysis of perfect competition, is to consider the limit as industry demands are proportionately and indefinitely expanded relative to the M locus. As this expansion proceeds (or as the M locus is proportionally contracted relative to market demands), the numbers of firms of each type expand proportionally in the associated fractional-firm competitive equilibria. Thus, a point must be reached beyond which the requirements of Proposition 11D1 are satisfied.

The second method is pertinent to the analysis of sustainable oligopolies with competitive properties. It is based on the following proposition, which utilizes trans-ray convexity to obtain its specific results.

Proposition 11D2   *There exists a sustainable configuration with competitive properties under the following conditions*:

(i) *All RAC curves are flat-bottomed.*
(ii) *If y_m is any output vector in M with minimum efficient scale, and if the quantity vector demanded under prices equal to marginal costs at y_m is a multiple of y_m, $Q[\nabla C(y_m)] = t y_m$, then $t \geq \lceil 1/k \rceil$.*
(iii) *The homogenized cost function $C^H(\cdot)$ is trans-ray convex.*
(iv) *The hypotheses of Proposition 9D10 are satisfied.*

Proof   Proposition 9D10 ensures the existence of a fractional-firm competitive equilibrium in this case. Condition (iii) and Proposition 9D9 imply that such an equilibrium can be composed of (fractional) firms with equal output vectors. Normalize this equilibrium so that it is comprised of $\tilde{a}$ (fractional) firms, each producing $\tilde{y}_m$, an output vector of minimum efficient scale. The corresponding prices are $\nabla C(\tilde{y}_m)$, and $Q[\nabla C(\tilde{y}_m)] = \tilde{a}\tilde{y}_m$. Then, condition (ii) implies that $\tilde{a} \geq \lceil 1/k \rceil$, and Proposition 11D1 implies that there is then an ordinary (integer firm) competitive equilibrium configuration that is equivalent to the fractional-firm competitive equilibrium. The sustainability of this configuration follows from its competitive equilibrium properties. Q.E.D.

Thus the assumption that RAC curves are flat-bottomed overcomes the integer problem and permits a proof of the existence of sustainable multiproduct oligopoly configurations with competitive properties. The technical condition (iv) of Proposition 11D2 is quite innocuous, as was discussed in Chapter 9. Condition (iii), trans-ray convexity of the homogenized cost function, as described in Chapter 9, can be weakened to an appropriate version of trans-ray supportability. However, a trans-ray cost condition (of some sort) cannot be dispensed with entirely, so far as we know, without introducing the possibility that the pertinent fractional-firm competitive equilibrium requires some small fractional number of normalized firms, with no counterpart in a sustainable configuration. Of course, the proposition does apply to industries offering a single homogeneous product, and in this case (iii) is unnecessary and vacuous.

The crucial assumption for industry structure is condition (ii), which asserts that the relevant market demands (at all possible equilibrium vectors of marginal-cost prices) are sufficiently large relative to minimum efficient scale to permit their supply by an integer number of firms producing at minimal ray-average cost. The economic import of this condition depends on the length of the flat bottoms of the RAC curves, as measured by k. For k extremely close to zero, condition (ii) requires that relevant market demands be very large relative to minimum efficient scale, and the proposition yields little beyond what we obtained from our analysis of perfect competition. However, for $k \geq 1$, the proposition assures us of the existence of a sustainable configuration with competitive properties as long as relevant market demands lie outside the region of increasing returns to scale. If they exhaust the economies of scale, but are less than twice minimum efficient scale, then the sustainable configuration is a natural monopoly earning zero profit at marginal-cost prices. If market demands are at least double minimum efficient scale, then this problem decreases rapidly as k rises. For example, if $k = 1/3$, it requires industry output to be only three times as large as minimum efficient scale. Moreover, a sustainable configuration is an oligopoly, comprised of at least two firms, that operate in a competitive manner.

In sum, the assumption that RAC curves are flat-bottomed effectively eliminates much of the integer problem for the existence of sustainable solutions under oligopoly. But, it must be emphasized, it does not remove it altogether. Where the ranges of the flat-bottomed portion of the firm's RAC curves are relatively small and where a small number of firms is required for sustainability, existence problems may still arise. This does not mean that in such cases sustainable solutions will not occur, but that they must then be ascribed to phenomena other than the flat bottoms of the RAC

curves. It is easy to think of examples of such supplementary mechanisms, but since they will play no role in our subsequent analysis, we will not pursue them.[4]

11E. Partial Monopoly

We have seen that industry structure in contestable markets may entail monopoly production of some goods (in particular, those characterized by global $DAIC$) and requires competitive pricing of goods offered by more than one firm. However, these two cases are compatible with sustainable operations by a wide variety of types of firms in a wide variety of structural patterns. Chapter 8 deals with firms that are natural monopolists for all products offered by their industries, while Chapter 9 and the preceding sections of this chapter relate to firms forced by at least some rivalry and by potential entry to price all their outputs at marginal costs. In this section, we analyze the simplest of mixed cases, that in which a firm is active in several contestable markets, some of which are naturally monopolistic while others are naturally oligopolistic. This is a case that has given rise to many difficulties in regulatory policy,[5] a subject to which our discussion will prove very relevant.

This scenario involves the production of (at least) one good (product 1) with global $DAIC$ that is linked by economies of scope to (at least) one other good (product 2) with an average incremental cost curve that is U-shaped or that rises globally. We know from Proposition 11B4 that good 1 must be produced monopolistically in any sustainable configuration. (Call the producer of this good firm A). It then follows that any other firms active in such a two-good industry must specialize in the production of good 2. The average cost curve of such a specialty firm, firm B, is depicted in Figure 11E1, along with its marginal cost curve. The figure also incorporates the (broken) AIC and marginal cost curves for firm A, which holds the

---

[4] For example, if the position of the demand hypersurface at the initial price vector leads to such an existence problem, we can expect prices to change, and quantities demanded will change as a consequence. On the Duesenberry (1949) hypothesis that the position of the demand function depends upon past consumption, that is, that tastes are affected by consumption experience, this will lead to shifts in the demand hypersurface that will continue so long as its position remains incompatible with a sustainable solution. Thus, this constitutes a mechanism that may work to eliminate such unsustainability problems. Moreover, if entry induced by unsustainability occurs in the form of the introduction of products differentiated from those of the incumbents, we find ourselves in the Chamberlinian case, which offers many types of sustainable solutions.

[5] This is precisely the situation in which the charge of cross subsidization is likely to be raised by the (partial) monopolist's rivals. See, for example, Kahn (1970, 1971) and references therein.

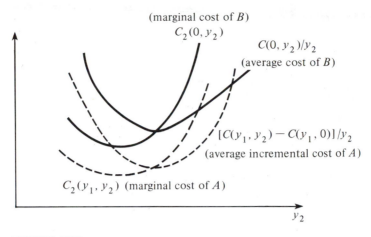

(marginal cost of B)
$C_2(0, y_2)$

$C(0, y_2)/y_2$
(average cost of B)

$[C(y_1, y_2) - C(y_1, 0)]/y_2$
(average incremental cost of A)

$C_2(y_1, y_2)$ (marginal cost of A)

y_2

FIGURE 11E1

monopoly in production of good 1. As indicated in the drawing, firm A's *AIC* of good 2 must lie everywhere below the corresponding average cost of the specialty firm B because of the assumed economies of scope. This follows because $C(y_1, y_2) < C(y_1, 0) + C(0, y_2)$ implies

$$\frac{C(y_1, y_2) - C(y_1, 0)}{y_2} < \frac{C(0, y_2)}{y_2}.$$

Consequently, any price for good 2 that enables a specialty firm B to cover its costs must exceed the *AIC* that the partial monopolist A would incur by equaling B's output. Such activity by firm A in the market for good 2 adds more to its revenues than it adds to its total cost. The added net revenues enable the partial monopolist to cover his total costs at a price for good 1 lower than would otherwise be required. Thus, to be sustainable, the monopolist of good 1 must also produce good 2, perhaps in competition with others, if it would otherwise be feasible for any specialty firm to produce good 2.

From what has been said so far, a monopoly by the two-product firm in both of its markets is not ruled out. However, because of the assumed absence of *DAIC* in the production of good 2, if the quantity of it demanded by the market in equilibrium is sufficiently large, industry cost minimization will require the supply of good 2 by specialty firms as well as by the partial monopolist.

In such a case, any sustainable configuration necessarily entails marginal-cost pricing of good 2 by all producers, and a price for good 1 that just enables its multiproduct supplier to break even. Further, that firm must then produce an output of good 2 at which increasing average incremental cost ($IAIC$) prevails, in order to avoid unsustainable cross subsidization of good 2 by consumers of good 1. Otherwise, if $DAIC$ prevailed, a price equal to the marginal cost of good 2 would necessarily be less than AIC_2: $p_2 = C_2(y_1, y_2) < [C(y_1, y_2) - C(y_1, 0)]/y_2$, so that a producer specializing in the production of good 1 could afford to offer it at a price lower than that at which it is sold by the partial monopolist.

If there were production complementarities between goods 1 and 2, the multiproduct firm's marginal cost curve for good 2 would lie below that of a specialized firm $[C_2(y_1, y_2) < C_2(0, y_2),$ for $y_1 > 0]$, as shown in Figure 11E1. Consequently, the multiproduct firm would have to produce a quantity of good 2 larger than that of a specialized firm in order for them both to equate their marginal costs to the same market price.

A clear example of a scenario of this kind is generated by the cost function

$$C(y_1, y_2) = y_1^{1/2} + y_2 + y_2^2 - y_1^{1/2}(y_2 + y_2^2)$$

for

$$y_2 + y_2^2 < 1, \qquad y_1 < 1.$$

This involves cost complementarities, global $DAIC$ in good 1, and global $IAIC$ in good 2. Because the marginal cost of good 2 increases with y_2, firms specializing in production of good 2 must be of infinitesimal size with their marginal and average costs equal to $\lim_{y_2 \to 0} C_2(0, y_2) = 1$. As a group, such firms have a horizontal supply curve at $p_2 = 1$, as shown in Figure 11E2. Also shown there are the AIC_2 and C_2 curves of the partial

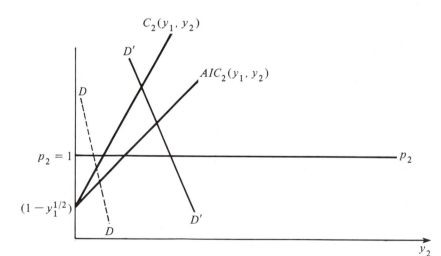

FIGURE 11E2

monopoly firm, for a given positive output of good 1. The slopes of these curves are 1 and 2, respectively, and their common intercept is $(1 - y_1^{1/2})$. Thus, they cross the supply curve of the specialized firms at quantities of y_2 that are greater, the larger is the monopolist's output of y_1. If the market demand for good 2 is represented by the broken line DD, then industry costs are minimized when the multiproduct firm holds a monopoly in both markets because its marginal costs for good 2 are lower than that of any specialized firm throughout the relevant range. If, instead, market demand were described by the higher solid line $D'D'$, then industry cost minimization would require that the multiproduct firm produce the output at which its marginal cost is 1 and that the remainder of market demand be supplied by the specialized firms. Here, the multiproduct firm would earn incremental net revenues in the market for good 2 equal to the area between the horizontal price line, p_2p_2, and its marginal-cost curve. It would be forced by potential entrants to use these gains to reduce the price of good 1. In this case industry structure in a contestable market must include a partial monopoly active in both a natural monopoly and a naturally competitive market. In other cases, the technological relationships will require only a handful of firms that are actively engaged in the production of a naturally oligopolistic good, together with a partial monopoly firm. Here, too, all producers of the homogeneous product must equate their marginal costs to the common price.

11F. Monopolistic Competition

The standard scenario of Chamberlinian monopolistic competition can be summarized in our terms as the case in which a multitude of (potential) products that are substitutes in demand are included in the product set, each such product having a U-shaped average-cost curve, and where there are no economies of scope among them.[6]

The equilibrium concept Chamberlin called the "large-group case" entails adoption by each of the single-product firms of the price that maximizes its profits, given the prices of its rivals which offer differentiated substitute products. Free entry of firms marketing additional product varieties then continues until the maximized profits of each active firm is driven to zero, as shown in Figure 11F1. Simultaneously, potential entrants are offered no profitable opportunity for entry, given the prices and product designs already supplied in the market. In this equilibrium, each active firm, of necessity, equates its price to its average cost in the region of increasing

---

[6] There is an illuminating recent and growing literature on the subjects of heterogeneous products, choice of product design, and the determination of industry structure. For example, see Salop (1981) and the references therein.

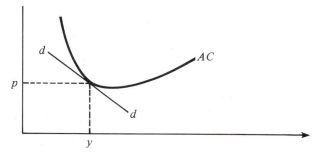

FIGURE 11F1

returns to scale, because of the requisite tangency between its downward sloping demand curve *dd* and the average-cost (*AC*) curve.

Plainly, then, by its very definition, this Chamberlinian equilibrium is a sustainable configuration over the set of potential products. However, other sustainable configurations are also compatible with the very same scenario. One such type, depicted in Figure 11F2, involves average-cost pricing (point *A*) by each single-product firm, but not the selection of those prices that maximize its profit on the (not always plausible) assumption that the prices of other active firms are given. Rather, each firm may choose to maximize its profit, given the behavior of potential entrants who would undercut it if it were to raise its price above average cost. Of course, such a configuration would not be sustainable if an entrant could earn a profit by supplying a new variety of product, given the prices and designs of those already offered for sale. However, a large number of sustainable configurations of the form illustrated will generally exist in monopolistically competitive markets, with an industry-wide trade-off between the number of available varieties and the price of each. Notice that a sustainable solution

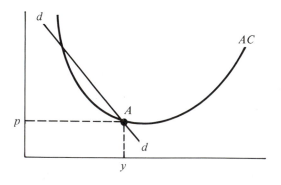

FIGURE 11F2

such as *A* does not exhibit excess capacity. That is, excess capacity in a monopolistically competitive market is not necessary for sustainability.

Another type of sustainable configuration in such a market would arise if the quantity of a particular product variety demanded in equilibrium were sufficiently large relative to the minimum efficient scale to require production by two or more firms for industry efficiency. Then, it is required that any such product variant coexist with any less popular variants (characterized by Figure 11F1 or 11F2) in equilibrium. Of course, the price of the popular variant must simultaneously equal marginal and average cost, while the others may conceivably be priced at average cost and above marginal cost. This type of sustainable configuration can be consistent with Chamberlin's assumption that firms choose prices to maximize their profits, given the decisions of the other active firms. As such, these configurations can perhaps provide the foundation for a substantive and relevant generalization of Chamberlinian monopolistic competition theory.

Chamberlin and his successors apparently never dealt with the incentive of a firm, in the standard framework, to produce *several* product varieties in order to coordinate their prices. In general, the profit-maximizing prices for a pair of substitute products are both higher than the prices that comprise a Nash equilibrium between two noncooperating price-setting vendors. However, in contestable markets, such coordinated prices are not sustainable in the absence of economies of scope, so that in the absence of such economies Chamberlin's disregard of the incentives for coordinated pricing can be justified.

Chamberlin suggested that outside the "large-group case," firms can usefully be assumed to set their own prices on the basis of anticipated price reactions of their rivals, rather than on the basis of the *dd* demand curve that takes rivals' prices to be fixed. The former yields what he called the "high-tangency equilibrium," pictured in Figure 11F3. Here, the demand

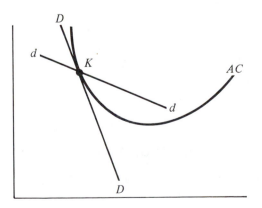

FIGURE 11F3

curve DD is based on the assumption that rivals are expected to react to any one firm's price moves by changing their own prices in exactly the same way. Hence, DD must be less elastic than the dd curve. The situation depicted in Figure 11F3 is purportedly a free-entry equilibrium of this kind, because the DD curve is tangent to the average cost curve, yielding zero profits that are maximal, given the assumed price reactions of rivals.

However, such a position cannot be an equilibrium in a contestable market. An entrant can closely or exactly duplicate the product design of the firm depicted, and enter at a lower price. In a contestable market, with an arbitrarily short lag in incumbents' price reactions, he would expect demand to behave in accord with curve dd during the period of the lag. Since dd is more elastic than the DD curve that is tangent to the AC curve, the dd curve necessarily cuts into and above the AC curve, as illustrated. Consequently, there exist (temporarily) profitable entry opportunities, and so "high-tangency equilibria" cannot be equilibria in contestable markets. It should be noted that Chamberlin's discussion implies strongly that the markets of which he was thinking satisfied the free-entry requirements of contestability.[7] Thus, it is of some significance for his analysis that his high-tangency solution turns out to be unsustainable.

It should be noted that, recently, important work on monopolistic competition by Dixit and Stiglitz (1977) and Spence (1976) has assumed Cournot quantity behavior among rivals supplying differentiated products. The resulting "free-entry equilibria" are analogous to those depicted in Figure 11F3. The assumption that rivals are expected to leave their outputs unchanged is equivalent, in a case of substitute products, to a conjecture that price decreases will induce rivals to respond by decreasing their prices, and to behave analogously when price increases occur. Thus, the corresponding demand curve is, like Chamberlin's DD curve, less elastic than the dd curve which takes rivals' prices to be fixed. It follows then, by the preceding argument, that quantity Cournot equilibria are unsustainable under free entry, and are not truly equilibria in contestable markets.

Finally, where economies of scope hold for some or all (potential) products, sustainable configurations in monopolistically competitive markets necessarily entail the presence of multiproduct firms. These are discussed below.

---

[7] So far as we have been able to determine, Chamberlin does not deal with sunk costs or costs of exit. But he *does* discuss freedom of entry in some detail (see, e.g., Chamberlin, 1962, pp. 199ff). He deals with cases in which "insofar as profits are higher than the general competitive level . . . new competitors will . . . invade the field and reduce them" and "wherever in the economic system there are profit possibilities they will be exploited so far as possible." All this is used to explain the zero-profit tangency equilibrium and clearly suggests something like contestability.

11G. Normative Analysis

Now that we have determined some of the types of industry structure for which sustainability is possible, so that they are consistent with equilibrium in contestable markets, it is appropriate to subject them to a normative analysis of their consequences for economic welfare. Should we find any sustainable oligopolistic or monopolistically competitive structures that necessarily produce optimal results, given the relevant economic circumstances and feasibility constraints, we could infer that in those cases the invisible hand holds sway outside its normal domain. On the other hand, where sustainable solutions are not necessarily optimal, or are even inconsistent with optimality, some degree of market failure must be expected. Such cases of market failure are not those attributable to entry barriers or to friction, because in our analysis they occur in the world of frictionless and reversible free entry that characterizes contestable markets. Rather, such market failures must be ascribed to the very nature of production technology, consumer preferences (as described by market demands), and, perhaps, to the system of nondiscriminatory simple prices that are taken throughout the analysis to constitute the only form of prices available.[8]

One of the principal lessons of Chapter 8 is that monopoly does not necessarily entail such welfare losses. Rather, the weak invisible hand theorem shows that under certain conditions sustainability and Ramsey optimality are consistent, so that the total of consumers' and producers' surpluses may well be maximized (subject to the constraint that firms be self-supporting) in the equilibrium of a monopoly that operates in a contestable market.

Even stronger results follow from the discussion in earlier portions of this chapter. Propositions 11B5 and 11D2 show that, under certain conditions, sustainability and a first-best solution are consistent in an oligopoly with a small number of firms. When minimization of industry cost requires that each good be produced by at least two firms, sustainability requires any equilibrium to satisfy the necessary conditions for a first-best allocation of resources. Thus, in these cases, the invisible hand has the same power over oligopoly in contestable markets that it exercises over a perfectly competitive industry.

However, in the other cases discussed in this chapter, there are systematic reasons indicating that unsustainability may plague the industry configurations that are Ramsey-optimal. Industry configurations involving firms that are partial monopolies and those that constitute generalized Chamberlinian structures may lie outside the set of market forms in which decentralized

---

[8] That is, we assume that there are no special prices for special customers, and no multipart tariffs or quantity discounts that are unrelated to costs. For demonstrations that the absence of nonlinear and discriminatory prices can be a source of social costs see Baumol (1979a) and Willig (1979b).

decision making, guided by the price system, can attain results as desirable as those achievable in theory by a perfectly informed and beneficent planner, even if constrained to permit firms to break even. Our results in these two cases turn out to be mixed. We find that, depending on the nature of the budget constraint and the standard of comparison considered relevant, there are three different variants of the Ramsey optimum, some of which are inconsistent and some consistent with sustainability of the market forms we are now considering.

As we know, under a price system and in the absence of subsidies, a first-best allocation in a decentralized market requires that the prices be equal to marginal costs and that they yield revenues sufficient to cover costs. As we have reconfirmed in Chapter 3, marginal-cost pricing cannot cover production costs where scale economies occur. Thus, in industries whose equilibria may entail outputs by particular firms at which there are increasing returns to scale, first-best allocations may not be feasible financially. In such cases, the relevant normative concept becomes Ramsey optimality—maximization of welfare subject to the constraint that firms be financially self-supporting.

The analysis of Chapter 8 examined the relation of Ramsey optimality to the behavior of a natural monopoly firm. Since the firm in question was taken to be the only enterprise actually in operation within the industry, the financial solvency constraint for the firm also required total industry revenues at least to cover total industry costs. However, in cases in which the industry may contain more than one active firm, a nonnegativity constraint upon industry profit does not guarantee that the profit of each firm will be nonnegative.

Ramsey Optimum I: The Viable-Industry Ramsey Solution

Let us consider as our first variant of the concept of Ramsey optimality that in which welfare maximization is subjected only to the constraint requiring *industry* profit to be nonnegative. We refer to it as the *viable-industry Ramsey solution*. Here, the objective is to

(11G1) $$\text{maximize } W(p) + \sum_{i=1}^{m} (p \cdot y^i - C(y^i))$$

subject to

$$Q(p) = \sum_{i=1}^{m} y^i \quad \text{and} \quad \sum_{i=1}^{m} (p \cdot y^i - C(y^i)) \geq 0.$$

For simplicity, $W(p)$ is taken to possess the following local properties of consumers' surplus: $\partial W(p)/\partial p_j = -Q_j(p)$, so that $W(p) + \sum_{i=1}^{m} (p \cdot y^i - C(y^i))$ can be interpreted as the portion of aggregate real income that can be affected, in this partial equilibrium analysis, by the industry variables in question. In (11G1), the maximization calculation is performed with respect to industry prices p, the number of active firms m, and the output vector of each firm $y^1, y^2, \ldots, y^m$. It is immediately clear that, whatever the optimal quantities of industry outputs $y^I = \sum_{i=1}^m y^i$, the maximization in (11G1) requires the number of firms and the output quantities of each to be those that produce y^I at least total cost for the industry.[9] This result may seem to offer some promise for the sustainability of such optima in light of the conclusion (Proposition 11B2) that sustainable industry configurations must also minimize the total industry cost of producing the industry's total output. This result also permits (11G1) to be rewritten:

(11G2)   maximize $W(p) + py^I - C^I(y^I)$

subject to

$$Q(p) = y^I \quad \text{and} \quad p \cdot y^I - C^I(y^I) \geq 0.$$

Because (11G2) has the same form as the standard Ramsey optimization problem for a monopoly, and because the cost function $C^I(\cdot)$ has (at least) the (weak) natural monopoly property of subadditivity (see Section 9B), this may also seem to suggest that the weak invisible hand theorem holds for the problems at issue.

While it is true that, in a formal sense, the theorem does hold in this manner for (11G1) and (11G2), unfortunately, for at least two reasons, the economic significance of the result is questionable at best. First, one must inquire about the economic meaning of the assumption that the requisite conditions for the weak invisible hand theorem pertain to the industry cost function $C^I(y^I)$. The only assumption known to imply that $C^I(\cdot)$ must exhibit decreasing ray-average costs is the premise that the underlying cost function for the firm $C(\cdot)$ has this property. Here, however, ten Raa (1980) has proved the following result:

Proposition 11G1   *If $C(\cdot)$ exhibits strictly decreasing ray-average costs, and if $C^I(\cdot)$ is trans-ray supportable at y^I, then $C(\cdot)$ is strictly subadditive at y^I.*

Thus, the assumption that $C^I(\cdot)$ has strictly decreasing ray-average costs and is trans-ray supportable, as is needed for the proof of the weak invisible hand theorem, is tantamount to the premise that $C(\cdot)$ is strictly subadditive

---

[9] This can be seen by rewriting (11G1) as the maximization of $W(p) + py^I - \sum_{i=1}^m C(y^i)$ subject to $Q(p) = y^I$ and $py^I - \sum_{i=1}^m C(y^i) \geq 0$. Then, for a given y^I and p, the objective function is maximized and the profit constraint is rendered least binding by the choice of the $m, y^1, \ldots, y^m$ that minimize $\sum_{i=1}^m C(y^i)$ subject to $\sum_{i=1}^m y^i = y^I$.

and that the industry is a natural monopoly. Consequently, the primary results of Chapter 8 are simply not relevant to the case of oligopoly.

Second, the solution of (11G1) will, in general, yield negative profits to at least one firm, and so it will be unsustainable. As a simple example, consider an industry that offers two products whose demands are independent. Then, the Ramsey optimum in (11G2) is characterized by the inverse elasticity rule (for a proof see Baumol and Bradford, 1970):

(11G3)
$$\left[\frac{p_1 - \partial C^I / \partial y_1}{p_1}\right]\varepsilon_1 = \left[\frac{p_2 - \partial C^I / \partial y_2}{p_2}\right]\varepsilon_2,$$

where ε_j is the (positive) price elasticity of demand for good j. To add simplicity, let costs be separable so that $C^I(y_1, y_2) = C^1(y_1) + C^2(y_2)$. Then (11G3) can be rewritten as

(11G4)
$$\left[1 - \frac{C^1}{R^1 S_1}\right]\varepsilon_1 = \left[1 - \frac{C^2}{R^2 S_2}\right]\varepsilon_2,$$

where R^j is the revenue derived from good j, $p_j y_j$, and S_j is the degree of scale economies in the production of good j. If there are increasing returns to scale, marginal-cost pricing is not feasible financially, so the profit constraint must be binding, and we must have $R^1 - C^1 + R^2 - C^2 = 0$ at the optimum. If both $R^1 - C^1 = 0$ and $R^2 - C^2 = 0$ were to hold at the optimum, (11G4) shows that necessarily

$$\left[1 - \frac{1}{S_1}\right]\varepsilon_1 = \left[1 - \frac{1}{S_2}\right]\varepsilon_2.$$

Of course, this will not generally be satisfied. In particular, $R^1 - C^1 > 0$ and $R^2 - C^2 < 0$ must characterize the optimum if good 1 has the less elastic demand and the smaller degree of scale economies. For such an optimum to occur in the market, one of the firms must be prepared to incur steady losses, or it must be in a position to use profits from the sale of good 1 to make up, by internal cross subsidy, the losses incurred by the sale of good 2. Of course, neither of these arrangements is sustainable. In particular, because the latter involves a cross subsidy, it permits an entrant to earn a positive profit by selling only good 1 at a lower price than that which is called for by the viable-industry Ramsey solution. If the preceding cost function were altered to $C(y_1, y_2) = C^1(y_1) + C^2(y_2) + y_1 y_2$, we see that diseconomies of scope would require the viable-industry Ramsey optimum to involve specialized firms, each producing only a single product, so that some of them must operate at a loss in perpetuity.

Thus, welfare maximization under an industry-wide profit constraint will generally require configurations that are infeasible in a decentralized market operating under a price system.

Ramsey Optimum II: The Viable-Firm Ramsey Solution

The obvious substitute for (11G1) is optimization subject to a nonnegativity constraint upon the profit of each firm in the industry. We call this second Ramsey concept *viable-firm* Ramsey optimality. Formally, it is described by

(11G5) $$\text{maximize } W(p) + \sum_{i=1}^{m} (p \cdot y^i - C(y^i))$$

subject to

$$Q(p) = \sum_{i=1}^{m} y^i \quad \text{and} \quad p \cdot y^i - C(y^i) \geq 0 \qquad \text{for } i = 1, \ldots, m.$$

Obviously, the solution to (11G5) will generally yield a lower aggregate real income than is offered by the solution to (11G1). For example, in the preceding illustration involving independent demands and diseconomies of scope, the solution to (11G5) must involve specialized firms, each producing an output at which its price is equal to its average cost. Generally, because this configuration is both feasible for (11G1) and distinctly different from the optimal solution to the latter, it must yield a lower value for the objective function. It is by no means clear whether this welfare loss should be considered a market failure. The loss occurs because, by assumption, transfers of revenues among firms in the market are ruled out. Where such transfers are not practically possible, the welfare loss is attributable to the nature of the free market and the necessary independence of its decision makers rather than to any imperfections in the structure of the product markets themselves. This situation is analogous to the relative welfare loss incurred by the Ramsey optimum under natural monopoly with increasing returns, in comparison with the welfare offered by a first-best solution. In the monopoly case, the welfare loss is attributable to the utilization of the price system and to the implicit assumption that a subsidy to the owners of a monopoly is unacceptable. Where this is so, it seems inappropriate to attribute the loss either to the technological property of increasing returns or to the presence of monopoly.

Curiously, however, the viable-firm Ramsey optimum (11G5) does require some implicit transfers among firms to minimize the damage to welfare incurred by each firm's compliance with its own nonnegative profit constraint. This accounts for the most surprising attribute of the viable-firm Ramsey optimum—the fact that it does not generally minimize total industry cost when it involves production of some good by more than one firm. This is shown by considering the first-order conditions for (11G5). Forming the Lagrangian for a given number of firms m and treating industry prices as the inverse demand functions of the industry output quantities $\sum y^i = y^I$ we obtain

$$\mathscr{L} = W(p) + \sum_i (\lambda_i + 1)(p \cdot y^i - C(y^i)).$$

Here, λ_i is the Lagrange multiplier for the constraint $p \cdot y^i - C(y^i) \geq 0$. Differentiating $\mathscr{L}$ with respect to y_j^k, the quantity of good j produced by firm k, we obtain the first-order condition

(11G6) $$\sum_h \frac{\partial W}{\partial p_h} \frac{\partial p_h}{\partial y_j^I} + \sum_i (\lambda_i + 1)\left(\sum_h \frac{\partial p_h}{\partial y_j^I} y_h^i\right) + (\lambda_k + 1)\left(p_j - \frac{\partial C(y^k)}{\partial y_j}\right) = 0.$$

Then, if at the optimum both firms k and r produce positive quantities of good j, subtraction of (11G6) for firm k from that for firm r gives

(11G7) $$(\lambda_k + 1)\left(p_j - \frac{\partial C(y^k)}{\partial y_j}\right) = (\lambda_r + 1)\left(p_j - \frac{\partial C(y^r)}{\partial y_j}\right).$$

Consequently, in an optimum, firms k and r can *only* be producing good j at equal marginal costs if the values of the multipliers for their profit constraints are equal. Hence, this unlikely occurrence is necessary for the equality of marginal costs, which is, in turn, necessary for industry cost minimization.

For example, if $\lambda_k > \lambda_r$, then at the optimum, firm k's marginal cost of producing j must be larger than that of firm r.

Intuitively, this surprising result—the inefficiency of the viable-firm Ramsey optimum—is not difficult to explain. Consider a viable-industry Ramsey optimum in which both firms k and r produce some of good j. In this situation, industry-wide costs are minimized, so that both firms produce good j at equal levels of marginal cost. Suppose, as will generally be the case where scale economies prevail overall, that the price of good j exceeds the common level of marginal cost.

Now, where this viable-industry Ramsey optimum permits firm r to earn positive profits but causes firm k to operate at a loss, the *viable-firm* Ramsey optimum forces us to seek modifications of these prices and firms' output levels which eliminate such financial losses with as little welfare loss as possible. A transfer of a small amount of production of good j from firm r to firm k is an effective modification of this sort. It adds to the profit of firm k because it yields incremental revenues that exceed the incremental costs. And it diminishes welfare relatively little, because such a transfer of production raises total industry costs relatively little as long as the marginal costs of the two firms remain close to one another. Consequently, in this example, the transformation from the viable-industry Ramsey optimum to the viable-firm Ramsey optimum will certainly entail the transfer of some amount of production of good j from firm r to firm k. And this transfer will move the allocation of production away from that required for minimization of industry cost.

Rearrangement of (11G7) shows that the transfer will proceed until the point at which

$$(11\text{G}8) \qquad \frac{\partial C(y^k)}{\partial y_j} - \frac{\partial C(y^r)}{\partial y_j} = \lambda_k\left(p_j - \frac{\partial C(y^k)}{y_j}\right) - \lambda_r\left(p_j - \frac{\partial C(y^r)}{\partial y_j}\right).$$

The left-hand side is the increase in industry cost caused by a marginal transfer of production from r to k. The right-hand side is the net social value of the implicit transfer of profits between the two firms that a marginal transfer of production effects. Thus, where $\lambda_k > \lambda_r$, the production transfer proceeds until the marginal costs of the firms are substantially different from one another: until a substantial increase in industry cost has been sacrificed to render firm k financially viable. That is why some inefficiency becomes part of the price that is paid for moving from a Ramsey optimum constrained only by an industry profit requirement to one involving a profit constraint for each and every firm.

This intuitive argument is, of course, not a proof, which requires merely an analytic example in which $\lambda_k \neq \lambda_r$ at the calculated optimum. Nevertheless, it does indicate the reason why industry costs are not generally minimized at the Ramsey optimum. The beneficent, all-knowing planner who would achieve the optimum must direct some of the firms in the industry to take actions harmful to their own interests and those of their customers in order to benefit other firms in the industry. This can take the form of incremental transfers of the output of a good from a lower cost to a higher cost producer, to enable the latter to achieve financial viability via larger sales at a positive price–cost margin. Or it can take other forms, such as a rise in the price charged by a profitable firm for a good with a relatively low elasticity of demand in order to stimulate the demand of a substitute good produced by an unprofitable firm, because the latter good is priced well above its marginal cost and has a relatively large cross elasticity of demand.

Intuitively, it should also be clear that such a viable-firm Ramsey optimum will generally not be sustainable. A firm which, in this optimum, is forced to accept this burden upon itself and/or its customers for the benefit of the customers of some other firm (who, of course, may be the same people demanding some other good) will generally find that at least some of its patronage is vulnerable to takeover by an entrant who is not subjected to such a disadvantage. While entry of this sort will benefit the former customers of the partially displaced incumbent, the customers of the other firms will, as a result, lose more than they gain. This must be true because of the Ramsey optimality of the original configuration. Thus, we cannot expect such Ramsey optima to be sustainable in contestable markets.

To take a simple and concrete example, consider the case of a mixed firm whose costs and demands are like those depicted in Figure 11E2 and redrawn here as Figure 11G1. This multiproduct firm is taken to have a natural monopoly in the production of good 1, while good 2 is naturally competitive. A host of small firms can produce good 2 at a constant cost of $1/unit. The multiproduct firm enjoys complementarities between the costs of the two goods, and its marginal costs of producing good 2 are those indicated in the figure by $C_2(y_1, y_2)$. The sustainable configuration here, as explained earlier, entails production by the multiproduct firm of the quantity y_2^1 of the competitive good at which its marginal cost, like that of any other of the good's producers, is $1/unit. The price of good 2 is $1/unit. The competitive firms supply the difference between the quantity demanded by the market and y_2^1, and the multiproduct firm charges the lowest feasible price for good 1.

If this price for good 1 is equal to the multiproduct firm's marginal cost, the situation is first-best and sustainable. However, if this price is above marginal cost, because of scale economies, this situation is neither first-best nor consistent with viable-firm Ramsey optimality. Instead, the Ramsey optimum calls for a higher price, p_2^* in Figure 11G1, for the naturally competitive good and an expanded output of good 2, y_2^{1*}, for the multiproduct firm, at which its new marginal-cost curve crosses the new price horizontal

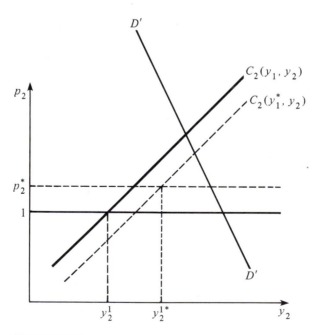

FIGURE 11G1

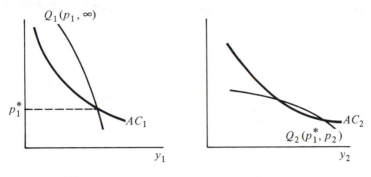

FIGURE 11G2

at p_2^*. The multiproduct firm uses its increased net revenues from good 2 to finance a reduced price for good 1, which drives up its output, thus shifting the $C_2(y_1, y_2)$ curve downward via the cost complementarity.

The reason viable-firm Ramsey optimality calls for $p_2^* > 1$, the competitive supply price, is that the marginal welfare loss, as p_2 is first raised from 1, is zero because the value of a unit of the reduced output is just equal to its marginal cost. However, the added net revenue this price increase yields to the multiproduct firm enables that firm to reduce the price of good 1, bringing it closer to its marginal cost, thereby stimulating the demand for good 1 whose unit value is greater than its marginal cost. The process of raising p_2 continues until the marginal welfare loss from an additional increase counterbalances the welfare gain from the concomitant reduction of p_1.

Thus, the viable-firm Ramsey optimum in this case requires the competitive suppliers of good 2 to raise their price from 1 to p_2^*, that is, to act in a manner disadvantageous to their customers in order to benefit the patrons of the multiproduct firm who purchase good 1. It is therefore hardly surprising that this Ramsey optimum is not sustainable.[10] With $p_2^* > 1$, entrants specializing in good 2 stand to earn profits by undercutting that price.

Normative Analysis of the Chamberlinian Case

Another concrete example provides a direct connection between the violation of optimality under monopolistic competition, and the unsustainability of the viable-firm Ramsey optimum. Figures 11G2 and 11G3 show the average-cost curves of two substitute products with scale economies and

---

[10] Braeutigam (1979) has already recognized this relationship.

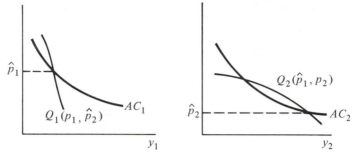

FIGURE 11G3

independent costs. Assume that one-half of the consumers find the two products to be close substitutes, while the other half are devoted to good 1 and do not find good 2 to be closely substitutable for it. The viable-firm Ramsey optimum is described in Figure 11G2. Here, good 2 is not offered, $p_2 = \infty$, and the large quantity of good 1 consequently demanded makes the low price p_1^* feasible for good 1. Yet, at that price, the $Q_2(p_1^*, p_2)$ demand curve intersects AC_2, so that profitable production of good 2 is possible. Such production is not consistent with viable-firm Ramsey optimality, however, because that reduced price for good 2 would divert considerable demand from good 1 and lead to a substantial rise in its equilibrium price to $\hat{p}_1$. The diverted demand would not yield any substantial benefits because good 2 is only slightly preferred to good 1 by those who switch their purchases. Yet, the diversion leads to the equilibrium prices $\hat{p}_1$, $\hat{p}_2$, depicted in Figure 11G3, where $\hat{p}_1$ is much higher than p_1^*, the corresponding value of p_1 in Figure 11G2.

In this scenario, the Ramsey prices $p_1 = p_1^*$ and $p_2 = \infty$ are not sustainable. At this optimum, the potential vendor of good 2 is, in effect, asked by a benevolent planner to refrain from profitable production because the consequent gain to new purchasers of good 2 is outweighed by the losses to purchasers of good 1 caused by the price increase. Yet, at p_1^*, there is an opportunity for profitable entry via production of good 2. Obversely, the equilibrium with zero profits for each firm that is illustrated in Figure 11G3 may be sustainable even though it is far from optimal. This is a simple example of the oft-noted[11] possibility that equilibrium under monopolistic competition may involve a product set that does not maximize welfare. Here, the equilibrium offers too many varieties that are insufficiently differentiated from one another.

_____

[11] See Chamberlin (1962), Willig (1973, 1979c), Spence (1976) and Salop (1979a).

This inefficiency of monopolistic competition may again be ascribed to the nature of the price system employed, because, as Spence[12] has shown, equilibrium with perfect price discrimination yields a product set that is first-best. However, since information problems and arbitrage incentives generally preclude perfect price discrimination, it seems inappropriate to attribute the difficulty to the price mechanism in question. Another view of the matter is that the problem is ascribable to the absence of incentives for each firm to take into account the effects of its decisions on other firms and their customers. This view is consistent with our general interpretation of the viable-firm Ramsey optimum in terms of a planner who must require firms to distort their own behavior in order to carry out, implicitly, interfirm transfers of net revenues. As we have seen, whether the interfirm transfers are implicit or explicit, they must lead to unsustainability.

Of course, the need for interfirm transfers disappears when the set of products in question is offered by a single firm. This is, however, an illusory solution, as the example with independent costs showed, because the same unsustainability problem reappears, this time in the guise of internal cross subsidization. Even where some economies of scope are present, the combination of product-specific scale economies and substitutability, which underlie the welfare problems of monopolistic competition, still causes unsustainability, as shown in Chapter 8. Only when the economies of scope are sufficiently large relative to the product-specific scale economies do these problems really disappear. This is one way of interpreting the weak invisible hand theorem for the case of monopoly, for trans-ray supportability is the measure of sufficiency of economies of scope for the consistency of Ramsey optimality with decentralized equilibrium in contestable markets.

Ramsey Optimum III: The Autarkic Ramsey Solution

We now propose a third variant of the concept of constrained welfare optimality that is designed to check for market failures over and above those that stem from the need for interfirm transfers. We call the concept *autarkic Ramsey optimality* to connote the requirement that each firm's operations be *independently* optimal, subject to its financial viability, given the behavior of the other firms in the industry. That is, in an autarkic Ramsey optimum, each firm maximizes its contribution to social welfare to the extent that is financially feasible, without taking account of the advantages or disadvantages its actions confer on other firms or their customers. Of course, such an optimum yields a smaller aggregate real income than the other two forms of Ramsey

---

[12] See Spence (1976).

optima. It may be appropriate to regard it as the best that can generally be hoped for in a decentralized market. It still remains to determine the structural features of an industry that can discipline its firms into playing their parts of an autarkic Ramsey optimum. Our conjecture is that the frictionless and reversible free entry of contestable markets can provide that discipline.

Definition 11G1   An industry configuration, $p, y^1, \ldots, y^m$, is an *autarkic Ramsey optimum* if each firm i selects prices and quantities that maximize $W(p) + (p \cdot y^i - C(y^i))$, subject to the constraints that the firm meets all consumer demands for its outputs and at least covers its costs, and given the prices set by the other firms.

We can make the meaning of this definition clearer with the aid of some of the examples and the properties of industry structures that have already been discussed. If more than one firm is actively producing a particular good in an autarkic Ramsey optimum, then each of these enterprises is faced with an infinitely elastic demand for this item. A small price increase will force the firm out of that market, while a small price decrease expands the quantity of the good demanded from that firm to the entire amount demanded by the market. Thus, in accord with the usual first-order conditions for a Ramsey solution, in each of these firms the marginal cost of the good must equal its market price. As a consequence, in an autarkic Ramsey optimum any two or more firms that produce the same good must have equal marginal costs that coincide with the market price. Hence, these optima are consistent with minimization of industry cost and yield first-best allocations when all goods are produced by two or more firms. Note, too, that for a partial monopoly like the one described in Figure 11G1 the sustainable configuration is an autarkic Ramsey optimum, though it is not a viable-firm Ramsey optimum. Note, also, that under monopolistic competition any sustainable configuration comprised of single-product firms is an autarkic Ramsey optimum, even though, as shown in Figures 11G2 and 11G3, it need not be a viable-firm Ramsey optimum. In each of these cases, the self-injuring acts by a firm that are required by 11G5 are ruled out in autarkic Ramsey optima.

It is noteworthy that all three Ramsey optima studied here yield identical results in three crucial cases: (i) natural monopoly industries, (ii) perfectly competitive industries, and (iii) oligopolies whose sustainable behavior is competitive—that is, each of whose products is sold by two or more firms. In the natural-monopoly case, this is so because there is no distinction between a constraint upon the firm and one upon the industry, so that the basis for the distinction among the optima disappears. In each of the other cases, any Ramsey optimum must be first-best, and so is any sustainable solution. Of course, these are the three cases in which sustainability is consistent with Ramsey optimality. Beyond these cases, we conjecture that only autarkic Ramsey optimality can generally be consistent with sustainability.

This may be interpreted to mean that even here contestable markets perform as well as can be hoped of decentralized markets employing a simple price system.

11H. Summary Comments

This chapter has taken us well beyond the conventional view of oligopoly equilibrium. Dealing with the special case of contestable markets, it has shown that the following results hold:

(i) Equilibrium is largely independent of any postulated or expected behavior patterns that characterize incumbents' responses to one another.
(ii) Equilibrium requires zero economic profits, minimization of industry costs, and absence of cross subsidies.
(iii) Any product offered by two or more firms must sell at a price equal to the marginal cost of each of its producers.
(iv) Except where each product is sold by more than one firm, equilibrium may generally be consistent only with Ramsey optimality of the autarkic variety described in the preceding section.

Of course, in reality many oligopoly markets are far from contestable. For an analysis of these cases, one must turn to the standard oligopoly models. But the model of oligopoly in contestable markets presented here provides a bench mark, based on the absence of entry barriers, with which both theoretical and actual performance of an oligopoly can usefully be compared. And our discussion also suggests that there is much to recommend any policy measures that can increase the contestability of particular oligopoly markets.

12

Powers of the Market Mechanism

Some of the preceding analysis has been rather formal and abstract. Yet it lends itself directly to the design of public policy toward the market structure and conduct of industries and firms. It argues for a significant modification in the traditional approaches employed by regulation and anti-trust programs. It suggests that, even in the absence of anything resembling pure competition, market forces are potentially a guardian of the general welfare far more powerful and effective than is commonly supposed. But, at the same time, it describes circumstances in which the market will not serve this role effectively, either if left entirely unconstrained or if constrained in the ways that traditional programs favor.

This chapter is divided into three parts. The first offers a brief review of the sorts of social damage one can hope to avoid either through the agency of the market or the exercise of public policy *vis-à-vis* industry structure and conduct. In the second, we discuss in general terms some salient features we have learned to expect of equilibria in *contestable markets*—an arrangement that includes perfect competition as a special case, while encompassing a broad variety of other market forms, even the case of monopoly. We discuss the implications of our finding that in a contestable market firms are constrained to behave in a considerable number of respects in a manner consistent with the requirements of optimality in resource allocation. Thus, in this second portion of the chapter, we describe the surprisingly broad realm of the invisible hand, as well as some limits upon what it can be expected to accomplish even within near-ideal contestable markets. The third portion of the chapter will discuss what public policy can do to make the markets of reality approximate more closely those of the contestable prototype and suggest the ways in which current policy actually impedes the attainment of this goal.

12A. Some Allocative Goals of Public Policy

Economists have long been concerned about imperfections in the workings of the market process, pointing out several forms of damage that imperfections in market arrangements can inflict. The list is familiar, but it will facilitate our discussion to recapitulate its main components, so we can simply refer to each problem by number as it arises.

1. *Inefficiency in the supply of products*, that is, supply of the industry's vector of outputs at a total cost (significantly) higher than the minimum figure permitted by the production techniques that are known and available. In real terms, this means a waste of resources, that is, production represented by a point inside the production possibility frontier. It may result from inefficient choices of inputs, inefficient use of those inputs, inappropriate pricing of inputs, failure to adopt new cost-saving production techniques, or an inefficient organization of the industry so that it is composed of a number or size distribution of firms that does not minimize total costs.

2. *Cross subsidization of particular products*, meaning that the buyers of some products do not bear the costs it takes to supply those products. This may mean that those prices fall short of the corresponding marginal costs or that the revenues of those products fall short of the corresponding total (incremental) costs incurred by the supply of those products. In either case one can expect inefficiency in resource allocation to result, with excessive demand for the products that receive a cross subsidy, inadequate demand (in terms of the social welfare) for products that bear the burden of the cross subsidies, and excessive demands for substitutes for the products that bear the burden, whether these substitutes are goods that would have existed in any event or are new products introduced in response to the opportunity created by the cross subsidy.

3. *Vulnerability of an industry to forms of entry that raise total production costs*, as when entry reduces the economies of scale and scope the industry might otherwise offer.

4. *The adoption of prices that conflict with optimal resource allocation*. This is illustrated by the prices of a profit-maximizing monopoly (unconstrained by potential competition or public policy) whose undesirable effects on resource allocation are described in all text books.

5. *The supply of a set of products or services inconsistent with optimal resource allocation*. This means that the variety and designs of the products

supplied do not strike the best balance between consumers' benefits and the social cost of their production.

6. *"Disorderly" evolution of industry*, that is, a time path of evolution of industry structure and utilization of capital that can be said to be inconsistent with intertemporal efficiency in resource allocation.

In this chapter we argue that, at least to some degree, contestable markets have the power to deal with the first five of these issues (and to some degree with the sixth) and that they do so automatically, without case-by-case intervention by any regulator. However, there are a number of issues with which even near-ideal contestable markets cannot deal directly or with which they can deal effectively only after the adoption of special measures, such as taxes and subsidies, involving magnitudes that must be determined case by case to suit the requirements of the particular circumstances. These less tractable issues include:

1. *Instability or unsustainability* of socially desirable prices and market structures.
2. *Diffuse externalities*, that is, externalities whose effects are spread among many individuals or groups so that no voluntary arrangement can be relied upon to bring them under control.
3. *Social goals in consumption*, that is, the supply of particular goods or services to particular consumers because their consumption is deemed socially desirable. Examples are the provision of nutritious foods to poor children, telephone service to the elderly, or postal services to isolated rural areas.
4. These social goals, in turn, may lead to the adoption of explicit or implicit subsidies, which constitute political as well as economic problems in themselves.

In addition, there are a number of other things that markets cannot do by themselves, as economists generally concede. They do not make for equality in the distribution of income, they do not guarantee full employment without inflation, and so forth. These last issues lie beyond the range of our discussion. However, later in this chapter we do offer some remarks on the issues of unsustainability, implicit and explicit subsidies, externalities, and social goals in consumption.

12B. The Nature of Equilibrium in Contestable Markets

A contestable market is one in which the positions of incumbents are easily contested by entrants. In brief, a perfectly contestable economic market is defined to be one into which entry is completely free, from which exit is costless, in which entrants and incumbents compete on completely symmetric terms, and entry is not impeded by fear of retaliatory price alterations.

Potential entrants may not fear such retaliation for one of several reasons: First, because entry and exit are so inexpensive that a profit opportunity, even one expected to be very short in duration, will attract new competitors; second, because incumbents are restrained by law or other impediments from undertaking retaliatory moves; or, third, because some potential entrants have "Bertrand-Nash" expectations, believing that entry will not lead to price responses by incumbents. (Such expectations may be rational, for example, if the entrants are sufficiently small relative to the size of the market.)

These markets are, then, open to entry by entrepreneurs who face no disadvantages *vis-à-vis* incumbent firms. The incumbents receive no subsidies that are not also freely available to newly entering firms. Potential entrants have available to them the same production techniques representing best current practice, the same input markets, and the same input prices as those before the incumbent firms. There are no legal restrictions on entry, and there are no special costs that must be borne by an entrant that do not fall on incumbents as well. Consumers have no preferences among firms except those arising directly from price or quality differences in firm's offerings.

In our analysis of these markets, entrepreneurs are profit seekers who take advantage of all profitable opportunities for entry. Potential entrants assess the profitability of their marketing plans by making use of the current prices of incumbent firms. Thus, for example, an entrepreneur will enter a market if he expects to earn a positive profit by undercutting the incumbent's price and serving the entire market demand at the new lower price. Potential entrants are undeterred by prospects of retaliatory price cuts by incumbents and, instead, are deterred only when the current market prices leave them no room for profitable entry. A constestable market may contain only a single monopoly enterprise whose as-yet unidentified competitors are nevertheless in the wings awaiting their entry cue.

We argue that, in theory, a perfectly contestable market offers many of the benefits heretofore associated only with perfect competition. Monopolists and oligopolists who populate such markets are sheep in wolves' clothing, for under this arrangement potential rivals can be as effective as actual competitors in forcing pro social behavior upon incumbents, whether or not such behavior is attractive to them. As we have seen throughout our formal analysis, this may be true where observed market phenomena are far from the competitive norm, and even where they superficially assume some pattern of behavior previously thought to be pernicious *per se*.

12C. Efficiency

We show (in Chapters 2, 8, 9, and 11) that an industry configuration can be sustainable only if the firms involved produce the total industry outputs at the least possible industry cost. Least-cost operation by a firm entails the purchases of inputs at the lowest available prices, no waste in use of inputs, and the choice of input combinations that keep cost to its minimum for the outputs produced. In a contestable market, firms must operate efficiently in order to survive. If a firm were to earn nonnegative profits while producing at costs greater than necessary, an entrant could undercut the firm's prices and earn a positive profit by operating more efficiently. Thus, potential competition from entrants who face no barriers guarantees that surviving, sustainable firms are cost minimizers.

Morevoer, in a sustainable industry configuration, total industry output must be distributed among firms in a way that minimizes total industry cost. If some rearrangement of output among firms could reduce total cost, then an entrant whose size was consistent with that rearrangement could earn a positive profit at prices below those that held previously. Thus, frictionless free entry has the power to enforce an efficient industry structure.

For the same reason, an incumbent firm is forced to adopt any new techniques capable of reducing costs, for failure to do so will invite entry by firms that do employ such cost-saving innovations, and those entrants will be in a position to take the markets away from the incumbents.

These general propositions have important consequences for the case of monopoly. A monopoly in contestable markets cannot find sustainable prices, and its monopoly cannot persist, unless it is the least costly provider of its array of outputs. We have defined an industry to be a *natural monopoly* if the cost-minimizing industry structure is a single firm. An enterprise that monopolizes an industry can be sustainable only if it is a natural monopoly. Frictionless free entry provides a market test of whether a monopoly enterprise is actually a natural monopoly; it is if it can succeed in remaining alone in contestable markets.

12D. Contestable Markets and Cross Subsidies

One of the major results of Chapter 8 (Proposition 8C9) is the idea that prices cannot be sustainable if they involve any cross subsidy. If any product or set of products of the incumbent firms does not yield incremental net revenues as great as its incremental net costs, then an entrant can cut prices and nevertheless earn more than the incumbent previously did. The entrant can do so by offering only the remunerative products, keeping the incumbent's subsidized products out of its own product line. Consequently, sustainable prices must all involve no cross subsidization.

As we have seen, this general result has several implications. First, sustainable prices cannot be below the associated marginal costs of production. Otherwise, the last units produced would not be compensatory and a competitor could enter profitably by scaling back production of the underpriced goods, provided he could still replicate the rest of the incumbent's operations. (See Propositions 8C7 and 11B1.) Thus, our analysis is consistent in spirit with the standard conclusion that prices below marginal costs indicate the presence of some form of cross subsidization.

Before proceeding with our discussion of the issue of cross subsidy, it is necessary to elaborate upon the rather technical definition of this concept given in Chapter 8. Quite simply, if the revenues collected from the sale of a subset of products, S, exceed the cost of providing the same quantity of those products independently, a profitable entry opportunity is offered to anyone willing to supply the same bundle at a slightly lower price and, in a perfectly contestable market, entry will occur. More formally, then, sustainability requires that

$$(12D1) \qquad\qquad \sum_{i \in S} p_i y_i \leq C(y_S).$$

Condition (12D1) has been referred to by Faulhaber (1975b) and others as the *stand-alone cost test*, and failure to pass it indicates that the set of services comprising S is in a significant sense subsidizing the remaining set of the firm's products. This is true because, *at current prices*, the users of these services will then be paying more than it would cost a separate firm to provide *only* those products at their current levels.

There is a natural symmetry between the stand-alone cost test, which can be used to determine whether products are *providing* a subsidy, and the *incremental cost test*, which indicates whether or not a group of services is *receiving* a subsidy. Formally, the incremental cost calculation tests whether the revenues from a group of products cover their incremental cost:

$$(12D2) \qquad\qquad \sum_{i \in S} p_i y_i \geq C(y) - C(y_{N-S}) \equiv IC_S(y).$$

Clearly, if a subset of services does not yield revenues sufficient to cover the additional or *incremental costs* incurred by providing them, they must be subsidized by the customers for the remainder of the firm's products, and a profit opportunity is created for a potential entrant offering only the latter. The correspondence between the provision of subsidy by some services and the receipt of subsidy by others, as we have defined the terms, arises from the requirement that, in equilibrium, firms in contestable markets earn only zero (normal) profits:

$$(12D3) \qquad\qquad \sum_{i \in N} p_i y_i = C(y).$$

To show the correspondence, let T be a product set that passes the stand-alone cost test, and subtract (12D1) from (12D3). Then we have

$$(12D4) \qquad\qquad \sum_{i \in N-T} p_i y_i \geq C(y) - C(y_T) \equiv IC_{N-T}.$$

Thus, if a product set T passes the stand-alone cost test, its *complement* product set, $N - T$, comprised of all other offerings, must automatically pass the incremental cost test. It is just as important to note that, if T *fails* the stand-alone cost test [so that the inequality in (12D1) is reversed], then its complement, $N - T$, must *fail* the incremental cost test. Therefore, we conclude that both tests must be passed for *all subsets of products*, $S \subseteq N$, for the firm's prices and outputs to be consistent with equilibrium in contestable markets. And, for a firm earning a normal rate of return, only one complete set of tests (either the stand-alone or incremental test) suffices.

The requirement that (either) the stand-alone or the incremental tests must be passed by *all subsets of products and services*, individually and in combination, may seem a mere technical complication, yet it can be crucial for the detection of cross subsidy. This is easily illustrated by a realistic example involving the application of the incremental cost test. Imagine a railroad that provides mainline service for a wide range of commodities and, in addition, has a spur line that carries only coal and iron ore, each carried on separate trains. It is tempting (and proper) to apply the incremental cost test to coal and iron ore separately. However, these tests are passed as long as the prices charged each yield revenues sufficient to cover the incremental costs of transportation of each commodity individually—the fuel, the crews on their trains, and so forth. But, the incremental cost of the coal service includes none of the roadbed cost of the spur line, because that line would have had to be built and maintained just to transport the iron ore. Similarly, the incremental cost of the iron ore service also includes none of the roadbed cost of the spur line. So, if the two services each were to contribute revenue just sufficient to pay for its own incremental cost, nothing would be available to finance replacement and maintenance of the spur, and it would make little sense to describe that set of rates as compensatory.

Consequently, in addition to testing the coal and iron ore services separately, it is necessary to test the two together to see whether their combined revenues cover the incremental cost of producing them both. This incremental cost figure does include the entire cost of the spur line, because, in our example, the spur would be unnecessary in the absence of the coal *and* iron ore services. On this view, it follows that, in order to pass the appropriate test of compensatory pricing, each product and every combination of products must contribute incremental revenues that equal or exceed the corresponding net incremental costs, including the cost of any associated incremental capital and the normal rate of return on that incremental capital. We should note that this test must, in particular, be passed by the totality of the firm's products. This last requirement ensures that the prices of the firm's products cover all of the company's costs, including a normal return on its capital. Thus, compensatory prices by definition mean that the firm as a whole is viable financially without any outside subsidy.

A yet more general view of cross subsidy, and its avoidance by compensatory pricing, is offered by the requirements of sustainability. Equilibrium in perfectly contestable markets requires that the revenues earned on any part of the total output of the industry be no more than the stand-alone production cost of that part. And it follows from this that each part of the industry's total output must earn revenues at least as large as its industry-wide incremental cost. Formally,

$$(12D5) \qquad C^I(y^I) - C^I(y^I - y) \le p \cdot y \le C^I(y), \qquad y \le y^I.$$

Of course, (12D1)–(12D4) are special cases of this relationship, in which the part of the industry output in question is comprised of all of the outputs of some product lines.

The significance of this generalization is that it applies to the part of industry output that is purchased by any consumer, or by any group of consumers. As such, (12D5) means that in perfectly contestable markets any and all consumers or groups of consumers will pay at least the incremental costs to the industry that their purchases cause. Yet, they will pay no more than they would have to if they were to provide their own products and services using the best available production techniques.

This property has aptly been referred to as *anonymous equity*,[1] because it holds regardless of the identities or the special purchasing patterns of consumers. In this connection, it should be noted that arguments before the courts and regulatory agencies against cross subsidy rarely are based on grounds of efficiency of resource allocation. Rather, cross-subsidy issues are discussed as matters of distributive equity among different groups of the supplier's customers.[2] A test of cross subsidy is intended to verify that the customers of each product bear the costs imposed by them and do not shift any portion of these costs to buyers of other products. In other words, by implication, each customer group then bears its contribution to total company costs.

However, this viewpoint should not be allowed to conceal the fact that that cross-subsidy tests are also directly relevant for efficiency. If revenues from some portion of the industry's output were less than incremental cost, then a potential entrant who was more efficient than the incumbent in producing that portion could thereby be excluded. However, such an innovator would be offered the proper incentive for entry if his costs should prove lower than the revenues that were obtainable, if those revenues were at least as large as the industry's incremental cost. Thus, the requirement that revenues be at least as large as the incremental cost floor is a necessary condition for efficiency over time.

---

[1] See Willig (1979d) and Faulhaber and Levinson (1981).

[2] For an extensive discussion of this issue, see Zajac (1978) and the references he cites.

Yet, the revenues from any portion of a firm's sales must not be prohibited from reaching that floor, especially when any higher prices would cause the firm to lose all (or a substantial portion) of those sales. As long as revenues exceed incremental costs, these sales must make a net contribution to the firm's profit. Then, where the overall profit of the firm is limited, either by regulatory constraint or by the competitive pressures in contestable markets, that net contribution permits other prices to be reduced, making other customers better off than they would otherwise be. That is, the requirement that there be a greater margin between prices and average incremental costs can lead to a substantial reduction in sales and work to the detriment of all consumers.[3]

In concluding our discussion of the issue of cross subsidy, it is also important to emphasize that, in contestable markets, situations often suspected to involve cross subsidy may in fact constitute efficient consequences of the working of market forces. The clearest example is that of a firm, regulated or not, that has a monopoly in one product and is active in the competitive market of another. Suppose, as is often true, that this firm has a larger share of the second market than any of its rivals. It is sometimes alleged that the source of the advantage that leads to this asymmetry in market positions is the multimarket firm's exploitation of its monopoly market to give it a better position in the competitive one. However, in a perfectly contestable market, the analysis of Chapter 11 has shown that this larger size is ascribable entirely to cost complementarities and that the customers of the monopoly service, far from being exploited, actually benefit, through a lower price, from the multimarket firm's participation in the competitive market. When the monopoly market is not perfectly contestable, regulation may be desirable; but regulatory policy should then be designed, insofar as possible, to replicate the results of a contestable market, thereby *encouraging* the monopolist's participation in the competitive market as a socially efficient means to take advantage of economies of scope, rather than adopting the socially wasteful policy that denies the monopolist that opportunity or impedes its use.

To summarize the conclusions of this section, in order to avoid competitive entry, an enterprise operating in contestable markets must set its prices so that all categories, subcategories, and groups of outputs are priced

---

[3] This argument goes back at least to the 1880s (Hadley, 1886, Chap. 6; Alexander, 1887, pp. 2–5, 10–11; Ackworth, 1891, Chap. 3; see also Lewis, 1949, p. 20ff., for an excellent discussion). It is one of the reasons why most economists oppose the propensity of some regulatory agencies, in response to pressures from competitors of the regulated firms, to force price above incremental cost, marginal cost, and the amount at which the regulated firm is willing to supply the product. The justification for such a decision is that the firm's price does not cover its "fully allocated cost," a figure which has no economic content and is totally irrelevant to rational decision making.

in a compensatory fashion, to yield incremental revenues between stand-alone and incremental costs. In this way, the mere possibility of competitive entry can prevent all cross subsidies.

However, it may not be considered socially desirable for all rates to be completely free of subsidy. Because such prices must reflect all cost differences, a sustainable price structure may be uncomfortably complex if costs vary among all customer groups and subgroups of services. Yet, rate averaging or any other means to simplify the structure of prices will be unsustainable if one of the resulting prices is below the corresponding marginal or average incremental cost. Services such as postal delivery and telephone directory assistance, which are often supplied without charge, render the prices of the enterprise that supplies them unsustainable in a perfectly frictionless market. Hence, there is a conflict between freedom of entry and deliberate cross subsidy, even where it is considered desirable. We shall return to this issue in Section 12H.

12E. Sustainability, Normal Profits, and Optimality of Industry Performance

We have seen that contestable markets dispose of Problems 1 and 2 (Section 12A), forcing its enterprises to operate and organize themselves efficiently and to avoid cross subsidy. These are necessary conditions for an industry's structure and rates to be sustainable.

In markets that are perfectly contestable, only an industry structure that is efficient can have sustainable prices. Suppose the firms in an industry can find and levy sustainable prices. In contestable markets such prices must yield each active firm a zero economic profit, or, in other words, a normal rate of return on capital. This follows because, by definition, sustainable prices are feasible in a free market without any subsidy, and because any positive profit would permit an entrant to undercut the incumbents' prices and still earn a positive, though smaller, profit.

Further, by definition, sustainable prices prevent the entry of entrepreneurs who propose to produce some or all of the same goods or services offered by the incumbents using techniques available to the latter. Thus, while sustainable prices do not necessarily protect firms from the entry of competitors offering improved products or production techniques, they do leave an industry invulnerable to profitable market incursions by those offering nothing new. Such incursions are, of course, wasteful socially in an efficient industry structure, since they accomplish nothing that the incumbents cannot themselves effect at a lower total cost to the industry.

Hence, sustainable prices for an industry in contestable markets solve Problem 3 in our list: Sustainable prices yield exactly normal profits to an industry's firms and protect them from competition involving socially wasteful duplication. Furthermore, in contestable markets, a configuration of firms that is not an efficient industry structure cannot find prices sustainable against any entry which can decrease total industry costs. Thus, since it seems reasonable to presume that an enterprise in contestable markets has the desire to survive and will therefore attempt to operate in a manner that is sustainable, and since equilibrium in contestable markets requires sustainability, one is tempted to conclude that contestable markets deal effectively with Problem 3, as well as with Problems 1 and 2 in our list.

There is a serious gap in the preceding arguments, however, since they have not yet dealt with the possibility that sustainable prices do not exist—even for a natural monopoly producing a set of socially valuable products. [This is a possibility first analyzed by Faulhaber (1975b). See Chapter 8, Proposition 8C5.] Where this is so, the natural monopoly enterprise in contestable markets will be unable to find equilibrium prices capable of deterring duplicative entry, and no industry structure producing the set of products or services most desirable to consumers can be stable in the presence of frictionless free entry. Thus, the unavailability of sustainable prices to an enterprise is not a proof that it is socially inefficient, and so contestable markets do not always guarantee social efficiency.

We have shown that the nonexistence of sustainable prices is more likely the stronger the degree of substitution in the demands for the various products of the enterprise, the weaker the degree of complementarity in production among its outputs, and the greater its product-specific economies of scale. Later in this chapter we offer some rough and ready suggestions for the design of public policy where unsustainability is a problem. However, where these influences are absent, contestable markets can be expected to yield an allocation of resources that is socially desirable.

Under appropriately felicitous cost and demand conditions even stronger efficiency results emerge. We have seen that under a set of conditions circumscribing the available production techniques and the structure of market demands, the weak invisible hand theorem describes how contestable markets may induce a natural monopoly enterprise to adopt Ramsey-optimal prices (see Chapter 8, Proposition 8E1). Given these conditions, sustainable prices for the natural monopolist do exist. To survive in equilibrium in perfectly contestable markets, the enterprise can and must seek to operate in a sustainable manner. The invisible hand thereby provides socially desirable guidance to a natural monopoly enterprise. The conditions that were just referred to guarantee the sustainability of the set of service offerings, product designs, and prices that are Ramsey-optimal, that is, socially optimal under the constraint that the enterprise's sales revenues cover its production costs.

Given the (global) qualitative properties of the cost and demand conditions under which the theorem holds, the enterprise can determine whether its prices are Ramsey-optimal using only economic information (marginal costs and demand elasticities) that pertains to its current levels of outputs. While prices and offerings that are not Ramsey-optimal may be sustainable, their sustainability cannot be verified without information about demands and costs at output levels that may be beyond the range of experience. Thus, the way that the enterprise can assure itself of its sustainability, using information observable directly from its current activities, is to operate in a manner consistent with Ramsey optimality.

In this case, the desire to survive can then induce a natural monopoly in contestable markets to adopt socially optimal prices and to offer the socially optimal set of services and product designs. The same invisible hand that forces Adam Smith's perfectly competitive firms, each small in its market, to operate in a manner that is socially optimal also guides a natural monopoly enterprise to a Ramsey optimum. Consequently, under the conditions of the weak invisible hand theorem, contestable markets do not dispose only of Problems 1–3. They also offer incentives to a natural monopolist that tend to eliminate Problems 4 and 5: arbitrary and socially inefficient prices, product designs, and variety of offerings.

However, it is important to note that the sustainable Ramsey prices will typically exhibit characteristics that are often viewed with some disquiet. The Ramsey price structure will generally involve prices that differ among groups of customers with varying elasticities of demand. Because they minimize the allocative distortion caused by the break-even constraint under increasing returns to scale, the Ramsey prices obtain proportionately larger contributions to fixed and unassignable common costs from those customers whose elasticity of demand is low—those who can least easily escape the required surcharge above marginal cost.

These features are also exhibited by the prices that maximize the profits of a discriminating monopolist who is unconstrained by regulation and has a market position that is protected from competitive entry. This may be the prime reason for which such a discriminatory price structure is considered undesirable. Yet, for an unsubsidized natural monopoly, which is constrained to earn no more than a normal return either by frictionless potential entry or by an overall profit constraint, such a pattern of prices is best for consumers in the aggregate. Of course, profit-maximizing prices are generally higher than Ramsey-optimal prices. However, both sets of prices follow the same qualitative pattern.

If the price structure that is required for Ramsey optimality is prohibited to a natural monopoly enterprise, then sustainability may be impossible. In such circumstances, the enterprise may require subsidies or legal protection to prevent duplicative and wasteful competitive entry. The weak invisible hand theorem only assures us that contestable markets can deal with Problems 1–5 if no constraints other than those implied by the contestable market arrangement are imposed upon the monopoly's rate structure.

Both overall increasing returns to scale and trans-ray convexity, the prime conditions for the weak invisible hand theorem, are features of the cost function that can be tested by means of econometric cost studies and by engineering simulations. Where these conditions are found to be satisfied, the weak invisible hand theorem suggests that contestable markets can indeed deal with Problems 1–5.

These problems can be solved equally well by perfectly contestable markets that are not natural monopolies. Of course, the classic form of a perfectly contestable market—a market that is perfectly competitive—in theory achieves the first-best resource allocation. However, perfect competition requires that each active firm produce only an insignificant portion of its industry's total output. Consequently, this model of industry structure has generally been considered decreasingly relevant, as actual industry structures have grown far more concentrated than those Adam Smith observed in the eighteenth century.

This observation adds considerable weight to the results that indicate that the behavior of highly concentrated oligopolies in perfectly contestable markets will have all the essential properties of competitive performance and will deal optimally with the Problems 1–5, yielding a first-best equilibrium. In particular, as was described in detail in Chapter 11, oligopolistic firms with only few rivals active in the production of each of their goods and services must adopt marginal-cost prices to survive in equilibrium in perfectly contestable markets. The socially optimal, first-best choices of product variety and product designs, and their optimal prices, comprise a sustainable equilibrium in contestable markets under a set of conditions far more generally relevant than the conditions required for perfect competition. These conditions require that a firm's costs be trans-ray convex; that, once its scale economies are exhausted, its unit costs remain constant for sufficiently large expansions beyond minimum efficient scale; and that quantities demanded by the market at optimal prices be greater than a firm's minimum efficient scale. Moreover, we know that these conditions can be generalized further, although to what extent is still an open question.

Thus, we conclude that Problems 1–5 can be solved, as it were, by the invisible hand in perfectly contestable markets in which the market is either a monopoly or an oligopoly with two or more active sellers offering each product and service. The more complex case of an oligopoly involving some firms that are the only sellers of some of their offerings is discussed later.

Problem 6 involves the intertemporal allocation of resources—the path of evolution of an industry, its patterns of investment, and the utilization and depreciation of its capital. These issues are analyzed in detail in Chapters 13 and 14, although in this area much remains for future research. There are two principal conclusions to be drawn from this analysis, one unfortunate and the other fortunate. The first of these asserts that unsustainability is the rule in growing markets where capacity construction costs are sunk and subject to increasing returns to scale that are sufficiently great. Here, it would seem that perfectly contestable markets cannot serve even as an ideal standard for socially desirable intertemporal performance. Rather, our theoretical analysis seems to indicate that industry has a tendency to evolve in a disorderly fashion, by fits and starts, perhaps even giving rise to substantial waste in the form of duplication of capital.

On the other hand, our more felicitous conclusion holds where capacity is fungible or where its construction costs are marked by scale economies that are smaller relative to market growth. Here, contestable markets can solve Problem 6 with remarkable efficacy. The threat of entry that hangs over them at all times can force incumbents, in intertemporal equilibrium, to invest in new equipment, adopt new techniques, salvage old equipment, depreciate capital to account for inflation, employ nondurable inputs, and set prices in socially optimal patterns. Thus, the analysis offers reason to believe that market forces can in some cases effectively guide the difficult decisions involved in intertemporal resource allocation. Furthermore, here perfectly contestable markets can serve as an effective bench mark against which the intertemporal performance of less perfect markets can be assessed.

12F. Policies that Move Markets toward Contestability

All the goals involved in solving Problems 1–6, then, can be attained under appropriate conditions more or less automatically in the sort of markets to which we have assigned the label "perfectly contestable." Of course, in practice many markets do not have the characteristics required for perfect contestability, and the issue, then, is to determine whether there exist straightforward policy measures capable of increasing substantially the degree of approximation of actual market arrangements to those of the contestable ideal. This issue is discussed in detail in Chapter 16, but a few pertinent remarks are appropriate here.

The two crucial characteristics of perfectly contestable markets are complete freedom of entry and exit, without handicaps upon the entrant; and, where entry and exist are costly, inhibition of strategic pricing responses by incumbents to moves by entrants. That is, firms must experience no impediments to movement in and out of the industry, or, where commencement or cessation of operation has some cost, entrants must have grounds for confidence that incumbents' prices will not be changed each time the entrant makes a move that promises to be profitable.

The first of these characteristics immediately suggest some appropriate changes in current public policy. Legal impediments to entry like those that, until recently, characterized almost all regulated industries, simply are the reverse of what is called for by the public interest. Special licensing requirements and other preconditions for the launching of operations must be eliminated unless fully justified by clear dangers to health or public welfare.

But elimination of legal barriers to entry is not always enough. Indeed, as in the theorem of the second-best, such a move by itself may well make matters worse. If, along with the dismantling of entry barriers, incumbent firms are required to adjust pricing and other policies in order to ensure the survival of entrants, or if incumbent firms are inhibited from attempting to reach sustainable prices, the net result may be the worst of all possible worlds: a mandatory cartel that offers neither the virtues of effective competition nor the benefits of efficient industry structure. Such a partial move toward an ideal market is certain to cost the consumer heavily and benefit only the inefficient entrant thereby protected from the rigors of true competition.[4]

The problem, then, is that of preventing punitive pricing responses from undermining effective freedom of entry. As indicated in detail in Chapter 10, if there are no sunk costs incurred in the establishment of a firm, the possibility of retaliation by the incumbent may not be terribly important. Where all of an entrant's plant and equipment is, as it were, on wheels, more can be introduced into and withdrawn from the industry quickly and cheaply. Then any profitable entry opportunity presented by unsustainable prices invites hit-and-run tactics. Entrants can gather in short-run gains and depart as soon as the incumbent's countermeasures threaten to do any damage.

But where there are significant costs of entry and exit, retaliation can be very damaging to entrants, who will thereby be inhibited from taking advantage of the entry opportunities dangled before them by unsustainable prices, unless something is done to prevent or discourage retaliation by incumbents.[5] Thus, where movement of capital is not cheap, contestability of the market requires some policy measures designed to inhibit retaliatory measures by incumbents. Although it is difficult to design such measures that do not impose ills worse than those they cure, one such proposal, called *Quasipermanence of price reductions*, is described in the Appendix. This policy would permit incumbents to adjust their prices in response to entry, but would freeze any such price reductions until exogenous changes warrant their upward readjustment. Thus, entrants would be freed from an unending succession of retaliatory price responses by incumbents.

12G. Explicit and Implicit Subsidies to Firms

As indicated in Section 12A, there are a number of issues with which even perfectly contestable markets cannot deal automatically—and a number of elements of socially desirable performance that they cannot be expected to

---

[4] Moreover, we have seen that in the case where no sustainable prices exist entry barriers may be necessary to prevent the waste inherent in duplicative entry.

[5] This possibility raises, once again, the theoretical problems associated with strategic entry deterrence and predatory pricing. The literature on this subject is enormous and growing. The reader is urged to consult relevant references cited above, for example, in Chapters 2, 8, 9, 10, and 11.

carry into effect without the help of special measures. In this and the next two sections we briefly discuss these elements and some of the special measures they may make appropriate.

It is well established that there may be circumstances in which firms require subsidies to enable them to perform certain delimited activities that are socially desirable, but which are fundamentally unremunerative. For example, there can, in principle, be services whose total benefits to consumers exceed the costs of providing them, but for which revenues obtainable from a price system cannot cover costs. Of course, activities of this kind cannot be expected to survive in contestable markets, in the absence of special arrangements, because they require firms either to operate at a loss or to undertake unsustainable cross subsidies.

In such circumstances, external subsidy may be warranted. However, unless subsidies are provided with great caution and only to activities selected with great care, they are apt to damage the general welfare as they are likely to encourage inefficiency in the operations of the firm by shielding it from the consequences of its own ineptitude. In addition, subsidies are likely to protect incumbent firms from competition, since they are artificial handicaps which must be overcome by entrants who are not eligible for them.

Such problems can be overcome, partially, by earmarking external subsidies to specific activities rather than to specific firms. If a given amount of subsidy is made available for a given level of performance of an activity, by any active firm and by any prospective entrant, then the market can remain contestable. In a system following this design, private demands for the goods or services in question are simply supplemented by the demands of the public sector that are expressed in the market through the subsidy scheme. Consequently, such a market can, in principle, operate as effectively as one not subject to public demands, although it provides the larger outputs that are the goal of the subsidies.

Explicit and implicit interfirm subsidies are far more difficult, if not impossible, for markets to handle, even in principle. Chapter 11 shows that viable-industry Ramsey optima require financial flows between firms, where the optima involve oligopolies with some firms that are the only sellers of some products. Of course, contestable markets do not permit such flows in equilibrium, unless special arrangements require them of the donor firms. The overwhelming difficulties caused by the imposition of financial flows lie in defining their size, the conditions under which a firm must pay a subsidy, and the conditions for eligibility for receipt of a subsidy. Determination of the optimal sizes of the subsidy flows requires complete information on all the economic relationships that bear on the industry, as well as exceedingly delicate calculations. The optimal sizes of the flows change with any alterations in the industry's circumstances. And the information on which the calculations were based would be subject to manipulation by firms in the attempt to reduce their obligations and to raise their receipts. Moreover, arrangements of this kind give firms the incentive to alter their operations so as to meet the eligibility requirements for subsidies and to avoid the conditions under which they would have to pay subsidies. Thus, for all these reasons, we conclude that markets, even contestable ones with whatever special arrangements are feasible, are probably unable to provide the interfirm subsidies necessary for the class of viable-industry Ramsey optima.

In the same circumstances, as discussed in Chapter 11, viable-firm Ramsey optima require what we described as implicit interfirm subsidies. These are pricing and production decisions by some active firms and by prospective entrants that are disadvantageous to themselves or to their customers, which they adopt to increase the profitability of other firms or the welfare of other firms' customers. Clearly, firms that are donors of such implicit subsidies cannot be in equilibrium in contestable markets because they can be replaced by entrants who do not impose such disadvantages upon themselves or their customers. Consequently, special arrangements are required if such a program is to be possible. But these, too, would be prey to all the difficulties that plague systems of explicit interfirm subsidies, and to one more in addition. Here, even the measures that are supposed to provide the implicit subsidies would be difficult for the relevant authorities to monitor and assess with any degree of precision. Hence, we conclude that neither contestable markets nor any other decentralized mechanisms are generally able to provide the implicit interfirm subsidies necessary for some viable-firm Ramsey optima.

It is this that led us to define autarkic Ramsey optima in Chapter 11 in terms of the socially efficient performance of each firm in an industry, taking as given the operations of all its other firms. Optimality of this sort seems to be the best we can hope for in the circumstances we are discussing. Virtually by definition, these optima require no interfirm subsidies. And, we conjecture that these optima are consistent with equilibrium in contestable markets. While autarkic optima coincide with both viable-industry and viable-firm optima for industry structures that are natural monopolies or oligopolies with at least two firms producing each output, in other circumstances their performance in welfare terms can be expected to be inferior. However, in general, contestability can achieve no better welfare performance than that provided by autarkic Ramsey optima where they contain firms some of which sell products in oligopolistic or competitive markets and some in markets in which the enterprise has a monopoly.

12H. A Policy Approach to Unsustainability

Without special measures, contestable markets are incapable of dealing efficiently with cases in which no sustainable prices and industry configurations exist. For as indicated in Chapter 8, one or even one hundred responses by incumbents will not bring stability to the industry, because each one will open some new opportunity for profitable entry.

The analysis of policy approaches to deal with endemic unsustainability or, what is more common in regulated activities, the absence of any sustainable prices deemed politically feasible, is still in its infancy. However, there is a promising approach, which we shall discuss briefly in this section. It operates on the basic principle that when the incumbent's obligation to serve the market leaves it no subsidy-free set of prices (or no politically acceptable or socially desirable set of subsidy-free prices), it may be possible to achieve many of the benefits of contestability by devising and imposing a burden equivalent to the requirement to serve all customer demands on potential entrants as well.

As an example, consider an industry with vertically related services, with the "upstream" (input) service exhibiting global product-specific scale economies so that it must be monopolized in any sustainable equilibrium.[6] As is well known, such a vertical structure may make it possible to carry out price discrimination in downstream markets via forward integration. This practice by an unfettered monopoly is usually deplored but, as we have seen, such price discrimination may be necessary for Ramsey optimality or even for the sustainability and/or viability of the upstream service. In any case, we know that in perfectly contestable markets such price discrimination can only serve, not harm, the public interest. However, such socially desirable price discrimination as well as possible economies of scope may be vulnerable to entry which, in effect, performs an arbitrage function that undermines the separation of the markets and the cost savings achieved by vertical integration.

One obvious policy option is an attempt to preserve these benefits by banning such entry. Unfortunately, this would also make innovative, cost-reducing entry impossible. However, in some circumstances it is possible to retain the advantages of vertical integration while leaving open the possibility of innovative entry. The solution involves retention of the socially desirable discriminatory price structure for the intermediate product, but requiring that the discrimination be based on categories of final *use* rather than the identity of final producers. That is, it requires potential entrants and the (perhaps hypothetical) downstream subsidiary of the upstream monopolist to pay the *same* price for an intermediate service needed for the provision of services competing in the *same* final product market.

This "symmetric burden" approach can also be used to remedy the prototype unsustainability situation depicted in Figure 8A3 by means of artfully selected taxes and subsidies that are applied identically to incumbents and potential entrants alike. This approach may also make it possible to

[6] See Panzar (1980) and, especially, Willig (1979b) for a formal treatment of this problem.

shift the region of sustainability to include previously unsustainable but socially desired price vectors. [See K.C. Baseman (1981) and the Comment by Panzar.] Additional research in this area will no doubt indicate still more options for policy.

12I. Diffuse Externalities and Social Consumption Goals

The economic literature tells us, at least in principle, how to deal optimally with diffused externalities in a contestable market. Either Pigouvian taxes and subsidies, applied identically to active firms and to prospective entrants, or a market in permits (open to all comers) for the generation of externalities can take care of the problem perfectly, at least in theory. This has long been recognized for a world of perfect competition. The contribution of our analysis is the extension of the result to *any* contestable market, even one which is oligopolistic or monopolistic. There is one required amendment, however. In a world of universally perfect competition, everything is priced at marginal private cost. Then a Pigouvian tax exactly equal to the marginal social damage can correct any misallocation produced by the externality. But, where the economy's equilibrium does not involve the equality of all prices to their associated marginal costs, the relevant welfare standard is the Ramsey optimum, with its optimal deviations from marginal cost pricing. Since every price in such an arrangement will differ from marginal cost, the Pigouvian tax, too, must assume its Ramsey value, which differs from the social marginal cost of the externality exactly in accord with the same relationship that holds for any other price.[7] If Pigouvian taxes and subsidies are set in any other way, they will undermine the Ramsey solution and prevent the attainment of welfare optimality, subject to the relevant financial constraints.

Similar comments apply to types of consumption the community wishes to sponsor because society considers them particularly meritorious. Consumption of nutritious food by children in improvished families, for example, can be considered to generate a benefit over and above what the parents are willing and able to pay. If society can determine a marginal value for this supplementary benefit, such consumption can be treated exactly as though it were an activity that generates an external benefit. An appropriate Pigouvian subsidy, determined exactly as in the preceding discussion, can take care of the problem in an ideal manner in contestable markets, provided the subsidies are offered to all suppliers, be they active firms or potential entrants.

Several of the programs to modify contestable markets that have been described in this chapter, including the Pigouvian tax and subsidies to types of consumption that are favored socially, can be considered special cases of

---

[7] This implication is at least implicit in Diamond (1973).

a more general approach. Under this general program, objectives not automatically taken into account by market forces may nevertheless be pursued. But this is done by a policy that relies upon general financial inducements, not case-by-case intervention and direct controls. The financial inducements are offered on a symmetric and even-handed basis to all active and potential market participants. In this way, market competitiveness and contestability can remain unimpaired. The costs of the pursuit of the special objectives are borne by society as a whole, not by particular classes of consumers who have the burden thrust upon them by cross subsidies either voluntarily adopted by or imposed upon firms by the authorities. Such cross subsidies, as we have seen, are sources of inefficiencies that can, at least in principle, be eliminated by the general approach discussed in this section.

12J. Concluding Comment

While it has gone beyond formal theory, this chapter nevertheless can be taken to deal primarily with the theory of policy on market structure and conduct. We have summarized the principles underlying the solutions that are offered by perfectly contestable markets to many basic problems of resource misallocation. We have offered some suggestions designed to move actual markets toward contestability. We have identified economic issues with which contestable markets cannot deal without the aid of special programs, and we have indicated in general terms how such programs can be structured to leave markets' contestability unimpaired. Undoubtedly there is much more to be said about these subjects. Indeed, there is an enormous literature on related topics, but most of it does not rest as directly as does ours on a systematic theoretical framework. It is not surprising, then, that our policy proposals differ from many offered in the earlier literature, and vary substantially from the policies pursued in practice.

The analysis needs to be extended in many directions, some of which have already been indicated in preceding chapters. It is certainly clear from Chapter 11 that the depths of general sustainable industry structures must be plumbed further, and that autarkic Ramsey optima warrant further study. Vertical industry structure in contestable markets is an important subject as yet uninvestigated. The policy measures suggested in this chapter require refinement and additional analysis. Extension of the framework to the intertemporal case is begun in the next two chapters.

APPENDIX

Quasipermanence of Price Reductions[1]

If the world were stationary, with no exogenous economic changes, it would, at least in principle, be simple to design an effective program to render markets contestable. Incumbents would be required to adopt prices for their products once and for all. These prices would be permitted and even encouraged to be designed in a way that minimizes the likelihood of successful entry. But once these prices had been adopted, the incumbent would never be permitted to change them again. In that way, consumers would be assured of the benefits of competition even though there were, as yet, no credible threat of entry. Entrants, too, would benefit by absolute protection from retaliatory and strategic price counter measures. Of course, once established, the entrants, too, would be subject to these restrictions. Such a program would also be easy to supervise by the authorities, who need merely assure themselves that no price was ever changed.

Of course, in our world, in which nothing stands still for very long, such a proposal is sheer fantasy. As demands shift, as inventions modify production techniques, as inflation changes price relationships elsewhere in the economy, absolute rigidity of prices in any industry becomes both impractical and undesirable.

In fact, there seems to be no measure capable of doing the job ideally. Granted that changing circumstances require changes in prices, the question becomes: Which price changes, if any, should be prohibited? In particular, how should entry affect the permitted range of freedom of pricing?

When entry occurs, the obvious alternatives for policy makers who are to pass on price responses proposed by incumbents seem equally distasteful: Rejection of proposals for competitive responses is tantamount to the grant of a protective shield to the entrant, the first step toward the establishment of a cartel and an invitation to inefficiency and, ultimately, to poor customer

---

[1] This approach to the issue of predatory behavior (Baumol, 1979b) was directed at the debate in the literature begun by Areeda and Turner (1975), with contributions from Scherer (1976) and Williamson (1977) with rejoinders from Areeda and Turner (1976, 1978).

service and perhaps to higher prices. On the other hand, approval of the proposed responses carries the danger that established firms will succeed in driving the entrant from the market and that the low prices offered by the established firms will then be quietly withdrawn after the competitive threat has passed.

Thus, the policy designer sails between Scylla and Charybdis, apparently with no safe middle course. Neither the creation of a cartel nor the sanctioning of the destruction of the entrant is an inviting prospect. And neither seems to get us closer to our near-ideal contestable market. But there *is* a third possibility that may prove effective under relatively stationary circumstances: The established firm can be left free to adopt prices that protect its interests without being permitted to *read*just those prices *in response to further moves by the entrant or after the entrant's demise.* In short, such price reductions can be made *quasipermanent.* Under such an arrangement established firms are told they can reduce their own prices as far as they wish; but these incumbents are also informed that, if the entrant should cease to operate or should change price tactics, they would not be permitted to withdraw these low rates in the future except in response to independent changes in costs and market demands.[2] Under such an arrangement the established firm is, in effect, put on notice that its decision to offer service at a low price is tantamount to a declaration that this price is compensatory and that (*in the absence of exogenous changes in costs or demands*) it can therefore be expected to offer the service at this price for the indefinite future.[3]

Inflation and other autonomous changes in costs and market demands pose no insuperable problem for the administration of such a program. The established firm which has cut a price by 50 percent can be permitted to adjust that price upward when, say, there is a 5 percent rise in fuel prices.

---

[2] Of course, if entry and exit are cheap and easy, or if the entrant is alive and well, there seems no reason to prevent the established firm from rescinding a price cut if that is obviously not a response to a new price move by the entrant. After all, such a rise in the established firm's price can only be to the advantage of the entrant.

[3] In practice, it may be necessary to set a finite time limit upon the period for which price increases are precluded. The period should not be so brief that it all but eliminates the cost of retaliation to the incumbent, but it should not be so long that it threatens to become a major source of price rigidity.

It probably would be desirable to enforce a quasipermanence program *ex post* rather than opening up the possibility of substantial regulatory delays by requiring advance approval of all price changes. The incumbent firm could be challenged under regulatory procedures or the antitrust laws on the basis of a questionable price increase, and would have to show in its defense that this price increase was, to a rough order of magnitude, commensurate with autonomous rises in cost, shifts in demand, or other exogenous influences. Alternative admissible defenses would be evidence that the costs of entry and exit in the industry are negligible, or that the entrant was alive and well at the time of the price rise.

But the firm is permitted a rise that brings in additional revenue sufficient to cover the rise in fuel cost, *starting from its new, low price*.

To see how this pricing arrangement is related to the requirements of a contestable market, it is convenient to recall the distinction between what we have referred to as *stationary* and *responsive* limit pricing. Responsive limit pricing refers to a policy under which the established firm continually readjusts its prices to meet each move of an entrant. Stationary limit pricing is a policy very close to the pricing that characterizes a contestable market. It refers to a set of prices that the established firm adopts "once and for all" (or changes only because of exogenous developments such as rises in input prices), those stationary prices being chosen so as to protect the established firm to whatever extent is possible from the incursions of entrants.

This is precisely the sort of price behavior that can facilitate the workings of a contestable market. For if incumbents' behavior indicates their commitment to the current set of prices, potential entrants will have no reason to fear retaliatory or responsive price moves if they decide to enter the market.

In addition, from the viewpoint of society, a policy that permits and encourages stationary limit prices offers the benefit that it leads the established firms to protect themselves from entry only by making themselves so attractive to consumers in the first place that they will have no motive to switch to entrants. In the simplest example, if the established firm offers products of high quality at low prices *before* entry takes place, there will be little inducement for entry to occur. Because under stationary pricing established firms must decide upon their prices in advance lest they be left vulnerable to entry later on, they are forced to offer consumers the price-quality benefits that would stem from competition in anticipation of the possibility of competition, even if it has not yet materialized.

The quasipermanent pricing proposal does not yield all the benefits possible when sustainable Ramsey prices are offered in an idealized contestable market. Ramsey prices render a case-by-case response to entry unnecessary while the former permits a price response but, so to speak, permits it only once. The established firm is not inhibited from reducing its price after entry occurs, whenever and wherever it occurs; it is only precluded from rescinding such a price cut, at least for a stipulated period. This implies that quasipermanence of price reductions may not be quite as effective in serving the public welfare as sustainable prices, since it does not offer consumers the advantages of competition *before* entry occurs. Yet the former may be about as close an approximation to the latter as we can expect to achieve through public policy, for in practice it will often be difficult to require established firms to prepackage their responses to any and all potential entrants, something that is necessary for sustainable prices but not essential for a policy of quasipermanence of price reductions.

On the other hand, it can perhaps be argued that the relative flexibility of the policy of quasipermanence offers more assurance that entry will have an effect upon prices when it does occur, an effect that might otherwise not be realized if the established firm lacks sufficient foresight.

It is, of course, difficult to judge the comparative validity of such impressionistic arguments whose relative force may, in any event, vary from case to case. It is sufficient for our purposes to reemphasize the claim of allocative efficiency that can be made for the broad class of pricing policies encompassing both sustainable pricing and the policy of quasipermanence of price reductions.

13

Intertemporal Sustainability

In previous chapters we ignored the passage of time. Here, and in the next chapter, we begin to consider intertemporal behavior patterns—patterns that must be explained principally in terms of the role of time in production relationships. We shall see that many of the concepts and tools presented earlier can still be used in the intertemporal analysis. And, although several special difficulties beset the achievement of optimality by the market mechanism, previous results such as the weak invisible hand theorem take on new significance where they nevertheless still hold.

Our basic analytic device takes the outputs of a given product or service at different dates as supplies of different goods. Viewed in this way, every industry that exists for more than a moment must be interpreted as a multi-product industry.

What does sustainability mean in the intertemporal context? We can define it formally, as before, with each distinct good in the ordinary sense now interpreted as a multiplicity of outputs, one at each of the dates under consideration. However, here, the interpretation runs into problems involving timing, anticipations, and information structure. For example, it is clear from our previous discussion that in a contestable market a firm with two product prices to select cannot choose price magnitudes that yield a positive profit without providing an incentive for entry. But if the two prices pertain to two different time periods, profitability need not always invite entry. For example, a potential entrant may infer from a high price in the first period that a low second-period price is to follow, making entry appear unprofitable. But if entry does not, in fact, occur for this reason, then the incumbent can adopt a high price in the second period too, thereby obtaining a positive net profit with impunity. Such manifestations of imperfect information and unfulfilled expectations can certainly be very significant and, as indicated in Chapters 1 and 10, they are the subject of much current research. We will return to these effects briefly in the following chapter, suggesting how they can serve as equilibrating forces and as instruments incumbents can use to make viable solutions possible. However, such phenomena lie outside the general scope of this volume. Thus, in this chapter, we will abstract from uncertainty and incorrectness of anticipations and analyze, as a benchmark, the implications of sustainability of intertemporal plans in the absence of these influences. Such intertemporal plans consist of sequences of price vectors, market-clearing output vectors, and choices of productive techniques and investments.

This raises another issue relating to the interpretation of the concept of sustainability in an intertemporal analysis. Sustainability is usually described as a Bertrand-Nash concept. It means that there are available *stationary* prices which can discourage entry—prices which can do this without their magnitudes' ever being changed in response to actual or threatened entry. Yet, it is important to recognize that, in an intertemporal analysis [in which the vector of prices, $p = (p_1, \ldots, p_h)$, may represent the price of a single good in h different periods, rather than the price of h different goods in some one period], the Bertrand-Nash premise need not preclude changes in prices over the course of time. That is, we may have $p_{t+1} \neq p_t$ for any or all periods t. However, what the Bertrand-Nash premise does mean in such an analysis is that, given any such intertemporal price vector p, the threat of entry which occurs at date s (where $1 \leq s \leq h$) must not lead to any change in the values of current and planned future prices, $p_1, \ldots, p_h$.

This chapter is devoted largely to the analysis of two important special cases. The first involves independence of the costs of production in different periods, and the second is that in which all the relevant cost relationships are linearly homogeneous. It is not entirely surprising to find that the first of these retains most of the features of the timeless analysis of the preceding chapters and that many of our earlier results on subadditivity, sustainability, and contestability continue to hold for this model. Nevertheless, it transpires that even in this case the influence of time does make a difference. For example, we find, paradoxically, that an invention that *can* be used to change production techniques may force a change in market prices by narrowing the range of sustainable prices, even if it is not adopted anywhere because it is not economically efficient to do so. After analyzing the simple world in which costs are intertemporally independent, we turn next to the case in which there is real intertemporal interdependence, but in which linear homogeneity still provides some analytic simplicity. Here matters still work out neatly, and sustainable prices will generally be available. In this case, it also proves possible to derive unique principles for the determination of Ramsey-optimal payments to capital, that is, depreciation plus return on investment. That is, such payments correspond to the vector of Ramsey-optimal prices over time, contributing a sufficient vector of surpluses of revenues above total variable costs, so that through the lifetime of a piece of capital equipment the discounted value of these surpluses is sufficient to cover the cost of that item. While the demand functions play a role in the determination of the optimal stream of depreciation payments, we show that the depreciation schedule can generally be bounded effectively just with the aid of cost information and knowledge about the dates at which some investment occurs. This optimal depreciation stream will be seen to be necessary and sufficient for sustainability of the corresponding price vector.

We conclude the chapter with a preliminary glance at the case of nonlinearity in the cost of construction of capital. There we see for the first time the nature of the severe problems for sustainability that accompany such cases. That is the issue the following chapter examines in detail.

13A. Intertemporally Separable Production

We begin our discussion with the particularly simple case of intertemporal separability, in which the influence of time is reduced to a minimum and which, therefore, can serve as a halfway house to a full intertemporal analysis. Indeed, we see that this case provides an interpretation of the single-period analysis of the preceding chapters for a multiperiod world.

Production is said to be *intertemporally separable* if the total cost function of a firm satisfies

$$(13A1) \qquad C(y^1, \ldots, y^T) = \sum_{s=1}^{T} C^s(y^s)\rho_s,$$

where y^s is the vector of outputs in period s, T is the time horizon, ρ_s is the discount factor that gives the present value of a dollar in period s, and $C^S(y^S)$ is the contemporaneous cost of producing y^s. If the cost function satisfies (13A1), costs during any one period are unaffected by the output quantities in any other period. There are no economies of scope among products supplied at different dates. There is, therefore, no natural monopoly encompassing the outputs in different time periods. Hence, if there are economies of scale, there cannot be trans-ray convexity over the entire dated set of goods $y^1, \ldots, y^T$. Yet, this relationship may be satisfied for the various outputs produced during each individual time period, taken by themselves.

If the demands for products at any time are independent of the prices of products produced at different dates, then the intertemporal sequence of prices, $p^1, \ldots, p^T$ is sustainable for C if and only if each period's price vector p^s is sustainable for C^s. This way of looking at the matter gives us a bridge to the atemporal analysis that occupies the early parts of the book. Further, in this case, anticipations cause no analytic problems as long as the production of potential entrants is also intertemporally separable. For in this case, decisions can be made period by period with no concern for interdependences with other periods.

Here, it is appropriate to offer a narrowed interpretation of the weak invisible hand theorem. Suppose each period's cost and demand functions satisfy the theorem's hypotheses. Then, we know that the Ramsey prices optimal for each individual period are sustainable within that period and, therefore, that the sequence of such price vectors is intertemporally sustainable. Moreover, because each period's cost function, by assumption, has declining ray average costs, the same must be true of the intertemporal cost function. But, because that function is not trans-ray convex, the intertemporal Ramsey-optimal prices need not be sustainable. This suggests the noteworthy conclusion that the sequence of Ramsey-optimal price vectors for the individual periods need not comprise a set of intertemporal Ramsey-optimal prices. To show this, suppose that the firm produces only one product in each period and that costs are stationary through time but that demands are not. Then, the Ramsey prices for all the individual periods will each be equal to the average cost incurred in its own period. But, suppose demands are independent and grow constantly less elastic, and so are more elastic in period 1 than they are in period 2. The intertemporal Ramsey prices must satisfy only one budget constraint: the intertemporal budget constraint, that is, the budget constraint pertaining to the discounted net receipts of all periods together. Then, in accord with the inverse elasticity rule, the intertemporal Ramsey prices must be lower in period one and higher in period two than the prices that are Ramsey-optimal for the two individual periods. The intertemporally optimal prices will therefore yield a negative profit in period 1 and a positive profit in period 2. This suffices to show that the intertemporal Ramsey prices will, in this case, differ from the Ramsey prices calculated period by period. It is also noteworthy that, because they may be profitable in some periods, the intertemporal Ramsey prices may be unsustainable since, if adhered to, they will permit profitable entry in any period with positive profits for the incumbent.

Other interesting effects arise if production technology is not stationary. Of course, if the entire set of available techniques changes, then one can expect the set of sustainable prices and the Ramsey-optimal prices to change concomitantly. However, it may be surprising to note that when some new techniques become available, even if it remains optimal to employ the same techniques as before, the set of sustainable prices may nevertheless be changed. Perhaps the most direct way in which this can occur is through the introduction of a new technique that reduces the stand-alone average production cost of some good below its initial price. The prices of such goods must then also decrease if cross subsidy and the resulting unsustainability are to be avoided, for the incumbent firm cannot sustain a price of a good higher than the average cost at which the item can be supplied all by itself. This situation is depicted in Figure 13A1. Here it can be recalled (see Chapter 8) that a point such as A on arc CD represents a pair of prices that are sustainable given the set of techniques corresponding to the solid curves, since at or below those prices neither good can be produced by itself without incurring a loss. But the price vector A must be changed to, say, B, when the stand-alone costs of producing good one are reduced so that the curve labeled $\pi_1 = 0$ shifts to the position of the broken curve.

Thus, despite continued use of the techniques corresponding to the curve labeled $\pi = 0$ (the techniques that produce the two goods together), the availability of the new technique, and the potential entry threat that it introduces, can force industry prices to change.

Such a situation can easily lead to misunderstanding in an antitrust examination. To see how this can occur, envision the following scenario: Assume entry is threatened by a specialized firm proposing to use a new production technique to supply good 1. The incumbent firm cuts its price of good 1 and simultaneously raises the price of its other product, good 2. Thereafter, entry does not materialize and the old techniques continue in use. It is easy, under these circumstances, to jump to the conclusion that the price cut was in some sense exclusionary—that it arose only from the incumbent's desire to prevent entry. But we see now that the mere *availability* of

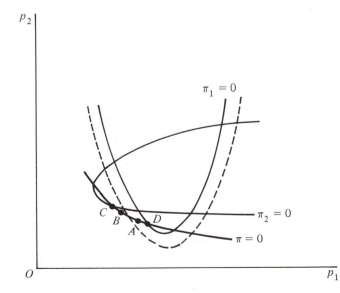

FIGURE 13A1

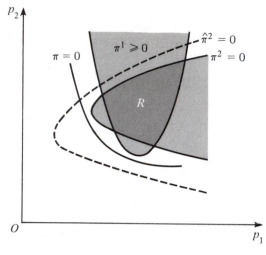

FIGURE 13A2

the new technique requires prices to change. By reducing the "stand-alone" cost of producing good 1 in isolation, it also increases the incremental cost of good 2, and this can account fully for price changes of the sort described. Thus, this price behavior can be entirely innocent, though, of course, it is not necessarily so.

The effect of an innovation of the sort we are considering can be even more extreme. As shown in Figure 13A2, it can even transform a sustainable situation into an unsustainable one, in which no fixed price vector can prevent an inefficient increase of the number of firms in the industry and in which no fixed price vector can sustain the full set of previously available products—despite their continuing social desirability.

Thus, even though the case of intertemporal separability that we have been discussing seems to assume away one of the most fundamental complicating features of production through time, we see that it already presents us with a variety of complexities absent from the analysis of single-period production.

13B. The Case of Perfect Fungibility

Before leaving the case of separability, it is worth pausing to inquire about the circumstances in which it can occur. If no capital (i.e., no durable input) were involved in production, intertemporal costs would obviously be separable since there would be no means by which one period's production decisions would transmit any effects to the cost relationships at any other date. A separable case that is far more interesting arises if all investments are perfectly reversible, that is, if there exist perfect rental or resale markets for all types of capital (human and physical). This requires *perfect fungibility* of all capital utilized in production. That is, the value of any item of capital equipment in its best alternative use must be equal to that in the firm which initially undertook the investment.

This is, perhaps, the best way to interpret the single-period analysis of the preceding chapters. In a world in which production does involve both capital and time, timeless cost functions such as those we have been employing can most usefully be taken as components of separable intertemporal cost functions involving capital costs that are determined in perfect rental or resale markets. For in such perfect rental or resale markets, each period's price for the use of a capital good is determined outside any particular industry, with no role played by allocative behavior within that industry. Moreover, there no decision maker is constrained by previous commitments in the form of capital expenditures. Anything previously done can be undone, and so rational decisions can be made considering only the period to which they apply and ignoring all other dates. For these reasons, in this case the cost relationship is intertemporally separable and can usefully be studied one period at a time.

Of course, in reality a case that is far more common (and more interesting) is that in which some portion of capital is imperfectly fungible, and constitutes a sunk cost once it is installed. At least some investment is then irreversible, and costs become intertemporally interdependent through the linkage provided by the nonfungible capital. And in this case, the determination of the implicit rental cost of capital, of the techniques to be used, and of the prices in different periods are inextricably linked and endogenous. The rest of this chapter focuses on the implications of sustainability for such processes.

13C. A Model of Intertemporal Interdependence

Significant sunk cost outlays on long-lived and nonfungible capital render production intertemporally interdependent because when such expenses are incurred in one period they affect the costs required to produce given output vectors in subsequent periods.

To analyze such effects, we construct a rather general model of production with sunk and imperfectly fungible capital, and define the associated intertemporal cost function.

(13C1) $$C(y^1, \ldots, y^T) = \min \sum_{t=1}^{T} V^t(y^t, K^t)\rho_t + \sum_{t=1}^{T} \beta^t I^t \rho_t$$

$$- \sum_{t=1}^{T} \alpha^t D^t \rho_t - \alpha^{T+1} K^{T+1} \rho_{T+1}$$

subject to

(13C2)
$$K^t = (1 - \delta^{t-1})K^{t-1} + I^t - D^t \geq 0,$$

$$K^0 = 0 \text{ (the capital formation constraint)}$$

and

(13C3)
$$I^t \geq 0, \qquad D^t \geq 0, \qquad \beta^t > \alpha^t.$$

Here, I^t is the vector of nonnegative gross additions (investments), at the beginning of period t, to the various nonfungible capital stocks. These are measured in physical units. Their prices (when new) are the time-dependent vectors β^t. Similarly, D^t is the vector of nonnegative disinvestments of the various capital goods that are undertaken at the beginning of period t. The vector α^t gives the scrap values of a unit of each of the various types of capital at that time. Of course $\alpha^t \leq \beta^t$. If α^t_j were zero, it would mean that capital stock j has no alternative use of any value once it is put into place. This is the case of perfect nonfungibility. At the other extreme, $\alpha^t_j = \beta^t_j$ indicates that investment in stock j is perfectly reversible in that scrap value and original cost coincide. We stipulate that $\alpha^t < \beta^t$ because all perfectly reversible investment costs are subsumed in the variable cost functions $V^t(\cdot)$. Note, also, that the "scrap value" α includes the saving in the capital-maintenance cost that must be incurred to keep the capital serviceable. The capitalized value of such costs can be included in the β's, and then taken to be saved upon disinvestment.

Equation (13C2) describes the process by which the vectors of capital stocks evolve, period by period. In that equation, δ^t is the vector of rates of physical deterioration in period t of the capital stocks K^t extant during that period. In equation (13C1), $V^t(y^t, K^t)$, expressed in money of period t, is the minimized cost of producing the vector of outputs y^t, during period t, when the available stocks of nonfungible capital are given by K^t. Since an increase in the available capital stocks does not increase variable cost, we have $V^t_{K_j} \equiv \partial V^t/\partial K^t_j \leq 0$. The V^t, then, are the intertemporally separable costs, and they incorporate the costs of all inputs completely utilized during period t, including the services of perfectly reversible stocks of capital; that is, those that entail no commitments or spillovers affecting other time periods. The intertemporal cost function $C(y^1, \ldots, y^T)$ is obtained by minimization, subject to (13C2) and (13C3), of the discounted sum of the intertemporally separable costs, plus the outlays on irreversible investments, after subtraction of the proceeds from the salvage values of disinvested capital stocks, and the proceeds from the salvage of all remaining capital at time $T + 1$, after the active life of the enterprise has ceased.

This model is sufficiently general to capture a variety of important intertemporal effects. As we see later, inflation can be represented by rises over time in the prices of new capital goods, β^t, in the levels of the variable cost functions, and in the interest rates r_t that underlie the discount factors

(13C4) $$\rho_t = \rho_{t-1}(1 + r_t)^{-1}.$$

Technological change can be taken to reduce, progressively, the levels of the variable cost functions. Technical advances that introduce new productive techniques can be represented by corresponding changes in the variable cost functions, for the date of the innovation and thereafter, and with new kinds of capital goods among their arguments.

As a prelude to our substantive economic analysis, it is necessary to characterize the solution to the cost-minimization problem inherent in (13C1). For this purpose, it is convenient to rewrite the conditions (13C2), which describe the formation of the capital stocks, as

(13C5) $$K^t = \sum_{s=1}^{t} (I^s - D^s) \cdot l^{st}.$$

This states simply that the capital stock vector at time t is the sum of all net investments, $I^s - D^s$, in all previous periods s, each such investment multiplied by a factor l^{st} representing the effects of deterioration between dates s and t. Thus, a representative element in vector l^{st}, say l_j^{st}, is the proportion of a unit of capital of type j installed at the beginning of period s that is still usable physically during period t. Hence, expressed in terms of the deterioration rates, we have $l_j^{tt} = 1$, and

(13C6) $$l_j^{st} = (1 - \delta_j^s)(1 - \delta_j^{s+1}) \cdots (1 - \delta_j^{t-1}), \text{ for } s < t.$$

Consequently,

(13C7) $$l_j^{st} = l_j^{sk} l_j^{kt}, \text{ for } s \leq k \leq t.$$

Thus, the constraints in (13C2) can be dispensed with formally by substituting the expressions (13C5) directly into (13C1). Then the Lagrangian for the minimization becomes

(13C8) $$\mathscr{L} = \sum_{t=1}^{T} V^t \left(y^t, \sum_{s=1}^{t} (I^s - D^s) l^{st} \right) \rho_t + \sum_{t=1}^{T} \beta^t I^t \rho_t - \sum_{t=1}^{T} \alpha^t D^t \rho_t$$
$$- \alpha^{T+1} \rho_{T+1} \left(\sum_{s=1}^{T+1} (I^s - D^s) l^{sT+1} \right).$$

The necessary (first-order) Kuhn-Tucker conditions are

(13C9) $$\partial \mathscr{L}/\partial I_j^s = \sum_{t=s}^{T} V_{K_j}^t l_j^{st} \rho_t + \beta_j^s \rho_s - \alpha_j^{T+1} \rho_{T+1} l_j^{sT+1} = 0$$

if $I_j^s > 0$, and $\partial \mathscr{L}/\partial I_j^s \geq 0$.

$$(13C10) \qquad \partial\mathscr{L}/\partial D_j^s = -\sum_{t=s}^{T} V_{K_j}^t l_j^{st} \rho_t - \alpha_j^s \rho_s + \alpha_j^{T+1} \rho_{T+1} l_j^{sT+1} = 0$$

$$\text{if} \quad D_j^s > 0, \quad \text{and} \quad \partial\mathscr{L}/\partial D_j^s \geq 0.$$

Condition (13C9) indicates that if any capital of type j is bought in period s, then such a quantity of j is purchased that, in present-value terms, its price $\beta_j^s \rho_s$, net of the horizon salvage value of whatever part remains usable, $\alpha_j^{T+1} \rho_{T+1} l_j^{sT+1}$, is equal to the sum, $-\sum V_{K_j}^t l_j^{st} \rho_t$, of the marginal savings in variable costs over time permitted by the surviving portions of the capital. Condition (13C10) indicates that if there is any disinvestment in capital of type j during period s, then salvage will be carried to the point at which the salvage value of capital in period s, net of the corresponding horizon salvage value thereby forgone, equals the sum of the savings in variable costs that would have been made possible over time by the additional portions of the capital that would have survived as the result of a marginal reduction in disinvestment.

One very reasonable implication of (13C9) and (13C10) and the premise $\beta_j^s > \alpha_j^s$ is that I_j^s and D_j^s cannot both be positive. That is, it cannot be efficient to invest and disinvest simultaneously in the same type of capital. To prove this, note that conditions (13C9) and (13C10) yield

$$(13C11) \qquad \alpha_j^s \rho_s \leq -\sum_{t=s}^{T} V_{K_j}^t l_j^{st} \rho_t + \alpha_j^{T+1} \rho_{T+1} l_j^{sT+1} \leq \beta_j^s \rho_s,$$

which states that in an optimal solution, $\beta_j^s \rho_s$, the cost of an additional unit of investment, must be no less than the net marginal saving in other costs that it permits, and that the opposite must be true of the yield from incremental disinvestment. The right-hand equality in (13C11) holds if investment takes place, the left-hand equality if disinvestment occurs, and both inequalities hold in any event. Thus, if I_j^s and D_j^s were both positive we would have $\alpha_j^s = \beta_j^s$, contrary to our assumption that the latter exceeds the former.

13D. Some Relationships for Pricing and Depreciation Calculations

We now derive some useful expressions for the (usual) case in which investment in capital of type j occurs during some two different periods, τ and τ', with $\tau' \geq \tau + 1$, and no such investment occurs between these dates. These results prove extremely useful below, in particular when we deal with the issues of pricing and depreciation.

With positive quantities of investment at two different dates, relation (13C9) must hold as an equality at each of them. Now, if the equality for τ'

is multiplied by $l_j^{\tau\tau'}$ and subtracted from the equality for τ, we obtain, making use of (13C7),

$$(13D1) \qquad \sum_{t=\tau}^{\tau'-1} V_{K_j}^t l_j^{\tau t} \rho_t + \beta_j^\tau \rho_\tau - \beta_j^{\tau'} \rho_{\tau'} l_j^{\tau\tau'} = 0.$$

This tells us that in an optimal solution the savings in other costs made possible by undertaking a marginal unit of investment at date τ, rather than postponing it to τ' (i.e., $-\sum_{t=\tau}^{\tau'-1} V_{K_j}^t l_j^{\tau t} \rho_t$), must equal the financial saving from postponement of (the surviving portion of) the investment from τ to τ', that is, $\beta_j^\tau \rho_\tau - \beta_j^{\tau'} \rho_{\tau'} l_j^{\tau\tau'}$. Next, if the equality form of (13C9) for τ' is multiplied by $l_j^{s\tau'}$ and substituted into (13C11) for periods s between τ and τ', we obtain the necessary optimality requirement

$$(13D2) \qquad \alpha_j^s \rho_s - \beta_j^{\tau'} \rho_{\tau'} l_j^{s\tau'} \leq -\sum_{t=s}^{\tau'-1} V_{K_j}^t l_j^{st} \rho_t \leq \beta_j^s \rho_s - \beta_j^{\tau'} \rho_{\tau'} l_j^{s\tau'}.$$

Multiplying (13D2) through by $l_j^{\tau s}$ and combining with (13D1) yields

$$(13D3) \qquad \beta_j^\tau \rho_\tau - \alpha_j^s \rho_s l_j^{\tau s} \geq -\sum_{t=\tau}^{s-1} V_{K_j}^t l_j^{\tau t} \rho_t \geq \beta_j^\tau \rho_\tau - \beta_j^s \rho_s l_j^{\tau s}.$$

One can provide an interpretation of (13D2) and (13D3) similar to that just given for (13D1). However, their primary economic significance will emerge later in this chapter.

Expressions analogous to (13D1), (13D2), and (13D3) can also be derived (and will later be found useful) for the cases in which disinvestment occurs at either dates τ and τ'.

13E. Intertemporal Cost Complementarities and Periods of Nonzero Investment

In our analysis the reason for the special significance of those periods in which some positive amount of investment (or salvage) occurs may at first seem rather mysterious. But it is really straightforward. The fact that some investment or salvage is undertaken at date t automatically transforms older capital of the same type that would otherwise be sunk into capital that is liquid *at the margin*. For if K^t units of such capital are sunk and I^t is expected to be invested in period t, then the freely available option to reduce planned investment by ΔI^t to $I^t - \Delta I^t \geq 0$ is equivalent to the option to reduce the otherwise-sunk capital stock from K^t to $K^t - \Delta I^t$. For this reason, at any date at which $I^t > 0$ (or $D^t > 0$), at the margin that type of capital becomes perfectly substitutable for investment (or salvage) and the two must, therefore, have exactly the same financial (rental) value. The value and cost of new investment are, of course, determined by the market. So that is why prices, rental values, and the implied economic depreciation payments for capital are all also rendered perfectly determinate over a stretch of time in which investment or salvage occurs throughout. Where there are gaps in investment, that is, dates at which no investment occurs, the process is not so straightforward. Nevertheless, for a time interval whose beginning and ending dates are both periods of positive investment or salvage, we can determine (in present-value terms) the sum of the depreciation payments for the interval as a whole because of the fungibility of capital (on the margin) at the two bounding dates. Viewed another way, these observations tell us that in this model the value of sunk investment in capital of a given type is intertemporally separable, on the margin, among groups of consecutive periods that are divided by an act of investment in or salvage of capital of that type. For even though the capital is sunk physically, it is not sunk economically on the margin at a date at which its owner chooses to augment its amount by means of gross investment, and we have seen that when capital is not sunk (economically) its intertemporal costs become separable. Thus, in the cost function defined in (13C1), if for all types of capital the optimal value of gross investment or of salvage is positive at the same date τ, then the intertemporal cost function is locally separable into a portion that pertains only to outputs before date τ and another portion pertaining only to outputs at and after date τ. In other words,[1] in this case there are no cost complementarities between outputs before and after τ.

13F. Propositions on Pricing under Constant Returns to Scale

It is easiest to derive substantive conclusions about pricing and industry structure from our general model in the special case in which each $V^t(y^t, K^t)$ function is linearly homogeneous in all its arguments. We therefore begin

---

[1] Yet, curiously, so long as at date τ there is some gross investment whose cost is greater than its immediate salvage value, there must exist economies of scope between outputs before and after τ. This must be true because the quantity of gross investment at τ that is optimal for current output levels would, in general, become zero or negative, optimally, if the later periods' outputs were to shrink to zero, as in the conceptual experiment that underlies the definition of economies of scope.

Thus, the nature of intertemporal cost complementarities becomes clear. They can be present without interruption only if the enterprise employs more than one type of (sunk) capital good. And the dates on which there is investment in or salvage of the various types of sunk capital goods must, so to speak, interlock. That is, there cannot be any date at which it is simultaneously optimal for investment or salvage to occur in each and every type of capital good.

with this case. First, we have the following plausible result:

Proposition 13F1[2]   *If $V^t(y^t, K^t)$ is continuously differentiable and linearly homogeneous in all its arguments, for all t, then $C(y^1, \ldots, y^T)$, as defined by (13C1), exhibits constant returns to scale. Marginal costs, in present value, satisfy*

$$\partial C/\partial y_i^t = \frac{\partial V^t(y^t, K^t)}{\partial y_i^t}\, \rho_t,$$

where K^t is the vector of cost-minimizing capital stocks for period t.

We also have the following crucial result:

Proposition 13F2   *The Ramsey-optimal prices in the model (13C1)–(13C3), with costs linearly homogeneous, satisfy $p_i^t = \partial V^t/\partial y_i^t$. These are the only prices that satisfy the necessary conditions for sustainability.*

Proof   With constant returns to scale, the first-best marginal cost prices satisfy the budget constraint and so are Ramsey-optimal. We know that no sustainable price can be less than the corresponding marginal cost. But, then, if any prices exceed the corresponding marginal costs, total revenues must

---

[2] *Proof*   With V^t linearly homogeneous, its first derivatives are homogeneous of degree zero. Then, if all outputs and the corresponding optimal values of investments, disinvestments, and capital stocks are multiplied by a common scalar factor λ, the Kuhn-Tucker conditions (13C9) and (13C10) continue to hold. Thus, for the production of $\lambda y^1, \ldots, \lambda y^T$, the efficient quantities of investments and disinvestments are $\lambda I^1, \ldots, \lambda I^T$ and $\lambda D^1, \ldots, \lambda D^T$, where $I^1, \ldots, I^T$ and $D^1, \ldots, D^T$ are efficient in the production of $y^1, \ldots, y^T$. The consequent capital stocks are $\lambda K^1, \ldots, \lambda K^T$, where $K^1, \ldots, K^T$ are optimal for $y^1, \ldots, y^T$. Then,

$$\begin{aligned}
C(\lambda y^1, \ldots, \lambda y^T) &= \sum_{t=1}^{T} V^t(\lambda y^t, \lambda K^t)\rho_t + \sum_{t=1}^{T} \lambda I^t \beta^t \rho_t \\
&\quad - \sum_{t=1}^{T} \lambda D^t \alpha^t \rho_t - \alpha^{T+1}\rho_{T+1}\lambda K^{T+1} \\
&= \lambda \sum_{t=1}^{T} V^t(y^t, K^t)\rho_t \\
&\quad + \lambda \left[\sum_{t=1}^{T} I^t \beta^t \rho_t - \sum_{t=1}^{T} D^t \alpha^t \rho_t - \alpha^{T+1}\rho_{\tau+1}K^{T+1} \right] \\
&= \lambda C(y^1, \ldots, y^T).
\end{aligned}$$

By the envelope theorem, $\partial C/\partial y_i^t = \rho_t \partial V^t(y^t, K^t)/\partial y_i^t$, because this is equal to $\partial \mathscr{L}/\partial y_i^t$, at the optimum, by (13C9).   Q.E.D.

exceed total costs, thus precluding sustainability. Thus, for sustainability, it is necessary (and, with appropriate cost properties, sufficient) that prices equal marginal costs.   Q.E.D.

13G. Optimal Pricing and Optimal Depreciation Policies

Economists generally recognize that accounting depreciation rules often bear little relationship to the underlying economic relationships. What is less clear is a set of depreciation rules that *is* consistent with the tenets of economic theory. We will show now that our analysis does yield specific rules for depreciation, rules which are not only consistent with recovery of invested resources, but which, in addition, satisfy the requirements of optimality in intertemporal resource allocation.

Our analysis of depreciation rests on the following fundamental result:

Proposition 13G1 *In period t, the revenues from the marginal-cost prices, $p_i^t = \partial V^t(\cdot)/\partial y_i^t$ are*

$$(13G1) \qquad p^t y^t = V^t(y^t, K^t) - \sum_j V_{K_j}^t K_j^t.$$

This follows immediately from Proposition 13F1 and the application of Euler's theorem to the linearly homogeneous $V^t(\cdot)$ functions:

$$V^t(y^t, K^t) = \sum_j V_{K_j}^t K_j^t + \sum_i (\partial V^t(\cdot)/\partial y_i^t) y_i^t.$$

Proposition 13G1 implies that, after discounting, the revenues from marginal-cost pricing during period t must cover the period's variable costs, $\rho_t V^t(\cdot)$, and in addition, make the contribution $-\sum_j V_{K_j}^t K_j^t \rho_t$ toward recovery of the present value of capital costs. Consequently, the contemporaneous value $-\sum V_{K_j}^t K_j^t$ may be interpreted as the period's payment to capital, including depreciation, with $-V_{K_j}^t K_j^t$ earmarked for capital of type j.

Proposition 13G2 *Since the total revenues derived from Ramsey prices just cover the present value of the stream of all costs incurred by the supplier, the discounted present value of all Ramsey-optimal payments to capital, $-\sum_j V_{K_j}^t K_j^t$, exactly covers the discounted value of all capital costs. In fact, this equality holds for each type of capital, j, that is employed:*

$$(13G2) \quad -\sum_{t=1}^T V_{K_j}^t K_j^t \rho_t = \sum_{t=1}^T \beta_j^t \rho_t I_j^t - \sum_{t=1}^T \alpha_j^t \rho_t D_j^t - \alpha_j^{T+1} \rho_{T+1} K_j^{T+1}.$$

For interpretation and application it is useful to relate our concept of Ramsey-optimal payments to capital to the more familiar concepts of depreciation and return on investment.[3] Denote by PC_t the payment made to capital at the end of period t. This sum can be divided into two portions: the depreciation charge for period t, denoted DEP_t, and the (interest and risk) return on investment, $r_t A_t$, where r_t is the rate of return and A_t is the value of the capital asset at the beginning of the period. Thus

(13G3) $$PC_t = rA_t + DEP_t.$$

The value of the capital asset can be assessed economically as the present value of the stream of future payments to capital, minus the present value of the stream of future costs of the net investments that, in part, make those payments possible. That is,

(13G4) $$A_t = \sum_{i=0}^{T-t} PC_{t+i}(1+r)^{-(i+1)} - \sum_{i=1}^{T-t} IC_{t+i}(1+r)^{-i},$$

where IC_j are the investment costs incurred at the start of period j. Algebraic manipulation of (13G4) reveals that

(13G5) $$A_{t+1} = (1+r)A_t + IC_{t+1} - PC_t.$$

Then, substitution of (13G3) yields

(13G6) $$A_{t+1} = A_t + IC_{t+1} - DEP_t.$$

Summing (13G6) over all periods, we have

(13G7) $$\sum_{t=1}^{T} DEP_t = A_1 - A_{T+1} + \sum_{t=2}^{T+1} IC_t.$$

If the discounted sum of all payments to capital is equal to the discounted sum of all investment costs, then (13G4) shows that $A_1 = IC_1$. If the firm salvages its capital stock at the horizon, or if that stock has zero value at that time, then $A_{T+1} = 0$. Finally, in this case (13G7) becomes

(13G8) $$\sum_{t=1}^{T} DEP_t = \sum_{t=1}^{T+1} IC_t.$$

This discussion shows that we have defined our categories of capital payments in a manner consistent with accounting usage. In particular, (13G8)

---

[3] Thanks are due to Frank Sinden, Sharon Oster, and Herbert Mohring for their help with this construction.

shows that the *undiscounted* sum of depreciation payments is equal to the *undiscounted* sum of investment costs. As shown by (13G6), the value of the capital stock is brought up to date by adding investment costs and subtracting depreciation. And the discounted sum of payments to capital equals the discounted sum of investment costs. This consistency with accounting usage is especially significant for economic analysis because it is derived from the economic assessment of the value of capital assets given by (13G4).

It is now easy to indicate how the stream of depreciation payments can be derived from the streams of investment costs and payments to capital. This is accomplished iteratively, through successive updating of the value of the capital assets. Initially, $A_1 = IC_1$, and $DEP_1 = PC_1 - rA_1$. Then, in the second period, $A_2 = A_1 + IC_2 - DEP_1$ and $DEP_2 = PC_2 - rA_2$. This process can be continued indefinitely by making successive use of (13G6) and (13G3).

Conversely, any stream of depreciation charges can be utilized to determine a corresponding stream of payments to capital. Here, $A_1 = IC_1$ and $PC_1 = rIC_1 + DEP_1$. Then, $A_2 = A_1 + IC_2 - DEP_1$ and $PC_2 = rA_2 + DEP_2$. This iterative process, too, can be continued indefinitely. Thus, we have

Definition 13G1   A set of payments to capital is *any* stream of differences between total revenues and variable costs whose present values add up to the discounted original cost of the capital in question, net of any returns from salvage. A set of depreciation payments is any stream of charges that add up to the sum of the original investment costs. For any stream of payments to capital there is an equivalent stream of depreciation charges, and vice versa, where the equivalence is based on the relationships (13G3), (13G4), and (13G6).

Thus, if the firm decides on a set of outputs, input purchases, and prices over time, it automatically decides, implicitly, on the streams of payments to capital and depreciation charges. Consequently, there is a direct relationship between the firm's pricing decisions and its depreciation decisions. For example, a decision to use accelerated depreciation for a piece of equipment is tantamount to the decision to charge a high price (relative to variable costs) for those products of the equipment which are supplied early in the life of that item, and then to reduce the products' prices (relative to variable costs) as the item ages.

With such a duality relation between pricing and depreciation decisions, it is clear that we can formulate

Definition 13G2   A set of *Ramsey-optimal payments to capital* may be defined as those payments that correspond to the stream of Ramsey-optimal product prices for the firm. This stream corresponds to *Ramsey-optimal depreciation charges*.

That is, among the infinite number of different streams of depreciation assessments that cover the cost of the capital equipment over its lifetime, we may select that stream of assessments which induces a Ramsey-optimal pattern of quantities demanded and supplied over the relevant time interval. This optimal set of depreciation assessments, and the corresponding prices, are those necessary for efficiency in the intertemporal allocation of resources. It is these optimal depreciation decisions that we are discussing here.

Later, we see more explicitly how the pattern of depreciation payments can induce efficient adaptations in intertemporal demand patterns in response to changes in interest rates, cost-saving innovations, and so forth.

Let $\sigma_j^t = -V_{K_j}^t K_j^t$ denote the payment for capital of type j that is made at the start of period t, but after the investments and asset evaluations for period t have been carried out. Then, $(1 + r_t)\sigma_j^t = -(1 + r_t)V_{K_j}^t K_j^t$ is the equivalent payment at the end of the period, which is the timing of the capital payments analyzed in (13G3)–(13G8). In any period in which investment or salvage does occur, the marginal optimality condition must be satisfied. If investment or salvage occurs in two adjacent periods, these equations will suffice to determine the relationship between the prices of capital in the two periods and the uses to which it can be put. This, in turn, determines the optimal rental value and the optimal depreciation assessment for the type of capital in question:

Proposition 13G3 *If there is an investment in the capital of type j in both periods t and $t + 1$, then the only value of σ_j^t consistent with Ramsey optimality and sustainability is*

(13G9) $$\sigma_j^t = K_j^t[\beta_j^t - \beta_j^{t+1}(1 - \delta_j^t)/(1 + r_t)] \equiv {}^*\sigma_j^t.$$

The corresponding depreciation charge is

(13G10) $$(1 + r_t){}^*\sigma_j^t - r_t\beta_j^t K_j^t = K_j^t((\beta_j^t - \beta_j^{t+1})(1 - \delta_j^t) + \delta_j^t\beta_j^t) \equiv {}^*DEP_j^t.$$

    Proof   By (13C4) and (13C6), $\rho_{t+1} = \rho_t(1 + r_t)^{-1}$ and $l_j^{tt+1} = (1 - \delta_j^t)$, so that the middle expression in (13G9) equals

$$K_j^t[\beta_j^t - (\rho_{t+1}/\rho_t)\beta_j^{t+1} l_j^{tt+1}].$$

But by (13D1) this, in turn, equals

$$-K_j^t V_{K_j}^t l_j^{tt} = -K_j^t V_{K_j}^t \equiv \sigma_j^t,$$

since $l_j^{tt} = 1$ by (13C6).   Q.E.D.

The optimal payment to capital, $*\sigma_j^t$, is the rental value of perfectly fungible capital, K_j^t, in a perfect capital market—it is the optimal rental value. Then (13G9) implies that the one-period rental value of a unit of such capital is its per-unit purchase price, β_j^t, at the beginning of the period, minus $\beta_j^{t+1}(1 - \delta_j^t)/(1 + r_t)$, the present value of the replacement cost of $(1 - \delta_j^t)$, the amount of the unit of capital that still remains at the end of the period. That is, it is the *net* cost incurred by purchasing the capital in period t, after deduction of the saving in $t + 1$ capital purchase costs permitted as a result.

Further, if the lessor of a unit of such capital had paid $*\sigma_j^t$ to its owner, at the start of the period, as a rental fee, and if the owner had invested the fee at a rate of return of r_t, at the end of the period the owner would have accumulated the sum $(1 + r_t)\beta_j^t - \beta_j^{t+1}(1 - \delta_j^t)$. This sum can be divided into three parts: $r_t\beta_j^t$, the rate of return applied to the value of the capital; $\delta_j^t\beta_j^{t+1}$, the sum needed to replace the deteriorated portion of the capital at the end of the period; and $(\beta_j^t - \beta_j^{t+1})$, minus one multiplied by the appreciation in the value of a unit of the capital.

Moreover, the depreciation charge associated with the optimal payment to capital, $*DEP_j^t$, as given by (13G10), is readily interpreted. It has two parts. The first, $K_j^t(1 - \delta_j^t)(\beta_j^t - \beta_j^{t+1})$, reflects the change in the value of the portion of the capital that survives the period. This portion of the depreciation charge is positive if obsolescence, deflation, or technological progress drives the price of the capital good (or its functional equivalent) in period $t + 1$ below the price that prevailed at the start of period t. The second part of the depreciation charge, $K_j^t\delta_j^t\beta_j^t$, is the cost, at the price prevailing in period t, of the portion of the capital stock that becomes unusable during the period. While this deterioration charge cannot be negative, the first portion of the optimal depreciation assessment can be negative, thereby offsetting the second, if inflation drives up the value of the capital goods.

Equation (13G9) is an extremely illuminating expression because it indicates how different economic influences affect the optimal payment to capital. Predictably, it indicates that if δ_j^t, the rate of deterioration of capital, increases, the optimal depreciation rate will increase correspondingly, since in (13G9) δ_j^t appears only in the term $\beta_j^{t+1}\delta_j^t/(1 + r_t)$. Similarly, if r_t, the rate of interest, rises, then $*\sigma_j^t$ must rise, as long as it is positive. In terms of optimal resource allocation, the explanation is simple. A rise in $*\sigma_j^t$ is tantamount to a rise in product price at period t relative to price at $t + 1$. This encourages postponement of consumption and, hence, of the construction of productive capital, a change which is obviously efficient when there is a rise in the opportunity cost of current resource use in terms of later use of resources.

Similarly, expected innovation which reduces the cost and hence the price of future investment β^{t+1} relative to its current price β^t will increase $*\sigma_j^t$. That is, if as a result of innovation the price of capital falls by the factor θ, where $0 \leq \theta \leq 1$, we may write $\beta_j^{t+1} = (1 - \theta)(1 + k)\beta_j^t$, where k is the rate of inflation. Then

$$*\sigma_j^t = K_j^t \beta_j^t \left(1 - \frac{(1-\theta)(1+k)(1-\delta_j^t)}{(1+r_t)} \right),$$

so that an increase in the value of θ (increased technological progress) will raise σ_j^t. This is, again, required for optimality because it induces postponement of investment to the later date when it will use up smaller quantities of resources.[4]

13H. Technological Progress Embodied in Sunk Capital

In our discussion of the optimal depreciation formulas, (13G9) and (13G10), in Section 13G we showed how one type of technological innovation can be taken into account in our analysis. Now we go on to consider technological change of a more general sort. The type of technological change just discussed, which for our purposes is the simplest kind to analyze, involves augmentation of a standard type of capital. That is, after date τ, one unit of new investment becomes equivalent in productivity to $\gamma > 1$ units of surviving stock of that type purchased prior to τ. In other words, it takes smaller quantities of inputs than it did before to achieve a given level of productive capacity. This is simply equivalent to a drop in the price of that type of equipment to $(1/\gamma)$ of the original price. Thus it can be dealt with in our model by a corresponding adjustment in the β^t vectors—that is, by a readjustment of the pertinent price parameters, leaving all other components of the model unchanged. For example, in (13G9) this reduces the value of β_j^{t+1} relative to β_j^t, increasing prices during period t and accelerating depreciation assessments in order to induce postponement of new investment until the later date, when cheaper capital becomes available.

---

[4] It is noteworthy that the rate of inflation *during* period t will not have a substantial effect on $*\sigma_j^t$. To see this, rewrite (13G9) in nominal terms, taking the price index at the start of period t to be unity, β_j^t and β_j^{t+1} to be real prices, $\bar{r}_t$ to be the real interest rate, and letting prices grow at the rate $(1 + k)$. Then (13G9) becomes

$$*\sigma_j^t = K_j^t[\beta_j^t - \beta_j^{t+1}(1+k)(1-\delta_j^t)/(1+\bar{r}_t+k)]$$
$$= K_j^t[\beta_j^t - \beta_j^{t+1}(1+k)(1-\delta_j^t)/[(1+\bar{r}_t)(1+k) - k\bar{r}_t]],$$

which, if we take $k\bar{r}_t$ to be negligible, is approximately equal to the value of $*\sigma_j^t$ in (13G9) without inflation. However, such inflation does raise the payment to capital, if it is made at the end of the period, from $(1+\bar{r}_t)*\sigma_j^t$ to $(1+\bar{r}_t+k)*\sigma_j^t$. Also, inflation preceding period t will raise $*\sigma_j^t$ correspondingly. That is, if we take the price index for period t to be some number $w \neq 1$, all terms in the preceding equation will just be multiplied by w.

More interesting, and at least equally relevant for obsolescence–depreciation issues is the advent of new productive techniques that require new capital goods, goods that were useless when only the old techniques were available.

Let $\tilde{K}$ be the vector of quantities of new capital goods, K be the vector of quantities of old capital goods, g be the vector of outputs produced using old techniques, $W(y, \tilde{K})$ be the variable cost function associated with the new techniques, and $\tilde{V}(y, K)$ be the old variable cost function. Now, the enterprise can and usually will find it efficient, at least for a while, to use a mixture of the old and new techniques, if only because it has sunken stocks of the old capital goods. But to do so it must buy some of the new capital goods. Efficiency requires that in a given period, with given stocks, the firm minimize the total variable costs of the production of its outputs with respect to the mixture of old and new techniques that it employs. This mixture gives rise to a new variable cost function:

(13H1) $\quad \underset{y \geq g \geq 0}{\text{Minimum}} \left[\tilde{V}(g, K) + W(y - g, \tilde{K}) \right] \equiv V(y, K, \tilde{K}).$

From this we obtain

Proposition 13H1 *The variable cost function obtained by mixing old and new techniques, as represented in (13H1), has the following properties:*

(i) *If neither old nor new capital goods, taken alone, are inferior inputs and do not raise marginal variable costs, then in their mixture, neither type of capital is an inferior input and each of them must (weakly) reduce marginal variable costs. That is*

(13H2) $\quad\quad\quad \partial^2 V / \partial y_i \, \partial K_j \leq 0 \quad\quad \partial^2 V / \partial y_i \, \partial \tilde{K}_j \leq 0.$

(ii) *Under the same conditions, the new and the old capital goods are substitutes;*

(13H3) $\quad\quad\quad\quad\quad \partial^2 V / \partial K_i \, \partial \tilde{K}_j > 0.$

(iii) *If the component techniques exhibit constant returns to scale, then so does the mixed variable cost function.*[5]

Thus, the advent of a new technique can be represented directly in the model with which we have been working. It is perhaps useful to take the

---

[5] *Proof* Conditions (i) and (ii) follow from straightforward comparative statics analyses of (13H1). Condition (iii) is shown by expanding all variables in (13H1) by the same scalar factor, and noting that the optimal mixture, g/y, is unaffected.

variable cost functions in (13C1) to embody the mixture of old and new techniques all along, even before the new techniques have actually been invented. But, when the new techniques are not, in reality, available, the prices of the required capital goods can be taken to be prohibitive (i.e., to be "infinite"). Then, when the new technique actually becomes available, the model can represent this as a decrease in the prices of the required new types of capital goods. So, once more the innovation can be dealt with by taking the prices of capital goods to be falling. As usual, when a capital good's price falls, prior investments are discouraged, raising prior periods' marginal costs and the associated sustainable prices of outputs; and subsequent investments are stimulated, reducing these periods' marginal costs and the corresponding prices. The effects on optimal depreciation assessments are then the same as those previously described in our discussion of (13G9) and (13G10).

13I. The Sporadic-Investment Case

The derivation of (13G9) is based on the assumption that investment occurs in the two consecutive periods, t and $t + 1$. More generally, with any pattern of sporadic investment, we obtain relationships analogous to (13G9) for the two most adjacent periods in which investment *does* occur.

Proposition 13I1 *If there is investment in capital of type j in both periods τ and τ', with no investment or disinvestment in between, then for consistency with Ramsey optimality and sustainability, the depreciation assessments must satisfy:*

$$(13I1) \qquad \sum_{t=\tau}^{\tau'-1} (\sigma_j^t)\rho_t = \sum_{t=\tau}^{\tau'-1} (*\sigma_j^t)\rho_t = K_j^\tau [\beta_j^\tau \rho_\tau - \beta_j^{\tau'} \rho_{\tau'} l_j^{\tau\tau'}]$$

$$(13I2) \qquad \sum_{t=\tau}^{\tau+s-1} (\sigma_j^t)\rho_t \geq \sum_{t=\tau}^{\tau+s-1} (*\sigma_j^t)\rho_t$$
$$= K_j^\tau [\beta_j^\tau \rho_\tau - \beta_j^{\tau+s} \rho_{\tau+s} l_j^{\tau\tau+s}], \quad s = 1, \ldots, \tau' - \tau$$

$$(13I3) \qquad \sum_{t=\tau}^{\tau+s-1} (\sigma_j^t)\rho_t \leq K_j^\tau [\beta_j^\tau \rho_\tau - \alpha_j^{\tau+s} \rho_{\tau+s} l_j^{\tau\tau+s}], \quad s = 1, \ldots, \tau' - \tau$$

The proof of these relationships is obtainable from (13D3) in the same way that (13G9) was derived from (13D1). Equations (13I1) tells us that between the dates when investment occurs it is no longer necessary for each

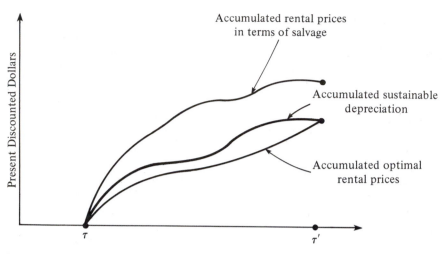

FIGURE 13I1

individual period's capital payment to equal the corresponding optimal rental price, because the capital good is, in effect, not perfectly fungible. That is, since investment in $\tau + 1$ is zero, if the inherited capital at the beginning of period $\tau + 1$ turns out to be excessive, it is no longer possible to offset the excess without loss by a corresponding decrease in investment in $\tau + 1$. However, even in this case the depreciation assessments do continue to be constrained to some extent. In particular, the discounted sum of payments to capital over the entire period from τ to τ' must be equal to the discounted sum of the optimal rental prices.

In addition, (13I2) tells us that the discounted sum of capital payments from the date of investment τ through any date $\tau + s - 1$ between τ and τ' must *at least equal* the discounted sum of the optimal rental prices, for otherwise it would pay to postpone some of the investment actually undertaken[6] from period τ to period $\tau + s$. In addition, that discounted sum must *not exceed* the discounted sum of the values that would be assumed by perfect rental payments if the capital were worth no more than its salvage value. For if this condition were violated, the solution could not be optimal. Rather, it would pay to expand investment at date τ and then scrap any excess at date $\tau + s$.

---

[6] To see why this is so, note that capital assessment is the marginal benefit of the investment to the firm, while the expression in brackets in (13I2) represents the cost of undertaking a unit of investment at date τ rather than postponing it to period $\tau + s$. If the benefit of early investment (the capital charge) were less than its net cost, postponement of some of the investment actually undertaken would have been profitable.

Of course, the interval between the upper and lower bounds on the accumulating sums of capital payments that are given in (13I2) and (13I3) will be smaller the closer the scrap value in each period is to current replacement cost. Figure 13I1 shows one potentially sustainable time path of accumulated capital charges (heavy curve). It indicates how at dates between τ and τ' (the dates at which investment takes place) the accumulated investment may lie anywhere between the two indicated bounds, but at date τ' it must rejoin the lower bounding curve.

Now, reasoning as we did to obtain Proposition 13I1, we obtain

Proposition 13I2   *If there is investment in capital of type j in period τ, and salvage of capital of type j in period $\tau' > \tau$, with no investment or salvage in between, then, to be consistent with Ramsey optimality and sustainability, the payments to capital must satisfy (13I2), (13I3), and*

(13I4)
$$\sum_{t=\tau}^{\tau'-1} (\sigma_j^t)\rho_t = K_j^\tau[\beta_j^\tau\rho_\tau - \alpha_j^{\tau'}\rho_{\tau'}l_j^{\tau\tau'}].$$

In contrast with Proposition 13I1, this asserts that if salvage, rather than investment, occurs in period τ', the capital payments between the dates of gross adjustments in the capital stock must have a discounted sum equal to the optimal rental price that would hold if the value of capital to everyone were to fall to its scrap value in period τ'. Nevertheless, the bounds encountered in Proposition 13I1 now also constrain the Proposition 13I2 capital charges accumulated between periods τ and τ'. Thus, in this case, the relationships are those depicted in Figure 13I2. The upper and lower bounding

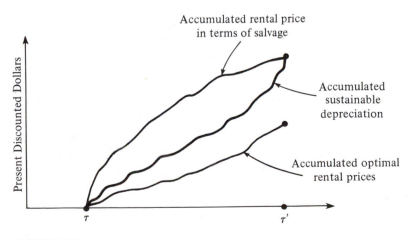

FIGURE 13I2

curves remain those of Figure 13I1. But, now, the actual value of accumulated capital charge assessments must attain the upper bound rather than the lower bound at date τ', the end of the pertinent time interval.

13J. Determinate Character of the Depreciation Assessments

It is important to note that although the figures and the relationships in (13I1)–(13I4) seem to suggest that there is a considerable range of indeterminacy for the payments to capital, σ_j^t, wherever there is a substantial gap between τ and τ' and between the values of β_j^t and α_j^t, this is not really true. The relationships in (13I1)–(13I4) are derived only from partial information—from information relating only to costs: capital purchase prices, capital scrap values, discount rates, and deterioration rates. And, as a matter of fact, the ranges of values for the capital charges obtained so far cannot be narrowed further without access to additional information.

However, the payments to capital really are determined completely by the full system that is comprised of the cost data that are summarized by (13C1) together with the demand functions and the relationships describing the objectives of the enterprise.

In the key case in which the objectives of the enterprise require it to adopt sustainable Ramsey prices, the implied equalities between marginal costs and prices, $p_i^t = \partial V^t / \partial y_i^t$, as indicated in Proposition 13F2, yield a system of equations which, together with the demand functions, determine the quantities of the outputs that are produced and sold in each period. The arguments of the marginal cost functions, $\partial V^t / \partial y_i^t$, are the outputs in period t, y^t, and the K_j^t, the magnitudes of the capital stocks in that period. In turn, the K_j^t values are endogenously determined by the minimization of cost, (13C1), as functions of the parametric cost data and the desired output levels. Suppressing the cost parameters, then, the implications of marginal-cost pricing can be summarized by the system of equations $\partial C(y^1, \ldots, y^T) / \partial y_i^t = p_i^t$ and $y = Q(p)$, where $Q(p)$ is composed of the full vector of intertemporal demands as a function of all intertemporal prices. With this system solved for the output levels $y^1, \ldots, y^T$, the cost-minimization problem, (13C1), can be solved for the optimal pattern of investments and for the corresponding magnitudes of the stocks of capital K^t. Then, the payments to capital $-V_{K_j}^t K_j^t$ defined in Proposition 13G1 can, in principle, be calculated from the V^t functions evaluated at the relevant quantities of the outputs and capital stocks. On a unit basis, that is $-V_{K_j}^t$, is just the rate of saving in variable cost in period t permitted by a unit increase in the quantity of the jth capital stock.

For another illuminating way of viewing this process, let us ignore the endogeneity of K_j^t and assume that demands are independent and that if the stocks of capital were fixed, the (short-run) marginal cost of each output would be independent of the quantities of all other outputs. Then we would

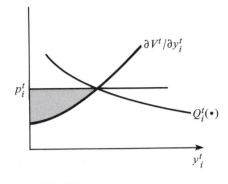

FIGURE 13J1

have the simple relationships depicted in Figure 13J1. Here, the optimal output of y_i^t is that at which the demand price is equal to the marginal variable cost. The shaded area is the contribution from product y_i^t to coverage of capital cost; that is, it is the excess of total revenue over variable cost. Summed for all the products sold in period t, these areas comprise the period's total payment to capital, and by Proposition 13G2, when discounted and summed over time, they recover the entire present value of the cost of the investment program.

This last result may appear curious. For it is clear that the capital payment in any period depends upon the position of the demand curve in that period. A shift in that curve would obviously change the size of the shaded area— the optimal payment to capital. How, then, can it happen that the sum of these demand-dependent magnitudes over the lifetime of capital always add up, precisely, to the total amount invested in that capital? What act of magic guarantees that such a sum of demand-determined magnitudes will be equal to a figure that is entirely cost determined?

The answer, of course, is that this is neither coincidence nor magic, but the result of the optimization process, in which the quantity of capital is determined endogenously. If, for example, the capital assessments as we define them were more than sufficient to cover the investment cost, the quantity of capital employed would not satisfy the requirement of Ramsey optimality. Its quantity should then be increased, and output should be increased along with it, depressing price and bringing total revenue closer to total variable cost. Only when the total capital costs are exactly covered by the optimal depreciation assessments can the size of the capital stock satisfy the Ramsey requirements.

In any event, it should be clear from this interpretation of the depreciation determination process, or from the interpretation offered a bit earlier, that the depreciation assessments are completely determined by the interactions between costs and demands, when there is a policy of marginal-cost pricing.

And, in principle, they can be calculated from complete information about the cost function and the demand functions.

Viewed in this way, Propositions 13I1 and 13I2 tell us that if the depreciation payments were calculated from complete information on demands and costs for the entire system, and if all investment decisions were optimal, then the depreciation assessments would satisfy the conditions (13I1)–(13I4). The great strength of these results, then, is that they make it possible to learn a great deal about depreciation assessments from relatively little information.

13K. Conditions (13I1)–(13I4) and Sustainability

Conditions (13I1)–(13I4), besides being necessary for optimality in resource allocation, are also necessary for sustainability. The connection is not difficult to see. Condition (13I1) is the requirement that the benefit of investment in capital stock j; that is $\sum_t (*\sigma_j^t)\rho_t = -\sum V_{K_j}^t K_j^t \rho_t$, equal the incremental cost of the investment for the time interval τ to τ'. But, if this incremental benefit were less than the cost, an entrant could outcompete the incumbent by reducing the quantity of investment j at date τ (and perhaps making up for the reduction later) but otherwise duplicating the incumbent's activities. On the other side, if the incremental benefit exceeded the incremental cost, an entrant could expand investment j at date τ and be in a position to outcompete the incumbent. Similarly, condition (13I2), as we discuss above, is the requirement that the incremental loss in benefit from postponement of investment be at least equal to the saving in present value of investment outlays permitted by postponement. Obviously, if this condition were violated, an entrant could operate more efficiently than the incumbent by duplicating all the other activities of the incumbent, but postponing the investment in question s periods from date τ. Finally, violation of (13I3) means that an entrant can outmatch the incumbent's efficiency by undertaking a quantity of investment larger than the incumbent's at date τ and then getting rid of any excess as scrap at date $\tau + s$.

A simple example will indicate more explicitly the connection among (13I1)–(13I4) and the issue of sustainability. In particular, we will show how violation of (13I2) must cause unsustainability. We deal with an asset whose life is exactly two periods, which does not deteriorate during the first period of its life, but whose value falls to zero at the end of the second. Market demands, asset prices, and the available set of techniques are assumed to be constant for at least two periods. The incumbent firm invests in the asset at the start of period 1 and depreciates it over periods 1 and 2 with capital assessments σ^1 and σ^2. Then, for a unit of this capital asset, (13I1) requires

$$\sigma^1 + [\sigma^2/(1 + r)] = \beta.$$

Since the capital does not deteriorate during the first period, $^*\sigma^1 = \beta(1 - 1/(1 + r))$ and $^*DEP_1 = 0$. Since it deteriorates completely during the second period, $^*\sigma^2 = \beta = {}^*DEP_2$. Thus, as required, $^*\sigma^1 + {}^*\sigma^2/(1 + r) = \beta$ and $^*DEP_1 + {}^*DEP_2 = \beta$.

Now, suppose σ^1 were to violate (13C2), so that $\sigma^1 < {}^*\sigma^1$ and therefore, by the preceding equations, $\sigma^2 > {}^*\sigma^2 = \beta$. Then an entrant could wait until period 2 to invest in a unit of the capital asset and to begin operation, and matching the incumbent's prices, earn a gross return on his investment of σ^2 in period 2. The entrant's net return, present value *circa* period 2, would thus be at least $\sigma^2 - \beta > 0$; that is, the entrant could slightly undercut the period-2 prices of the incumbent and yet make a profit for himself. Obviously, then, prices yielding $\sigma^1 < {}^*\sigma^1$ are unsustainable, as we want to show.

13L. Fixed Cost, Depreciation, and Sustainability

Much of our later analysis deals with the case in which the firm's overall activities incur a fixed cost, $F > 0$, which does not have to be repeated when the firm undertakes an investment after it is already in operation. For this case we have

Proposition 13L1 *Let the cost function $\tilde{C}(y^1, \ldots, y^T) \equiv F + C(y^1, \ldots, y^T)$, where $C(y^1, \ldots, y^T)$ is as defined in (13C1) and is linearly homogeneous. Then $\tilde{C}$ exhibits declining ray average costs. Moreover, it is trans-ray convex, provided that C exhibits weak cost complementarities.[7] If $F = 0$, then (as shown before) marginal-cost pricing is necessary and sufficient for sustainability. With positive fixed costs, F, sustainable prices cannot be lower than marginal costs, and the corresponding payments to capital and depreciation assessments must not be less than those characterized above. Given the assumptions about demands needed for the weak invisible hand theorem, Ramsey prices are sustainable. These prices provide revenues in each period that cover variable costs in that period, as well as the optimal capital charges just discussed and, in addition, some positive portion of the common fixed cost, F.*

---

[7] This follows from the fact that a twice-differentiable, linearly homogeneous function, $C(\cdot)$, with $C_{ij}(\cdot) \leq 0$ for $i \neq j$, is convex over the positive orthant. By linear homogeneity, $\sum_j C_{ij}(y)y_j = 0$, so that $C_{ii}(y)y_i = -\sum_{j \neq i} C_{ij}(y)y_j \geq 0$ and $|C_{ii}(y)|y_i = \sum_{j \neq i} |C_{ij}(y)|y_j$. Thus, the Hessian of $C(\cdot)$ has a (weakly) dominant diagonal and is positive semidefinite. Consequently, $C(\cdot)$ is convex for $y \geq 0$ and $\tilde{C}(\cdot)$ is trans-ray convex. As discussed earlier, outputs in different time periods have weak cost complementarities as long as capital stocks are normal inputs in each period's production process. This suffices if only one output is produced in each period. Otherwise, to show that $C(\cdot)$ exhibits weak cost complementarities among all outputs, we must assume in addition that $V^t(y^t, K^t)$ exhibits weak cost complementarities between y_i^t and y_j^t, $j \neq i$, for given capital stocks K^t.

13M. Increasing Returns to Scale in Variable Costs and Construction Costs

Until now, we have assumed that the variable cost functions exhibit linear homogeneity. This imparts constant returns to scale to $C(y^1, \ldots, y^T)$. Now, suppose instead that the degree of returns to scale of the V^t function is $S_v^t > 1$. Then $\sum_i y_i^t (\partial V^t / \partial y_i^t) + \sum_j K_j^t V_{K_j}^t = V^t / S_v^t$, and the revenues derived from marginal cost pricing, $\sum_i y_i^t (\partial V^t / \partial y_i^t)$, must fall short of covering the period's total variable costs V^t after subtraction from those revenues of the capital assessment $-\sum_j K_j^t V_{K_j}^t$. Nevertheless, the latter remains the optimal value of economic depreciation; that is, it still remains optimal for the capital charges to satisfy conditions (13I1)–(13I4), and their discounted sum must still equal the discounted total net payments for investment.

However, because variable costs are not covered, here, Ramsey prices must generally exceed the corresponding marginal costs and, as we shall see, the overall intertemporal cost function need not be trans-ray convex. It follows that intertemporal unsustainability is no extraordinary phenomenon. This is explored carefully in the next chapter for a special class of cases. But, as an introduction to the issue, it may be useful to indicate graphically the implications of either increasing or diminishing returns in capital construction for the requirements of sustainability.

For simplicity, we deal with the single-product, two-period case in which the only costs are construction costs, and in which constructed capacity lasts two periods, without deterioration.

Let y^t be output at time t

k^t be quantity of construction at t, measured in units of output-producing capacity

$C(y^1, y^2)$ = the output cost function, with $C_1 \geq 0$ and $C_2 \geq 0$.

$K(k^1, k^2)$ = the construction cost function with $K_1 > 0$ and $K_2 > 0$.

Here, because k^1 and k^2 represent investments in capacity, feasibility requires

(13M1)
$$k^1 \geq y^1$$
$$k^1 + k^2 \geq y^2.$$

Thus, $C(y^1, y^2)$ is the minimum of $K(k^1, k^2)$, subject to (13M1).

Let us now examine the effects of increasing or diminishing returns in the *construction* cost function $K(\cdot)$ upon the trans-ray behavior of the production-cost function $C(\cdot)$. For this purpose, we work with the cross section of the cost surface above the symmetric trans-ray hyperplane in output space:

(13M2)
$$y^1 + y^2 = a, \qquad y^1 \geq 0, \qquad y^2 \geq 0.$$

On this hyperplane, $y^1 \geq y^2$ for $a \geq y^1 \geq a/2$. Clearly, in this interval, efficiency requires that $k^2 = 0$ and $k^1 = y^1$, so that

(13M3) $\qquad\qquad C(y^1, y^2) = K(y^1, 0) \qquad$ for $a \geq y^1 \geq a/2$.

Thus, the curvature of the cross section of $C(\cdot)$ in that interval will be the same as that of $K(k^1, 0)$. Specifically, $C(\cdot)$ will be concave or convex as $K(k_1, 0)$ is concave or convex in the interval $a/2 \leq k^1 = y^1 \leq a$.

Figure 13M1(a) is a construction cost curve, and we see now how it is used to determine the corresponding symmetric trans-ray cross section of the cost function $C(y^1, y^2)$, which is shown in Figure 13M1(b). For this purpose consider how any particular point, say point W, is derived. Point W clearly corresponds to the output vector $V = (y_1, y_2) = (a3/4, a/4)$. Now, in accord with (13M3), with this pattern of outputs, cost minimization requires the construction of capacity $k^1 = y^1 = (\tfrac{3}{4})a$ in the first period, with with no construction taking place in the second period. From Figure 13M1(a) we see that the cost of such a construction program (point v) is vw. Hence $C(y_1, y_2) = C((\tfrac{3}{4})a, a/4) = K(k_1, 0) = vw = VW$ where point W is found by moving horizontally from point w in Figure 13M1(a) to the point above V in Figure 13M1(b). The other points in the trans-ray cross section SR are derived in exactly the same way from arc rs in Figure 13M1(a). A moment's thought confirms from the method of construction that arc SR in the right-hand diagram is simply the mirror image of arc rs in the left-hand diagram, as (13M3) requires.

Because of the concavity of curve Os, Figure 13M1 clearly depicts a case in which there are decreasing marginal costs (and increasing returns) in

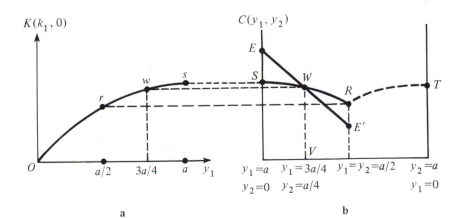

a $\qquad\qquad\qquad\qquad\qquad\qquad\qquad$ b

FIGURE 13M1

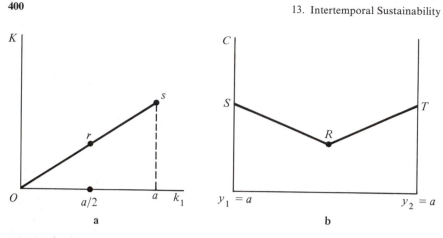

FIGURE 13M2

construction. By the symmetry between rs and SR, since the former is concave the latter must also be concave. So we see at once

Proposition 13M1 *Decreasing marginal construction costs imply that the output cost function $C(y_1, y_2)$ cannot be everywhere trans-ray convex.*

Figures 13M2 and 13M3 show, respectively, the corresponding relationships for the cases of constant and increasing marginal (and average) costs of capacity construction. Here we see that both cases do yield trans-ray convexity of $C(\cdot)$ over the arc SR.

For the case of scale economies in construction, Figure 13M1(b) shows, however, that whatever behavior of cost and whatever construction practices prevail in the right-hand interval of the trans-ray cross section, where $y_2 > y_1$,

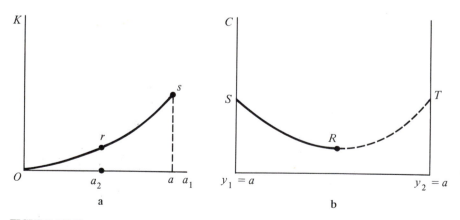

FIGURE 13M3

we already have a nonconvex region where $y_1 \geq y_2$, yielding points like W which are not trans-ray supportable. That is, it is impossible to draw a revenue hyperplane under the cost surface whose (linear) trans-ray cross section EE' covers total cost at W without crossing the cost surface elsewhere, hence offering the possibility of positive profits (as in interval EW). Thus, for at least some patterns of demand, there will be no sustainable prices.

Other demand patterns, of course, make the right-hand interval of the cross section the relevant one. To show, briefly, how one can construct this portion of the cost surface, where $y_2 \geq y_1$, we deal here only with anticipatory construction; that is, the (analytically simple) case in which capacity is built all at once and on a scale sufficient to handle the output growth in the second period.[8] Here, $k^1 = y^2 \geq y^1$ and $k^2 = 0$, with capacity k_1, therefore, being sufficiently large to produce output in either period. Then,

$$(13M4) \qquad C(y^1, y^2) = K(k^1, 0) = K(y^2, 0), \qquad \text{for } a/2 \leq y_2 \leq a.$$

Comparison with (13M3) confirms that this right-hand region of the trans-ray cross section is completely symmetric with the left-hand region ($a/2 \leq y_1 \leq a$). This yields the curves RT, symmetric with SR, in Figures 13M1(b), 13M2(b), and 13M3(b) as the relevant right-hand arcs when all construction is anticipatory.

However, where the construction cost function exhibits decreasing returns to the scale of construction in each period, it is not efficient to engage in anticipatory construction. Rather, postponement of the construction of the capacity needed only for period 2 reduces total costs by decreasing the scale of construction in period 1 and by reducing the (implicit) interest costs of financing the capacity construction. Thus, in this case,

$$(13M5) \qquad C(y^1, y^2) = \begin{cases} K(y^1, 0) & \text{for } y^1 \geq y^2 \\ K(y^1, y^2 - y^1) & \text{for } y^1 < y^2 \end{cases}$$

In particular, suppose $K(k^1, k^2)$ has the simple but plausible form

$$(13M6) \quad K(k^1, k^2) = c(k^1) + c(k^2)(1/1 + r), \qquad c''(\cdot) > 0 \text{ and } c'(\cdot) > 0.$$

Here, construction costs are intertemporally separable and stationary, and exhibit increasing marginal costs. Then, it is straightforward to show that the production cost function, (13M5), is strictly trans-ray convex. Moreover,

---

[8] In the next chapter we see that where it pays to postpone the construction of some or all of the capacity needed to handle future output growth, matters become somewhat more complicated and the sustainability problems of the increasing returns case may become far more serious. But all of this is best postponed for systematic discussion in the following chapter.

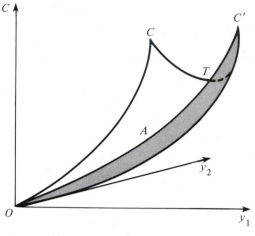

FIGURE 13M4

it is also strictly ray convex, and globally strictly convex, having the form illustrated in Figure 13M4. Consequently, an industry comprised of such firms is naturally competitive, and the competitive equilibrium entails an infinity of firms with infinitesimal output quantities. By the results of Chapter 9, since (13M5) exhibits economies of scope between outputs in the two

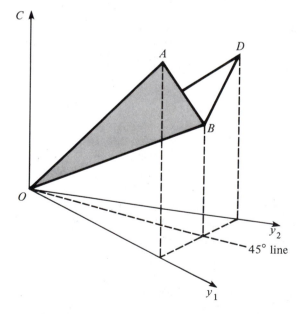

FIGURE 13M5

periods, as well as strict trans-ray convexity, all of the firms in competitive equilibrium must produce actively in both periods and must have the same combination of intertemporal outputs.

Where the construction cost function exhibits constant returns to the scale of construction in each period, the production cost function will also be trans-ray convex. Figure 13M2(b) depicts this case under the simplifying assumption that construction costs are separable and identical (in present value) in the two periods. Separability alone ensures that the production cost function is linearly homogeneous and trans-ray convex, as illustrated in Figure 13M5. This is, then, a special case of the model defined in (13C1) and used throughout the subsequent sections of this chapter. Consequently, sustainable intertemporal prices exist in this scenario, and they embody the depreciation assessments described and analyzed here.

13N. Concluding Notes

This chapter has approached the intertemporal sustainability issue in stages, starting from the case of complete separability of cost among periods, the case in which the formal analysis comes closest to one from which time is entirely abstracted. From that simple case, the chapter's analysis gradually moved on to one involving full intertemporal interdependence. The case of linear homogeneity was particularly rich, offering us a variety of insights into intertemporal pricing and depreciation assessments for capital goods and their relation to technical change, variation in interest rates, and inflation. And these fundamental insights were shown to be equally applicable in cases far more general than just that of linear homogeneity.

Perhaps the most important lesson of this chapter is that deviations of firms' pricing decisions from the precepts of capital theory can put their market positions at risk. In contestable markets, depreciation assessments that deviate substantially from Ramsey-optimal depreciation figures give latent rivals incentives to enter at some time and displace the incumbent firm. Thus, the invisible hand is hard at work in some contestable markets, guiding investment, depreciation, and intertemporal pricing decisions along socially beneficial paths.

However, toward the end of the chapter, it became clear that in intertemporal models the existence of sustainable prices cannot be assumed lightly. In the absence of fixed costs, at least in theory, the slightest scale economies in capacity construction can permit unsustainability. In such a case, even though it is most efficient for a single firm to continue to provide output over a succession of dates, such an assignment of the task of production simply may not be possible because entry may prevent it.

In the next chapter, unsustainability under economies of scale in construction is examined in detail, and its implications are considered with some care. We find reason to believe that the problem can be real and significant, and that in intertemporal analysis there may be other significant sources of unsustainability. One seems to end up inescapably with the conclusion that in the intertemporal case unsustainability may come closer to being the rule than the exception, though how serious a problem this poses for resource allocation, equilibrium, and stability is still far from settled.

14

Intertemporal Unsustainability

Parable: A succulent, chubby lamb stands in a narrow passageway before a line of ravenous lions, all of whom have smelled the lamb's presence. No lion near the beginning of the line knows how many lions are lined up behind him. The first lion is tempted to gobble up the lamb but recognizes that if he does so he may be eaten by the lion immediately behind him, who has no other way of getting at the lamb. But that second lion knows that, if he eats the lion in front of him to get at the lamb, he, in turn, is apt to be eaten by the lion who is next in line. The same is true of the third lion, the fourth, and so on.

Question: Who, if anyone, will be eaten?

In this book and in other recent writings the possibility that no sustainable prices may exist for a particular firm or industry has been demonstrated. However, the static examples that have been constructed are frequently characterized by an aura of artificiality. Some observers have consequently been led to conjecture that unsustainability is a pathological phenomenon which has more interest as an analytic curiosum than as a phenomenon in the world of reality.[1]

In this chapter we find that all this changes drastically in an intertemporal setting. Unsustainability can come closer to being the rule rather than the exception. Put another way, one of the main implications of the earlier portions of the book is that in a timeless world the invisible hand is more powerful than we had reason to expect. It can exert a powerful and beneficent influence even upon monopolistic and oligopolistic industries, forcing them to be efficient and innovative and, perhaps, even to adopt Ramsey-optimal prices.

In the intertemporal case, on the contrary, we will find that the domain of the invisible hand can be far more limited than intuition may have led

---

[1] See, for example, Joskow and Noll (1981).

us to suspect. Often there may well be no sustainable solutions and the market mechanism may well produce an intertemporal allocation of resources that is patently inefficient.

We begin our discussion with the source of the difficulties that appear to be most general and most serious—economies of scale in sunk construction costs. At the end of the preceding chapter we already obtained an inkling of the problem that arises in this case. But this chapter shows that the problems actually can be considerably more substantial than even the diagrams at the end of the chapter suggested.

We find that, as long as efficiency does not require construction to occur at several different dates, the analysis, though containing various novel features, yields no startling results. However, where efficiency requires the construction to be spread over time and there are scale economies in sunk costs for the construction of capacity—the case in which one usually expects the market to be supplied by a single firm, intertemporally as well as within a period—matters do not work out quite that way.

We prove that with a fixed vector of prices the firm that at one period of time was the sole supplier of the market will find itself *vulnerable* to being driven from the market precisely at the date when efficiency calls for its expansion.[2] Even in the absence of technological change or superior efficiency, given the incumbent's prices, an entrant will *always* have an opportunity to take at least part of the market away from the established firm if the scale economies in construction are sufficiently strong, and to do so before the incumbent can recoup its investment. But then, that entrant's market in turn becomes vulnerable to takeover in the same fashion, before it can recover its outlays.

We usually ascribe the decline of a firm or a national economy that previously dominated the world market to a combination of technical obsolescence and deteriorating workmanship or entrepreneurship. However, we see now that the triumph of one such firm or one such country over another, and still another after that, may be possible even if no one falls behind either in efficiency or in pace of innovation. As it were, each master of the market may be fated to become vulnerable to a successor, even if that successor is superior only in the larger scale of its operations. Moreover, such a process, if it occurs, will be wasteful, causing unnecessary duplication, very plausibly on a vast scale.

The real moral of this intertemporal unsustainability analysis is not that incumbents are inevitably doomed to early failure, but that they may only be able to protect themselves, and the interests of the economy, by the very sort of strategic pricing behavior economists have tended to deplore.

---

[2] A somewhat analogous effect was noted in a more specialized model in Faulhaber (1971).

The threat of responsive pricing and its deterrent effect upon potential entrants, far from frustrating the invisible hand, may be the only effective pricing implement available to it.[3]

We end the chapter with a discussion of two special sources of intertemporal unsustainability, learning by doing and postponability of investment. For reasons we will describe, each of these can also lead to violation of trans-ray convexity and, therefore, may well increase the possibility of sustainability problems.

Part I
The Role of Scale Economies in Construction

14A. Basic Unsustainability in Multiperiod Production

In this section we will prove the fundamental result of this chapter: that where demand is growing, if capital construction costs are sunk and are the determining cost component, and there are declining average costs in such construction, then no set of fixed product prices for the different dates involved will enable the firms that comprise the cost-minimizing intertemporal arrangement to prevent entry at the time when it is efficient for the incumbents to expand their capacity. In later sections the source of this result will be examined, and it will be contrasted with the case in which there are constant unit costs in construction—the basic case of the preceding chapter.

We focus on the case in which the unit cost of construction of capacity (for the production of a nonstorable good or service) declines with the amount of capacity constructed. That is, letting $K(y)$ be the cost of producing capacity sufficient for output y, we assume that $K(y)/y$ declines as y increases and that $K(0) = 0$. We also assume that the constructed facilities have no other valuable use outside the industry in question, so that once built, these facilities are sunk. To focus our attention upon the effects of such construction scale economies, we ignore the costs of operating the plant[4] and assume that the plant is sufficiently durable to operate with undiminished capacity for the two periods of our analysis.[5]

---

[3] Here we must be careful. On balance we still believe that the social costs of that sort of pricing behavior should be presumed usually to outweigh its benefits. In other words, this passage is certainly *not* intended as automatic apology for anticompetitive pricing, or for the threat to carry it out.

[4] However, note that constant unit operating costs can be taken into account simply by interpreting the prices that appear below as being net of such costs.

[5] Later we will examine the case of capacity that deteriorates over time.

However, we allow for the possibility that, in addition to the costs of producing capacity, there are fixed costs, $F \geq 0$, incurred for the establishment of any firm.[6] Then, in present value terms, the cost $C(y_1, y_2)$ of producing y_1 and y_2 units in periods 1 and 2, respectively, is

(14A1) $$\min_{k_1, k_2} \left[F + K(k_1) + K(k_2)/(1 + r) \right],$$

subject to

$$k_1 \geq y_1, \qquad k_1 + k_2 \geq y_2, \quad \text{and} \quad k_i \geq 0,$$

where k_i is the quantity of capacity constructed in period i. Note that (14A1) embodies the assumption that, once constructed, capital facilities are sunk. Otherwise, the cost function would comprise what are essentially the rental costs of capacity facilities, well tailored to each period's output level; for example,

$$C(y_1, y_2) = F + \frac{rk(y_1)}{1 + r} + \frac{rk(y_2)}{(1 + r)^2}.$$

In contrast, with our sunkedness assumption, and using k_1^* and k_2^* to denote the efficient levels of capacity construction in the two periods, (14A1) can be rewritten as

(14A2) $$C(y_1, y_2) = F + K(k_1^*) + K(k_2^*)/(1 + r).$$

Here, if $k_1^* > y_1$, which can only occur if $y_2 > y_1$, we say that the firm engages in *anticipatory construction*, building in the first period capacity that will not be used until the second. If $k_2^* > 0$, the firm is said to engage in a program of *capacity expansion*. This can only occur if $y_2 > y_1$, in which case $k_2^* = y_2 - k_1^*$. In general, it may be efficient for the firm to undertake both some anticipatory construction and, later, some capacity expansion.[7]

For the case of expanding output, $y_2 > y_1$, it is efficient to undertake *some* later capacity expansion, rather than completely anticipatory con-

---

[6] These may include outlays on the legal process, the initial acquisition of managerial and technical skills, and the collection of required marketing and technical information.

[7] However, if $K(\cdot)$ is concave, both of these programs cannot simultaneously be efficient. In this case, efficiency requires either the construction of all required capacity in the beginning of the first period, or construction only of the capacity immediately needed at that time. This can be seen by noting that otherwise $y_1 < k_1^* < y_2$, $k_2^* = y_2 - k_1^*$, so that k_1^* must then be the unconstrained (interior) minimizer of $K(k_1) + K(y_2 - k_1)/(1 + r)$. But the necessary second-order condition for the minimization contradicts the assumed concavity of $K(\cdot)$.

struction, if, for some k_1 with $y_1 \leq k_1 < y_2$, $K(k_1) + K(y_2 - k_1)/(1 + r) < K(y_2)$, or, rearranging, if

$$(14A3) \qquad K(y_2 - k_1) + K(k_1) - K(y_2) < r[K(y_2) - K(k_1)],$$

for some k_1 with $y_1 \leq k_1 < y_2$. Thus, maximal anticipatory construction is wasteful if the right-hand side of (14A3), the cost of holding the initially unused increment in capacity ($y_2 - k_1$), exceeds the savings [the left-hand side of (14A3)] from full exploitation of the scale economies of construction obtainable by simultaneous construction of the entire capacity necessary to produce y_2.

In general, (14A3) will hold for given $K(\cdot)$, y_1, and y_2 if the discount factor r (which is an increasing function of the length of time between the two dates in question)[8] is sufficiently large. And, given $K(\cdot)$ and r, it will hold only if y_2 is sufficiently large relative to y_1.

We have defined a firm as an intertemporal natural monopoly if its intertemporal sequence of outputs cannot be produced at lower cost by *any* combination of two or more firms. It can be shown by a straightforward but tedious argument that a firm with the cost function $C(y_1, y_2)$, which is defined by (14A1) or (14A2), must be an intertemporal natural monopoly, unless $F = 0$ and it increases its capacity at some date after that on which it first began to operate. This last case is not a natural monopoly since total industry costs will not be affected if, instead of the incumbent firm, it is a second firm that builds capacity $k_2^* = y_2 - k_1^*$ at the beginning of period 2. That is, here we have $F = 0$, $C(y_1, y_2) = C(y_1, k_1^*) + C(0, y_2 - k_1^*)$. Yet, in this case, as in the intertemporal natural monopoly case, minimized industry cost is given by $C(y_1, y_2)$ as a result of the preceding equality.

Making use of the industry cost function, which is also the cost function of any one firm, let us consider an intertemporal industry configuration that is efficient; that is, it is least costly in terms of the present value of the cost of producing whatever output vector is involved. Such a configuration is defined to consist of a pair of market prices, p_1 and p_2, one for each period, and the associated market demands, y_1 and y_2, such that, in present-value terms, total industry revenues cover the associated minimal industry costs. Further, for a configuration to be sustainable, it is necessary that

$$(14A4) \qquad \begin{array}{ll} \text{(i)} & p_1 y_1(p) + p_2 y_2(p)/(1 + r) = C[y_1(p), y_2(p)]; \\[2mm] \text{(ii)} & (p_1, p_2) \text{ is undominated.}[9] \end{array}$$

---

[8] That is, if t is the number of units of time that intervene between our two dates and i is the interest rate for one unit of time, then $1 + r = (1 + i)^t$.

[9] In the sense defined in Chapter 8.

Condition (i), of course, means that the firm's discounted revenues just cover its costs, while (ii) means that there exists no different pair of prices, p_1', p_2', such that $p_1' \leq p_1$ and $p_2' \leq p_2$, with (p_1', p_2') yielding a positive profit.

Such a configuration calls for capacity expansion rather than completely anticipatory construction if and only if (14A3) is satisfied. Which of these cases is relevant depends on the forces determining p_1 and p_2, on the time pattern of market demand functions, and on the size of the discount factor r, which in turn grows with the interval of time between the two dates we are considering. The fundamental properties of the two cases are quite different, and here we shall focus exclusively on the case in which capacity expansion is the less costly alternative. Thus, throughout the next few sections of this chapter, we assume that (14A3) holds for all candidate sustainable configurations, that is, for all configurations that satisfy the basic necessary conditions (14A4).

We show that, unless F is large, any efficient industry configuration that involves capacity expansion is unsustainable. Note that, to prove that a monopoly has no sustainable prices available to it, it is only necessary to select some *particular* set of entry strategies and to show that the monopoly will be unable to protect itself against those strategies by the selection of any time pattern of prices. Using this observation we now derive the basic theorem of this chapter:

Proposition 14A1   *When there are declining average costs in construction of sunk capacity for the production of a nonstorable good or service, any efficient intertemporal industry configuration that involves capacity expansion is unsustainable, unless the firm incurs fixed costs F sufficiently large to satisfy*

$$\text{(14A5)} \qquad \frac{F}{y_2} > \frac{K(y_2 - k_1^*)}{y_2 - k_1^*} - \frac{K(y_2)}{y_2},$$

where k_1^* is defined in (14A2); that is, unless the fixed cost averaged over the larger capacity, y_2, is sufficiently great to make up the difference between the average construction cost of the incremental capacity, $y_2 - k_1^*$, and that of the larger capacity, y_2.

Proof   Let p_1, p_2 and y_1, y_2 satisfy (14A4) and (14A3), so that they characterize an efficient break-even configuration involving expansion of capacity. It follows from (14A3) that $y_2 > y_1$, and

$$\text{(14A6)} \qquad C(y_1, y_2) = K(k_1^*) + K(y_2 - k_1^*)/(1 + r) + F.$$

For the configuration to be sustainable, p_1 and p_2 must yield revenues that do not exceed cost for *any* potential entrant, in particular for one who plans to construct only k_1^* units of capacity in period 1, and to sell y_1 and

k_1^* units in periods 1 and 2, respectively. This requires

(14A7) $$p_1 y_1 + \frac{p_2 k_1^*}{1 + r} \leq F + K(k_1^*)$$

Subtraction of this inequality from (14A4)(i) yields, in view of (14A6),

(14A8) $$\frac{p_2(y_2 - k_1^*)}{1 + r} \geq \frac{K(y_2 - k_1^*)}{1 + r}.$$

Thus, (14A8) is necessary for sustainability.

Sustainability also requires that an entrant at the beginning of period 2 not be able to expect to earn a positive profit if it chooses to supply the entire quantity y_2. That is, another necessary condition for sustainability is

(14A9) $$p_2 y_2 - K(y_2) - F \leq 0.$$

But, rearrangement of (14A8) gives

$$p_2 y_2 - K(y_2) - F \geq K(y_2 - k_1^*)(y_2/(y_2 - k_1^*)) - K(y_2) - F,$$

where $K(y_2 - k_1^*)(y_2/(y_2 - k_1^*)) - K(y_2) > 0$ by the assumption of declining cost of construction, or

(14A10) $K(y_2 - k_1^*)/(y_2 - k_1^*) > K(y_2)/y_2,$ for $y_2 > k_1^* > 0$.

Thus, unless this is offset by fixed costs which are sufficiently large, that is, unless condition (14A5) holds, the two necessary conditions for sustainability, (14A8) and (14A9), are inconsistent and, consequently, the configuration is unsustainable.[10]   Q.E.D.

Let us look once more at the critical condition (14A5). First, as has just been noted, the right-hand side is positive if average construction costs are

---

[10] It has been suggested to us by Tim Cooke of Johns Hopkins University that the proof can be made more intuitive in the case that $y_1 = k_1^*$ by rearranging (14A8) to read

(14A8′) $p_2 \geq K(y_2 - y_1)/(y_2 - y_1),$

meaning that, unless period 2's price covers the incremental cost of expansion, period 1's price will have to be sufficiently high to attract entry. Similarly, (14A9) can be rewritten as

(14A9′) $K(y_2)/y_2 + F/y_2 > p_2,$

meaning that the average cost of a second-period entrant must exceed his price. But (14A8′) and (14A9′) together imply $K(y_2)/y_2 + F/y_2 > K(y_2 - y_1)/(y_2 - y_1)$, which is only possible if F is sufficiently large to satisfy (14A5).

decreasing, because of (14A10). In (14A10), $K(y_2 - k_1^*)/(y_2 - k_1^*)$ is the unit incremental capacity cost borne by the incumbent in expanding his capacity. $K(y_2)/y_2$ is the lower unit capacity cost expected by a potential entrant in evaluating the profitability of building capacity sufficient to serve the entire market at the beginning of period 2. The value of the right-hand side of (14A5) is the difference between what the cost of constructing y_2 units of capacity would be at the unit cost borne by the incumbent and the actual construction cost for a second-period entrant.

If no general fixed costs were borne by the firm, then (14A5) could not be satisfied, and Proposition 14A1 would establish the unavailability of any sustainable prices for an efficient industry configuration which at some date expands its capacity. Such a configuration can involve a single firm which builds k_1^* units of capacity at the beginning of period 1 and $y_2 - k_1^*$ units at the start of period 2. Or it can involve one firm which carries out the first construction project and another firm which carries out the second. In either case, any fixed vector of industry prices must inevitably offer incentives for entry by an additional firm whose presence in the market must destroy the efficient configuration.

At the opposite extreme, it can be shown that sustainability of an efficient industry configuration can be ensured, under reasonable assumptions, by fixed costs that are sufficiently large and that must be incurred for any activity of the firm to be possible (see Chapter 10). In contrast, (14A5) shows that, given the *fixed* costs incurred in the establishment of a firm, and so long as capacity expansion is sunk, efficient, and subject to scale economies, the higher are the *sunk* costs involved in the construction of added capacity the less likely is sustainability. Thus, Proposition 14A1 again demonstrates the crucial distinction between costs sunk in capacity construction and fixed costs incurred in the establishment of a firm. The former militate against sustainability while the latter tend to support it.

In the midrange of F, between 0 and the bound given by (14A5), the industry will be an intertemporal natural monopoly that does not have any sustainable prices available to it. That is, production of the intertemporal output vector will be less costly if it is done by a single firm than if it is carried out by any multiplicity of firms. Yet it will be impossible for such a single firm to select any financially viable vector of prices for the different periods in its history which can prevent entry. Thus, the efficient single-firm arrangement inherently involves an element of instability in price, industry structure, or both.

Proposition 14A1 becomes more surprising (and more disturbing) as one considers it further.

(i) At first it may appear to have a simple explanation: With low fixed costs, the second-period entrant is able to undercut the incumbent firm by building later simply because increased demand permits him to build a larger plant whose unit cost is, by assumption, lower. But there is much more to the matter. The older firm will have an offsetting advantage on its side. It will have been using its more costly plant for some time before second-period entry threatens, and will thereby have been able to write off some of its construction cost by selling output from that plant before the appearance of the second-period entrant. The surprising part of the theorem is that, if fixed costs are low, this earlier revenue cannot be sufficient to permit the incumbent to select prices that meet the competition of the later entrant (provided that the first-period price is not so high as to provide an incentive for duplicative entry in the first period.) In other words, despite the substantial cost complementarity between the two periods' outputs, and the competitive advantage it offers to established enterprises, we have found that the incumbent firm can be driven from the market in the second period.

(ii) Moreover, Proposition 14A1 implies more than the risk of demise and replacement of the incumbent. It implies that, in the absence of large fixed costs, the incumbent will be vulnerable to displacement before it is able to recover its initial investment cost. If its first-period price is sufficiently high to permit it to recover its costs in the absence of entry, it must leave itself open to successful entry (and takeover) in that first period. On the other hand, if the first-period price is low enough to prevent entry at that time, the incumbent firm cannot set the second-period price sufficiently high to recover capital cost without permitting the second-period entrant to drive it from the market. The implications about the profitability of firms and the investment process in general are more than a little disquieting.

14B. Repeated Demsetz Auctions of Monopoly Franchises

Another way of looking at the problem indicated by the preceding analysis is to see whether similar difficulties arise under the arrangement suggested by Demsetz (1968) for the control of monopoly. He has proposed that where an industry is a natural monopoly it be run under government franchise. The franchise is to be put up for auction periodically and awarded to the entrepreneur making the best bid. But at the next auction date the current holder of the franchise will lose it unless he is once again prepared to make the best bid. In our model, at the beginning of period 1 the franchise is auctioned, the winner being the firm that promises to supply all of the quantity demanded by the market during period 1 at the lowest price offered by any of the bidders. Similarly, at the beginning of period 2, the franchise for period 2 is auctioned off, with the winner being the entrepreneur who promises to serve the market at the lowest price during period 2.[11]

---

[11] Successive auctions of this type have recently been studied in Sinden (1979).

This case is important for our analysis because in an auction with many potential bidders, each with free access to the required productive techniques, the crucial Bertrand-Nash assumption is necessarily satisfied. That is, every participant knows that the lowest bid must win out and that losers cannot then reopen bidding with revised prices. It is important to note here that the bidding takes place *ex ante*, so that at the time of the competition, no bidder other than the incumbent has committed any capital (i.e., has incurred any sunk costs), while the incumbent's capital will be salable at its true economic value and thereby become fungible *within* the industry.

We now show that

Proposition 14B1   *Unless the firm's operations as a whole incur fixed costs sufficiently large to satisfy condition (14A5), if minimization of cost requires construction of sunk plant at two different dates, the winning prices of the repeated auction make it profitable for a firm to offer quantities that fail to satisfy the entire market demand in period 2. Alternatively, it will pay a subset of the population to withdraw from the overall consumer group that runs the repeated auction.*

This means that for the auction to yield an equilibrium, the law must proscribe private provision of the good outside the auction, and the terms of the auction must explicitly require the winner to satisfy market demand completely at the price it has set. In other words, to permit an equilibrium, the auction must offer an exclusive franchise, it must enforce its exclusivity, and it must also enforce a "common-carrier obligation" upon the holder of the franchise to serve the market completely at the winning price.

Proof   Let p_1^m, p_2^m be winning prices of the auctions, and let y_1^m, y_2^m be the associated market demands. Let $C(y_1^m, y_2^m) = F + K(k_1^m) + K(y_2^m - k_1^m)/(1 + r)$, with $y_1^m \leq k_1^m < y_2^m$; that is, let minimum cost require construction in both periods.

Case 1: The same enterprise wins the auction in both periods. The enterprise must then earn zero profit because it would not have bid prices p_1^m, p_2^m if they would have yielded a negative profit, and it would not have won if profit were positive. Thus we have

(14B1) $$p_1^m y_1^m + \frac{p_2^m y_2^m}{1 + r} = F + K(k_1^m) + K(y_2^m - k_1^m)/(1 + r).$$

Moreover, price p_2^m would not have won period 2's auction for the firm unless

(14B2) $$p_2^m y_2^m \leq F + K(y_2^m),$$

because, otherwise, a new firm could have been started at the beginning of period 2 and profitably bid a price lower than p_2^m. As in the proof of Proposition 14A1, unless F were large enough for (14A5) to hold, (14B1) and (14B2) together imply

(14B3)
$$p_1^m y_1^m + \frac{p_2^m k_1^m}{1 + r} > F + K(k_1^m),$$

so that a firm could profitably sell y_1^m and k_1^m units in periods 1 and 2, respectively, at lower prices than those that won the auctions. This completes the proof for Case 1.

Case 2: Different firms win the auction in the two periods. Here if the k_1^m units of capacity are not transferred between the two periods' winners, then profit must equal zero for period 1's winner, so that

$$p_1^m y_1^m = F + K(k_1^m).$$

Similarly, the profit of period 2's winner must be zero; that is,

$$\frac{p_2^m y_2^m}{1 + r} = \frac{F}{1 + r} + \frac{K(y_2^m)}{1 + r}.$$

Addition of these equalities yields

$$p_1^m y_1^m + \frac{p_2^m y_2^m}{1 + r} = \frac{2F + rF}{1 + r} + K(k_1^m) + \frac{K(y_2^m)}{1 + r} > C(y_1^m, y_2^m),$$

so that this cannot be an equilibrium. It can be undermined by a single firm that undertakes to produce at minimum cost and can consequently earn a profit at prices somewhat lower in both periods.

More interesting is

Case 3: The winners in the two periods are, again, different firms, but k_1^m units of capacity are sold by period 1's winner to period 2's winner for V dollars. Then for period 1's winner to earn a zero profit, we must have

(14B4)
$$p_1^m y_1^m + \frac{V}{1 + r} = K(k_1^m) + F.$$

For period 2's winner to earn zero profit,

(14B5)
$$p_2^m y_2^m = V + K(y_2^m - k_1^m) + F.$$

Rationality on the part of period 1's winner requires

(14B6)
$$V \geq 0.$$

On the other hand, rationality on the part of period 2's winner requires his purchase of k_1^m units of capacity to be worth its cost so that

(14B7) $$V \leq K(y_2^m) - K(y_2^m - k_1^m).$$

[Note that (14B5) and (14B7) together again imply (14B2).]

We can now prove that the revenues from the sale of y_1^m and k_1^m in periods 1 and 2, respectively, at the auction-winning prices must again exceed production cost. Multiplying (14B5) by $(k_1^m/y_2^m)(1/(1 + r))$, we add the resulting equation to (14B4), and rearrange the result, to get

(14B8) $$p_1^m y_1^m + \frac{p_2^m k_1^m}{1 + r} - [F + K(k_1^m)] = \frac{F k_1^m}{(1 + r)y_2^m}$$

$$+ \frac{K(y_2^m - k_1^m)k_1^m}{(1 + r)y_2^m} - \frac{(y_2^m - k_1^m)V}{(1 + r)y_2^m}$$

which, by (14B6) and (14B7),

$$\geq \frac{(y_2^m - k_1^m)}{1 + r}\left[\frac{F k_1^m}{y_2^m(y_2^m - k_1^m)} + \frac{K(y_2^m - k_1^m)}{(y_2^m - k_1^m)} - \frac{K(y_2^m)}{y_2^m}\right].$$

This expression is positive because of the scale economies in sunk capacity construction and because $F \geq 0$, and it follows once again that (14B3) holds; that is,

$$p_1^m y_1^m + \frac{p_2^m k_1^m}{1 + r} > F + K(k_1^m) \geq C(y_1^m, k_1^m). \qquad \text{Q.E.D.}$$

14C. Trans-Ray Convexity in the Multiperiod Model[12]

If instead of being composed of n products $y_1, \ldots, y_n$, the output vector y is taken to refer to the outputs of one nonstorable product over n periods, the definitions of declining ray average costs and trans-ray convexity translate themselves immediately into multiperiod terms, as we have seen. The question that naturally arises then is whether these properties are plausible in these circumstances.

It seems clear that declining ray average cost is at least as plausible for one good produced in n periods as for n goods produced in one period.

---

[12] For simplicity, throughout this section fixed costs are assumed to be zero.

Specifically, if in every period taken by itself a k-fold increase in output leads to a less-than-k-fold increase in total cost, the same is apparently to be expected if output in *every* period is increased k-fold. Moreover, intertemporal economies of scope only serve to magnify such intertemporal scale economies.

Intertemporal trans-ray convexity, however, is a more problematic concept, as we have seen in the last section of the preceding chapter. To examine its plausibility, one must consider the character of intertemporal complementarity in production. The existence of some degree of intertemporal complementarity seems quite obvious. For if equipment is durable, it will be cheaper for one firm to produce period after period rather than switching from one firm to another. Indeed, because of this relationship, which was represented in the models of Chapter 13, the case for the presence of some degree of complementarity among the outputs of a given product in a sequence of different periods seems stronger than that for multiproduct output in a given period.

But will this complementarity be sufficiently strong to guarantee trans-ray convexity? We have seen in Chapter 13 that this condition will not be satisfied when construction involves scale economies and construction is anticipatory, and we show next that this problem is likely to be aggravated *when optimality requires some construction in (at least) two different periods.* When this is so, scale economies in construction of sunk capacity yield a curious and rather surprising cost surface, which offers some insights into the absence of sustainable prices. We show that the graph must be of the general form depicted in Figure 14C1. This total cost surface consists of two concave wings with a valley between them just over the $45°$ line. The two wings are asymmetric, with the concavity of the wing nearer the y_2 axis far sharper than that of the other wing.

The asymmetry follows from the difference in cost function for the cases in which all capacity is constructed at the start of period 1 and that in which capacity is expanded at the start of period 2. To see why the surface has the shape depicted, we show the derivation of cross section $RSWUQ$ in Figure 14C1, using the procedure first laid out at the end of Chapter 13. In Figure 14C2(b), we see two curves showing $K(y)$, total construction cost, as a function of capacity. The upper curve, Ow, shows current construction costs, which require no discounting. Its concavity corresponds to the assumed scale economies.[13] The lower curve, Ow', corresponds to the discounted present value of costs of construction in the second period. For simplicity, we again deal with a symmetric trans-ray cross section. This means that we select the points R and S on the two axes of Figure 14C1 to be equidistant from the origin, so that they, respectively, represent outputs $(y, 0)$ and $(0, y)$. Thus, in Figure 14C2, we have $RQ = yw$ and $SW' = yw'$.

---

[13] Actually, as we know, concavity is a stronger assumption than declining average cost, but the case of concave construction costs brings out the issue more clearly.

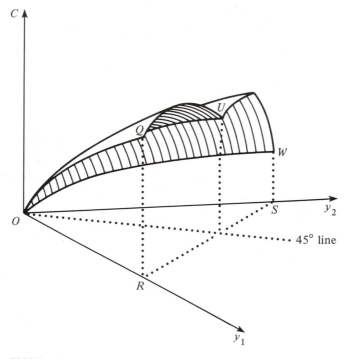

FIGURE 14C1

At the midpoint, U, of the cost surface, with output $y/2$ in either period, one only needs capacity $k_1 = y/2$, which will be used in both periods and which is built entirely in period 1. The cost at that point is seen from Figure 14C2(b) to be given by the height of u, or U in Figure 14C2(a). Next, we examine the capacity cost at point A in Figure 14C2(a), which lies $\frac{1}{4}$ of the way from the left-hand end of the diagram. Here we have $(y_1, y_2) = (3y/4, y/4)$ so we must initially build capacity $k_1 = 3y/4$, whose cost is given by the height of v in Figure 14C2(b), or V in 14C2(a). Because of the concavity of the curve in 14C2(b), V is more than halfway between the height of Q and U. Similarly, we have seen that the remainder of the curve VU in Figure 14C2(a) must also be concave since QVU in 14C2(a) is merely the mirror image of uw in 14C2(b).

Finally, we come to the point D in Figure 14C2(a) which lies $\frac{3}{4}$ of the way toward S and so represents the output vector $(y_1, y_2) = (y/4, 3y/4)$. Here, suppose capacity $k_1 = y/4$ is constructed in period 1 with the residual capacity $k_2 = y/2 = (3y/4 - y/4)$ produced in period 2. In accord with (14A2) this will have the cost given by the sum of the heights of t and the (discounted) height u' in Figure 14C2(b), which gives us point M' in Figure 14C2(a). The other points on the curve $UM'KW'$ are derived similarly. Thus, all points

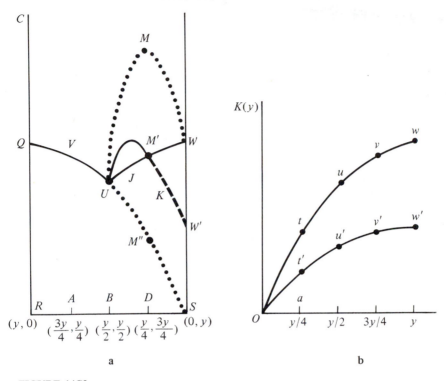

FIGURE 14C2

on the curve represent minimum anticipatory construction, that is, the delay of as much of the necessary construction as time permits. Suppose, on the other hand, that all necessary construction were to be carried out in the first period. It is easy to verify (see Chapter 13) that then the curve $UM'KW'$ would be replaced by $UJM'W$, the mirror image of QVU. Since the firm will choose between delayed and anticipatory construction on the basis of cost, it will always operate on the lower of the two curves. Hence the relevant portion of the cost function will be the lower envelope of the two curves, that is, the thick curve $UJM'KW'$.

It will be noted that the shape of the delayed construction cost curve depends on the discount rate. As the discount rate falls, the curve Ow' in Figure 14C2(b) rises. At the one extreme, where the discount rate is zero, Ow' becomes identical with Ow and the delayed construction cost curve in Figure 14C2(a) rises to UMW, never cutting below UJW. Then it will never pay to delay construction. At the other extreme, when the discount rate is infinite, the curve Ow' in 14C2(b) falls to the horizontal axis, and the delayed construction cost curve in 14C2(a) falls to $UM''S$. It then pays to delay construction to whatever extent is possible.

Now it is clear that the sharp concavity of the wing OUW in Figure 14C1 (the wing corresponding to output growth) means that a revenue hyperplane which touches the cost surface at any output point y^* in that region must also cut the cost surface for some output $y < y^*$ and, hence, there will be smaller output vectors for which total revenue exceeds total cost. This means that an entrant will be able to produce such a smaller output vector and more than cover his costs; that is, there are no sustainable prices for such an output vector, y^*.

14D. An Intuitive Explanation of Intertemporal Unsustainability

The geometric discussion of the preceding sections may be helpful in pointing out what goes wrong formally in the intertemporal equilibration process. But it does not, as it stands, offer an intuitive grasp of the source of the unsustainability problem. There is, however, an alternative way of describing the matter which may lend itself more readily to intuition.

As already mentioned, the basic results of this chapter seem paradoxical because strong complementarity in production is the key influence making for the production of a multiplicity of outputs by a single firm. Surely, there are few cases offering complementarities as strong as those between the production of output today and the production of output tomorrow, when both are supplied by a single firm using much of the same plant in both periods. One would expect that to give this firm a commanding advantage over any firm choosing to supply output in only one of the two periods. Yet, we see that this is not always so—the firm that produces only in the second period may have a strong advantage *vis-à-vis* the firm that produces in both.

What has gone wrong with the intertemporal complementarity argument? The answer is that it has left out of consideration a second element in multiperiod production—one which has what we may call, for want of a better term, an *anti*complementarity effect. This element is the opportunity cost of the resources tied up by the intertemporal process, which we may interpret as the interest cost, as the cost of time, or any of the other usual forms in which the phenomenon is described. If the interval between the two crucial periods is sufficiently long, this cost rises, without let-up, and does so at a compounded rate. True, the complementarity advantage of being able to use the same plant in producing the outputs of both periods does not disappear. But it is overtaken and overwhelmed by the growing opportunity cost over the time during which resources are tied up.

The precise way in which this occurs can be seen by returning to the condition that anticipatory construction is unprofitable, under which Proposition 14A1 holds. It will be recalled from Section 14A that the unsustainability result holds when the interval between the two periods (call it t units of time) is so long that it does not pay to expend $K(y_2)$ initially to build capacity in anticipation of the growth of demand from y_1 to y_2. Rather, it is cheaper to expend $K(k_1^*)$ in the first period and $K(y_2 - K_1^*)$ in the second, with $y_1 \leq k_1^* < y_2$, so that (letting i be the interest rate per unit of time)

$$(14D1) \quad K(y_2) > K(k_1^*) + K(y_2 - k_1^*)/(1 + r) \equiv K(k_1^*) + K(y_2 - k_1^*)/(1 + i)^t.$$

Here, by the assumption of economies of scale in production, we must have

$$(14D2) \qquad\qquad K(y_2) < K(k_1^*) + K(y_2 - k_1^*),$$

and this expresses the complementarity advantage derived from simultaneous construction of capacity to produce k_1^* and $\Delta y = y_2 - k_1^*$. But (14D1) now tells us that

$$(14D3) \quad [(1 + i)^t - 1][K(y_2) - K(k_1^*)] > K(y_2 - k_1^*) + K(k_1^*) - K(y_2).$$

That is, it tells us that the cost of time, represented by $(1 + i)^t$, eventually grows more than sufficiently large to offset the complementarity advantage represented by (14D2). And it is only when the time interval involved is long enough for this to be true (i.e., when the net complementarity of construction for the future is negative) that the firm offering to produce in both periods finds itself necessarily vulnerable at its break-even prices to takeover of some or all of its market by a firm producing only in period 2.

14E. Extension: Deterioration of Capital

Our discussion may appear to suggest that growth in demand lies at the heart of the problem of intertemporal unsustainability. After all, in the preceding treatment it is rising volume that permits the later entrant to enjoy the economies of large-scale construction and offers him his advantage over the incumbent firm. However, it is easy to show that the same problem arises when volume is constant or even declining, so long as deterioration of capital with the passage of time reduces capacity still faster. If from time to time the incumbent must still construct to supplement that portion of initial capacity that still remains in order to keep abreast of the stationary or shrinking market, the same sustainability problem must entrap him. That is, we have

Proposition 14E1 *In an industry with declining average cost in production of sunk capacity, if capacity deteriorates with the passage of time and for this or any other reason future construction to augment remaining capacity is efficient,* *then, unless the general fixed costs of the firm are sufficiently large, any inter-temporal efficient industry configuration will have no vector of sustainable prices available to it even if quantity sold remains constant or declines with time.*

Without going through the details of the argument, we note that this result can once more be shown in a two-period model, with plant of capacity k_1^* constructed in the first period declining in capacity to $\lambda k_1^* > 0$ in the second, and then augmented by construction of $k_2^* > 0$. It follows exactly as before that unless the general fixed costs are sufficiently large, the associated pair of break-even prices cannot simultaneously protect an incumbent (a) from an entrant who builds k_1^* units of capacity only in the first period and (b) from the alternative entrant who begins operation only in the second period by constructing $\lambda k_1^* + k_2^*$ units of capacity.

14F. Forms and Effects of Unsustainability in Practice

Even if the preceding analysis proves to be unassailable in terms of its formal logic, it leaves many questions unanswered. First, viewing the issue as an international trade phenomenon, we have taken no account of the role of variations in exchange rates. Second, we have so far ignored the possibility of bankruptcy and its effects on the valuation of the assets of the incumbent firm. We will argue that either of these may affect substantially our interpretation of the problem of intertemporal unsustainability.

The observations in this section may help us to deal with two more crucial issues to which we will return in the next section. First, casual empiricism does not indicate that there is continuous replacement of one dominant firm by another, as the analysis would seem to suggest. Second, and more important, the analysis seems to imply that, in the long run, in an industry where there are economies of scale in sunk construction, no firm can ever hope to recoup its investment. Why then, with any reasonable foresight, should any firm ever choose to enter such a field? Obviously, firms do in practice and, apparently, some of them do prove successful.

Let us first offer a few remarks upon the modifications in the unsustainability process caused by exchange-rate variations when the entrant and the incumbent firm operate in different countries, that is, where the process involves international trade. If, as was suggested in the introduction, unsustainability can be used to help explain the series of successions of one country by another as master of the international market place, it becomes inappropriate to analyze the issue with the aid of the assumption that the exchange rate is fixed. Suppose that in country A a number of industries more or less simultaneously find themselves threatened by replacement by foreign entrants in country B; the law of comparative advantage tells us that under freedom of trade a country cannot normally be driven out of *every* economic activity. What may happen is that A's exchange rate will fall sufficiently to prevent that takeover of the market by B's enterprises. It is even possible that there will be little change in the pattern of specialization. This will occur if all industries in B increase their absolute advantage more or less proportionately so that there is little change in the pattern of comparative advantage.

But a real and significant change may nevertheless result from the unsustainability mechanism. For whichever of the old lines of activity A is able to retain for itself under this scenario will have been kept at the expense of real wages and real standards of living. Country A's economy will have suffered a real decline in the form of higher real payments for its imports and lower real receipts for its exports. This is not an unfamiliar pattern for countries whose hold on the market is effectively challenged by foreign entrants.

We turn next to the role of bankruptcy. Our comments on the subject are intended to be preliminary because we are far from clear in our own minds about the way in which matters will work out. Bankruptcy can change our story in at least two ways. First, it may reduce or eliminate the wastefulness of the process. Suppose the absence of sustainable prices drives the incumbent firm into insolvency, having been expelled from the market by a later entrant. Then the incumbent's assets become available at a much reduced price: indeed, in the absence of other users, at whatever price is sufficiently attractive to lead the entrant to purchase them rather than to build his own. Thus, at least part of the social waste induced by intertemporal unsustainability—the abandonment of the incumbent's capacity and its replication by the entrant— need never occur.

Second, the phenomenon of bankruptcy may mean that the only effect of unsustainability is a transfer of ownership of the assets whose operation continues, precisely as it would have if there were no such problems. Firm X continues to operate at the same premises using the same labor force as it would have otherwise, except that its name is perhaps changed to firm Y.

There are, however, two flies in the ointment. First, the process of bankruptcy may transform the social risk into a purely private risk. It now takes the form of the danger to the investor that today's successful firm will become tomorrow's financial casualty, whose assets will be transferred to someone else—someone who will enjoy their fruits at knocked-down prices. As with other differences between private and social costs, this can have very real efficiency costs by leading to a level of investment in the industry lower than is socially optimal.

Second, the sequence of events may, in this case, prevent bankruptcy from achieving its prime salutary effect—the prevention of wasteful duplication of plant. For with foresight imperfect, as it is in reality, the failure of the incumbent is likely to occur *after* the entrant has already come into the business and demonstrated its ability to take the market away from its predecessor. But if this is so, the entrant will already have built his new and duplicative plant *before* the former incumbent is forced to make his old plant available at bargain prices. If this is the real sequence of events, the process of bankruptcy will contribute little to the elimination of the wastefulness of the process.

Let us examine in somewhat greater detail the possibility of transfer of assets via the bankruptcy process. Suppose, under the assumptions of Proposition 14A1, that p_1 and p_2 satisfy (14A7) so that a potential entrant planning to market y_1 and k_1^* in periods 1 and 2, respectively, is deterred from doing so. Then, the proof of the proposition shows that, given (14A4) and unless (14A10) holds, an entrant can hope to earn positive profits by the construction and operation of y_2 units of capacity during period 2. Such entry can result in the duplication and waste of the general fixed costs F of the firm, as well as of k_1^* units of capacity in the second period, and in the inability of the incumbent to recover its capital costs. But, the possibility arises that the social waste of k_1^* units of capacity can be averted by their sale to the entrant by the incumbent, though the wasteful replication of F may still occur.

The value of the purchase of k_1^* units of capacity to the entrant is $K(y_2) - K(y_2 - k_1^*)$, for this is the resulting saving in his capacity construction costs. Such a transfer of assets would indeed avoid duplication of plant. However, the maximal purchase price of these assets, $K(y_2) - K(y_2 - k_1^*)$, must still leave the incumbent firm unable to recover its capital costs. That is, (14A4), (14A6), and (14A7) imply that

(14F1) $$p_1 y_1 + \frac{K(y_2) - K(y_2 - k_1^*)}{1 + r} < F + K(k_1^*).$$

Thus, the revenues of the incumbent firm from the sale of y_1 in the first period and from the sale of his assets at the beginning of the second period do not cover the incumbent's costs of launching the firm and of the building of k_1^* units of capacity at the outset.

Proof of Inequality (14F1)   Together, (14A4) and (14A6) yield

(14F2) $$p_1 y_1 + \frac{p_2 y_2}{1 + r} = F + K(k_1^*) + \frac{K(y_2 - k_1^*)}{1 + r}.$$

Multiply (14A7) by y_2, subtract from it (14F2) multiplied by k_1^*, and divide through by $(y_2 - k_1^*)$, obtaining

(14F3) $$p_1 y_1 \leq F + K(k_1^*) - \frac{K(y_2 - k_1^*)}{(y_2 - k_1^*)} \frac{k_1^*}{(1 + r)}.$$

A bit of straightforward manipulation confirms that

$$-\frac{K(y_2 - k_1^*)}{(y_2 - k_1^*)} \frac{k_1^*}{1 + r} < \frac{K(y_2 - k_1^*) - K(y_2)}{1 + r}$$

follows from

$$\frac{K(y_2 - k_1^*)}{y_2 - k_1^*} > \frac{K(y_2)}{y_2},$$

so that decreasing unit construction costs of capacity together with (14F3) implies (14F1).   Q.E.D.

Thus, we have shown that while a sale of assets at the end of period 1 can avert duplication of capacity, any transfer price that is rational for the entrant will not cover the incumbent's sunk costs. Consequently, if the incumbent were to foresee these events at the beginning of period 1, before his costs were sunk, and if he considered them certain, then he would not enter the industry at all.

14G. Inducements to Entry and Profitability of Incumbents in Practice

We come next to two important issues raised in the preceding section. If the intertemporal unsustainability model is widely applicable, what induces firms to enter into industries characterized by decreasing unit costs of sunk capacity construction in the first place? And how does one explain the apparent fact that firms in such industries, even if they do not last forever, seem frequently to prosper for considerable periods, usually long enough to recoup their original investments and more?

We will leave until last what many microeconomists will consider the prime explanation: the likelihood of violation of the premise of Bertrand-Nash behavior. There are many other plausible explanatory elements, some of which can only be suggested informally. One, which is perhaps not of major importance, has been suggested already. If the incumbent and the potential entrant are in different countries, then a declining exchange rate may preserve some of the former's market. One way to interpret the effect of a fall in exchange rate is to view it as a means to distribute the effects of unsustainability among *all* the inhabitants of the incumbent's country, thereby shielding the firms directly involved from much of the damage they would otherwise suffer. That country's transformation into a cheap labor economy reduces its standard of living but prevents its exporters from being driven from the world's markets. Thus, the old firms may be protected from the bulk of the unsustainability problem, continuing to earn a profit well after foreign entry might have become profitable.

We have also seen that substantial fixed costs for the firm can protect sustainability. Barriers to entry may also be an important element helping to explain the survival and profitability of incumbents. Here, it is sufficient to note that where governmental regulation and licensing inhibit entry, the profitability of the incumbent is no mystery at all. But, even where the barriers are much weaker, it is clear that they can transform a case of mild unsustainability into one in which the incumbent can choose among many price vectors, all capable of preventing profitable entry.

Imperfect knowledge may also go far to account for the profitability of the incumbent firm, and its initial willingness to enter, despite the unsustainability problem that clouds its future. At least two types of information gaps are pertinent: ignorance on the part of the incumbent, who probably was unaware of the threat the future held for him at the time he opened his business, and ignorance on the part of later potential entrants, who may be unaware of the opportunities for profitable entry that unsustainability may offer them.

For the potential entrant, there is one source of imperfect knowledge that is absent from the intratemporal sustainability issue. In the intratemporal analysis, the entrant can usually observe the current prices of each of the industry's goods directly. But in the intertemporal case, the potential entrant can only conjecture about the pertinent future prices on which the profitability of entry is entirely dependent. This can cause potential entrants to miss opportunities provided by intertemporal unsustainability and thus help to protect the incumbent in ways that may not be entirely obvious.

A diagram for a simple specialization of the two-period case used to derive Proposition 14A1 will help to bring out the issue. Figure 14G1 shows in terms of price space the constraints imposed by the requirements of sustainability when quantities demanded are fixed,[14] $F = 0$, and $k_1^* = y_1$. Line $\pi_i \pi_i$ is the locus of prices p_1 and p_2,

$$p_1 y_1 + \frac{p_2 y_2}{1 + r} \geq K(y_1) + K(y_2 - y_1)/(1 + r).$$

Any point below the line representing this relationship will correspond to a pair of prices at which the incumbent cannot cover its costs.

Sustainability requires the incumbent to be able to protect himself from an entrant (whom we can call an entrant of type 1), among others, proposing to sell y in each of the two periods. Line $\pi_1 \pi_1$ is the outer boundary

---

[14] For simplicity we deal here with the case where demands are zero-elastic. This accounts for the linearity of the three loci. One can easily draw in the corresponding loci for any other specified demand functions.

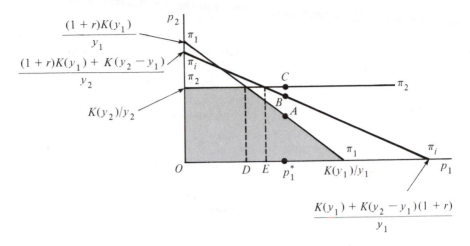

FIGURE 14G1

of the set of all price vectors which are unprofitable to such an entrant, that is, of price vectors which satisfy

$$p_1 y_1 + \frac{p_2 y_1}{1 + r} < K(y_1).$$

We also considered in our discussion of Proposition 14A1 the incumbent's need to protect himself against an entrant (of type 2) who proposes to supply nothing in the first period and y_2 in the second, that is, the late entrant who proposes to take over the entire business from the incumbent in the second period. The prices that can prevent this must lie in the region with

$$p_2 y_2 < K(y_2),$$

whose outer boundary is represented by the horizontal line $\pi_2 \pi_2$. It is clear that only points in the shaded region represent price vectors that can prevent both types of entry, and since $\pi_i \pi_i$, the incumbent's break-even line, does not enter the shaded region, no sustainable prices are possible.[15]

However, suppose the incumbent selects price p_1^* in the initial period.[16] An entrant of type 1 knows that it will lose money *unless the price set by the*

---

[15] Tim Cooke has also shown us that it is easy to modify Figure 14G1 to indicate how fixed costs that are sufficiently large can permit sustainability, by shifting the curves so that a region of prices emerges which satisfies the three inequalities that are necessary conditions for sustainability.

[16] The same argument clearly applies for any values of p_1 such that (in terms of the figure) $E < p_1 < K(y_1)/y_1$, that is in the range where the break-even value of p_2 for the incumbent firm is less than that for an entrant of type 2.

incumbent in the second period will be above A. But of this the potential entrant cannot be sure. If it decides to stay out because it is not worth the risk, in the second period the incumbent will find itself free to select a price between *B* and *C*, which gives it a healthy profit, while keeping out any entrant of type 2. True, at this price the entrant of type 1 can see in retrospect that it could have prospered in the industry, but by then it is too late. Period 1, its time of entry, will have come and gone, leaving the entrant of type 1 only the option of regret.

There is a final and very critical reason why the implications of our unsustainability theorem may not be as drastic as they seem. A fundamental limitation of sustainability analysis is its implication that entrants' expectations are those of the Bertrand-Nash models. That is, each entrant is implicitly assumed to expect that after entry occurs the incumbent will keep his pre-entry prices unchanged for a period sufficiently long to make entry profitable. For only on this premise can the entrant be assumed to take advantage of any profitable entry possibilities presented by the incumbent's initial prices. However, if the potential entrant fears pricing changes after entry, then he may decide to ignore the apparently profitable entry opportunities and simply stay out of the industry. The entry threat may therefore evaporate even before it can be noticed.

In intertemporal analyses, neither the Bertrand-Nash assumption nor its denial is completely convincing in general. The questions that can be raised about the Bertrand-Nash premise have just been suggested. But in the multiperiod model, if entry occurs in period n, prices in the preceding $n - 1$ periods can no longer be changed. The potential entrant can act on the conviction that at least those prices will not be affected by entry. And this, as we have just seen, may be enough to leave the former incumbent with two unpleasant choices: the adoption of prices that prevent entry but guarantee his own bankruptcy, or the retention of prices that would be profitable if no entry were to occur, but leave him completely vulnerable to replacement by a rival firm.

In practice, given these circumstances, virtually any scenario is possible. Potential entrants may all be cowed from entry forever by fear that they in turn will suffer the fate that their entry imposes upon the incumbent, or that the latter will retaliate so sharply and dramatically that anyone who has the temerity to enter will soon have cause to regret it. Such a scenario would depict a world of equilibrium with industry structure and ownership held unchanged by the forces of fear.

At the other extreme, potential entrants may grasp recklessly at every entry opportunity and so produce constant upheaval and perpetual unprofitability for every firm.

There is an intermediate possibility, which seems much more plausible. When entry opportunities arise, they are not recognized and used at once.

Rather, imperfect information, some degree of timidity, and other frictions suggested in the preceding discussion introduce substantial lags into the process. These lags do not protect incumbents forever. Ultimately, their territory will be invaded, and finally they are likely to succumb to the invaders. But typically the lag will be sufficient to permit many incumbents not only to recoup their investments, but to earn handsome profits over their lifetimes. Casual empiricism seems to confirm the reality of this last scenario.

Finally, where truly sunk costs constitute a high proportion of the incumbents' expenses, entrants should expect incumbents to respond to entry with sharp price reductions, even if they will preclude recovery of investment.[17] Responsive pricing patterns, and rational threats of such behavior, may then effectively prevent replacement of the incumbent, and such a price pattern becomes the instrument that stabilizes the industry and prevents wasteful duplication of facilities. It may, of course, lead to all the undesirable consequences usually attributed to it—excessive profits, protection of inefficiency, and the like. But it may have *some* virtues: In particular, under the circumstances we have studied, it may be the only avenue to some sort of intertemporal equilibrium in the structure of the industry.

If this is the reality of the matter, things do not work out as the most extreme interpretation of the unsustainability analysis might suggest. However, that analysis may nevertheless help us to understand the underlying dynamic of the process of evolution of industry structure and ownership. It certainly suggests that we cannot be comfortable with a standard equilibrium analysis of industry structure, or feel confident that the invisible hand has matters here firmly under control.

Part II
Other Sources of Intertemporal Unsustainability

14H. Intertemporal Sustainability and Learning by Doing

We turn now to some other potential sources of unsustainability in multiperiod models. In the preceding chapter we saw that increases in productivity induced by technological change do not, inherently, constitute a threat to sustainability. Learning by doing is another source of expansion

---

[17] It should be noted that in our model the capital outlays need only be entirely sunk *vis-à-vis* use outside the industry. The discussions of the Demsetz auction and of the transfer of assets through bankruptcy show that the incumbent may be able to realize something from those assets inside the industry, and will then be unwilling to reduce prices to zero.

of productivity. We see now how this phenomenon can be incorporated into our formal analysis.[18] From this extension of our model we can conclude

Proposition 14H1   *Learning by doing results in intertemporal natural monopoly. Where diminishing returns to the learning prevail, the weak invisible hand theorem can apply formally. However, unsustainability may result from increasing returns to learning.*

For simplicity, we deal with a two-period production process in which returns to scale would be constant if learning effects were absent and in which the general state of technological information is fixed. However, for us the essential new feature of the process is that in this model the second period's productivity is raised by the experience derived from production in the first period. Thus, our cost function assumes the form

(14H1) $$C(y_1, y_2) = c_1 y_1 + c_2(y_1)y_2, \qquad c_2'(y_1) < 0$$

where $c_2(y_1)$ is the second period's marginal cost. Here it is easy to show that $C(\cdot)$ in (14H1) manifests economies of scope. For $C(y_1, 0) = c_1 y_1$ and $C(0, y_2) = c_2(0)y_2 > c_2(y_1)y_2$. Consequently,

$$C(y_1, y_2) = c_1 y_1 + c_2(y_1)y_2 < c_1 y_1 + c_2(0)y_2 = C(y_1, 0) + C(0, y_2),$$

as is required by the definition of economies of scope. $C(\cdot)$ also exhibits economies of scale since, for $t > 1$,

$$C(ty_1, ty_2) = c_1 ty_1 + c_2(ty_1)ty_2 < tc_1 y_1 + tc_2(y_1)y_2.$$

Moreover,

$$\frac{\partial C}{\partial y_1} = c_1 + c_2' y_2 < c_1 = \text{first-period out-of-pocket costs.}$$

For the seller it would seem tempting to offer low prices in the first period in order to sell a great deal early, thereby reducing later costs. But if he selects a price below

$$\frac{C(y_1, y_2) - C(0, y_2)}{y_1} = c_1 + y_2 \left[\frac{c_2(y_1) - c_2(0)}{y_1} \right],$$

---

[18] The classic reference on learning by doing is Arrow (1962). Our approach is similar to that of Rosen (1972).

then that price will be unsustainable. For a second-period entrant can then undercut the initial seller's break-even level of p_2, even though the entrant's (second-period) costs are at the high level $c_2(0)$.

However, this potential unsustainability is, generally, surmountable. By charging a first-period price that is sufficiently high, the break-even price in the second period will not invite entry in general. The issue to be investigated is whether and under what conditions such sustainable price vectors exist. Let us deal, in turn, with the cases of diminishing and increasing returns to learning by doing.

Case 1: Diminishing Returns in Learning.   Suppose learning is characterized by weakly diminishing returns, so that $c_2' < 0$ but $c_2'' \geq 0$. We show that $C(\cdot)$ must then be trans-ray convex. Consider the trans-ray cross section $y_2 = k - ay_1, a > 0$. Then, writing $C(y_1, k - ay_1) = \phi(y_1)$, we have

$$\phi(y_1) = c_1 y_1 + c_2(y_1)(k - ay_1)$$

$$\phi'(y_1) = c_1 + c_2'(y_1)(k - ay_1) + c_2(y_1)(-a)$$

$$\phi''(y_1) = c_2''(k - ay_1) - ac_2' - ac_2' > 0;$$

that is, the trans-ray cross section is convex, as was to be shown. It follows from this and the economies of scale which have already been demonstrated that, by the results of Chapter 7, costs are strictly subadditive.

As an interesting special case, let

$$c_2(y_1) = \beta - \alpha y_1 \qquad \alpha > 0, \beta > 0, \beta - \alpha k > 0.$$

Then,

$$C = c_1 y_1 + (\beta - \alpha y_1)y_2 = c_1 y_1 + \beta y_2 - \alpha y_1 y_2.$$

This cost function, which is trans-ray convex, is the only cost function with $S_1 = S_2 = 1$ everywhere. That is, for any value of y_2, $\partial C/\partial y_1$ is constant and (for $y_2 < c_1/\alpha$) positive, and a similar relation holds for $\partial C/\partial y_2$. As before,

$$\phi(y_1) \equiv C(y_1, k - y_1) = c_1 y_1 + \beta(k - y_1) - \alpha y_1(k - y_1)$$

$$= \beta k + y_1(c_1 - \beta - \alpha k) + \alpha y_1^2.$$

An example of such a trans-ray cross section of the cost function is depicted in Figure 14H1. The sharper the diminishing returns to learning, the smaller the value of y_1 at which costs are minimized on a given trans-ray cross section. This is the result of the tradeoff between increased learning derived from increased period-1 production, and the volume of future production to which the learning will later be applied.

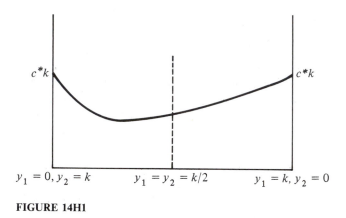

FIGURE 14H1

In the case of diminishing returns, with the resulting trans-ray convexity, Ramsey prices will be sustainable for the intertemporal natural monopoly, and the firm will not be driven to adopt so low a price in the learning phase that profitable entry will become possible for an inexperienced firm in the second period. However, the prices during the learning phase may be lower than the period's out-of-pocket costs, and, at the same time, prevent entry. As a result, well-intentioned authorities, or less-well-intentioned competitors who found themselves excluded, may be too quick to claim that these prices are anticompetitively low. They may also be disturbed by the fact that the incumbent is a monopolist, even though his costs in each period seem to exhibit no scale economies. Yet, as has been shown, in this scenario the true marginal costs are well below the out-of-pocket costs during periods of learning by doing, and both Ramsey optimality and sustainability may require prices that lie between these two figures, and are thus likely to be objects of suspicion, though they are really benign. Moreover, in this scenario, the learning induces economies of scope and economies of scale, and confers an intertemporal natural monopoly upon an incumbent, so that the monopoly and the absence of entry should not necessarily be considered undesirable.[19]

Case 2: Increasing Returns in Learning. Suppose the learning effect increases with production, so that $c_2''(y_1) < 0$ over some range (Figure 14H2).

---

[19] However, it must be noted that in this oversimplified model, with its limitation to two time periods, an incumbent who avoids entry in the first period may have the opportunity to earn positive rents overall by charging a high price in the second. Here, the second-period price will be immune from entry as long as it remains below the costs of a firm that had accumulated no learning. This effect will be attenuated substantially in a model involving many time periods.

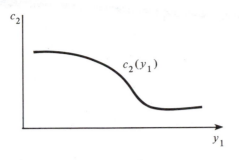

FIGURE 14H2

First, let us consider the returns to scale in each period's product by itself. We have

$$S_2 = \frac{C(y_1, y_2) - C(y_1, 0)}{y_2 C_2(y_1, y_2)} = \frac{c_1 y_1 + c_2(y_1) y_2 - c_1 y_1}{y_2 c_2(y_1)} = 1;$$

that is, there will be constant returns to scale in y_2. This, of course, will be true whether or not our increasing returns assumption holds. However,

$$S_1 = \frac{C(y_1, y_2) - C(0, y_2)}{y_1 C_1(y_1, y_2)} = \frac{c_1 y_1 + c_2(y_1) y_2 - c_2(0) y_2}{y_1 [c_1 + c_2'(y_1) y_2]}$$

$$= 1 + \frac{y_2 [(c_2(y_1) - c_2(0)) - y_1 c_2'(y_1))]}{y_1 (c_1 + c_2'(y_1) y_2)}.$$

Thus, since $[c_2(y_1) - c_2(0)]/y_1 - c_2'(y_1)$ is the average minus the marginal returns in learning, we will have $S_1 < 1$ for diminishing returns in learning, but $S_1 > 1$ for increasing returns in learning. That is, $c_2'' < 0$ or increasing returns in learning are equivalent to increasing returns to scale in y_1 alone. Therefore (Chapter 4), with $c_2'' < 0$, trans-ray convexity does not necessarily hold. But we have $S_2 = 1$, $S_1 > 1$, which (Chapter 7) with the economies of scope implies subadditivity, that is, natural monopoly. Thus, in this case although the conditions for natural monopoly are satisfied, the intertemporal monopoly may not be sustainable.

More specifically, $\phi'' = c_2''(k - a y_1) - 2 a c_2'$. Consequently, the cross section will generally be convex for large y_1, because the $y_2 = k - a y_1$ term will then be small. But it can be locally concave for small y_1, as for example, when $C = c y_1 + (w - y_1^2) y_2$. Here, the trans-ray cross section will be like that in Figure 14H3. So, Ramsey pricing will then be sustainable if it entails a large value of y_1 (placing it in the trans-ray supportable region). But the Ramsey solution need not meet this requirement. The danger of inefficient entry will

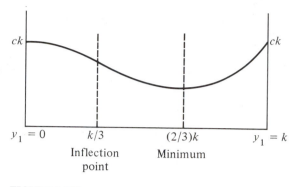

FIGURE 14H3

obviously be increased if the Ramsey solution entails an output level in the first period that is relatively small and that leaves substantial learning effects unexploited.

14I. Postponability of "Fixed" Outlays

Although it is traditional to consider the irreducible costs required to inaugurate a firm, a plant, or a project to be *fixed*, there is a basic reason why, viewed in intertemporal terms, no cost that enters the decision process is really fixed. The reason is simply that only *ex ante* costs are relevant for economic decisions, and before decisions have been made, the dates on which outlays are to occur still can normally be varied. The inaugural date of the enterprise can be delayed (moved forward), and thereby its setup cost will be reduced (increased) by an amount dependent on the discount rate. That is, if F is the setup cost expressed in present value of period 1, and if it is decided to delay this outlay to the second period, then the setup cost is reduced to $F/(1 + r)$, where r is the discount rate. Viewed in this way, the outlay is not really fixed. That is, the incremental cost of starting business in the first period must include the incremental setup cost, $F - F/(1 + r) = rF/(1 + r)$. Similarly, a two-period delay yields an incremental saving equal to $F - F/(1 + r)^2$. Unfortunately, we see

Proposition 14I1   *The lumpy portion of incremental cost introduced by postponability of "fixed" investment is inconsistent with trans-ray convexity.*

Thus, if the sustainable cost functions, $C(\cdot)$, constructed in the preceding chapter are simply amended to $\hat{\hat{C}}(y^1, \ldots, y^T) = F/(1 + r)^{z-1} + C(\cdot)$, where z is the first period in which any positive amount of production occurs, unsustainability can result.

The reason that such variability in the firm's "fixed" setup costs violates trans-ray convexity is easy to see. Consider once more a two-period world in which y_1 and y_2, the outputs of the two periods, are taken to constitute distinct products. Then the incremental setup cost of y_1, the amount $rF/(1 + r)$, becomes a product-specific fixed cost. That is, letting $F(1)$ represent the incremental setup cost for period 1 we have

$$F(1) = \begin{cases} 0 & \text{if } y_1 = 0 \\ rF/(1 + r) & \text{if } y_1 > 0. \end{cases}$$

Thus, the cost surface jumps by the amount $rF/(1 + r)$ anywhere along the y_2 axis as y_1 goes from $y_1 = 0$ to $y_1 = \varepsilon > 0$. This is shown in Figure 14I1 by the shaded vertical strip $JHBF$ above the y_2 axis, the height of that strip being $rF/(1 + r)$, the rise in fixed cost from $F/(1 + r)$ (at point J) to F when y_1 increases by the slightest amount above zero. We see that the typical trans-ray cross section, that above ab, is ABH, which exhibits a nonconvexity in the neighborhood of B. The resulting cost function may be described as semi-transylvanian because it has just one batlike wing.

Next, we turn to some specific examples, illustrating the disappearance of trans-ray convexity and showing the surprising fact that effective entry threats under Ramsey pricing by the incumbent need not be confined to entrants who cut cost by delaying as long as possible the launching of their firms.

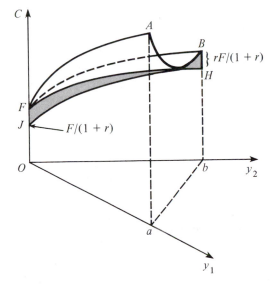

FIGURE 14I1

We will first offer a three-period example, proving this. Then we will deal with an important special case which may be described as that of the intertemporal public good.

Case 1: A Three-Period Case with Second-Period Entry Opportunities. Consider the cost function

$$C(y^1, y^2, y^3) = \sum c_i(T) y^i,$$

where the marginal costs c_i depend on the set T of products of which positive amounts are produced. We know that as long as $c_i(R) \geq c_i(T)$ for $R \subseteq T$, C is trans-ray convex. Then, of course, increasing returns and trans-ray convexity are guaranteed by the addition of a fixed cost so that the cost function becomes $\hat{C}(y) = F + C(y)$. But this is *not* true of a cost function in which F can be incurred earlier, if that is desired. That is, $\tilde{C} = F/(1 + r)^{z-1} + C(y)$ is not generally trans-ray convex. To explore the consequences, consider the example

$$F = 3.9, \qquad r = 1,$$

$$c_i(1) = c_i(2) = c_i(3) = 12.1$$

$$c_i(1, 2) = c_i(1, 3) = c_i(2, 3) = 12.05$$

$$c_i(1, 2, 3) = 10.7,$$

so that $C(\cdot)$ is clearly subadditive. For appropriate demand functions,[20] the Ramsey quantities and prices can be taken to be $y^1 = y^2 = y^3 = 1$ and $p^1 = 10.75$, $p^3 = 10.75$, $p^2 = 14.5$, where the prices are expressed in present-

---

[20] Given marginal costs, C_i^*, and the appropriate second-order conditions, choose a set of demand functions yielding any set of preselected Ramsey prices and quantities. Thus, let

$$y_i = a_i p_i^{-E^i}$$

be the demand function for product i. Then the demands for the different products are independent, and E^i is the elasticity of demand for product i. Given the vector of prices and outputs (p^*, y^*), we can substitute these values into the Ramsey expressions

$$\sum p_i^* y_i^* = C, \qquad (C_j^* - p_j^*)/(C_j^* - p_j^*) = E^j p_i^* / E^i p_j^* \qquad (j = 1, \ldots, n),$$

which gives us the required values, $E^{1*}, \ldots, E^{n*}$, of the elasticities and then we can obtain the remaining set of unknown parameter values, the a_i^* satisfying

$$y_i^* = a_i^*(p_i^*)^{-E^{i*}}.$$

These demand functions, then, give us as the Ramsey prices and outputs the desired vector (p^*, y^*).

value terms. Then by direct substitution we obtain

$$\tilde{\tilde{C}}(1,0,0) = 16 \qquad \tilde{\tilde{C}}(0,1,0) = 3.9/2 + 12.1 = 14.05$$
$$\tilde{\tilde{C}}(0,0,1) = 3.9/4 + 12.1 = 13.075$$
$$\tilde{\tilde{C}}(1,1,0) = \tilde{\tilde{C}}(1,0,1) = 28 \qquad \tilde{\tilde{C}}(0,1,1) = 26.05$$
$$\tilde{\tilde{C}}(1,1,1) = 36.$$

The prices are all greater than the corresponding marginal costs, $C_i(1,2,3) = 10.7$. Moreover, the prices cover these incremental costs:

$$p^1 y^1 = 10.75 > IC_1 = \tilde{\tilde{C}}(1,1,1) - \tilde{\tilde{C}}(0,1,1) = 9.95$$
$$25.25 = p^1 y^1 + p^2 y^2 > IC_{1,2} = \tilde{\tilde{C}}(1,1,1) - \tilde{\tilde{C}}(0,0,1) = 22.925.$$

At these Ramsey prices entry is not possible in either the first or third periods since $p^1 y^1 = 10.75 < \tilde{\tilde{C}}(1,0,0) = 16$ and $p^3 y^3 = 10.75 < \tilde{\tilde{C}}(0,0,1) = 13.075$.

However, we have

$$p^2 y^2 = 14.5 > \tilde{\tilde{C}}(0,1,0) = 14.05.$$

Thus, at the Ramsey prices, it is possible for a new firm to enter in the second period and to operate profitably in that period alone because of the saving in setup costs made possible by delaying its inauguration from the first period to the second. Indeed, this second-period entrant cannot continue to operate beyond the second period without losing money since

$$p^2 y^2 + p^3 y^3 = 25.25 < \tilde{\tilde{C}}(0,1,1) = 26.05.$$

This proves what we desired to show—that postponability of setup costs may lead to unsustainability of the Ramsey solution, which can invite profitable entry at an intermediate period rather than in the period that constitutes the relevant horizon when the savings from postponement of investment are greatest.

14J. Pure Public Goods with Postponable Purchase Date

We come now to the last potential source of intertemporal unsustainability considered here—the interesting special case of investment upon entry which constitutes a public good, in the sense that it permits as large an increase in output as desired without incurring additional costs. That is, it is a situation involving purely "fixed" costs, with zero marginal costs, which (as we saw in Chapter 10) is formally equivalent to the case of a pure public good which can serve additional persons without any rise in total cost. We want to see now what happens when, as in the preceding section, it is possible to reduce the pure fixed cost (in present-value terms) by deciding to incur it at a later date.

Consider such a "pure public good" provided in a two-period world and sufficiently durable to last the entire two periods. All money figures will be expressed in present-value terms. As before, the good costs F if it is constructed at the beginning of the first period, while it costs $F/(1 + r)$ if built at the beginning of the second period. Let the demand function for use of the item in the first period be $x^1(p_1)$, and $x^2(p_2)$ be the demand function in the second period. Assume x^2 is large enough so that it definitely pays to build the item at some time. Then, as before, the incremental cost of use in the first period is

$$C(x_1, x_2) - C(0, x_2) = F - F/(1 + r) = rF/(1 + r).$$

Suppose that the good is socially worthwhile if used in both periods; that is,

(14J1)
$$\int_0^\infty x^1(t)\, dt + \int_0^\infty x^2(t)\, dt > F.$$

Further, assume that the first period's surplus covers the first period's incremental cost so that

(14J2)
$$\int_0^\infty x^1(t)\, dt > rF/(1 + r).$$

Hence it is socially optimal to build the item at the beginning of period 1. Notice that this is equivalent to a much-simplified case in which it is optimal to build capacity early, in anticipation of future demand. In that more general case, with increasing returns in the activity of capacity building, we obtain cost functions such as

$$C^1 = F + k \max(x_1, x_2) + c_1 x_1 + c_2 x_2$$

$$C^2 = F + k x_1 + F/(1 + r) + \frac{k(x_2 - x_1)}{1 + r} + c_1 x_1 + c_2 x_2, \qquad \text{for } x_2 > x_1 > 0,$$

where $F =$ fixed costs of capacity construction,

$k =$ unit cost of capacity,

$c_t =$ running costs in period t,

$C^1 =$ total cost if all construction occurs in period 1,

$C^2 =$ total cost if some construction occurs in each period, and

$$C(x_1, x_2) = \min(C^1, C^2), \qquad x_1 > 0$$
$$C(0, x_2) = F/(1 + r) + [k/(1 + r)]x_2 + c_2 x_2.$$

The public good model is simply the preceding model with $k = c_1 = c_2 = 0$, an assumption that just contributes simplicity to the discussion. This re-establishes the fact that, where there are increasing returns, a case with setup costs is formally equivalent to a public good, as already discussed in Chapter 10.

Now, suppose the first-period demand curve $x_1 = x^1(p_1)$ everywhere lies below the corresponding value of the average incremental cost function $AIC_1 = x_1^{-1} rF/(1 + r)$. Then, despite the efficiency of construction to serve in period 1, by (14J2), for all x_1, $p_1(x_1) < AIC_1(x_1)$. It follows that, here, no sustainable prices can exist since

$$p_1 x_1 + p_2 x_2 = F$$

and

$$p_1 x_1 < rF/(1 + r)$$

imply

$$p_2 x_2 > F - rF/(1 + r) = F/(1 + r).$$

Consequently, someone can always build *another* public good-producing facility at the beginning of period 2, offer a price below p_2, and make a profit.

Next, suppose there are values of x_i for which the demand curve lies above the curve of average incremental costs. Here, $p_1 > AIC_1$ is possible and sustainable since

$$p_1 x_1 \geq rF/(1 + r)$$

and

$$p_1 x_1 + p_2 x_2 = F$$

imply

$$p_2 x_2 < F/(1 + r),$$

so that no entry can occur in the second period. Also no entry can occur in the first period alone as long as $p_1 x_1 < F$.

Nevertheless, the Ramsey-optimal price, p_1, may well be below AIC_1, if $x^1(p_1)$ is more elastic than $x^2(p_2)$. We now provide a detailed example involving this problem. For simplicity, we use the linear demand functions

$$X_1 = A_1 - B_1 p_1 \qquad X_2 = A_2 - B_2 p_2.$$

Then

$$\int_0^\infty X_i(t)\,dt = \int_0^{A_i/B_i} X_i(t)\,dt = \left[A_i t - \frac{B_i t^2}{2} \right]_0^{A_i/B_i} = \frac{A_i^2}{2B_i}.$$

Conditions (14J1) and (14J2) are then easily represented, since the former here becomes

$$\frac{A_1^2}{B_1} + \frac{A_2^2}{B_2} > 2F,$$

while the latter is

$$\frac{A_1^2}{B_1} > \frac{2Fr}{(1+r)}.$$

With the postulated demand functions, the maximal revenue obtainable from market i is max $p_i(A_i - B_i p_i)$, requiring

$$p_i(-B_i) + A_i - B_i p_i = 0 \qquad p_i = A_i/2B_i$$

so that

$$\max R_i = \frac{A_i}{2B_i}\left(A_i - B_i\left(\frac{A_i}{2B_i}\right) \right) = \frac{A_i^2}{4B_i}.$$

Hence, the two-period system can cover costs if

(14J3) $$\frac{A_1^2}{B_1} + \frac{A_2^2}{B_2} > 4F.$$

The period-1 market can cover its incremental cost if

(14J4) $$\frac{A_1^2}{B_1} > 4rF/(1+r).$$

Of course, (14J3) implies (14J1), and (14J4) implies (14J2). Yet if (14J3) and (14J2) hold but (14J4) does not, then the best social solution is unsustainable, although if entry in the second period is prohibited, total costs can, of course, be covered.

As shown in Baumol and Ordover (1977) the first-order Ramsey condition for such a public good case is[21]

(14J5)
$$\frac{\partial x^1}{\partial p_1}\frac{p_1}{x^1} = \frac{\partial x^2}{\partial p_2}\frac{p_2}{x^2}$$

or, here

$$\frac{B_1 p_1}{A_1 - B_1 p_1} = \frac{B_2 p_2}{A_2 - B_2 p_2},$$

which is equivalent to

(14J6)
$$p_2 = \frac{B_1 A_2}{B_2 A_1} p_1.$$

For zero profit we must have

$$p_1(A_1 - B_1 p_1) + p_2(A_2 - B_2 p_2) = F$$

or

$$B_1 p_1^2 - A_1 p_1 + B_2 p_2^2 - A_2 p_2 + F = 0.$$

Substituting (14J6) into this equation we obtain

$$p_1^2\left(B_1 + B_2\left(\frac{B_1^2 A_2^2}{B_2^2 A_1^2}\right)\right) - p_1\left(A_1 + A_2\left(\frac{B_1 A_2}{B_2 A_1}\right)\right) + F = 0$$

$$B_1 p_1^2\left(1 + \frac{B_1 A_2^2}{B_2 A_1^2}\right) - A_1 p_1\left(1 + \frac{B_1 A_2^2}{B_2 A_1^2}\right) + F = 0.$$

---

[21] The proof is trivial. We are again assuming independence of demands. Consequently, Ramsey optimality, as usual, requires

$$(p_i - C_i)/p_i = k/E^i, \qquad (i = 1, \ldots, n).$$

But since for the public goods case marginal cost, $C_i = 0$, this becomes

$$K/E^i = 1 \qquad (i = 1, \ldots, n)$$

or

$$E^i = E^j, \qquad \text{all } i, j.$$

That is,

(14J7)
$$p_1(A_1 - B_1 p_1) = F \Big/ \left[1 + \frac{B_1 A_2^2}{B_2 A_1^2}\right].$$

Note that if (14J3) holds, some prices must cover costs, and so (14J7) must have a real solution. But (14J7) characterizes the Ramsey value of p_1 (actually it gives two values, of which only the smaller is Ramsey-optimal) and directly yields as the expression for revenue under Ramsey pricing in period 1

$$F \Big/ \left[1 + \frac{B_1 A_2^2}{B_2 A_1^2}\right].$$

At this Ramsey optimum, since the two periods' market revenues together just cover total cost F, then $p_1 x^1 < F =$ period 1's stand-alone cost, so that entry with operation in period 1 alone can never be profitable. However, if this revenue is also less than period 1's incremental cost, $rF/(1 + r)$, then the Ramsey prices are unsustainable, with profitable entry in period 2 being possible. This will be so if and only if

$$\frac{F}{1 + B_1 A_2^2/B_2 A_1^2} < \frac{rF}{1 + r}$$

which is equivalent to

$$1 + \frac{B_1 A_2^2}{B_2 A_1^2} > \frac{1 + r}{r} = 1 + 1/r$$

or

(14J8)
$$\frac{A_1^2}{B_1} < \frac{r A_2^2}{B_2}.$$

Thus, suppose (14J3), (14J4), and (14J8) hold. Then the system can cover its costs, there exist sustainable prices, but the Ramsey-optimal prices are *not* sustainable.

Finally, the assumption that the system can cover its costs and that the Ramsey prices are unsustainable [i.e., that (14J3) and (14J8) are satisfied] together imply

$$\frac{r A_2^2}{B_2} + \frac{A_2^2}{B_2} > \frac{A_1^2}{B_1} + \frac{A_2^2}{B_2} > 4F$$

implying in turn

$$\frac{A_2^2}{4B_2} > \frac{F}{1+r}.$$

This means that period-2 consumers can, all by themselves, cover costs of building just for themselves.

And, similarly, the assumptions that the system can cover its costs, although period one cannot cover its incremental costs [i.e., that (14J3) is satisfied but (14J4) is not] imply

$$\frac{A_2^2}{B_2} > 4F - \frac{A_1^2}{B_1},$$

$$\frac{A_1^2}{B_1} < \frac{4rF}{(1+r)},$$

which, in turn, yields,

$$\frac{A_2^2}{B_2} > 4F\left(1 - \frac{r}{1+r}\right) = \frac{4F}{1+r}.$$

This implies

$$\frac{A_2^2}{4B_2} > \frac{F}{1+r}.$$

Therefore, here too, period 2 can itself cover its own costs.

Consequently, if the system as a whole can cover its costs, in both generic cases of unsustainability it is the period-2 consumers who can profit by violating the conditions for intertemporal optimality by wasteful replication of plant, with the new duplicate plant constructed only when second-period consumers need it for themselves.

14K. Concluding Comment

This chapter has undoubtedly raised more questions than it has answered. Despite the attempts at intuitive explanation that are offered at several points, we cannot pretend to understand completely why intertemporal relationships seem so vulnerable to unsustainability, and why this unsustainability can be introduced from such a variety of sources—a set of sources our discussion has undoubtedly not exhausted. One conclusion is clear. The discussion certainly does not lend confidence to the belief that intertemporal markets are as readily subject to the benign control of the invisible hand as are single-period structures. Much, clearly, remains to be explored here, including the implications of all this for public policy.

15

Toward Empirical Analysis

One clear lesson from all the preceding chapters is that much can be learned about actual and potential industry structure and performance, and about policies that improve the latter, from appropriate information about market demands for the industry's products and about the nature of the productive techniques available to the industry's firms. Because everything novel that has been said here relates to the latter, and because there is already an enormous literature on methods of estimation of market demand relationships, we confine ourselves here to a discussion of the means appropriate for empirical analysis of the production technology of the firms in an industry.

Throughout the book we have found it most convenient and natural to conduct our analyses in terms of firms' cost functions and their properties. Indeed, the analysis has driven us to this approach, because our results show that it is the properties of the cost function that are directly related to the elements of industry structure and performance with which we have been concerned.[1] This focus on cost functions is fortunate since recent advances in duality theory and in techniques of empirical analysis have guided the profession away from the problem-ridden task of direct estimation of technological relationships (production functions and multi-output production transformation functions) and toward the estimation of cost functions

---

[1] These results are recapitulated in Chapter 16, which emphasizes the lessons that can be derived from empirical analysis.

instead (see Varian, 1978).[2] Thus, we find ourselves felicitously aligned with mainstream trends. Nevertheless, it is appropriate to offer some recommendations for adaptation of work on the estimation of cost relationships to the requirements of the analysis of industry structure.

15A. Earlier Approaches: Single-Product Techniques

The first generation of cost function estimation studies virtually all dealt with single-output production, not because the industries offered only one output, but because they were attempting to provide a useful simplification—and because the investigators either did not recognize, or care to focus on, the policy issues that arise only in the presence of multiproduct activities. Their methods involved aggregation of the different outputs of the firms into a single measure, the "scalar output," using value weights, value-added weights, or weights related to some physical characteristic of the various outputs, such as size. Then, typically, the behavior of the average cost of this "scalar output" was studied and its implications for industry structure and public policy were examined.[3]

It is useful to reconsider this rather common approach in light of what we have learned about multiproduct cost functions and the theoretical analyses with which they are associated. If the true cost relationship for an industry is the multiproduct function $C(y_1, \ldots, y_n)$, then the construction of an aggregate "scalar output" $Y \equiv \sum a_i y_i$ and the estimation of the parameters of the associated scalar-output cost function is tantamount to the imposition of a special form upon $C(y_1, \ldots, y_n)$. That is, one thereby implicitly requires $C(y_1, \ldots, y_n) \equiv \tilde{C}(Y) = \tilde{C}(\sum a_i y_i)$. Now, in some cases this may actually be an appropriate approximation, one that we ourselves have found it convenient to use as a simple example from time to time. But the imposition of such a special functional form, without any evidence of its appropriateness, must *generally* introduce serious statistical biases that render suspect any inferences derived from it.[4]

Statistical problems aside, let us consider what can and what cannot properly be inferred from estimates of $\tilde{C}(\sum a_i y_i)$. If there were reason to believe that the true multiproduct cost function does have this form, then all the required cost measures could be calculated readily and reliably from the simplified function. But in the absence of such evidence, virtually nothing can be inferred with any degree of confidence from the estimates. Yet, there

---

[2] The literature on duality is vast. Perhaps the best source for those interested in an exposition of the theory and its empirical implementation is Fuss and McFadden (1978).

[3] See, for example, Walters' (1963) classic survey. Even as this volume goes to press, another survey focusing on the single product approach has appeared; see Gold (1981).

[4] For discussion of econometric tests of aggregation methods, see Blackorby, Primont, and Russell (1977) or Denny and Fuss (1977).

are special cases in which some things can be learned from an estimate of $\tilde{C}(\sum a_i y_i)$.

For example, suppose that in the sample from which $\tilde{C}$ is estimated all the observations involve output vectors that lie very close to the same ray in output space, so that they all can be written approximately as $t y^0$ for various scalar values of t. In this special case, the estimated scalar cost function will exhibit the scale properties that truly characterize the behavior of multiproduct costs *along the ray in question*. In particular, the calculated single-product measure of economies of scale, $\tilde{S} \equiv \tilde{C}(y)/[(Y)(d\tilde{C}(Y)/dY)]$, would correctly measure the degree of multi-output scale economies, $S_N = C(y)/\sum y_i C_i(y)$, regardless of the weights a_i used to define the scalar aggregate[5] Y. However, it must be emphasized that only the returns to scale along the single ray, $y = t y^0$, can be indicated by the properties of $\tilde{C}$. This function cannot accurately tell us anything about the shape of the RAC cross section along any other ray. Moreover, nothing can logically be inferred from $\tilde{C}$ about the industry's cost-minimizing structure—whether the industry is naturally competitive, a natural oligopoly, or a natural monopoly.[6]

15B. Hedonic Cost Functions: An Intermediate Approach

Hedonic cost functions constitute a recently introduced intermediate step between scalar cost functions of aggregated multiple outputs and full-blown multiproduct cost functions. A hedonic cost function expresses production cost as a function of (i) input prices, (ii) some aggregated measure of the size of the multiple outputs, and (iii) some "hedonic" variables that describe, in a condensed form, the mixture of output quantities produced. This technique was pioneered by Spady and Friedlaender (1978), in their path-breaking analysis of trucking firms, as an ingenious method of achieving economy in the number of parameters whose values must be estimated, while introducing at least some indicators of the multiproduct behavior of costs. For example, instead of estimating costs as a function of the amounts of the vast number of types of freight carried by a trucking firm over its myriad origin–destination pairs, Spady and Friedlaender estimated cost as a function of ton-miles carried and, simultaneously, as a function of such hedonic descriptors as the average length of haul. A hedonic cost function can be described formally

---

[5] To see this, recall that over the sample, $y = t y^0$, $Y = \sum a_i y_i = t \sum a_i y_i^0$, and $\tilde{C}(Y) = \tilde{C}(t \sum a_i y_i^0) = C(t y^0)$. By the latter equality and the chain rule, $dC(t y^0)/dt = d\tilde{C}(t \sum a_i y_i^0)/dt = (\sum a_i y_i^0)(d\tilde{C}(Y)/dY) = t^{-1} Y(d\tilde{C}(Y)/dY)$. Further, $dC(t y^0)/dt = \sum y_i^0 C_i(t y^0) = t^{-1} \sum y_i C_i(y)$. Hence, $\sum y_i C_i(y) = Y(d\tilde{C}(Y)/dY)$, so that $S = \tilde{S}$.

[6] For examples of such unsupported inferences, see Christensen and Green (1976) who imply that they can derive such conclusions from their study of the costs of electric power generation as a function of total kilowatt-hours produced, irrespective of the distribution of this production among peak and off-peak times of day and seasons of the year.

with the aid of the expression $\tilde{C}(Y, Z_1, \ldots, Z_k)$, where $Y = \sum a_i y_i$ is a scalar measure of aggregate output, and where Z_i, $i = 1, \ldots, k$, are hedonic descriptors of the output mixture. Each descriptor can be expressed as some function of the output vector that is homogeneous of degree zero, so that it remains constant along a ray. Of course, if enough linearly independent descriptors were included in $\tilde{C}$, then the entire output vector y could be reconstructed from information given by $Y, Z_1, \ldots, Z_k$. In that case, there would exist a hedonic cost function perfectly equivalent to the true multi-output cost function, but then they would both necessarily require estimation of the same number of parameters.

If fewer hedonic descriptors are used, estimation of $\tilde{C}(Y, Z_1, \ldots, Z_k)$ instead of $C(y)$ is necessarily restrictive to some degree, if more economical. If the descriptors are explicitly written as functions of y, $\tilde{C}(\sum a_i y_i, Z_1(y), \ldots, Z_k(y))$, it becomes clear that a hedonic cost function is no more and no less than a multi-output cost function of a special and restrictive form. Where data limitations make necessary the imposition of some restrictive form, the hedonic approach permits informal information on the behavior of multi-product costs to be used to select restrictions that are more appropriate than arbitrary ones. Viewed in this way, a special hedonic cost function can be tested econometrically in comparison with more general forms, and, if it is shown to be at least a reasonable approximation, it can be used to evaluate all the parameters corresponding to the properties of multiproduct cost functions described in earlier chapters. Clearly, the decision to use a hedonic cost function should rest on the reliability of the information employed in selecting the hedonic variables and on the power of the data set to yield useful estimates of a multiproduct cost function without restrictions.

With this discussion as background, we now turn to the main task of this chapter: to discuss the forms of multiproduct cost functions that are appropriate vehicles for empirical analyses of industry structure. We confine ourselves to a discussion of functional forms for multiproduct cost functions because this seems to be the prime shortcoming of empirical work in the area and because we are in no position to offer new insights on appropriate econometric techniques.

15C. Desiderata for Multiproduct Cost Functions

What are the desiderata for the form of a multiproduct cost function? What properties should it possess? First, it should be a *proper* cost function—one consistent with minimization of the outlay on inputs, at constant input prices, subject to the technological feasibility of the outputs' production with the aid of the input quantities used. It is well known [see Varian (1978) for a clear discussion] that to qualify as a proper cost function, $C(y, w)$ must be nonnegative and nondecreasing, concave, and linearly homogeneous in the input prices w. Further, $C(y, w)$ must be nondecreasing in its outputs (particularly if the possibility of free disposal is assumed).

Second, for the purposes of an analysis of a multiproduct industry, the cost function should yield a reasonable cost figure for output vectors that involve zero outputs of some goods.[7] For, as we have seen, industry configurations compatible with efficiency or equilibrium may well include firms that do not produce all of the industry's products. This is a desideratum violated by several of the functional forms often used in statistical studies of cost functions. For example, the Cobb-Douglas functional form,

$$C(y, w) = b y_1^{a_1} y_2^{a_2} \cdots y_n^{a_n},$$

where $b, a_1, \ldots, a_n$ may be functions of the input prices, must be rejected for our purposes because of its property that $C(y, w) = 0$ if the magnitude of *any* of the outputs, $y_1, \ldots, y_n$, is zero. The translog cost function,

$$\ln C(y, w) = a_0 + \sum a_i \ln y_i + \sum \sum a_{ij} \ln y_i \ln y_j,$$

where the a's are functions of input prices, is also condemned by this property, despite its great recent popularity.

Third, the functional form of $C(y, w)$ should not, in itself, prejudge the presence or absence of any of the cost properties that play an important role in the analysis of the industry. Rather, the functional form should be consistent with either satisfaction or violation of any such properties so that the data set and not the mere choice of functional form will determine the substantive findings. For example, any function that, like the Cobb-Douglas or the translog, takes the value zero wherever the output of any product in the product set is zero automatically precludes the possibility of economies of scope or of subadditivity if $C(y) > 0$ for other relevant values of y. For then the costs of an industry can always be driven (ostensibly) to zero by dividing its outputs among specialized firms, none of which produces every one of the industry's products. Similarly, differentiation of the Cobb-Douglas functional form shows that for $i \neq j$, $\partial^2 C / \partial y_i \partial y_j = a_i a_j C / y_i y_j$. Then, as long as marginal costs are positive and all outputs are positive, $\partial^2 C / \partial y_i \partial y_j > 0$. Hence, use of the Cobb-Douglas functional form itself preimposes the conclusion that weak cost complementarities are always absent and that, in fact, diseconomies of scope prevail in the absence of fixed costs. Because an empirical analysis using the Cobb-Douglas form must always reach these conclusions, regardless of the truth of the matter and regardless of the data, the Cobb-Douglas cost function is automatically ruled out as a functional form sufficiently flexible for the analysis of industry structure.

---

[7] The material in this paragraph and in some other portions of this chapter first appeared in unpublished work by Dietrich Fischer.

Thus, our third and most novel desideratum, to which we refer as *substantive flexibility*, is that the form of the cost function should be consistent with both economies and diseconomies of scope, with both cost subadditivity and superadditivity, and that it not prejudge the shapes of the RAC curves, the shapes of the AIC curves, and the crucial properties of the trans-ray cross sections. For these are the properties of cost functions that the preceding chapters have shown to be most important for the analysis of industry structure.

Fourth, for the sake of empirical practicality, the cost function should not require estimation of the values of an excessive number of parameters.

A fifth desideratum, which many empirical analysts have begun to pursue, is that the cost functions assume what is called a flexible functional form, that is, a form that imposes no restrictions on the values of the first and second partial derivatives.

To illustrate what can be done about these desiderata and to offer some suggestions that may, perhaps, prove useful for empirical analysis, we now discuss, in turn, three promising functional forms.

15D. Specific Forms (i): The Hybrid Translog Function

We begin with a functional form that may be called the hybrid translog cost function. It has already been used in empirical studies of costs, and it has the advantage of relative familiarity because of its similarity to the ordinary translog function. As we show, it also satisfies our desiderata fairly well, though it does leave something to be desired in terms of the property we have called substantive flexibility. The hybrid translog cost function can be written as

(15D1)
$$\ln C(y, w) = \alpha_0 + \sum_i \alpha_i \ln w_i + \sum_k \beta_k \left[\frac{y_k^\theta - 1}{\theta} \right]$$
$$+ \tfrac{1}{2} \sum_i \sum_j \gamma_{ij} \ln w_i \ln w_j$$
$$+ \tfrac{1}{2} \sum_k \sum_l \delta_{kl} \left[\frac{y_k^\theta - 1}{\theta} \right] \left[\frac{y_l^\theta - 1}{\theta} \right]$$
$$+ \sum_i \sum_k \rho_{ik} \left(\ln w_i \right) \left[\frac{y_k^\theta - 1}{\theta} \right],$$

where $\gamma_{ij} = \gamma_{ji}$ and $\delta_{kl} = \delta_{lk}$.

This functional form was proposed by Caves, Christensen, and Tretheway (1980), and apparently independently by Fuss and Waverman (1981), as a generalization of the translog function.[8] It was explicitly designed to overcome the basic shortcoming of the translog function for our purposes—its violation of desideratum two by yielding the value $C(y) = 0$ whenever the firm does not produce one item in the industry's set of products. It is clear that in (15D1) costs need not be zero even if some output is zero. Further, (15D1) is a true generalization of the translog because

$$\lim_{\theta \to 0} \left[\frac{y_k^\theta - 1}{\theta} \right] = \ln y_k,$$

so that (15D1) can approximate the translog function arbitrarily closely, with sufficiently small values of the parameter θ.

There are restrictions on the parameters of (15D1) that are needed to ensure that it is a proper cost function, as defined in the discussion of our first desideratum. In particular,

$$\sum_i \alpha_i = 1, \qquad \sum_i \gamma_{ij} = 0, \qquad \text{and} \sum_i \rho_{ik} = 0$$

are necessary for the cost function to be linearly homogeneous in input prices. In practice, restrictions of this kind are usually imposed explicitly in the process of estimation. However, the other requirements of the first desideratum are not. Instead, the parameter values are estimated without restrictions and then the available data are substituted into the resulting function to determine whether requirements such as nonnegativity are satisfied.

As for our third desideratum, it can be seen that the hybrid translog is substantively flexible at least to some degree. Direct calculation shows that for (15D1) *defined on p. 21 & p. 50*

$$(15\text{D}2) \quad 1/S = \sum_i y_i^\theta \left(\beta_i + \sum_j \rho_{ji} \left(\ln w_j \right) - \sum_k \delta_{ki}/\theta \right) + \sum_i \sum_k y_i^\theta y_k^\theta \delta_{ki}/\theta.$$

It is easy to infer from (15D2) that if $\theta > 0$, then RAC is U-shaped with $S > 1$ for sufficiently small y and $S < 1$ for large y. If $\theta < 0$, then the RAC curves have inverted U shapes. One can obtain the equation of the M locus by equating the expression in (15D2) to unity. That is, since $S = 1$ where RAC is at a minimum, this equation yields the relationship among the output levels, for any vector of input prices, that characterizes the M locus.

The analytic characterizations of the trans-ray properties of a hybrid translog cost function are awkwardly complicated and difficult to work with.

---

[8] Our discussion of (15D1) generally follows that of Fuss and Waverman (1981).

Nevertheless, numerical examples show that (15D1) can exhibit economies or diseconomies of scope, $DAIC$ or $IAIC$, and cost complementarities or their opposite ($C_{ij} \gtreqless 0$). Further, it may or may not involve subadditivity or trans-ray convexity of costs. Moreover, Fuss and Waverman have been able to use this form to construct and carry out tests of hypotheses relating to overall scale economies, economies of scope, and product-specific scale economies at given output levels, using data on the operations of Bell Canada. However, what remains unknown is the extent to which (15D1) imposes rigidities upon the various combinations of properties that can simultaneously hold at particular output combinations.

The property of both the hybrid translog (15D1) and the translog forms that makes them particularly convenient for estimation is the statistically tractable set of equations that can be derived with the aid of Shephard's lemma for the shares of total cost efficiently expended on the various inputs. These equations are

$$\frac{\partial \ln C(w,y)}{\partial \ln w_i} = \frac{\partial C(w,y)}{\partial w_i} \cdot \frac{w_i}{C(w,y)} = \frac{X_i w_i}{C(w,y)} \quad \text{by Shephard's lemma}$$

(15D3)
$$\frac{w_i X_i^*}{C} = \alpha_i + \sum_j \gamma_{ij} \ln w_j + \sum_k \rho_{ik}\left[\frac{y_k^\theta - 1}{\theta}\right],$$

where X_i^* is the efficient quantity of the ith input. In fact, one usually estimates the values of the parameters of (15D1) by fitting (15D1) and the system of equations (15D3) simultaneously to the data, taking account of the resulting restrictions upon the equations.

The hybrid translog is a "flexible functional form," in that the magnitudes of C and of each of its first and second derivatives can assume any values at any one input price and output vector. As a consequence, it is not entirely surprising that it is consistent with either the presence or absence of the cost properties that concern us, at any given point.[9] However, we are left somewhat uneasy about the use of the hybrid translog for two reasons.

First, it is not clear that it is sufficiently flexible for the study of the effects of variations in input prices on industry structure. As was shown in the analysis of Chapter 6, these effects depend on the manner in which optimal input uses vary with the magnitudes of *all* the outputs, and the comparison between these variations and the corresponding variations in total costs. Yet, equation (15D3) shows that for the hybrid translog, interactions among the magnitudes of different outputs can only affect input utilization through their effects upon the total cost. However, it is probably unrealistic to expect statistical input demand functions to employ forms as flexible and as independent of one another as pure theory permits. While flexible input demand functions

---

[9] However, in principle, flexibility of a cost functional form need not ensure substantive flexibility, because many of the relevant properties of costs entail more than the magnitudes of the first and second derivatives at one point. Thus, as mentioned before, the translog is a flexible functional form that is not substantively flexible.

are desirable in themselves, they may undoubtedly entail an unacceptably vast number of parameters with which the econometrician would therefore be forced to deal.

A second shortcoming of the hybrid translog function is that it does not seem to permit representation of the relevant cost properties as tractable expressions of its parameters. It is this intractability problem that prevents us from judging just what, if any, constraints (15D1) does impose upon combinations of the cost properties and upon the manner in which the cost properties can vary through output space. Of course, sufficient experience with numerical computations can, perhaps, serve as a substitute for analytic relationships. However, at this early stage of inexperience in their use, there is much to be said for tools whose quirks are more readily visible. It is this concern that induces us to investigate the two functional forms that are considered in the remainder of this chapter.

15E. Functional Forms (ii): The Quadratic Cost Function

We turn now to a form, the quadratic cost function, that may be the most obvious one to use for our purposes. It is shown to perform well in terms of all of our desiderata. Its only shortcoming is the absence of any explicit theoretical foundation for use of this form in preference to any other. The quadratic cost function can be written[10]

(15E1)
$$C(y, w) = F + \sum_i a_i y_i + \tfrac{1}{2} \sum_i \sum_j a_{ij} y_i y_j,$$

$\left(a_i, a_{ij} \text{ are funct. of input prices} \right)$

where $F \geq 0$ and $a_{ij} = a_{ji}$. This function is a quadratic form in the magnitudes of the outputs, augmented by linear terms and the fixed cost parameter F. To deal more clearly with the effects of changes in outputs, we ignore the role of input prices w_i and simply take F, a_i, and a_{ij} to be unspecified functions of the vector w. The quadratic function is a flexible form in the output levels. The marginal costs are

$$C_i = a_i + \sum_j a_{ij} y_j,$$

$\downarrow y_i$ because there will be 2 of each term in the summation $a_{ii} y_i^2 \cdot a_{12} y_1 y_2 \cdot \ldots \left(a_{11} y_1 y_{11} \right. \ldots$ $\overset{\text{same}}{+ \left(a_{41} y_4 y_1 \right)}$

and these must be restricted to nonnegative values throughout the relevant domain. Our second desideratum is also satisfied by (15E1); that is, if some components of y are zero, the value of $C(y, w)$ in (15E1) is not necessarily

---

[10] Equation (15E1) was first constructed and used by Braunstein (see Baumol and Braunstein, 1977). He also employed a useful modification of (15E1), the form $C(y, w) = F + \sum_i a_i y_i + \sum_i \sum_j a_{ij}(y_i y_j)^{b_{ij}}$ where flexibility is increased by the parameters b_{ij}. One can, of course, reduce their number by setting $b_{ij} = b$ for all i, j.

zero. In fact, the behavior of C in relation to the output levels of the goods actually produced is identical with that exhibited by (15E1) if it had been applied to only those goods in the first place.

The performance of (15E1) in relation to our third desideratum, substantive flexibility, is readily investigated through calculations that are only a bit more extensive than those just concluded. For this purpose, we examine the conditions under which the various pertinent cost properties are satisfied by (15E1). For example, direct calculation shows that

$$S_N = [F + \sum_i a_i y_i + \tfrac{1}{2} \sum_i \sum_j a_{ij} y_i y_j] / [\sum_i a_i y_i + \sum_i \sum_j a_{ij} y_i y_j]$$

so that

See p. 157

(15E2) $S_N \gtreqless 1$ as $F \gtreqless \tfrac{1}{2} \sum_i \sum_j a_{ij} y_i y_j.$

Consequently, RAC is U-shaped if $F > 0$ and if $\sum_i \sum_j a_{ij} y_i^0 y_j^0 > 0$ for any point y^0 on the ray in question. If $\sum_i \sum_j a_{ij} y_i^0 y_j^0 \le 0$, economies of scale hold throughout that ray, unless $\sum_i \sum_j a_{ij} y_i^0 y_j^0 = 0 = F$, in which case RAC is constant.

The degree of product-specific returns to scale is given by

$$S_i = [a_i y_i + \tfrac{1}{2} a_{ii} y_i^2 + \sum_{j \ne i} a_{ij} y_i y_j] / [a_i y_i + a_{ii} y_i^2 + \sum_{j \ne i} a_{ij} y_i y_j].$$

Consequently

(15E3) $S_i \gtreqless 1$ as $0 \gtreqless a_{ii}.$

Thus, the average incremental costs of product i are either globally declining, constant, or rising, as a_{ii} is negative, zero, or positive, respectively. This may perhaps be considered overly restrictive, particularly as it rules out U-shaped average incremental cost curves.

There is however a simple way in which (15E1) can be modified to eliminate this restriction, albeit at the cost of an expansion (which can be considerable) in the number of parameters whose values must be estimated. For this purpose, instead of taking F to be fixed, it can be replaced by $F(T)$, where $T = \{i \in N \mid y_i > 0\}$, is the set of goods of which positive quantities are produced. Thus, F becomes the value of a "fixed cost" whose magnitude may depend on the set of items actually produced. For purposes of estimation, the values of $F(T)$ may be regarded as dummy variables, for the various product sets T that are represented in the sample.

This modest modification is perhaps the simplest, most direct step toward the improvement of the substantive flexibility of *any* empirical cost function. It involves only the addition of a string of dummy variables, the most basic and readily interpretable econometric means to capture structural differences within the data. While it is true that, *in theory*, this may result in the introduction of $2^n - 2$ additional parameters (the number of proper nonempty subsets of N), *in practice n* is usually small and the number of *distinct* subsets of products actually observed is, even then, usually much smaller than the theoretical maximum. While this also means that it will not be possible to estimate $F(T)$ completely, it is generally worth trying this technique since acquisition of some of the information to be gleaned from the presence of different product sets in the data is better than none. One can also employ various special forms for $F(\cdot)$, say $F(T) = \sum_{i \in T} f_i$ or $F(T) = (\#T)f$, among which one can choose on the basis of convenience or on other grounds. Viewed in this way, the most restrictive of the candidate assumptions is that fixed cost is a constant, immutable parameter, F.

Employing this modification of (15E1), the term $F(T) - F(T - \{i\})$ is added to the numerator of S_i, and we now conclude that

(15E4) $\qquad S_i \gtreqqless 1 \quad \text{as} \quad F(T) - F(T - \{i\}) \gtreqqless \tfrac{1}{2} a_{ii} y_i^2.$

Thus, the average incremental cost curve for product i can be U-shaped if $a_{ii} > 0$ and if some setup costs can be eliminated by exclusion of output i from production.

The degree of economies of scope between product sets T and $N - T$ is given by

(15E5)
$$\frac{[F(T) + F(N - T) - F(N)] - \sum_{i \in T} \sum_{j \in N-T} a_{ij} y_i y_j}{F(N) + \sum a_i y_i + \frac{1}{2} \sum_i \sum_j a_{ij} y_i y_j}$$

Clearly, then, there can be either economies or diseconomies of scope, depending on the signs of the a_{ij} and on the sizes of y_i and y_j, for $i \in T$ $j \in N - T$, as well as on the product-specific fixed costs. Since there are cost complementarities between products i and j if $a_{ij} < 0$, it is not surprising that the degree of economies of scope is larger the smaller is a_{ij}, and the larger are the outputs of goods i and j for which a_{ij} is negative.

To investigate trans-ray convexity, we first recall that any product-specific fixed costs, that is, any variation in $F(T)$, will impart an element of trans-ray concavity (as per the transylvanian cost function discussed in Chapters 4 and 7) near the relevant points with $y_i = 0$. For the case in which there are no product-specific fixed costs, $F(T) \equiv F$, it is very useful to recall the following standard result:

Proposition 15E1 *A function $C(y)$ that is twice-differentiable on the hyperplane segment $H_{TR} = \{y \in R_+^n | h \cdot y = k\}$ is convex on H_{TR} if and only if,* *throughout H_{TR}, the bordered Hessian*

(15E6)

$$\begin{bmatrix} C_{11} & C_{12} \cdots C_{1n} & h_1 \\ C_{21} & C_{22} \cdots C_{2n} & h_2 \\ \cdots\cdots\cdots\cdots\cdots\cdots \\ C_{n1} & C_{n2} \cdots C_{nn} & h_n \\ h_1 & h_2 \cdots h_n & 0 \end{bmatrix}$$

has border-preserving principal minors whose determinants are all nonpositive.

Corollary   *With two outputs, costs are trans-ray convex on*

$$H_{TR} = \{y \in R_+^2 \mid h \cdot y = k\}, h \geq 0,$$

if $C(y)$ is twice-differentiable on H_{TR} and if $C_{12}(y) = C_{21}(y) < 0, C_{11}(y) \geq 0$, and $C_{22}(y) \geq 0$ for all $y \in H_{TR}$.

For the quadratic cost function, $C_{ij} = a_{ij}$, so that the bordered Hessian in (15E6) is constant in the interior of every output subspace. This makes the conditions for trans-ray convexity that are given by the proposition easy to check, because they are independent of the output point. However, the proposition also reveals a restriction imposed by the quadratic functional form. If costs are trans-ray convex on the hyperplane $h \cdot y = k$, then they must also be so on any parallel hyperplane, $h \cdot y = k'$.

For example, in the two-output case, the bordered Hessian is

(15E7)

$$\begin{bmatrix} a_{11} & a_{12} & h_1 \\ a_{21} & a_{22} & h_2 \\ h_1 & h_2 & 0 \end{bmatrix}$$

The relevant minors are

(15E8)

$$\begin{bmatrix} a_{11} & h_1 \\ h_1 & 0 \end{bmatrix} \begin{bmatrix} a_{22} & h_2 \\ h_2 & 0 \end{bmatrix},$$

and the full-bordered Hessian itself. The determinants for the smaller matrices (15E8) are $-h_1^2 \leq 0$ and $-h_2^2 \leq 0$, and the determinant of (15E7) is

(15E9) $a_{12}h_1h_2 + a_{21}h_1h_2 - a_{11}h_2^2 - a_{22}h_1^2.$

Thus, costs are trans-ray convex along any and all hyperplanes $h \cdot y = k$ if and only if the expression in (15E9) is nonpositive. The economically relevant hyperplanes are positively sloped, with $h_i \geq 0$. Thus, if $a_{12} = a_{21} \leq 0, a_{11} \geq 0$, and $a_{22} \geq 0$, then trans-ray convexity holds in all positive directions. Otherwise, trans-ray convexity may hold in some directions and not in others. For example, if a_{11} is positive, then costs are trans-ray convex along hyperplanes with h_2/h_1 sufficiently large. It can be seen that if both a_{11} and a_{22} are negative, but if $a_{12} < -\sqrt{a_{11}a_{22}}$, then costs are trans-ray convex for $h_1 = -a_{12}$ and $h_2 = -a_{22}$, as well as for other values of h.

Since trans-ray convexity and economies of scale between the origin and the hyperplane suffice for subadditivity, by Proposition 7E1, the preceding straightforward tests can shed much light on the region over which the cost condition for natural monopoly holds. In particular, if the conditions for trans-ray convexity are satisfied for some $h \geq 0$, then subadditivity holds at any point for which the parallel hyperplane that contains it lies inside the M locus. It should be noted, however, that the quadratic cost function possesses sufficient substantive flexibility to permit violation of subadditivity inside the M locus. In particular, in the two-output case, (15E2) and (15E5) show that there are diseconomies of scope along with economies of scale if

$$a_{12}y_1y_2 > F > a_{12}y_1y_2 + \tfrac{1}{2}a_{11}y_1^2 + \tfrac{1}{2}a_{22}y_2^2.$$

Thus, we have explored and confirmed the substantive flexibility of the quadratic cost function and derived straightforward relationships between the critical cost properties and the parameters of this flexible functional form.[11]

15F. Input Prices and the Quadratic Cost Function

So far, we have ignored the role of input prices and the corresponding behavior of input demands when the cost function is quadratic. As mentioned earlier, the parameters in (15E1), F, a_i, and a_{ij}, may be considered to be functions of the input prices w. And, to ensure that (15E1) is a proper cost function, these functions should be nondecreasing, linearly homogeneous, and concave in w. Then, the cost-minimizing quantity of input k becomes, by

---

[11] Aside from its usefulness in empirical work, there is another important area in which the quadratic cost function seems highly promising, though it has not yet been employed for the purpose. In a theoretical analysis, it can be used together with a set of simple (perhaps even linear) market demand functions to characterize analytically the behavior of an industry in contestable markets. In such a model, one could explicitly determine the values of the parameters of the cost and demand functions that yield the various types of sustainable industry configurations as well as those for which no sustainable solutions exist. One could thereby carry out what is, in effect, a simulation exercise that can shed light on the working of the entire analysis.

Shepard's lemma,

$$X_k^* \equiv \frac{\partial C}{\partial w_k} = \frac{\partial F(w)}{\partial w_k} + \sum_i y_i \frac{\partial a_i(w)}{\partial w_k} + \tfrac{1}{2}\sum_i \sum_j y_i y_j \frac{\partial a_{ij}(w)}{\partial w_k}$$

Practicality aside, it is most desirable to take $F(w)$, $a_i(w)$, and $a_{ij}(w)$ to be distinct and independent functions. Then, little restriction is imposed on the response patterns of the different input demands to variations in output quantities. As was indicated in our discussion of the hybrid translog function, such an unrestricted form is particularly desirable for the study of the effects of changes in factor prices on industry structure. However, this requires a highly impractical proliferation of the parameters whose values must be estimated.

At alternative that achieves the greatest economy in the number of parameters entails the imposition on F, a_i, and a_{ij} a common functional dependence on w. This can be represented as

$$F(w) \equiv Fg(w), \qquad a_i(w) \equiv a_i g(w), \quad \text{and} \quad a_{ij}(w) \equiv a_{ij} g(w)$$

where $g(w)$ is a linearly homogeneous, concave, and nondecreasing function. Then, (15E1) becomes

$$C(y, w) = g(w)\big[F + \sum_i a_i y_i + \tfrac{1}{2}\sum_i \sum_j a_{ij} y_i y_j\big],$$

a cost function in the family suggested by Braunstein and Pulley (1980) [who also discuss some of the methods they have used to estimate the parameter values of a member of that family, for which $g(w)$ was assumed to take Cobb-Douglas form]. Cost functions of this kind, with w and y multiplicatively separable, require all input demands to vary with outputs in the same fashion. In fact, input ratios and cost shares are thus assumed to be independent of output levels, and input demand elasticities with respect to each output become equal and independent of w. Consequently, in view of the results in Chapter 6, multiplicatively separable cost functions are not well suited for investigation of those properties of input demand functions that relate to the effects of input prices on industry structure.

Thus, we have another example of the tradeoff between tractability in empirical analysis of the functional form chosen for the cost relationship and its usefulness in testing the many properties that theory suggests are important for industry analysis.

We turn next to the last functional form to be considered here. This form is suggested by a particular and detailed model of joint production. The cost function's structure is considerably more rigid and it has fewer parameters than those forms which are designed for functional flexibility. This is, of course, not necessarily a drawback because one must be prepared to employ relatively rigid structural forms whose economic meaning is clear, in circumstances in which their economic content appears to be appropriate.

15G. Function Forms (iii): The CES Function

The "CES" cost function[12] is given by

(15G1) $C(y, w) = F + [\sum a_i y_i^{b_i}]^\rho,$ where $a_i, b_i, \rho > 0$ and $F \geq 0$

This special, rather inflexible form takes its name from the similar functions with constant elasticity of substitution, which have frequently been used in analyses of other economic issues.[13] As we did earlier, we first ignore the input prices to focus our attention upon the relation between cost and output quantities. After discussing the relationships between the parameters of (15G1) and our basic cost properties, we show the connection between (15G1) and the model of shared inputs discussed in Chapter 4. This connection permits the theory to guide us toward incorporation of input prices into (15G1) in a way that is reasonable and illuminating. It also suggests empirical tests of the hypothesis that some actual set of productive relationships is consistent with the shared input model.

The CES cost function does satisfy our second desideratum, for its behavior when some output quantities are zero is identical with what it would have been if the function had originally been formulated to pertain only to the relevant subspace.

The degree of overall scale economies of the CES function is

(15G2) $$S_N = \frac{F + [\sum a_i y_i^{b_i}]^\rho}{\rho(\sum a_i y_i^{b_i})^{\rho - 1}[\sum b_i a_i y_i^{b_i}]}$$

It can be seen[14] from (15G2) that on a given ray, $y = ty^0$,

$$\lim_{t \to \infty} S_N = 1/\rho \max_{y_i^0 > 0} (b_i).$$

---

[12] CES multiproduct cost functions have been proposed by Fischer and by Panzar and Willig in unpublished manuscripts and explored in Keeler (1974).

[13] The CES functional form was first constructed by Arrow, Chenery, Minhas, and Solow (1961). Note however that, here, we use the term "CES" to describe the way the outputs enter the cost function, not the way input quantities relate to production. This cost function is *not* dual to a CES representation of production.

[14] The method of analysis entails division of numerator and denominator of (15G2) by t raised to the power $\rho \max(b_i)$ and, then, calculation of the limit as $t \to \infty$. Similarly, the limit of S_N as $t \to 0$ is obtained by division by t raised to the power $\rho \min(b_i)$.

Similarly, if $F = 0$,

$$\lim_{t \to 0} S_N = 1/\rho \min_{y_i^0 > 0} (b_i),$$

while, if $F > 0$,

$$\lim_{t \to 0} S_N = \infty.$$

Consequently, the CES cost function is consistent with U-shaped ray average costs, with global increasing returns to scale, or, for $F = 0$ and $b_i = b$, with homogeneity in outputs of any degree. Note, too, that the shape of RAC curves may vary with the set of products of which positive·quantities are produced.

Differentiation of (15G1) shows that

$$C_{ij} \gtreqless 0 \quad \text{as} \quad \rho \gtreqless 1, \quad \text{for } i \neq j.$$

Thus, complementarity must hold for each and every pair of outputs if $\rho < 1$, and for none of them if $\rho > 1$. Obviously, if $\rho = 1$, the variable costs of all the outputs are independent. The requirement that all outputs must be complementary if any of them are, is, of course, highly restrictive. However, we shall soon see that this restriction can be attributed to the underlying model from which (15G1) can be derived.

It follows that economies of scope are present if $\rho < 1$. However, if $\rho > 1$, there may nevertheless be economies of scope if the magnitude of F is sufficiently large relative to that of ρ and relative to the output quantities. Average incremental costs of output i are decreasing if $b_i > 1$ and $\rho > 1$. In particular, it can readily be seen that

$$\lim_{y_i \to \infty} S_i = 1/\rho b_i.$$

If there are product-specific fixed costs, so that F is replaced by $F(T)$, then the curves of average incremental costs will be U-shaped with $\rho b_i > 1$.

In the absence of product-specific fixed costs, trans-ray convexity holds in all directions if $\rho \geq 1$ and $b_j \geq 1$ for all j, since, in this case, the variable costs are a convex function of y. However, for two outputs, the Corollary of Proposition 15E1 shows that $b_i \geq 1$ and $1 \geq \rho \geq 1/b_i$ for all i also suffice for trans-ray convexity in nonnegative directions, since, in this case, $C_{ii} \geq 0$ and $C_{12} \leq 0$.

To illustrate some of the shapes cost function (15G1) can take, consider the case of two products where $b_1 = b_2 = b$ and $a_1 = a_2 = 1$. Then

$$C(y_1 y_2) = F + (y_1^b + y_2^b)^\rho.$$

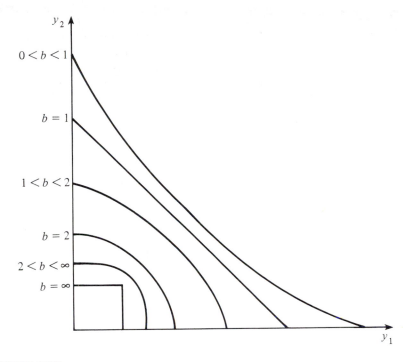

FIGURE 15G1
Isoquants of the CES cost function in two variables for $b_1 = b_2 = b$.

Figure 15G1 shows iso-cost curves for various values of b. It is easily seen that the cost function is quasiconvex, that is, it has iso-cost curves that are concave to the origin, if $b > 1$. The function is ray-concave if $b\rho < 1$. A sufficient (but not necessary) condition for subadditivity is quasiconvexity and ray concavity everywhere. These conditions are satisfied when $\rho < 1/b < 1$. If $0 < b < 1$, the cost function is quasiconcave. If $b\rho > 1$, it is convex on segments of rays bounded away from the origin. It can therefore assume a full variety of combinations of ray and trans-ray behavior, a necessary property for a cost function to be sufficiently flexible for consistency with either economies or diseconomies of scale and scope.

Now that we have investigated the substantive flexibility of the CES cost function, let us turn to its derivation from a shared input model of production. Suppose that the cost of producing y_i entirely by itself is

$$(15G3) \qquad y_i^{b_i} g_i(w) Z^{-e}, \qquad e \geq 0, \qquad b_i > 0,$$

where $g_i(w)$ is a positive function of input prices that is nondecreasing, linearly homogeneous, and concave, and where Z is the available quantity of the services of a process or some inputs which can be shared without congestion in the production by a single firm of the items that comprise the product set N. Let the cost of providing the quantity of shared services, Z, be given by the proper cost function $Zh(w)$. Then, a firm that produces the output vector y at minimum total cost will incur the cost

$$C(y, w) = \min_{Z} \left((\sum y_i^{b_i} g_i(w)) Z^{-e} + Zh(w) \right).$$

Calculation of the cost-minimizing value of Z is a straightforward matter. We obtain

$$Z = [h/e \sum y_i^{b_i} g_i]^{-1/(e+1)},$$

so that the associated cost function is

$$(15G4) \quad C(y, w) = (e^{-e/(e+1)} + e^{1/(e+1)}) h(w)^{e/(e+1)} \left[\sum g_i(w) y_i^{b_i} \right]^{1/(1+e)}.$$

Clearly, after addition of some fixed costs, the derived cost function in (15G4) has the CES form (15G1). In particular, the parameters in (15G1) and (15G4) have the following relationships:

$$\rho = 1/(1 + e), \qquad b_i = b_i, \quad \text{and} \quad a_i = g_i(w)h(w)^e (e^{-e/(e+1)} + e^{1/(e+1)})^{1+e}.$$

Consistency with this shared input model requires several properties to hold for the CES (15G1). First, and most striking, ρ must lie between 1 and 0, because e is nonnegative. As has been noted, for ρ in the interior of this range, there are cost complementarities as well as economies of scope among all outputs. In this model the reason is clear. For $e > 0$, the shared input services contribute to the production of each and every output. If, instead, e were zero, then there would be no connection between the production processes so that ρ would equal one and there would be no cost complementarities. Thus, the most direct test of the presence or absence of shared input services that do contribute to production in (15G1) is whether or not $0 < \rho < 1$.

Other tests of consistency deal with the character of the interactions between output variables and input prices. Prices of inputs that are used only in the production of output i must affect only a_i in (15G1), for this relationship to be consistent with (15G4). Perhaps more noteworthy is the requirement that prices of inputs used only in the production of the shared services must affect all the a_i's in (15G1) similarly, and the corresponding elasticities must all be the same. Thus, if we add a conjecture about the identity of the shared inputs to the hypothesis that the shared input model correctly describes the technological relationships of the firm or industry under study, then the combined hypothesis can be tested by determination of the value of ρ and the behavior of the a_i functions. On the other hand, if one maintains the hypothesis that the presence of some shared input services underlies (15G1), then one is helped in determining which inputs are shared by the estimates of the a_i functions.

The preceding discussion has been offered primarily as an example of an analysis of the relationship between the character of the productive process and the form of the cost function. The shared input model is, of course, hardly very general, and so our specific results are not likely to prove widely applicable. Much more work remains to be done in the generation of multi-product cost functions from production models that are more flexible and realistic—models that permit a variety of shared inputs, some that are partially congestable, and inputs that are only shared by proper subsets of the production processes. [The work of Scherer *et al.* (1975) will undoubtedly prove very useful in providing guidance for this endeavor.] More basic theoretical work is also needed to design methods that permit the investigator to obtain from a multiproduct cost function information about the structure of input utilization by the basic production processes. [See Fuss and McFadden (1978) for some relevant techniques.]

Perhaps the clearest lesson to be drawn from this chapter is that, while the required empirical techniques are only in their beginnings, these beginnings are auspicious. In the next chapter, we discuss the ways in which policy analyses can be helped by effective empirical investigation of the sort we have been discussing.

16

Toward Application
of the Theory

We have now gone as far as we can, within the confines of one volume, in providing a coherent theory of industry structure and an analysis of its implications for the performance of the economy in terms of the general welfare. Yet, it is clear that this is hardly the end of the analysis. Every chapter seems to suggest subjects that call for further investigation, and in many cases the way to begin seems clear.

It is also clear that the theory offers a considerable number of policy implications. Although some of these have already been discussed in Chapter 12, it seems appropriate, in concluding the book, to examine somewhat more systematically what the analysis suggests for the role of government in relation to market behavior. This becomes particularly important in light of the conclusion that deviations from the classical competitive norm need not prevent the invisible hand from doing its work. Thus traditional *per se* indicators of market performance such as concentration, price discrimination, conglomerate mergers, or vertical and horizontal integration do *not automatically* call for government intervention in contestable markets. Therefore, it is incumbent upon us to attempt to provide some alternative guidelines.

We will start by considering the contribution to policy analysis that can be hoped for from empirical investigation, continuing the subject matter of Chapter 15. Then we will proceed to examine briefly what can be inferred for public policy from the analysis of each of the major subjects dealt with in this volume. Finally, we will offer a few broader observations about the applications of the theory to regulation of industry, interpreted in its most general sense. Here, too, while we can offer some conclusions that seem reasonably firm, in several cases we are forced to leave the reader only with questions, because we have no clear-cut answers.

16A. Empirical Evidence for Policy Design

In order to summarize the factual materials that are needed by a decision maker to make effective use of the analysis of this book, it is useful to start with a brief outline of its general policy implications. By and large, we have concluded that if an industry is structurally contestable and is behaving accordingly, and if, in addition, it has sustainable configurations available to it, and if none of these involve any significant welfare problems, then that industry is best left to its own devices with no government interference, *even if it is composed of a very small number of firms.*

Where an industry is not behaving in a way that is to be expected in a contestable market, even though there seem to be no inherent structural impediments to contestability, then the most promising course for public policy is identification and removal of any artificial obstacles to contestability (such as governmental inhibition of entry or other such constraints imposed by public policy, as well as other entry barriers) and the prevention, so far as possible, of predatory acts or deterring threats by incumbents. If the industry is structurally not contestable, it may be possible to seek measures that decrease "natural" entry barriers to some degree—a subject that will be discussed later in this chapter. Or it may be possible to isolate the portion of the industry's activities that cause the uncontestability and regulate that portion, leaving the remainder of the industry's activities free from governmental interference. Where markets are not contestable or where no sustainable solutions are available, direct regulation or attention by the antitrust authorities is likely to be justified, though even here, as we will note later, some modifications of the usual criteria and procedures are likely to be warranted.

In any event, the criterion for the appropriateness of intervention is *not* the degree of departure from the attributes of perfect competition (a goal that is obviously often unrealistic and excessively demanding) since contestability is so much more applicable a benchmark. This general outline of the tenor of the policy implications that can be drawn from our analysis immediately indicates the nature of the factual information useful in performing such policy analysis.

In addition to such obvious indicators of contestability as a history of frequent entrance and exit, pertinent information one can hope to obtain from empirical investigation can usefully be described in terms of the following seven tasks:

1. Determination of the cost-minimizing industry structure for the industry being examined. Here one should identify which, if any, of the industry's products can be produced most inexpensively by a monopoly, and which are "naturally competitive."
2. Determination of the degree of contestability of the market, and of the proximity of the behavior of the industry to that which would be expected under perfect contestability.
3. Determination of obstacles to contestability and evaluation of the difficulty of their reduction or elimination. If the market is not reasonably contestable it becomes important to determine whether measures to increase contestability are a practical option, and for this purpose it is necessary to identify those obstacles and the degree to which they can be affected by public policy.
4. Determination of whether or not sustainable configurations for the industry exist.
5. Qualitative and quantitative description of sustainable configurations.
6. Identification of any substantial welfare problems, such as externalities or second-best difficulties, associated with the sustainable configurations.

   Finally, we may add, in light of our discussion of intertemporal sustainability,

7. Description of any institutional inhibitions to the adoption of efficient intertemporal price patterns.

Undoubtedly, this list is not exhaustive, but it clearly encompasses the essence of the analysis and its relationships to public policy. We may note, incidentally, that these tasks extend well beyond the econometric estimation of cost and demand functions discussed in Chapter 15, though that sort of work may be assigned a leading role. Besides such technically sophisticated analysis, the relevant empirical work includes the gathering of several types of information of more traditional varieties. To assess the behavior of the industry, it will be useful to collect descriptive statistics, calculated in accord with requirements of economic analysis, which may differ significantly from those of conventional accounting practices. It will also be essential to gather qualitative information, describing the structure and behavior of the industry and of the institutions that affect it. The potential utility of such information is, of course, hardly a novel idea. The only novelty in the discussion of such descriptive materials that follows lies in the indication of the precise character of the information required to apply the theoretical results described in this volume.

   Let us now proceed to some observations about each of the tasks in our list.

16B. Task 1: Determination of Cost-Minimizing Industry Structure

The theoretical analysis of cost-minimizing industry structure in Chapters 5–7 provides the criteria needed for the corresponding empirical analysis. For this purpose it is necessary to determine the potentially profitable region, T, of industry output vectors, y^I, that is, the values of y^I that can be judged, in light of evidence on the cost and demand functions, to be consistent with nonnegative profits for all incumbent firms. It is also necessary to determine the set of output vectors that comprise the M locus, since the cost-minimizing industry structure depends on the relationship between the positions of y^I and M. For each individual product, it is necessary to determine the range of outputs over which average incremental costs (AIC) are decreasing, in order to find out whether or not y^I lies within the region where $DAIC$ holds for some of the various products. If it does, then we know that efficiency requires production of each of those products by just a single firm.[1]

This is rendered particularly significant by the result in Chapter 11, which shows that a sharp increase in the degree of approximation to competitive behavior can be expected in a contestable market once the number of firms producing a good equals or exceeds two. For then, under perfect contestability, each such good must be priced at its marginal cost, which will be the same for all of its producers. And if all of the industry's goods are produced by two or more firms, then each firm must operate on the M locus under conditions of locally constant returns.

In this connection it is also useful to apply the result of Chapter 5 about the number of firms needed to produce any particular good in a cost-minimizing industry structure. It will be remembered that this is determined by the maximal point m^* of the projection of the M locus on the axis representing the good in question. If the industry output of that good is more than twice as large as m^*, two or more firms will be required for efficient production of that item. Where this is true of every good in the industry's product line, if it operates in contestable markets the industry will have to behave competitively, selling competitive output quantities at marginal-cost prices.

It is important also to determine the groups of products over which the cost function exhibits economies of scope. This can help in assessing whether the horizontal structures of firms are dictated by productive efficiency or whether they have been introduced by motivations less consistent with the public interest. In such an assessment, it may be important to investigate

---

[1] Even if an industry contains many firms, and each of them seems to be producing an output combination on the M locus, they cannot constitute a competitive equilibrium if any of them produces an output vector at which $DAIC$ holds for any single product or for any set of its products. For if this is true and the goods are priced at marginal cost, then their revenues must be less than their total incremental costs, and so some cross subsidy must occur if the firm is not to lose money overall.

whether any economies of scope that are present can be ascribed to shared inputs whose services can be offered by upstream firms supplying the enterprises competing within the industry, without any increase in costs above those that would be incurred by a multiproduct firm.

It may also be useful to determine whether or not the cost function has the property of trans-ray convexity or trans-ray supportability. As shown in Chapter 7, either of these properties, together with increasing returns to scale, implies that the industry is a natural monopoly. And as shown in Chapter 9, these trans-ray properties can imply that the cost-minimizing industry structure requires all firms to produce the same output mix.

Most of these issues can be resolved with the aid of the cost function of the industry's firms and the demand functions for their products. Consequently, Task 1 is the focus of the econometric work discussed in Chapter 15.

16C. Task 2: Contestability of the Market

Determination of the structural contestability of a market requires evaluation of the costs of entry and exit and of the magnitude of unavoidable sunk costs. In particular, the availability of resale markets for durable inputs and their usability in other activities (fungibility) must be investigated since, clearly, the less the financial loss incurred in such a transfer, the lower will be the costs that are truly sunk and the smaller will be the costs of exit. It will also be essential to have a description of the sources of entry and exit costs as well as of the sunk costs to determine whether these costs can readily be reduced by appropriate policy measures. For example, if sunk costs are increased by legal inhibitions to the formation of resale markets, appropriate remedial measures may suggest themselves. Finally, it will be desirable to determine how entry costs are affected by the size of the potential entrant, that is, to estimate the $E(y)$ function roughly.

If $E(y^I)$ is large (where y^I is the industry's output vector), market forces may do little to constrain an incumbent monopolist's profitability, and so some regulation of profit may be justified. Yet, if at the same time $E(y)$ is much smaller for smaller values of y, or for specialized production, the incumbent may be vulnerable to entry by many firms, each planning to operate on a modest scale or in a specialized manner. Such latent competition may still suffice to force the incumbent to operate efficiently, to adopt useful innovations without delay, and to adopt an optimal vector of prices. In such a case, it may be essential to avoid any public policy that imposes additional impediments to entry or that imposes inappropriate pricing constraints upon incumbents, preventing them from adjusting prices in accord with competitive pressures. More will be said later about the importance of avoiding such pricing restrictions.

16D. Task 3: Comparison with Behavior in a Contestable Market

The purpose of a description of industry behavior and its comparison with that to be expected in a contestable market is clear enough. The methods by which this can be done do, however, require a few comments.

A test widely employed for purposes of this kind entails examination of the profits of the firm or the industry under consideration, on the assumption that returns that exceed the cost of capital for any substantial period of time constitute evidence of noncompetitiveness or absence of contestability. However, there are perils in such a procedure. First, the profit data usually available are those supplied by accountants, and, as is well known, accounting profits can differ substantially from economic profits for a variety of reasons, including the fact that accountants consider a normal rate of return to constitute a positive profit. Second, rents to scarce inputs and to innovative firms are very difficult to distinguish from supernormal profits. Third, as our intertemporal analysis in Chapters 13 and 14 shows, zero economic profit should be expected to hold in contestable markets only in terms of the discounted present value of the stream of expected net returns. Optimal behavior may well call for losses in some years that are offset by positive net returns in others. Finally, as is clear from the discussion in Chapter 13, optimal economic depreciation rates are very different from accounting depreciation rates, in part because the figures called for by economic analysis can only be calculated with the aid of information and projections not easily verifiable by straightforward and mechanical tests.

The simple accounting test for profits is not the only obvious criterion ruled out as a mechanical indicator of the consistency of observed behavior patterns with contestability. For example, while simple scalar output models suggest that under perfect contestability we should expect all firms to be similar in size in that they all operate near minimum efficient scale, as we have seen in Chapter 5, the shape of the M locus is unpredictable in the multiproduct case. As a result, if the firms in the industry produce outputs in different proportions, their cost-minimizing sizes (for example, as measured by employment or by capitalization) may differ radically. Thus, such wide variation in size is no evidence that behavior is inconsistent with contestability.

One useful test of the consistency of behavior with contestability is whether the resulting industry structure satisfies, at least approximately, the requirements of industry-wide cost minimization, as described in Chapters 5 and 7. We have already referred to these in our discussion of Task 1. A comparison of the attributes of the cost function and the industry structure can be very informative here. The following checklist can be helpful for this purpose:

(a) if the industry output vector is large relative to the convex hull of the M locus, there must be many firms in the industry;

(b) if the industry output of a particular good is large relative to the amounts of that good throughout the M locus, there must be many firms producing that good in the industry;

(c) if two or more firms produce product i, both must produce it at the same marginal cost, which is equal to the price charged for it;

(d) if each of the industry's outputs is produced by two or more firms, each firm must be operating at an output vector on the M locus, at which returns to scale are constant, at least locally;

(e) if the industry cost surface exhibits decreasing ray average cost and trans-ray convexity, it must operate as a monopoly;

(f) if output i involves decreasing average incremental cost, it must be produced by only a single firm;

(g) no obvious inefficiencies must persist—all cost-saving opportunities must be used and all opportunities for realization of economies of scope must be seized;

(h) Chamberlinian firms (i.e., sole producers of slightly differentiated products), must be myopic takers of the prices of rival firms, for otherwise, as shown in Chapter 11, they can be displaced in contestable markets unless they are protected by special arrangements such as patent and copyright laws;

(i) intertemporal price patterns must satisfy the rules of economic depreciation, with no contribution toward recoupment of fixed or sunk costs in periods of excess capacity.

This is only a sample of the attributes that can be used to test whether observed behavior is consistent with contestability. Obviously, the list can be extended considerably, but it illustrates the sorts of tests that are valid and that may prove to be tractable in practice in at least some cases.

16E. Tasks 4–6: Sustainability and Unsustainability

It is convenient to discuss the next three tasks on our list simultaneously.

There is relatively little to be said in general terms about cases in which sustainable solutions exist and can be calculated at least approximately from the cost and demand relationships. It should be noted, however, that in an economy with other types of imperfections, a sustainable configuration, even if it constitutes a Ramsey solution, need not produce a true optimum. It may, for example, involve detrimental externalities of every familiar sort including the generation of toxic wastes and other serious environmental effects. On this, the only remark we can offer that goes beyond the standard analysis is that the appropriate corrective Pigouvian taxes[2] must apply to

---

[2] See Pigou (1920).

new entrants in the same manner that they fall on incumbent firms. Then, and only then, is the configuration sustainable under the tax mechanism assured consistency with industry-wide productive efficiency. Aside from externalities, sustainable solutions may raise problems because of imperfections of capital or labor markets, and they may (or may not) contribute to inequality of the distribution of real income. Once again, here we have only one distinct comment that emerges from our analysis.

In the past, at least in regulated industries, distributive concerns have explained (in part) the deliberate adoption of sets of cross subsidies; for example, urban-area consumers have been forced to help to defray the costs of transportation or communication to rural communities, younger customers have, in effect, financed the supply of services to retired persons, and so forth.[3] To render such systems of cross subsidy financially viable for a regulated supplier, entry was deliberately impeded or ruled out altogether. But, as we have seen repeatedly in this book, in a contestable market any cross subsidies are incompatible with sustainability of prices. They always invite profitable entry into the subsidizing portion of the business, and so they cannot persist. Perhaps one of the more attractive features of sustainable configurations is their preclusion of hidden cross subsidies. In a contestable market, subsidies may still be desired, say, to help particularly deprived social groups. But they must come from the outside, presumably from government or private philanthropy, and be granted or levied through the operations of all suppliers, on an equal basis. They are therefore more likely to be provided openly and as a consequence of deliberate decision rather than, as seems to have happened under regulation, as the result of historical happenstance, covert pressures, and as a by-product of the pursuit of other objectives.

Some special welfare problems may also beset the sustainable configurations associated with monopolistic competition, the configurations described briefly in Chapter 11. It is plausible that these may involve an undesirably large or an undesirably small set of product varieties. But that is mere conjecture. We are not in a position to offer any substantial new insights on these issues. We hope that future work by others will shed more light on this subject.

Beyond the problems that may stem from particular sustainable solutions, there is the problem that arises when no sustainable configurations exist. As we have seen, this problem may arise in single-period analysis, as well as in an intertemporal setting. Two policies immediately suggest themselves for such circumstances. First, it may be considered appropriate to adopt programs that inhibit or prevent entry into the affected markets. Second, it may be desirable to revise received attitudes toward such entry-deterring measures as strategic pricing—the use of the threat of responsive

---

[3] On this subject, see Posner (1971).

prices. Rather than constituting an instrument of predation, such strategic behavior may be the only means open to the market mechanism to maintain any sort of order and any approximation to equilibrium.

Yet, one must proceed with great caution. As long as any doubt remains about the unavailability of sustainable solutions, one must hesitate before bowing to pressures for the encouragement of barriers to entry. It is understandable and natural for the incumbent firms in an industry who are fearful of enhanced competitive pressures to seek the erection or toleration of protective umbrellas against entry. But those who have the task of protecting the interests of society must resist such demands until the evidence for them is all but incontrovertible. We have seen again and again the sorts of benefits that unrestricted freedom of entry can bring. It is dangerous to risk those benefits on the basis of imperfect evidence indicating that, in a particular case, the market mechanism is likely to function badly.

16F. Task 7: Analysis of Intertemporal Pricing and Allocation

The issues raised by our analysis of intertemporal efficiency in resource allocation are most imperfectly understood at this point. The unsustainability problems seem to arise from increasing returns to scale in the production of durable capital, and from the opportunity cost of tying up resources by building spare capacity in anticipation of the demand for capacity in the future. Empirical investigation can certainly provide material on both these issues. It can determine the areas in which capital construction is characterized by significant economies of scale, and the time periods for which construction in anticipation of future demands is justifiable economically. One can also investigate whether, in practice, anticipatory construction patterns approximate those called for by efficiency considerations.

In addition, one can examine and compare the economically efficient depreciation schedules and those actually used under the influence of taxation, regulation, and accounting practices. Where the two diverge, some analysis of the resulting welfare losses and sustainability problems may prove useful.

While our analysis has suggested that intertemporal pricing and resource allocation may give rise to special efficiency problems, it is noteworthy that in other respects contestable markets eliminate some of the difficulties associated with intertemporal phenomena that are normally considered sources of efficiency problems. Learning by doing is such a phenomenon. Thus, suppose larger firms derive an intertemporal advantage from such a learning process, with the discounted value of the benefits increasing with the scale of the firm's activities throughout the relevant range. It is clear that this can be a source of natural monopoly in the same way that fixed costs contribute to subadditivity, even in a single-period analysis.

Nevertheless, just as contestability deprives a natural monopoly of its power to cause welfare losses in the single-period case, where sustainable intertemporal prices are possible, contestability can eliminate any welfare losses that might otherwise derive from any advantages to natural monopoly flowing from learning by doing. In contestable markets, because there can be no cross subsidies, all activities must pass Areeda-Turner tests of predatory behavior; that is, prices must all equal or exceed both marginal and average incremental costs. In an intertemporal setting, the marginal costs to which prices must be compared must include an allowance from any future savings that result from any increase in current output. These are the true marginal costs of that rise in output, and any prices that exceed that marginal cost cannot be considered predatory in Areeda-Turner terms. In that sense, the notion that learning by doing is a potential source of legitimate antitrust problems seems to have no defensible basis in a contestable market with sustainable intertemporal prices.

16G. Policy for Unsustainability

Cases in which no sustainable prices exist are troublesome, and, as we have seen, no easy policy recommendations suggest themselves for such circumstances. While this list is by no means exhaustive, there are five phenomena that seem to be prime causes of unsustainability. The first is the obligation, typically imposed upon a firm accorded the status of a public utility, to supply any quantities demanded, at the prevailing prices, which are not supplied by others. This obligation to serve, which is not imposed upon entrants, can place incumbents at a disadvantage and, as we have seen in the chapters on monopoly, can sometimes cause unsustainability. A second source of unsustainability is the U-shaped average cost curve in a natural oligopoly, which leads to unsustainability if demand is such as to rule out a market clearing configuration of integer multiples of vectors that minimize ray average cost. Third, unsustainability may occur where economies of scope are insufficiently strong and economies of scale for individual substitute products are sufficiently strong that specialized firms can take advantage of favorable demand conditions despite the cost advantages of the multiproduct supplier. Fourth, unsustainability can be ascribed to a combination of expanding demand for capacity utilization and economies of scale in the production of capacity, as in the exploration of intertemporal unsustainability problems in Chapter 14. Finally, unsustainability can result from a variety of public policies that impose special impediments upon incumbents.

As already indicated, the solution to such unsustainability problems is by no means an open-and-shut matter. The first of the sources of unsustainability, the incumbent's obligation to meet all unsatisfied demand, can perhaps be dealt with by imposing a similar requirement upon entrants. That is, if the entrant chooses to undercut the incumbent's price, one may, perhaps, consider requiring the entrant to satisfy all demands its prices elicit. It is not clear at this point whether this constitutes an excessive burden upon potential entrants. But in raising this issue we come to the heart of the problem, which is to devise rules (where the free market's operations are not completely satisfactory) that do not impose an unjustifiable burden upon entrants, and yet do not give them an unjustifiable advantage *vis-à-vis* the incumbents. We will return to this point at the end of this section.

We have nothing to add here to our protracted discussion in Chapter 11 of the integer problem, its unsustainability consequences and the phenomena that can act as countervailing influences. We can, however, say more about the third source of unsustainability, the presence of weak economies of scope in combination with strong economies of scale in the production of individual substitute products. In practice, this problem seems likely to be more serious and more widespread than the integer problem to which we have just alluded. The U.S. Postal Service seems to provide a striking example of this problem, which may plague it increasingly in the future, as electronic mail becomes increasingly inexpensive in comparison with first class mail. The fact that the two are close substitutes may mean that firms specializing in electronic communication can draw large numbers of customers away from the U.S. Postal Service. The latter may find itself without any prices sustainable against such competition, even if the two types of mail do offer sufficient economies of scope to make it efficient for one enterprise to supply them both.[4]

A partial remedy for such a problem may involve the adoption of more subtle and adaptable pricing rules, even though some of them may be vulnerable to charges of price discrimination under conventional legal standards. For the postal system, the availability of a special express-mail service at higher prices, of discounts for presorted mail, of charges for any delivery services going above and beyond what can reasonably be expected to constitute normal service, can all be helpful. Of course, economists have long recognized that price discrimination should not automatically be viewed as an undesirable practice. After all, Ramsey pricing is, strictly speaking, discriminatory, even though it is designed to maximize the total of consumers' and producers' surplus. Moreover, without discrimination we can never be sure that a product whose benefits exceed its costs will necessarily be viable financially.

---

[4] On the other hand, other models may more accurately describe this situation. For example, if electronic mail is naturally competitive, despite economies of scope with other mail services, the model of mixed firms in Section 11K is applicable. In addition, Owen and Willig (1980) argue that the rate structure of the U.S. Postal Service involves substantial cross subsidies.

In any event, where the multiproduct supplier runs into sustainability problems of the sort we have just been describing, there is certainly no automatic justification for intervention by the public sector to *protect* production of the goods by separate specialized firms. If there are economies of scope, even if they are fairly weak, such induced specialization can impose unnecessary inefficiencies and excessive costs upon the general public.

We may comment briefly upon intertemporal unsustainability problems and note that these must, in some ultimate sense, be ascribable to the presence of sunk costs. In this case, the sunk costs are composed of investment in durable plant. If there were no such sunk costs, then, when an entrant proposed to take over an expanded market using plant that is larger and cheaper (per unit of output) than the incumbent's, the latter could simply dispose of any inherited plant without loss and start up again on terms identical to those available to any entrant. Thus, here too, sunk costs can be singled out as the villain. We will presently return to consider what, if anything, can be done about them.

Finally, we come to a variety of artificial sources of unsustainability which result from special disadvantages imposed by public policy upon incumbents over entrants, either intentionally or unintentionally. Examples include regulatory rules on depreciation policy, which force prices for some periods to fall below the pertinent marginal costs; deliberate imposition of cross subsidies designed to benefit groups considered particularly meritorious (e.g., "lifeline rates" for the elderly or geographic rate averaging that benefits isolated communities which are particularly expensive to serve); environmental regulations, if they are more severe for incumbents; and rules against price discrimination, which prevent adoption of sustainable Ramsey prices. Any of these measures, as we have seen, can lead to unsustainability.

The way to deal with such artificial sources of unsustainability is fairly obvious. One simply avoids the measures that give rise to them. For one thing, incumbents and entrants should always be treated similarly in any cost-imposing rule. For example, Pigouvian taxes upon detrimental externalities should always apply to incumbents and entrants alike. Similarly, internal cross subsidies (such as those for the elderly) should be replaced either by direct government grants or by funds obtained from the imposition of a tax that falls on an equal basis on all firms in the industry.

16H. Appropriate Areas for Public Intervention

On balance, contestability analysis leans on the side of those who advocate extension of the domain of *laissez-faire*. Since the analysis shows that a variety of market forms far removed from perfect competition may, nevertheless, perform commendably, it follows that the set of circumstances recognized to require no public intervention is expanded considerably.

By and large, what emerges is a two-part test for the desirability of intervention somewhat analogous to that recently advocated by Joskow and Klevorick (1979):

Step 1: Examine the market's contestability. If it satisfies the criteria of contestability, interference with the market mechanism should be ruled out;

Step 2: Even if the contestability criteria are violated, proposals for intervention should be approved only on the basis of an evaluation of costs and benefits.

Perhaps the most noteworthy implication of contestability theory is that a wide difference in appearance between a particular market and the form of perfect competition need not deprive the invisible hand of its power to protect the public interest. With abandonment of the unrealistic standard of perfect competition as the model for market behavior, many old rules of thumb which have served as guides for the antitrust agencies must be permitted to fall by the wayside. Whatever the appearances, as already noted, we can no longer accept as *per se* indicators of poor market performance evidence such as concentration, price discrimination, conglomerate mergers, or vertical or horizontal integration. It is true that sophisticated analysts in the field of industrial organization have long taken exception to blanket proscription of such practices; the literature contains a profusion of isolated models showing, for each of these phenomena, circumstances under which they serve legitimate business purposes, contribute to the general welfare and, sometimes, even promote effective competition.[5]

However, we have moved one step further and provided contestability as a benchmark that is generally far more applicable. In contestable markets, all of the ostensibly questionable phenomena that have just been listed *can* be desirable and should, indeed, be presumed so, with the burden shifted to those who in any particular case maintain the contrary.[6]

But the analysis certainly does not rule out government intervention altogether. In fields where technological conditions and other unavoidable circumstances impose heavy sunk costs and other obstacles to exit and entry, markets will not be contestable and the market mechanism cannot always be trusted to produce benign results. In such circumstances, one may,

---

[5] See Scherer (1980), Chapters 19–21 and references cited therein.

[6] This position is, of course, not dissimilar to that held by George Stigler, Harold Demsetz, Richard Posner, and their associates. However, our work differs from theirs in two important respects. First, our analysis does not rely for its results on either the perfectly competitive model or on special institutions, such as bidding for franchises, and in this respect we believe that ours is more general. Second, we probably go further than they do in arguing that there are legitimate roles for antitrust activities and regulation in markets in which there are substantial departures from contestability and in which contestability cannot readily be stimulated.

for example, still not wish to preclude single-firm production in an industry that is clearly a natural monopoly. But this monopoly will be a legitimate candidate for regulation or antitrust scrutiny.

Where intervention continues to be called for, our analysis also has some new insights to offer. One example pertains to the multiproduct firm, some of whose outputs are sold on a monopoly basis while other are sold in competition with similar goods supplied by other firms. Though such firms have often been suspected of extending their monopoly powers into otherwise competitive markets, our analysis shows that where substantial economies of scope are present, it may be costly to require a divorce between the supply of monopolistic and competitive outputs. Moreover, as was shown by the analysis of Chapter 11, it is to be expected that a firm which operates in both a competitive and a monopoly market will have a larger output in the former than any of the competitive firms which sell no products on a monopoly basis. This will be true even if there is absolutely nothing that can be considered remotely predatory or otherwise objectionable in the behavior of the partial monopoly. Rather, its large share in the competitive market will result from the cost advantages it derives from the economies of scope among its monopoly and competitive products. If all the markets in which the partial monopoly operates are perfectly contestable, it will be forced to pass along to consumers any cost advantages it derives from these economies of scope, and, as a result, it will have no option but to sell more in the competitive market than the firms enjoying no such economies. The analysis of Chapter 11 shows that such a market structure may be the only one that satisfies the requirements of efficiency.

Yet problems may arise here if the monopoly market is characterized by substantial entry costs so that the partial monopoly is protected from the full discipline of potential competition. Excess profits, inefficiency, and cross subsidies then become very real possibilities, and regulation may therefore be a legitimate response. But, in such a case, regulation can introduce problems of its own. In an attempt to prevent cross subsidy by the partial monopoly which gives it an unacceptable advantage in the competitive market, legitimate freedom of pricing decisions may be undermined, causing misallocation of resources and offering protected enclaves in which inefficient entrants are safe from the normal consequences of their incompetence. Moreover, if regulation proceeds, as is usually done, via the imposition of a ceiling on return to capital, new difficulties may easily arise. In an inflationary world, the rate of return ceiling may be set below the real cost of capital because money illusion makes the nomial cost of capital seem unconscionably high. The unfortunate consequences are disincentives for the maintenance of an adquate capital stock by the regulated firm. On the other hand, if the rate of return ceiling is set too high—if it is above the cost of capital, then Averch-Johnson (1962) consequences may follow. The regulated firm will be offered an inducement to acquire excessive capital and, consequently, to expand excessively in its competitive as well as its monopolistic market. Thus, both the possibility of cross subsidy and that of an Averch-Johnson effect imply that the relatively large size of the competitive outputs of the partial monopoly may not be entirely justified either in terms of fairness of competition or the public interest. As has just been said, some excess of the output of the partial monopoly over that of its competitive rivals is expected and legitimate and will occur even in perfectly contestable markets. But the difference may well go beyond this. Unfortunately, it is not always easy to distinguish what portion of the difference in the outputs of the two types of firm is in fact legitimate and efficient, and what portion is not.

We must leave policymakers to wrestle with these difficult decisions. We have only been able to provide a framework appropriate for the analysis of industry performance that is applicable generally. Before, only the competitive model was available as a standard for industry performance. Yet, the logical foundations of this model preclude it from offering guidance for the analysis of industry structures such as natural monopoly and the partial monopoly that has just been discussed. In contrast, perfect contestability does provide a benchmark that is appropriate and applicable to virtually all industry structures. Where markets are contestable, the presumption should be that intervention is unwarranted. Where markets are not contestable, intervention should be guided by the performance perfect contestability can be expected to produce.

16I. Stimulation of Contestability

Since contestability is ultimately the hero of our piece, it is natural to ask what can be done to advance its cause. Part of the answer is fairly clear, but another portion is far from obvious. The desirability for this purpose of reducing artificial impediments to entry and exit, most notably those imposed by deliberate public policy, is evident enough, though even here there are some complications, as we will see. But there are also "natural" barriers, imposed by technological conditions and the state of knowledge, which are not readily eliminated. The need for sunk investments immediately suggests itself as an example, and one may well be skeptical about the policymaker's ability to remove or even reduce such impediments to entry. As Dr. Salop so felicitously put it [in his discussion of Bailey (1981)], "we cannot legislate away sunk costs." Yet, as we will argue, even here public policy is far from powerless.

On any rational basis, it seems difficult to understand the extent to which regulatory policy in the United States devoted itself in the past to the inhibition of entry. Elizabeth E. Bailey, Member of the Civil Aeronautic Board, writes in her paper on policies to stimulate contestability (1981, p. 179):

> The traditional licensing policies of the Civil Aeronautics Board (CAB) and the Interstate Commerce Commission (ICC)... restricted entry whether by new suppliers into the industry or by established suppliers into routes already served by others or not served by anyone. Entry was restricted both in dense markets which were structurally competitive and in thin markets where there might be expected to be only a single supplier. Authority was only conferred if it was likely to be used. There was no value placed on the benefit of having a pool of potential competitors who could respond to a potential profit opportunity by entering the market.
>
> In contrast, the current policies of both the ICC and the CAB are to confer substantial new authority, whether actually used or not, thereby enhancing the degree of both actual and potential competition. The new policies are based on the theory that both trucking and aviation markets are, in the absence of regulatory intervention, naturally contestable.

Impediments to exit are equally traditional. They have ranged from direct acts of intervention by the ICC to prevent abandonment of highly un-profitable railroad routes, to more subtle measures such as prevention of mergers which would permit one of the merging firms to unsink its invest-ment, that is, to divest itself of its assets with its finances presumably more or less intact. Here too, as we have seen, contestability theory sheds a different light upon such prohibitions. This is again brought out by Dr. Bailey (1981, p. 181):

> Another step in altering antitrust policy to reflect contestability theory was taken by my colleagues and myself at the CAB when we refused to use traditional market share measures to preclude merges . . . . The Depart-ment of Justice recommended disapproval [of a proposed merger] based in large part on the market share data ... [which indicated that the expected share of the merged firm] was greater than comparable figures in mergers declared unlawful by the Supreme Court. The CAB countered by arguing that concentration ratios were not instructive in this case since, with the passage of the Airline Deregulation Act of 1978 (Pub.L. 95–504), there was now relative ease of entry, even for small carriers into such markets.[7]

---

[7] Note, however, that there are exceptions to the trends toward policies consistent with the implications of contestability theory. A striking example is the proposal by Representative William Ford, a Democrat from Michigan, that any plant whose yearly sales exceed $250,000 must, if it closes, provide 52 weeks of severance pay to its employees, pay the community 85 percent of a year's taxes, and pay the federal government 3 years taxes if it goes abroad! Up to 2 years advance notice is required, to give employees a chance to buy the plant or to seek govern-ment assistance to keep it open. The bill is reported to have the backing of more than 60 Congressmen as well as that of Ralph Nader. (*New York Times*, August 26, 1980, p. A19).

It should be noted, incidentally, that removal of impediments to international trade are also important for the promotion of contestability. We have seen in recent years how the entry of foreign competitors can enhance the contestability of an oligopoly market with a small number of incumbents. The U.S. automobile makers have found to their sorrow that German and Japanese firms can enter into their markets, and their entry has surely served as an effective disciplinary force upon the behavior of the incumbents.

There is an important caveat that should accompany a statement indicating the desirability of enhanced ease of entry. The gains from such a change can be eliminated, and even reversed into losses, unless the elimination of entry barriers is accompanied by suitable reductions in restrictions upon price setting. Freedom of pricing and freedom of entry are beneficial only if they are permitted to arrive together.

There has long been awareness of the dangers of complete freedom of pricing in markets in which incumbents are effectively shielded from any threat of entry. Such freedom of pricing is an open invitation to exploitation of consumers by sellers who can then extract monopoly profits, constrict output, and degrade output quality without fear of inviting new competition.

But freedom of entry accompanied by the absence of freedom of pricing can be equally detrimental. Where, as a result of tradition, legal, or regulatory influences, relative prices of incumbents for their different services are not closely tailored to demand conditions and to relative costs, entry opportunities will naturally be most attractive in those lines of activity whose incremental revenues are highest relative to incremental costs. It is entirely legitimate and desirable for entrants to seek out and exploit such "cream skimming" opportunities, even though it is invariably painful to the incumbent. But from the viewpoint of society, one of the main things to be gained from such entry, actual and threatened, is readjustment of the prices charged by the incumbents who are now forced to align them more closely with costs and with conditions of demand.

Naturally, just as the incumbent can be expected to denounce "cream-skimming entry," entrants can be relied upon to denounce any reduction in price by the incumbent as "predatory." Yet if one does not permit reductions of prices that are high relative to costs, and the elimination of the associated cream-skimming opportunities, a protected haven for inefficient entrants is thereby provided. Such entrants are permitted to survive not through the quality of their products or the efficiency of their operations, but through the protection conferred upon them by the regulators. And this all occurs at the expense of the consuming public, which is deprived of the price reductions which are presumably one of the prime benefits of freedom of entry.

Our proposal for quasipermanence of price reductions, described in the appendix to Chapter 12, is designed to help overcome the dangers of both overprotective prevention of price reductions and predatory responses to entry. The proposals of Ordover and Willig (1981) for modifications in antitrust policy toward predation are also designed to protect entrants who would be viable if they had entered a contestable market. It should be noted, however, that in perfectly contestable markets in which entry is fully reversible, the danger of predatory actions is negligible. First, as we have seen repeatedly, perfect contestability rules out pricing below marginal cost or any other form of cross subsidy, and so goes well beyond the requirements of the Areeda-Turner criterion of absence of predation. Besides, if entry is really free and exit is really costless, predation can have no payoff. By most definitions,[8] a predatory act is one that is not in the firm's normal business interests, and so it must incur at least an opportunity cost for the enterprise that undertakes it. But where the entrant can run away costlessly—to return when the circumstances become propitious once again, there is little to be gained by using costly predatory measures to drive him out.[9]

16J. On Policy toward Sunk Costs

As has already been noted, sunk costs, which are a prime impediment to contestability, cannot be wished away. Yet there are things that can be done about them. Here we turn once more to Dr. Bailey's (1981, pp. 179, 182) remarks on the subject:

> Rules must be devised to handle sunk cost problems. These may include encouraging technical changes that replace technologies involving large sunk costs with technologies that offer more opportunity for mobility or shared use. They may also include a careful look by policymakers at access rules to sunk facilities. For example, access problems can arise when airport authorities attempt to meet slot or noise constraints by banning new entry while allowing incumbent carriers to expand their operations at will. They can also arise under long-term lease arrangements which allocate airport space to particular carriers, and give these carriers the power to determine when, if, to whom, and at what price to sublease space to their competitors . . . . (p. 182)
>
> The single most important element in the design of public policy for monopoly should be the design of arrangements which render benign the exercise of power associated with operating sunk facilities.

---

[8] See in particular, Ordover and Willig (1981).

[9] This point has been made by Posner (1976, p. 87): "If the resources he [the entrant] uses to compete with the predator are perfectly mobile, the predator cannot hurt him; at the first sign of below cost pricing he will withdraw from the market without suffering any loss." Curiously, in markets where there are very great entry barriers, the likelihood of predatory practices is also small. There, they are simply unneeded, since the unaided entry barriers will offer the incumbent the protection he desires. Also, where virtually all capital costs are sunk, exit is unlikely because it would save the firm little (see Ordover and Willig, 1981). Thus, predatory behavior is most likely to be a pertinent issue where entry costs are neither very low nor very high.

> One way to avoid the exercise of monopoly power is to have the sunk costs borne by a government or municipality, as they are in U.S. highway systems or airports, or by mandating that sunk costs be shared by a consortium, as is to some extent true of international broadcasting satellites, rather than to have the sunk costs incurred by the firm that is supplying the services. Virtually any method will do as long as there are contractual or other arrangements that are nondiscriminatory and permit easy transfer or lease or shared use of these cost commitments. (p. 179)

One of the main points that emerges from the discussion is that sunk costs are often not pervasive in an industry, but rather are centered in a particular sector of its operations, such as airports in air transportation. By isolating the activities with which the heavy sunk costs are associated, their damaging consequences can be quarantined. By placing relations with the remainder of the industry at arm's length, to the extent that is permitted by economies of scope, it may be possible to leave the operations of the bulk of the industry safely to the free market, drawing a regulatory net over only the segment of the activities of the industry that are inextricably associated with heavy sunk costs.

Aside from this, one can think of other means to reduce sunk costs. Tax advantages for rapid depreciation, tax reductions for retooling of plant, and tax incentives for reuse of old plant in new activities can all be helpful. No doubt, other devices will occur to the reader. This is not the appropriate place for an attempt at a compendium. Our objective, rather, has been only to suggest directions which are promising for the design of policy, given the issues that the analysis has raised.

16K. Afterword

We wish to conclude by reemphasizing that this book largely represents work in progress, rather than a report on a more or less completed body of doctrine. There is a great deal of room for contributions by others, and, happily, we understand that many others have indeed started to carry on the work whose beginnings this book describes. Its positive models, its normative analysis, and its policy discussions leave many obvious questions unanswered, though we believe it suggests ways in which the answers can fruitfully be pursued. We hope, then, that many others will join us in this work.

17

Contestability

Developments since the
First Edition

The ideas encompassed in the theory of contestable markets have entered both the academic literature and the discussions of practitioners with surprising speed. At hearings before regulatory agencies in antitrust cases, terms such as "economies of scope," "sustainable prices," and "contestable markets" are interjected casually as though long usage has made their connotations familiar to everyone. This is in part attributable to the steady stream of papers, published and about to be published by authors offering careful theoretical extensions and critiques, empirical and interpretative applications, as well as heated reactions to contestability theory.

The central focus of this chapter is a summary of substantive developments relating to contestability since this book was originally published. We will discuss four subjects: theoretical advances provided by others, econometric studies using the multiproduct cost concepts accompanying contestability analysis, market simulation studies testing various attributes of contestability, and policy developments, covering both policy studies in the economic literature and some remarkable decisions in the regulatory and antitrust arenas, that rely explicitly on the theory of contestability.

However, before reaching these central areas, in order to deal with any remaining misunderstandings we shall begin with a brief restatement of some of the things our work does not and was not intended to imply and the things we instead hope it accomplishes. Specifically, we will deny emphatically that it offers carte blanche to mindless deregulation and dismantling of antitrust safeguards. On the contrary, as far as policy is concerned, contestability theory provides guidance in ascertaining where intervention is warranted socially, and it provides a more widely applicable benchmark to guide regulatory agencies and the courts in those arenas where intervention is called for by considerations of economic welfare.

17A. Contestability and "Libertarian" Ideology

Contestability theory does not, and was not intended to, lend support to those who believe (or almost seem to believe) that the unrestrained market automatically solves all economic problems and that virtually all regulation and antitrust activity constitutes a pointless and costly source of economic inefficiency. In a market that approximates perfect contestability, it is true, we believe matters can be left to take care of themselves. Small numbers of large firms, vertical and even horizontal mergers, and other arrangements which have traditionally been objects of suspicion of monopolistic taint and worse are rendered harmless and perhaps even beneficent by the presence of contestability. But that observation is no whitewash and establishes no presumption, one way or the other, about the desirability of public sector intervention in any particular market of reality. For before anyone can legitimately use the analysis to infer that virtue reigns in some economic sector and that interference is therefore unwarranted, that person must first provide evidence that the arena in question is, in fact, highly contestable. The economy of reality is composed of sectors which vary widely in the degree to which they approximate the attributes of contestability. Thus, the conclusion that *perfectly* contestable markets require no intervention claims little more than the possibility (which remains to be proven, case by case) that *some* markets in reality may automatically perform in a very acceptable manner despite the small number of firms that inhabit them.

Thus, it is simply incorrect to associate our writings on contestability with an all-pervasive laissez-faire position on the role of regulation and antitrust (for such a characterization of our position see, e.g., Shepherd (1984)). We disagree vehemently with such a view of the world. On the other hand, we reject with equal conviction the position of those who hold that mere large size of a firm means that it *must* serve the economy badly, that high concentration ratios are sufficient to justify governmental restrictions upon the structure or conduct of an industry, or that a horizontal merger merits automatic condemnation if it increases concentration ratios substantially. For it is true that a high degree of contestability, where and if it does happen to hold sway, should remove the undesirable consequences which might otherwise stem from these phenomena. Thus, contestability theory supports neither extreme interventionists nor extreme noninterventionists. We believe that antitrust and regulation have valuable roles to play, and that contestability theory can help to identify and sharpen those roles and thereby benefit the public.

17B. Objectives of Contestability Research

Contestability theory offers an analytic framework within which the fundamental features of demands and production technology determine the shape of industry structure and many of the characteristics of industry prices. The theory accomplishes this via a process of simplification; by stripping away through its assumptions all barriers to entry and exit, and the strategic behavior that goes along with them both in theory and in reality.

The model of perfect competition has also served an analogous simplifying role, and in so doing has provided the foundation for the most elegant and well-articulated portions of economic theory. However, because the model of perfect competition prespecifies industry structure by its very construction, it cannot serve as a useful benchmark for the study of the determinants of industrial structure. Perfect competition is a special case of perfect contestability, and perfect contestability applies with equal force to circumstances where perfect competition is impossible because economies of scale are present. Because of this fact, and because contestability theory encompasses an endogenous determination mechanism from which any industry structure may emerge (depending on circumstances), we feel that it is an extension of the competitive model appropriate for use in the theory of industrial organization.

In addition to this broad purpose in theoretical analysis, we feel that research on contestable markets has two objectives related to policy. The first is the establishment of an improved set of guidelines for appropriate government intervention in the structure and conduct of firms and industries, that is, of the rules to be followed by the regulators and antitrust authorities in those cases in which their intervention is called for. The second objective for policy analysis is the determination of criteria distinguishing between those cases in which intervention by the public sector is warranted and those in which it is not.

With regard to the first policy objective, it is widely understood that the basic guideline for intervention should be replication of the consequences of effective market forces in those cases in which competition, actual or potential, is insufficient to do the job. That is, intervention should aim to induce or compel an industry in which monopoly power is present to perform as it would if effective competitive pressures were available. But, for such cases perfect competition is often unsuitable as a standard, particularly in circumstances common to regulatory and antitrust issues where scale economies and related attributes dictate the presence of only a small number of firms, at least some of them relatively large.

Here, as we will see from recent developments in regulatory practice that will be reviewed presently, contestability theory was, indeed, able to suggest answers where other bodies of theory did not. Moreover, it was able to eliminate curious gaps in the standard theory of policy. Perhaps the most noteworthy of these was the lack of a defensible criterion for regulatory ceilings on prices, though one would have thought that to be the first order of business for social control of firms with monopoly power. Yet, to our knowledge, neither regulators nor courts previously had available to them from the theoretical literature any defensible test to determine just when a particular price set by one of the multiproduct firms of reality was excessive. We will also see later how contestability theory determines how high is enough and will observe that regulatory agencies have begun to adopt the criterion that emerges, with explicit acknowledgment of its source.

In saying this, we do not mean to claim that the theory sheds light on the welfare economics of markets that are highly uncontestable. The theory does not even try to contribute to the analysis of behavior in such cases or to the evaluation of the social cost of any resulting departures from optimality. All the analysis pretends to say about such cases is that if they entail behavior which is deemed socially undesirable and if government undertakes to change this behavior, then it is appropriate to require the individuals to act as they would have if the threat of potential competition had constrained them effectively. If, for example, firms charge prices and earn profits that are deemed excessive, it is appropriate to offer consumers the protection that the threat of barrier-free entry would have provided them.

Finally, we feel that contestability theory has helped to show that in *some* cases where regulatory or antitrust intervention was previously common, it was in fact undesirable. Contestability theory, we think, does help to clarify which arenas are the proper candidates for deregulation and cessation of other forms of intervention. But the theory neither calls for anything like universal deregulation nor for its automatic extension without careful study of the pertinent facts, case by case.

17C. Theoretical Developments

There has been an outpouring of theoretical work on contestable markets and allied subjects, as there has been recently throughout the field of industrial organization (as this very book indicates). Our brief overview of the recent developments cannot do justice to the full range of valuable contributions and is, unfortunately, bound to miss important items. This review is organized into four topics: (i) conditions for sustainability of prices set by a monopoly incumbent; (ii) theoretical applications of contestability; (iii) the characteristics of production cost that determine market structure in contestable markets; and (iv) the relationship between strategic firm behavior and contestability.

In our book, we identified one set of conditions sufficient to assure the existence of sustainable prices for a multi-output natural monopolist, and those conditions also were shown to guarantee that the Ramsey optimal prices are sustainable. Among those conditions are the stipulations that the cost function exhibit overall economies of scale and the property of trans-ray convexity. We showed that the sustainability of Ramsey prices could tolerate some deviation from these properties, such as product-specific fixed costs of bounded magnitude. Faulhaber and Levinson (1981) developed the idea of anonymously equitable prices which offered no cross-subsidies to any possible groups of consumers, and connected the properties of such prices to sustainability. Sharkey (1981, 1982) and Spulber (1984) presented new sets of conditions sufficient for the sustainability of monopoly prices, and showed that the Ramsey prices need not be among them. Ten Raa (1983, 1984) sharpened the connections among subsidy-free, market-clearing, and sustainable prices. Mirman, Tauman, and Zang (1985) further clarified these connections, established novel sets of sufficient conditions for the existence of sustainable prices, and identified conditions under which the Aumann-Shaply prices would be sustainable.

Brown and Heal provided the valuable result that where Ramsey prices are unsustainable, the government can solve the consequent policy conflict between open entry and optimal prices via a system of excise taxes and subsidies to incumbent and any entrants alike. Working in a general equilibrium context with independent demands for the outputs of the natural monopolist, they proved the existence of such a tax system, with zero net effect on government revenues, that permits Ramsey optimal consumers' prices and sustainable producers' prices.

Panzar and Postlewaite (1985) demonstrated the sustainability of Ramsey optimal non-linear prices where costs are linear, and provided counterexamples for other circumstances. In contrast to the usual results for contestable markets, Perry (1982) showed that a natural monopolist can at once earn positive profits and deter entry by committing itself to the sale of some units of homogeneous output at prices lower than those that apply to marginal purchases.

Our second category of new theoretical work comprises applications of contestability to various issues. Quirmbach (1982) studied vertical integration by an upstream monopolist into a competitive downstream industry. While the welfare effects are, in general, ambiguous, Quirmbach showed that profitable integration would raise social welfare if the upstream market were perfectly contestable.

In their pathbreaking book, Helpman and Krugman (1985) provide a sketch of a new theory of international trade in which industries in the trading nations are perfectly contestable rather than perfectly competitive, as in conventional trade theory. This construction permits them to study the effects of increasing returns to scale in the production processes internal to firms, without the need to specify particular forms of oligopolistic and strategic behavior. Among their conclusions are the result that the sustainable factor-price-equalization equilibrium is a highly useful construct for trade analysis; that even in the presence of economies of scale comparative advantage shapes trade patterns with unequal factor rewards; and that there is a strong presumption that there will be gains from trade when national and international markets are contestable.

Recent work on properties of production costs has been stimulated by the significance they have for the character of industry structure in contestable markets. Weitzman (1983) showed that constant returns to scale must hold when demand can be met by the instantaneous accumulation of inventories created by production that is turned on and off repeatedly without additional costs. He concluded that the hit and run entry opportunities said to be a necessary part of contestability must then be inconsistent with increasing returns to scale. In reply, we pointed out that quantities of services cannot be accumulated in inventories, no matter how short their duration, and that contestability does not require the ability to start and stop production costlessly—only the ability to sell without vulnerability to incumbents' responses for a time long enough to render all production costs economically reversible.

Teece (1982) has contributed illuminating work on the sources of economies of scope, both in theory and in specific applications. Gorman (1985) has recently provided theoretical results on economies of scope where there are complementarities or anticomplementarities in marginal costs and also product specific fixed costs that may be subadditive or superadditive. Bittlingmayer (1985) has provided a theoretical analysis that builds multiservice cost functions for airlines upon the foundation of the costs of operating aircraft. He finds economies of scope among routes arising from the hub and spoke architecture of the cost efficient configuration of routes. He goes on to study the characteristics of rate structures that are sustainable and that are Ramsey optimal.

The final category of theoretical work, the relationships between contestability and strategic firm behavior, is most difficult to review succinctly because it is intertwined with the bulk of new research in industrial organization concerning strategy and oligopoly solution concepts. Perfect contestability is a theoretical benchmark that is by its very construction immune from considerations of strategic behavior by dint of its assumption of the absence of economically sunk costs and irreversible commitments necessary for entry. Nevertheless, as pointed out by our distinguished reviewers, Brock (1983) and Spence (1983), and by our constructive critic Schwartz (1985), it is worthwhile to investigate what forms of game and what models of dynamic strategic oligopoly yield outcomes consistent or inconsistent with contestability. Here, Knieps and Vogelsang (1982) and Brock and Scheinkman (1983) have contributed models of quantity sustainability in which outcomes are quite different from those in contestable markets because potential entrants take incumbents' quantities rather than prices as given. In view of the enlightening work of Kreps and Scheinkman (1983), it is now clear that quantity-taking behavior is a reflection of precommitments to quantity-determining capacity by firms that later simultaneously announce prices.

In contrast, contestable outcomes are yielded by Nash equilibrium in a price setting game without precommitments (Mirman, Tauman, and Zang, 1983), and in a game in which supply schedules comprise the strategy sets (Grossman, 1981). We showed that contestable outcomes are a necessary feature of equilibrium in the limit as sunk costs approach zero, using a model in which incumbents' prices are sticky for at least a short period of time (Baumol, Panzar, and Willig, 1983). Maskin and Tirole (1982) analyze a model of dynamic oligopoly involving firms with fixed costs and show that the solution approaches the contestable outcome as strategic advantages fall to zero. In contrast to these results, Dasgupta and Stiglitz (1985) provide models in which incumbents' strategic advantages over a single potential entrant, no matter how small those advantages may be, permit the incumbents to earn economic profits without danger of entry. And Applebaum and Lim (1985) have contributed a model in which the degree of contestability is endogenous, determined by the incumbent's incentives to commit to quantity-determining capacity in view of demand uncertainty and the pattern of costs over time.

In short, there is a flood of exciting research activity in the area of strategic firm behavior, and one of the ways in which the various models are differentiated is the relationship between the games' solutions and the outcomes that would emerge in perfectly contestable markets. At present, that relationship seems to be highly sensitive to the fine grained structure of the models' game forms, so much so that one suspects that empirical reality embodies relationships more robust and stable than does oligopoly theory in its current tumultuous state.

17D. Experimental Market Studies of Contestability

The fruitful and growing body of research using experimental simulation of market behavior has produced a number of papers investigating contestability issues. This work uses human subjects (usually students), real money payoffs and experimental rules corresponding to preselected cost and other pertinent functions that permit direct calculation of the results predicted by theory and their comparison with those that are observed from the behavior of the experimental subjects.

Taking off from earlier studies on monopoly behavior, Coursey, Isaac, and Smith (1984) undertook the first of the experimental studies of contestability. They sought to determine whether complete freedom of entry and exit in a market which could most cheaply be served by one firm but where at least two (potential) participants were available would yield prices closer to the competitive than the monopoly level (they call this the "weak contestable markets hypothesis") or to prices actually equal to the competitive level (the "strong contestable market hypothesis"). Participants were given money which they were required to spend in accord with a strictly decreasing and known marginal cost function, the amount spent depending on how much of the good they decided to "produce." The consumers' demand function was also given but was not known by the sellers (except by experience derived from repeated play) who kept for themselves any excess of sales revenues over costs. A participant made a sale if that person's price offer was not undercut by a rival. Both sellers were required to post their price and quantity offers privately, *and* at the same time. Any seller was permitted (but not required) to sell less than the amount the market demanded at the posted price. The authors concluded that "the ... experiments strongly support the contestable markets hypothesis, namely, that to observe approximately competitive behavior by a single producing firm with substantially decreasing costs, it is sufficient that (a) sunk costs are zero and (b) there are two contesting firms acting noncooperatively in the sense that there is no explicit nonprice communication between them that leads to excessive restriction of supply" (Coursey, Isaac, Luke, and Smith, 1984, p. 69).

Harrison (1987) reaches a rather weaker conclusion from these results and those that emerged from a replication of the experiments (Harrison and McKee, 1985). He uses as a test criterion an "index of monopoly effectiveness" $M = (\pi - \pi c)/(\pi m - \pi c)$, where π, πc, and πm are, respectively, the total profit that emerges from an experiment, the competitive, and the monopoly profit levels. He notes that four of the six experiments did attain the $M = 0$ value expected of perfect contestability after 10 periods (i.e., 10 replications of the experiment by a given set of players). However, over all replications M averaged 29 percent in the original set of experiments and 19 percent in the second set, which while not very far from a competitive result, " ... is significantly positive." Thus, Harrison concludes, clearly, we can reject this strong form of the CMH (contestable markets hypothesis)" (p. 196).

The Harrison study also concludes from experiments using essentially the same procedure that " ... the discipline of contestability does serve to mitigate the monopoly [significantly] when we compare it to an unregulated monopoly ... [but] enlightened monopoly regulation vastly outperforms contestability in terms of reducing monopoly effectiveness" (Harrison, pp. 196–197). (Here the regulator's role is simulated by the offer of a subsidy payment equal to the consumers' surplus from any output exceeding the monopoly level, as calculated from the demand curve.)

In another experiment, Coursey, Isaac, Luke, and Smith (1984) evaluate the effects of sunk costs imposed equally on "incumbents" and "entrants."

The sunk cost takes the form of a $2 entry permit, valid for five periods. The incumbent is one of two sellers chosen randomly and permitted to operate as a monopolist for five periods. Thereafter, the other seller is permitted to enter and post prices that compete with the incumbent, who is required to purchase an entry permit in period 5 in order to continue operation in periods 6–10. The results, according to Harrison's reanalysis, exhibit "... a complete absence ... of any attempt by the incumbent to limit price in periods 1 through 5 ... despite the threat of future entry" (p. 222, n. 6). He also notes that the sunk costs constitute a differential incremental risk to the entrant (and hence an entry barrier) only in periods 6–10. Accordingly, he considers it noteworthy that "For all eighteen periods there is a slight and statistically insignificant effect of sunk costs reducing efficiency. However, there is a significant decline in efficiency in periods 6–10 due to sunk costs" (pp. 196).

Finally, Harrison tested the effects of imposition of the Bertrand–Nash (or, perhaps more accurately, the Stackelberg leader–follower) assumption that entrants take the incumbents' prices to be given—a premise widely associated with contestability. This was done by letting all participants post prices simultaneously in the first period, designating the winner as the next period's incumbent, who is then *required to post a price before anyone else does, and to announce it publicly*, with the following periods handled in the same way. Both two- and three-participant experiments were run. The author concludes, " ... satisfaction of the Bertrand–Nash assumption is associated with a dramatic decline in monopoly trading effectiveness. ... Moreover, we find support ... for a *strong form* of the (contestable market hypothesis) that claims that observed prices will converge to and attain competitive predictions" (p. 207).

These results would appear to offer strong support for the conclusions of contestability theory but, unfortunately, they still leave a major gap with which future research will have to deal. As Marius Schwartz, in his latest paper criticizing contestability theory, rightly observes, "Harrison's experiments basically do constitute a fair test of behavior under perfect contestability: the incumbent faces a price response lag and the entrant can hit and run costlessly. The real question, however, is not whether competitive results will emerge under perfectly contestable conditions but how often such conditions are likely to exist and what happens when they do not" (Schwartz, 1985, p. 20, footnote omitted). Put another way, the critical issue that remains is the determination of the circumstances under which the Bertrand–Nash assumption holds or at least is assumed by the participants to hold approximately. It is noteworthy to have it confirmed that *imposition* of Bertrand–Nash behavior leads to the results predicted by the theory. However, that result still leaves work for future statistical or experimental investigation, which Harrison and his colleagues are pursuing.

17E. Econometric Studies: Multiproduct Cost Functions and Industry Structure

In the past three years there has been an outpouring of empirical studies using the cost concepts derived from the contestable markets literature (see the survey of earlier work by Bailey and Friedlaender (1982)). They provide estimates of the shapes of ray average cost curves, of economies of scope, of the M-locus (the locus of points of minimum ray average costs on each ray in product-quantity space). Many such studies use the functional forms for the multiproduct cost function that were recommended in our book. The empirical studies encompass a wide range of industries, including banking, rail, truck, and water transportation, insurance, hospitals and a number of others. Here we can only recapitulate briefly a sample of these studies which, predictably, show that industries differ markedly in their scale and scope attributes and in their degree of approximation to contestability.

Cost Function Estimates

Perhaps the most consistent message of the "new" multiproduct cost studies is that the results of previous work using Cobb–Douglas or CES functional forms cannot be trusted. The more general translog form used by most of the multiproduct literature contains, as special cases, the Cobb–Douglas and CES forms. The appropriateness of these forms can be tested within the translog framework, and in all such tests reported, those forms are rejected in favor of the general translog form.[1]

Banking and Related Activities   Banking institutions seem a natural subject for multiproduct analysis. Gilligan, Smirlock, and Marshall (1984) and Gilligan and Smirlock (1984) divide output into deposits and loans, using data on many banks provided by the Federal Reserve System. They find that there are large economies of scope "across the balance sheet." They suggest banking theory explicitly consider that fact, and that bank regulation intended to affect one output also be analyzed in multiproduct terms. They encounter decreasing ray average costs (RAC) for small banks, concluding that some small single-bank markets may indeed be subadditive.

---

[1] Note that the translog form is itself unreasonable for cases involving zero output of one product since then the translog form implies that the total cost of the remaining products must be zero, no matter what their output levels. One method used to avoid this problem is recourse to a Box–Cox transformation of the outputs. (See Caves, Christensen, and Tretheway [1980] for a discussion of the translog form). As an alternative, Friedlaender, Winston, and Kung Wang (1983) use a quadratic approximation to a hedonic function to deal with the problem. The quadratic form also offers a simple measure of economies of scale and scope as well as marginal cost.

Overall scale economies are estimated to dissipate as bank size increases, and banks with deposits of $100 million or more manifest increasing RAC. Product specific diseconomies of scale also appear and the authors conclude that policy that precludes entry and encourages mergers cannot be justified by any cost savings.

Murray and White (1983) examine the investment side of British Columbian credit unions. They find significant economies of scope between mortgages and consumer lending and no economies of scope between all lending and investments. Unlike the U.S. studies, returns to scale are observed in most credit unions in the study. The authors conclude that regulation to "inhibit growth and diversification" hampers market efficiency.

Using the same output categories (mortgages, other lending, and investments) for California Savings and Loan data, Mester (1985) finds both overall and product specific constant returns to scale, suggesting that large assets offer no cost advantage. She also finds no advantage to a network of branches, given asset size, and absence of economies of scope between outputs. She rejects cost convexity, although finding pairwise transray convexity, and ray subadditivity along some rays.

Kellner and Mathewson (1983) provide a multiproduct analysis of the life insurance industry. They use a four-output industry model. Scale effects are rejected for the average firm while economies of scope appear in some product lines. In contrast to previous work, they find no evidence of "natural monopoly tendencies" in the cost structure.

Transportation Bailey and Friedlaender (1982) discuss work done on the trucking and railroad industries. Network configurations are reported to offer large economies of scope in both industries. These bias upward the measures of scale effects in single product analyses. In the case of trucking, apparent scale economies disappear for large firms when multiproduct analysis is used. Thus, single product analysis of scale effects may improperly reflect scope effects.

Stimulated by the examination of the regulation of bus transport by the Interstate Commerce Commission (ICC), Tauchen, Fravel, and Gilbert (1983) examine the cost structure, taking regular travel, charter, local service and school busing as the outputs of the industry. Using data on 950 privately owned intercity bus firms in the United States, they find that economies of scale are exhausted at low output levels while economies of scope occur throughout. They point out that price and entry regulation is usually meant to deal with economies of scale which are minor here. Economies of scope can be enhanced by fostering of service cooperation among firms in terms of schedules, interlining of tickets and baggage and use of common terminals.

Friedlaender, Winston, and Kung Wang (1983) examine the American auto industry's cost surface for the individual firm. Output is divided into small cars, large cars, and trucks. They find that the size of the firm has little to do with the returns to scope or to scale and that the cost function is not convex; some regions exhibit economies of scale and scope and others exhibit diseconomies. This demonstration of the variability of the industry's cost structures indicates the danger of using the neighborhood of a particular output vector as a basis for the empirical analysis of costs. Generalizations based on studies of aggregate output data are, thus, unreliable.

Caves, Christensen, *et al.* used the multiproduct framework to analyze cost attributes other than economies of scope and scale. In Caves, Christensen, and Herriges (1984) they construct models of consumer reactions to peak load electricity rate structures. They also examine productivity growth in U.S. railroads employing multiproduct estimates (Caves, Christensen, and Swanson, 1981). And they analyze the effects of deregulation on airline productivity—distinguishing between local service and trunk lines (Caves, Christensen, and Tretheway, 1983).

Telecommunications   Much work has been done on the telecommunications industry. Its output is readily divided at least into local and long distance service, and so constitutes a classic multiproduct case. Evans and Heckman (1984) perform an empirical test of the subadditivity of the industry's costs. To avoid the need for global cost information Evans and Heckman derive a local test for subadditivity by restricting the measurement of subadditivity to a subregion of the observed data. For each year of data they calculate the cost savings offered by a single firm relative to the minimum two-firm cost. Finding all such savings to be positive, they conclude that cost was subadditive during the period 1958–1977.

On the other side, Charnes, Cooper, and Sueyoshi (1985) use a goal programing/constrained regression method (derived in operations research studies) and a multiproduct framework, data set, and functional form identical with those of Evans and Heckman to obtain opposite results. Basically, they drop the standard economic assumption that the firm always operates on the efficient frontier and seek to derive an envelope from observed costs. While Evans and Heckman found the savings were positive, Charnes, Cooper, and Sueyoshi found the maximal savings to be negative. They question the validity of standard econometric procedures, but accept the multiproduct nature of the issue and the test of subadditivity. They stress the importance of use of methods derived from several disciplines, particularly when large issues of policy depend on the outcome.

Other Industries   Cowing and Holtmann (1983) divide hospital care into five care types to analyze the efficiency of general-care hospitals in New York state. They reject the use of a single aggregated measure of output and find the economies of scope "may be at least as important as scale effects in designing more efficient hospital service." They urge policy makers to take scope effects into account when deciding on mergers and encourage future health care research to employ multioutput specifications.

Scott (1982) examines the effects of concentration on manufacturing firms. He investigates the effects of multiproduct competition, or "contact," between firms upon excess profits and resource mobility. He constructs a measure of multimarket contact and then shows that an increase in contact in markets with high sales concentration permits larger profits. However, when sales concentration is low, multiproduct contact can drive profits toward costs.

Vertical integration theory suggests that the presence of economies of scope in an industry need have no categorical implications for the scope of individual firms. Teece (1980) examines this issue and concludes that as long as the economies of scope derive from proprietary information or a specialized indivisible physical asset, a multiproduct enterprise is most efficient. He takes the energy industry as an example of this, where petroleum search and removal techniques can be applied to coal.

Mayo (1984) examines data to pursue Teece's ideas. He uses a cost relationship with coal and petroleum as the outputs. Economies of scope, presumably attributable to similarity of extraction techniques, are found. He also shows that ray average costs decline and, near the axes, declining incremental costs and transray convexity yield evidence of output specific subadditivity. This suggests that divestiture of large laterally integrated petroleum firms may increase costs.

Water supplies are generally run as monopolies in the United States, and Kim and Clark (1983) analyze this industry's cost structure with supplies to residential and nonresidential customers as outputs. Using gallons of water per day as the metric, they find overall constancy in returns to scale. But nonresidential output offers product specific economies while diseconomies appear in residential supply. By analyzing ray average costs for many output mixes, they plot the M-locus for the industry and find it to be concave to the origin. This yields a wide range for the optimal number of firms in a market. In addition, by looking at an "average" ray (79 percent residential), they find that most firms produce near the essentially flat portion of the average cost curve. They conclude that market concentration is generally unnecessary for scale economies and may hinder potential competition in water supply.

Tests of the Contestability of Industries

There have been a number of industry studies seeking to test whether the area of the economy under investigation can legitimately be classed as "contestable." Several, but not all, have yielded results that are somewhat surprising. For example, as we will see below, some studies of the airline industry conclude that the industry is less close to the model of perfect contestability than has sometimes been suggested. On the other hand, one very recent study indicates that the aluminum industry is closer to contestability than might have been expected. Froeb and Geweke (1984) describe the U.S. aluminum industry as an area in which the long term threat of entry drives profits downward. Allowing for short run effects of entry barriers, they test the long run behavior of firms from 1949 to 1972 and find it consistent with what one would expect in an imperfectly contestable market.

On the other hand, the results of some of the studies we are now discussing are less surprising. Extensive work by Davies (1986) for the Canadian Transport Commission on the liner shipping industry concludes that contestability theory may offer the most appropriate model to describe that industry since the transferability and resalability of capital characterize oceanic shipping. The theory "accounts for the condition of entry to the industry, the cost characteristics and structure of the industry, the existence of loyalty ties [to cartels], the reported absence of supernormal profit and the pricing structure operative in the industry. No other body of theory can explain the existence of all these different phenomena in such a coherent and elegant manner." These conclusions are supported in an independent study of ocean transport by Peter Cassidy (1982).

17F. Contestability and Public Policy

As has already been said, contestability theory aspires to offer two types of guidance to policy makers: first, it undertakes to provide criteria to distinguish the cases in which government intervention is desirable from those in which it is not and, second, it seeks to offer tools to the regulator that will increase the public welfare benefits of this intervention. There have recently been significant applications of contestability theory on each of these fronts. On both issues we will examine some of the recent discussions in the academic literature and devote the bulk of our attention to pertinent developments in recent regulatory and antitrust decisions.

Contestability as a Guide to Arenas Inappropriate for Intervention   Contestability theory follows the lead of Bain, Sylos-Labini and others in stressing that potential competitors, like currently active competitors, can effectively constrain market power, so that when the number of incumbents in a market is few or even where only one firm is present, sufficiently low barriers to entry may make antitrust and regulatory attention unnecessary. Indeed, their costs and the inefficiencies they cause may then offer little or no offsetting benefit.

Since this viewpoint thoroughly antedates contestability theory it is not surprising that it has appeared in a variety of official policies. For example, the 1982 Merger Guidelines of the U.S. Department of Justice define market power as the ability to profit by raising (and maintaining) price above the competitive level, not referring to market share in that definition. It includes in its market definition firms whose entry would be attracted should an elevation of price be attempted. It asserts, moreover, that mergers will go unchallenged by the Department if they affect only markets subject to potential competition that is sufficiently strong. It proposes to take concentration into account only where potential competition is inadequate.

Similarly, the Federal Communications Commission has shown its awareness of the role of potential competition. In a recent Notice of Inquiry (FCC, 1985) it adopts a market definition for the analysis of long distance telecommunications similar to that of the Department of Justice. It asserts that " . . . satellite carriers readily can *shift* their capacity from one area to another and terrestrial carriers can readily *expand* their service areas through new facilities, interconnection, or resale." "There may be numerous large *potential* entrants in common carrier or private systems that further check AT & T's market power to some degree." And, "The absence of entry does not show that an existing firms possess market power; existing firms may be charging competitive prices because of competition among themselves or the *threat* of *potential* entry, and thereby make entry unattractive." Thus, the FCC implies that a large market share need not confer any monopoly power.

Contestability theory makes direct appearances in a recent decision of the Federal Trade Commission (FTC, 1985).[2] The case involved the acquisition in 1982 by the Echlin Manufacturing Corporation of the automotive aftermarket divisions of the Borg Warner Corporation, including its carburetor kit activities which overlapped with some activities of Echlin. The FTC staff challenged the acquisition under Section 7 of the Clayton Act and Section 5 of the Federal Trade Commission Act on the grounds that the merger would add significantly to the concentration of this market.

However, testimony before the Administrative Law Judge indicated that the field was not beset by any significant entry barriers and that potential entry was sufficient to deprive the merged firm of any monopoly power over the assembly and sale of carburetor kits. The Commission agreed, and permitted the acquisition. In its decision the Commission asserted (explicitly citing contestability theory and other writings on potential competition):

> An attempt to exercise market power in an industry without entry barriers would cause new competitors to enter the market. This additional supply would drive prices back to the competitive level. Indeed, the threat of new entry can be as potent a procompetitive force as its realization. As the Supreme Court has recognized, the presence of potential entrants on the fringe of a market

---

[2] It should be made clear that the authors of this book cannot claim to be disinterested reporters, having testified in several of the cases described here.

*can prevent the exercise of market power by the incumbent firms
even if the potential entrants never actually enter the market. Thus,
in the absence of barriers to entry, incumbent firms cannot exercise market power, regardless of the concentration in the nominal
"market," and indeed even if that market has been "monopolized"
by a single firm. (Pp. 9–10).*

The decision goes on to espouse the same definition of entry barriers as
that adopted (from Stigler and others) in the contestability literature " . . . as
additional long-run costs that must be incurred by an entrant relative to the
long-run costs faced by incumbent firms" (p. 12) and, again citing the contestability literature, it concludes that "if sunk costs are considered an entry
barrier, it must be because they create a difference in the risk confronting
the incumbent firms who have already committed their resources and potential entrants who have yet to make that decision" (p. 17). The decision finds,
along with contestability analysis, that " . . . we cannot agree that economies
of scale and declining markets necessarily create barriers to entry" (p. 18),
and that "the absence of past entry, however, does not prove the existence
of entry barriers because it is equally consistent with alternative explanations,
such as a declining industry or competitive prices" (p. 19).

Deregulatory Experience: Airlines, Buses, and Trucks   The intellectual foundations of the deregulation of airlines, trucking, and buses included recognition of the power of potential entry. While it was clear that in many
transportation markets efficiency is not inconsistent with the operation of
enough active carriers to make the replacement of regulation by competition
appropriate, it was also clear that on many routes efficient operation is incompatible with the presence of several carriers. Nevertheless, deregulation
proceeded in the expectation that potential competition could adequately
protect consumers of transportation services in such arenas.

In the initial enthusiasm with which we described contestability analysis
we agreed with this assessment, and more than once cited the airline industry as a case in point, using the metaphoric argument that investments
in aircraft do not incur any sunk costs because they constitute "capital on
wings." Reconsideration has led us to adopt a more qualified position on
this score. We now believe that transportation by trucks, barges, and even
buses may be more highly contestable than passenger air tranportation.
Barges and trucks have business firms rather than individual consumers as
their primary customers, and that facilitates the provision of service via contracts on which potential entrants can effectively bid against incumbents.
Where the contracts apply to long run relationships, as they often do in
transportation, even capital costs that are physically irreversible are not
economically sunk in the pertinent time period. Moreover, trucks and buses
do not face the heavy sunk costs involved in the construction of airports or
the shortage of gates and landing slots at busy airports such as that which
prevented People Express from acquiring even a single gate of its own at

Denver's Stapleton International Airport, so that it was forced instead to lease gates from other carriers, catch as catch can, during a year of flying to that airport.

In fact, post-deregulation experience in the airlines industry has revealed several elements of the structure of supply that conflict significantly with the conditions necessary for the pure theory of contestability to apply without modification. While these structural elements may be transitory, they nevertheless appear to have influenced the performance of the industry in important ways since the advent of deregulation.

First, as the previously noted difficulties of People Express exemplify, there have been constraining shortages of facilities and services of air traffic control at several pivotal airports. These constrain flights in and out of the affected airports and, in addition, they also restrict the prospects of entry and expansion on routes that would otherwise interconnect efficiently with such flights. Second, technological advances, changes in the relative prices of jet fuel and equipment, and changes in the desired configurations of route networks have significantly altered the types and mix of aircraft demanded by the industry. As a result, there have been shortages in the availability of the aircraft demanded, with delivery lags frequently stretching to three years. Third, newly certificated airlines have been able to avoid the costly labor contracts that pervaded the industry before deregulation, so that their labor costs have been substantially lower than those facing the older, established carriers. Recently, the older carriers and their unionized work forces have been adapting themselves to this new competitive reality with the aid of more flexible wage contracts, dual wage structures, and less costly contract settlements. Nevertheless, substantial differences in labor costs between older and newer carriers persist.

These conditions make it easy to see why the airline industry does not conform perfectly to the contestability model, even if aircraft are "capital on wings." Thus, it remains to analyze both quantitatively and qualitatively the degree to which the performance of the industry does, or can be expected to, reflect the predictions of contestability theory. This should come as no surprise since most industries can be expected to depart in some important respects from the model of perfect contestability, and it will therefore generally be necessary in applying the theory to assess the economic significance of the deviations.

Several econometric studies have confirmed the imperfection of the contestability of the airline markets (see, e.g., Call and Keeler, 1984). They have shown, for example, that there is a significant positive correlation between profits and concentration in airline markets. Thus the threat of entry does not by itself suffice to keep profits to zero, as perfect contestability would require. Moreover, when new entry does occur, established carriers do reduce their fares in response, something one would expect in a conventional oligopolistic market other than one that is perfectly contestable. Even the study of Morrison and Winston, which concludes that potential entry does constrain price significantly, finds that the coefficient describing the influence of potential entrants does not become significant until the number of such prospective entrants exceeds three.

The econometric study by Graham, Kaplan, and Sibley (1983) permits comparison among the effects on 1981 route-by-route prices of various influences that lie outside the predictions of contestability theory. They find that prices deviate from costs an additional 10 percent on the average and, other things being equal, for routes that utilize one of the major slot-constrained airports in New York, Chicago, or Washington, D.C. In contrast, prices are reduced some 22 percent relative to costs on routes where a newly certificated carrier operates. And prices are some 18 percent lower relative to costs if four carriers fly the route rather than only one.

These results indicate the significance of the physical constraints on entry and expansion in two ways. First, the direct influence of the slot constraints on prices shows up in the data. Second, the substantial effect of the presence of new carriers, with their low labor costs, demonstrates how much lower fares would generally be if those carriers were able to expand their operations to cover more routes. Finally, the significant effect of the number of carriers on a route shows that active competition plays a role in holding price down toward cost. However, since the size of that effect is so much smaller than would be predicted by most theories of oligopoly that focus on the role of active competitors, it can be concluded that the forces of potential competition still play an important role, despite the structural conditions that impede their workings.

Similar conclusions seem to us to follow from qualitative evaluation of the behavior of the airline markets under deregulation. As we have just seen, the responsive price cuts of incumbents when faced with incursions of low priced entrants is certainly not compatible with the predicted qualitative behavior of contestable markets in the long run. However, there are other qualitative properties that are highly pertinent. In a recent study, Elizabeth Bailey (1986) sets out to examine these. She undertakes to analyze

> the consequences that have emerged under deregulation in terms
> of the following behavioral properties predicted by contestability
> theory: a variety of products will emerge, each of which will yield
> zero economic profit; the revenues from any subset of the products
> must exceed the incremental costs of those products, so that no
> cross-subsidy can exist; prices for each product will equal or exceed
> marginal costs; and an equilibrium market structure will minimize
> costs of the industry. Thus, if the theory has some degree of validity,
> more of these properties should be displayed after deregulation than
> before. Cross-subsidy, which was pervasive in the regulatory era,
> should be significantly eroded. Prices should move nearer to costs.
> Products capable of producing zero economic profits, but which were
> formerly excluded, should now appear. (Pp. 2–3, footnote omitted)

Here evidence leads her to conclude that, while the performance in each respect has not been absolutely clear-cut, on balance the patterns predicted by contestability theory have indeed emerged. She writes, in summary of the consequences of deregulation of the airlines, that

> *Prices in the cheaper-to-serve long-haul and dense markets were substantially lowered (by about forty percent), whereas prices in the more expensive short-haul and thin markets went up somewhat. A diversity of price-service options arose. Individuals could select between low-service/low-price discount carriers and full-service national carriers. Even among the full-service carriers, prices were lower for customers willing to improve load factors by traveling in off-peak periods or by taking one-stop rather than non-stop flights.*
>
> *The encouragement to entry under air transport deregulation brought with it a variety of contributions to efficiency. One involved delivery systems. It quickly became clear that hub-and-spoke operations rather than the mostly linear systems imposed by regulation offered savings to the airlines. Hub-and-spoke systems also substantially improved service for consumers. A second efficiency contribution involved input productivity. The post-deregulation period has been characterized by pressures to reduce pay scales toward those in unregulated economic sectors, to increase productivity through changes in work rules, and to choose a more efficient fleet configuration. (P. 22)*

In light of these observations and similar evidence drawn from the opening of brokerage commissions to market determination, the deregulation of trucking, and the opening of telecommunications markets to entry, Bailey concludes that performance in these areas has shown no significant inconsistencies with the predictions of contestability analysis.

In short, in terms of the airlines case we can infer that market forces through the pressures of competition, both actual and potential, have done a commendable if imperfect job in protecting consumer interests. This is suggested by a number of developments following deregulation: the decrease in real average prices; the reduction in average time spent in traveling from point to point; the falling real costs of airline operation; the erosion or disappearance of cross-subsidies, with the elimination of financing of sparsely traveled routes by those that are heavily traveled, and of peak travelers by off-peak flyers, and the considerable improvements in efficiency through computerized routing and hub-and-spoke operations—approaches that had evaded both executives and the Commission during the era of regulation. (For a more measured and highly intelligent evaluation of the contestability of the airlines, see Levine, 1987.)

Contestability Theory as Guide for Regulation We come, finally, to the arena in which the viewpoint of contestability may make its main contribution—as a guide for regulation, rather than as an argument for its elimination. How does contestability theory help in this domain? After all, perfect competition has long served usefully as the ideal for government intervention to follow; that is, as the model of performance to which it should seek to make the regulated firm adhere. What does contestability analysis have to add to this? The answer is that in some circumstances, notably in the presence of substantial economies of scale and scope, the standard of perfect competition is totally inappropriate. For example, where economies of scale and scope are present society no longer is sure to benefit if firms are required to be small, as they would be under perfect competition. Similarly, a rule that price must be set equal to marginal cost is a prescription for financial disaster. Of course, Ramsey pricing theory is of considerable help here, but in some circumstances its usefulness in practice is limited, particularly where there are no reliable data on elasticities and cross-elasticities of demands for the considerable number of products at issue. It is then that contestability theory can come to the rescue (and several times already has). For it can propose, for example, to offer consumers in markets with unavoidable entry barriers just the same sort of protection from excessive pricing that they would have derived from perfect freedom of entry, if such freedom had been possible. This is precisely how the stand-alone cost, criterion for price ceilings, which will be described below, emerged from contestability theory and could not have been deduced from the model of perfect competition.

To see just how contestability theory can be used to guide regulation we can do no better than to follow the outlines of a remarkable decision of the Interstate Commerce Commission (1985), a decision that encompasses the foundations of its current policies toward those elements of railroad activities in which competitive pressures are judged to be inadequate (that is, in regulatory terminology, in areas in which a railroad possesses "market dominance," or its shippers are "captive").

Early in the discussion of its economic framework the decision provides a section headed "Contestable Markets." However while, as we will see presently, much is made of the logic of contestability, the decision asserts flatly and quite appropriately that the pertinent arena is *not* contestable: "the railroad industry is recognized to have barriers to entry and exit and thus is not considered contestable for captive traffic" (p. 10).

The question, then, is what is best done to control the pricing terms on which such traffic is served. The Commission adopted a set of rules which it termed "constrained market pricing." Before getting to the details of those rules it is desirable to take note of two fundamental attributes underlying the approach: rate of return-rate base regulation of overall earnings, and acceptance of differential (non uniform) pricing. On the first of these, the Commission undertakes to avoid decisions that preclude a railroad from earning in the long run what it refers to as "adequate revenues." Following long regulatory tradition, these are defined in terms of a permitted rate of return on the railroad's rate base (its total invested capital). What is new here is that the Commission adopts for this purpose most of the criteria called for by economic analysis, for example, determining that "'adequate' returns are those that provide a rate of return on net investment equal to the current cost of capital (i.e., the level of return available on alternative investments)" (p. 18).

Second, the Commission recognizes that solvency of the railroads is likely to require differential pricing:

> *Most importantly, railroads exhibit significant economies of scope and density. Economy of scope refers to the fact that the rail plant is indivisible and can produce numerous services at less cost than those services could be produced by separate rail plants for each service. Economy of density refers to the fact that greater use of the fixed plant results in a declining average cost. Thus, the marginal cost of rail service is less than the average cost, because the fixed plant is used in a progressively more efficient manner. The differential between marginal costs and average costs cannot be assigned directly to specific movements by any conventional accounting methodology. Hence, we refer to it as the "unattributable costs." These are the costs which must be covered through differential pricing. (Pp. 7–8, footnotes omitted)*

The decision goes on to point out that where unattributable costs cannot be covered by marginal cost pricing, then demand considerations as well as cost data must enter into decision making, both in order to permit adequacy of revenues and in order to achieve efficiency.

> *Any means of allocating these costs among shippers other than actual market demand is arbitrary and may not permit a carrier to cover all of its costs. This is because non-demand-based cost apportionment methods do not necessarily reflect the carrier's ability (or inability) to impose the assigned allocations and cover its costs. Thus, they frequently "over-assign" or "under-assign" the carrier's unattributable costs to particular services. If a carrier sought to apply the formula price to all of its traffic, it would lose that traffic for which the demand could not support the price assigned. In that event, the remaining shippers might be required to pay a larger portion of the carrier's unattributable costs because they would lose the benefit of sharing these costs with the lost traffic.*
> *"Ramsey pricing" is a widely recognized method of differential pricing, that is, pricing in accordance with demand. Under Ramsey pricing, each price or rate contains a mark-up above the long-run marginal cost of the product or service to cover a portion of the unattributable costs. The unattributable costs are allocated among the purchasers or users in inverse relation to their demand elasticity. Thus, in a market where shippers are very sensitive to price changes (a highly elastic market), the mark-up would be smaller than in a market where shippers are less price sensitive. The sum of the mark-ups equals the unattributable costs of an efficient producer. (P. 8, footnote omitted)*

Nevertheless, the Commission comments,

> *Ramsey pricing is based on a mathematical formula which re-*
> *quires both the marginal cost and the elasticity of demand to be*
> *quantified for every movement in the carrier's system. Thus, the*
> *amount of data and degree of analysis required seemed over-*
> *whelming. We concluded that while formula Ramsey pricing is*
> *useful as a theoretical guideline, it is too difficult and burdensome*
> *for universal application. In setting flat rates, that is, rates which*
> *do not vary with the volume shipped, Ramsey pricing, in principle,*
> *yields the least inefficient price structure. However, even under*
> *pure Ramsey pricing, output levels are less than they would be if*
> *rates were set at marginal costs. This results in an economic ineffi-*
> *ciency because the value of the lost output to the shipper is greater*
> *than the value of the resources saved by reducing output. In such*
> *a situation, it may be feasible for the parties to negotiate a contract*
> *which will leave both parties better off than at the [flat] Ramsey*
> *price. Rail freight differs from financing governement services (for*
> *which shippers are relatively large and few in number.) Thus, the*
> *feasibility of contracting is more evident for rail freight than for*
> *these other services.*
>
> *As an alternative to pure Ramsey pricing, we proposed Con-*
> *strained Market Pricing. (P. 9, footnotes omitted)*

The key issue to be faced by constrained market pricing is the formulation of a criterion to be used in setting a ceiling over the price to be charged
for traffic over which a railroad possesses market dominance. This is an issue
with which the Commission had been struggling only since 1978. Before that,
curiously, the Commission had been preoccupied primarily with the setting
of floors beneath rail prices, to prevent railroads from undercutting barge
and truck competitors (the Commission had once described its role as that
of a "a giant handicapper").

In the early 1980's, in its search for a tenable price ceiling rule, it turned
initially to the tools it had designed long before in its floor setting endeavors.
Here, at least for the bulk of the postwar period, it has been using an
accounting concept which it called "fully allocated cost." In essence, this
assigned to each part of a railroad's traffic a figure intended to approximate
the incremental cost of that traffic, plus a share of the remaining (unattributable) cost proportionate to some criterion of "relative use" such as number of carloads, weight, or monetary value of the commodity in question
carried by the railroad. The apportionment of the unattributable costs was
admittedly arbitrary, bore no necessary relation to the cost data (such as
marginal costs) that an economist would consider pertinent, and allowed for
no adjustments for variations in demand conditions. Even earlier than the
landmark Ingot Molds Case (1967) in which the Supreme Court threw up
its hands on the matter, economists had begun to argue before the ICC that

the use of a fully distributed cost floor was pernicious and a source of substantial inefficiency, and that marginal or incremental costs were the only defensible cost data for use in the calculation of price floors, a view that gradually acquired acceptance among regulatory agencies, including the ICC.

It seems odd, then, that in 1978, when the Commission began its search for a rate ceiling formula (ICC, 1978) it turned to fully allocated cost, now discredited as a cost floor, as its candidate for a viable cost ceiling. Initially, it proposed for this purpose to allocate a railroad's unattributable costs in proportion to the variable costs (as a proxy for the incremental costs) of the various portions of its traffic. Because this formula soon proved fatally restrictive, particularly in ignoring demand considerations, the Commission permitted the railroads leeway to charge prices as much as seven percent above fully allocated costs where demand conditions permitted. The courts agreed to the legitimacy of this move toward differential pricing but rejected the use of the arbitrary seven percent range of tolerance (see, e.g., ICC, 1979). The Commission made one more attempt to salvage the fully allocated cost approach to price ceilings, but this time adopted weight and distance traversed (calling it the "ton/ton-mile" method) as its basis of allocation (ICC, 1980). However, in response to careful arguments by the railroads and others, it withdrew from this approach altogether in 1981 (ICC, 1981), and early in 1983 offered constrained market pricing as its alternative.

Here, the ceiling proposed by the Commission was *stand-alone cost*, a concept it acknowledged to have derived from the contestability literature.

> ... *Stand-alone cost (SAC) test ... is used to compute the rate a competitor in the market-place would need to charge in serving a captive shipper or a group of shippers who benefit from sharing joint and common costs. A rate level calculated by the SAC methodology represents the theoretical maximum rate that a railroad could levy on shippers without substantial diversion of traffic to a hypothetical competing service. It is, in other words, a simulated competitive price. (The competing service could be a shipper providing service for itself or a third party competing with the incumbent railroad for traffic. In either case, the SAC represents the minimum cost of an alternative to the service provided by the incumbent railroad.)*
>
> *The theory behind SAC is best explained by the concept of contestable markets. This recently developed economic theory augments the classical economic model of pure competition with a model which focuses on the entry and exit from an industry as a measure of economic efficiency. The theory of contestable markets is more general than that of pure competition because it does not require a large number of firms. In fact, even a monopoly can be contestable. The underlying premise is that a monopolist or oligopolist will behave efficiently and competitively where there is a threat of losing some or all of its markets to a new entrant. In other words, contestable markets have competitive characteristics which preclude monopoly pricing. (P. 10)*

Here it is worth reviewing the logic of the stand-alone cost ceiling a bit more closely. The first-best lesson of the perfect competition model, calling for prices to be set equal to marginal costs, has no doubt contributed to the common regulatory ethos which *equates* price to *some* measure of cost. This doctrine has been used frequently where it is completely inappropriate and without logical foundation, that is, in cases where prices should be based on demand as well as cost considerations, because of the presence of economies of scale and scope. Such arbitrary measures as fully distributed costs cannot substitute for marginal cost measures as decision rules for proper pricing, and the search for a substitute is a remnant of inappropriate reliance on the model of perfect competition for guidance in regulation.

In contrast, contestability theory suggests cost measures that are appropriate guideposts for regulated pricing—incremental and stand-alone costs. The incremental cost of a given service is, of course, the increment in the total costs of the supplying firm when that service is added to its product line. In perfectly contestable markets, the price of a product will lie somewhere between its incremental and its stand-alone cost, just where it falls in that range depending on the state of demand. One cannot legitimately infer that monopoly power is exercised from data showing that prices do not exceed stand-alone costs, and stand-alone costs constitute the proper cost-based ceilings upon prices, preventing both cross-subsidization and the exercise of monopoly power (see Faulhaber, 1975, for tests of cross-subsidy and their equivalence). A simple example will show why this is so.

First, suppose that a firm supplies two services, A and B, which *share no costs* and that each costs 10 units a year to supply. The availability of effective potential competition would force revenues from each service to equal 10 units a year. For higher earnings would attract (profitable) entry, and lower revenues would drive the supplier out of business. In this case, in which common costs are absent, incremental and stand-alone costs are equal to each other and to revenues, and the competitive and contestability benchmarks yield the same results.

Next, suppose instead that of the 20 unit total costs, 4 are fixed and common to A and B, while 16 are variable, 8 of the 16 being attributable to A and 8 to B. If, because of demand conditions, at most only a bit more than 8 can be garnered from consumers of A, then a firm operating and surviving in contestable markets will earn a bit less than 12 from B. These prices lie between incremental costs (8) and stand-alone costs (12), are mutually advantageous to consumers of both services, and will attract no entrants, even in the absence of any entry barriers. In contrast, should the firm attempt to raise the revenues obtained from B above the 12 unit stand-alone cost, it would lose its business to competitors willing to charge less. Similarly, the same fate would befall it in contestable markets if it priced B in a way that earned more than 8 plus the common cost of 4, less the contribution toward that common cost from service A.

Thus, the forces of idealized potential competition in perfectly contestable markets enforce cost constraints on prices, but prices remain sensitive to demands as well. Actual and potential competition are *effective* if they constrain rates in this way, and in such circumstances regulatory intervention is completely unwarranted. But if, in fact, market forces are not sufficiently strong, then there is likely to be a proper role for regulation, and the theoretical guidelines derived from the workings of contestable markets are the appropriate ones to apply. That is, prices must be constrained to lie between incremental and stand-alone costs.[3,4] This is the approach adopted by the ICC to determine maximum rates for railroad services, and the method has already withstood appeals to the Federal courts. (More recently—on

---

[3] Note that, properly applied, the SAC criterion should hold for all subsets of a firm's product and not only for its individual products. Thus, suppose a firm produces three items, A, B, and C, and that a widget of limitless capacity must be used in order to produce any A, any B or any combination of the two. The cost of the widget will then not enter A's or B's incremental cost, but it will clearly constitute part of the incremental cost of supplying them both. Then, the combined total revenue of the two items must suffice to cover the cost of the widget as well as any other incremental cost incurred individually or in common. Note also that efficiency requires that a supplier never set price below either marginal *or* per unit incremental cost, because if it is set so low it may lure away customers from a more efficient supplier, i.e., one who can supply the product at lower marginal or per unit incremental cost, and hence supply either one unit of the product or its entire amount at lower resources cost.

[4] It is easy to prove that if a firm's total earnings are exactly equal to its cost of capital, i.e., if it earns exactly zero economic profits, then if all of its revenues equal or exceed the corresponding incremental costs they are *automatically guaranteed* not to exceed their stand-alone costs and *vice versa*. That is, the passing of either test (the SAC ceiling or the incremental cost floor) automatically demonstrates that the other must also be passed by the prices at issue. To show this simply we deal with the case of two products, though the proof is perfectly general. Let

$X_1, X_2 =$ the quantities of the outputs supplied by a firm

$P_1, P_2 =$ the prices of the products

$C(X_1, X_2) =$ the total cost function

$C(X_1, O) =$ the stand-alone cost of product 1 and

$C(X_1, X_2) - C(X_1, O) =$ the incremental cost of product 2.

Then, zero economic profit requires

(1) $P_1 X_1 + P_2 X_2 = C(X_1, X_2)$

and the incremental cost test for product 2 requires

(2) $P_2 X_2 \geq C(X_1, X_2) - C(X_1, O)$.

Subtraction of (2) from (1) immediately yields the stand-alone cost criterion for product 1,

(3) $P_1 X_1 \leq C(X_1, O)$.

Similarly, subtraction of (3) from (1) immediately yields (2).

August 17, 1987—the FCC proposed adoption of a similar approach to regulation of telecommunications.)

17G. Concluding Remarks

Whatever one's attitude toward contestability theory or the policy recommendations that have derived from it, it must surely be agreed that it has evoked a flood of imaginative and valuable research and writing in opposition, in extension, and in application. To see how rapidly it and associated concepts have spread one need only recall that as recently as 1970 the concept of Ramsey pricing was unknown to most economists, though the analysis, of course, had appeared in 1927. Indeed, the term "Ramsey pricing" was coined by the present authors less than a decade ago. Terms such as "economies of scope," "stand-alone cost," and "contestable markets" were coined in the 1970's as well. Yet today they are used routinely not only in professional journals, but in hearings before U.S. courts and regulatory agencies. Clearly, propagation of the substance rather than the terminology is what really counts, but this, too, is being achieved.

Thus, whatever contribution the future will judge contestability theory to have made, it will surely conclude that the analysis has succeeded in stimulating thought in both the realm of academic research and policy formulation.

Bibliography

Ackworth, W.M. (1891). *The Railways and the Traders*, London: 1891.

Alexander, E.P. (1887). *Railway Practice*, New York: 1887.

Apostol, T. (1957). *Mathematical Analysis*, Reading, Mass.: Addison-Wesley, 1957.

Applebaum, E., and Lim, C. (1985). "Contestable Markets Under Uncertainty," *Rand Journal of Economics*, **16**, No. 1, Spring 1985, pp. 28–40.

Areeda, P., and Turner, D.F. (1975). "Predatory Pricing and Related Practices under Section 2 of the Sherman Act," *Harvard Law Review*, **88**, 1975, pp. 637–733.

Areeda, P., and Turner, D.F. (1976). "Scherer on Predatory Pricing: A Reply," *Harvard Law Review*, **89**, March 1976, pp. 891–900.

Areeda, P., and Turner, D.F. (1978). "Williamson on Predatory Pricing," *Yale Law Journal*, **87**, June 1978, pp. 1337–1352.

Arrow, K.J. (1962). "The Economic Implications of Learning by Doing," *Review of Economic Studies*, **29**, No. 80, June 1962, pp. 155–173.

Arrow, K.J. (1964). "Optimal Capital Policy, the Cost of Capital, and Myopic Decision Rules," *Annals of the Institute of Statistical Mathematics*, **16**, Nos. 1–2, Tokyo, 1964, pp. 21–30.

Arrow, K.J., Chenery, H.B., Minhas, B., and Solow, R.M. (1961). "Capital Labor Substitution and Economic Efficiency," *The Review of Economics and Statistics*, **43**, No. 3, August 1961, pp. 225–250.

Averch, H., and Johnson, L. (1962). "Behavior of the Firm under Regulatory Constraint," *American Economic Review*, **52**, No. 5, December 1962, pp. 1052–1063.

Bailey, E.E. (1981). "Contestability and the Design of Regulatory and Antitrust Policy," *American Economic Review*, **71**, No. 2, May 1981, pp. 178–183.

Bailey, E.E. (1986). "Price and Productivity Change Following Deregulation: The U.S. Experience," *Economic Journal*, March 1986 (forthcoming).

Bailey, E.E., and Baumol, W.J. (1984). "Deregulation and the Theory of Contestable Markets," *Yale Journal of Regulation*, **1**, 1984, pp. 111–137.

Bailey, E.E., and Friedlaender, A.F. (1982). "Market Structure and Multiproduct Industries," *Journal of Economic Literature*, **20**, September 1982, pp. 1024–1048.

Bailey, E.E., and Panzar, J.C. (1981). "The Contestability of Airline Markets during the Transition to Deregulation," *Journal of Law and Contemporary Problems*, **44**, Winter 1981, pp. 125–145.

Bailey, E.E., and White, L.J. (1974). "Reversals in Peak and Off-Peak Prices," *Bell Journal of Economics*, **5**, Spring 1974, pp. 75–92.

Bain, J.S. (1954). "Economies of Scale, Concentration and Entry," *American Economic Review*, **44**, March 1954, pp. 15–39.

Bain, J.S. (1956), *Barriers to New Competition*, Cambridge, Mass.: Harvard University Press, 1956.

Baseman, K.C. (1981). "Cross Subsidies and Open Entry in Regulated Markets," *in* G. Fromm, ed., *Studies in Public Regulation*, Cambridge, Mass.: MIT University Press, 1981, pp. 329–360.

Bassett, L.R., and Borcherding, T.E. (1970). "The Relationship between Firm Size and Factor Price," *Quarterly Journal of Economics*, **84**, August 1970, pp. 518–522.

Baumol, W.J. (1971). "Optimal Depreciation Policy: Pricing the Products of Durable Assets," *Bell Journal of Economics and Management Science*, **2**, Autumn 1971, pp. 365–376.

Baumol, W.J. (1975). "Scale Economies, Average Cost, and the Profitability of Marginal-Cost Pricing," *in* R.E. Grieson, ed., *Essays in Urban Economics and Public Finance in Honor of William S. Vickrey*, Lexington, Mass.: D.C. Heath, 1975, pp. 43–57.

Baumol, W.J. (1977). "On the Proper Cost Tests for Natural Monopoly in a Multiproduct Industry," *American Economic Review*, **67**, December 1977, pp. 809–822.

Baumol, W.J. (1979a). "Quasi Optimality: The Price We Must Pay for a Price System," *Journal of Political Economy*, **87**, No. 3, 1979, pp. 578–599.

Baumol, W.J. (1979b). "Quasi Permanence of Price Reduction: A Policy for Prevention of Predatory Pricing," *Yale Law Journal*, **89**, November 1979, pp. 1–26.

Baumol, W.J., and Bradford, D.E. (1970). "Optimal Departures from Marginal Cost Pricing," *American Economic Review*, **60**, June 1970, pp. 265–283.

Baumol, W.J., and Braunstein, Y.M. (1977). "Empirical Study of Scale Economies and Production Complementarity: The Case of Journal Publication," *Journal of Political Economy*, **85**, October 1977, pp. 1037–1048.

Baumol, W.J., and Fabian, T. (1964). "Decomposition, Pricing for Decentralization and External Economies," *Management Science*, **11**, No. 1, September 1964.

Baumol, W.J., and Fischer, D. (1975). "On the Optimal Number of Firms in an Industry," Discussion Paper No. 75–51, New York University Center for Applied Economics, 1975.

Baumol, W.J., and Fischer, D. (1978). "Cost-Minimizing Number of Firms and Determination of Industry Structure," *Quarterly Journal of Economics*, **92**, August 1978, pp. 439–467.

Baumol, W.J., and Ordover, J.A. (1977). "On the Optimality of Public-Goods Pricing with Exclusion Devices," *Kyklos*, Fasc. 1, **30**, 1977, pp. 5–21.

Baumol, W.J., Panzar, J.C., and Willig, R.D. (1983). "Contestable Markets: An Uprising in the Theory of Industry Structure: Reply," *American Economic Review*, **73**, June 1983, pp. 491–496.

Baumol, W.J., and Willig, R.D. (1980). "Intertemporal Unsustainability," Princeton University Economic Working Paper, 1980.

Baumol, W.J., and Willig, R.D. (1981a). "Intertemporal Failures of the Invisible Hand: Theory and Implications for International Market Dominance," *Indian Economic Journal*, 1981.

Baumol, W.J., and Willig, R.D. (1981b). "Fixed Cost, Sunk Cost, Entry Barriers and Sustainability of Monopoly," *Quarterly Journal of Economics*, **95**, August 1981, pp. 405–431.

Baumol, W.J., Bailey, E.E., and Willig, R.D. (1977). "Weak Invisible Hand Theorems on the Sustainability of Prices in a Multiproduct Monopoly," *American Economic Review*, **67**, June 1977, pp. 350–365.

Baumol, W.J., Fischer, D., and ten Raa, T. (1979). "The Price Iso-Return Locus and Rational Rate Regulation," *The Bell Journal of Economics*, **10**, Autumn 1979, pp. 648–658.

Beckenstein, A.R. (1975). "Scale Economics in the Multiplant Firm: Theory and Empirical Evidence," *Bell Journal of Economics*, **6**, Autumn 1975, pp. 644–660.

Berge, C. (1963). *Topological Spaces, Including a Treatment of Multivalued Functions, Vector Spaces and Convexity*, New York: Macmillan, 1963.

Bertrand, J. (1883). "Review of *Théorie Mathematique de la Richesse Sociale* and *Recherches sur les Principles Mathematique de la Théorie des Richesses*," *Journal des Savants*, 1883, pp. 499–508.

Bittlingmayer, G. (1985). "The Economics of A Simple Airline Network," mimeo, 1985.

Blackorby, C., Primont, D., and Russell, R.R. (1977). "On Testing Separability Restrictions with Flexible Functional Forms," *Journal of Econometrics*, **5**, March 1977, pp. 195–209.

Blackorby, C., Primont, D., and Russell, R.R. (1978). *Duality, Separability and Functional Structure: Theory and Economic Applications*, New York: North-Holland, 1978.

Body, F.M. (1939). "The Influence of Costs on Production and Price Policy in a Joint-Product Industry," unpublished Ph.D. dissertation, University of Minnesota, 1939.

Bohm, P. (1967). "On the Theory of 'Second Best'," *Review of Economic Studies*, **34**, 1967, pp. 301–314.

Boiteux, M. (1960). "Peak-Load Pricing," *The Journal of Business*, **33**, No. 2, 1960, pp. 157–179.

Boiteux, M. (1971). "On the Management of Public Monopolies Subject to Budgetary Constraints," *Journal of Economic Theory*, **3**, September 1971, pp. 219–240.

Braeutigam, R.R. (1979). "Optimal Pricing with Intermodal Competition," *American Economic Review*, **69**, No. 1, March 1979, pp. 38–49.

Braunstein, Y.M., and Pulley, L.B. (1980). "Market Structure in Information Industries: An Empirical Study," Economics Working Paper, Brandeis University, 1980.

Bresnahan, T.F. (1981). "Duopoly Models with Consistent Conjectures," *American Economic Review*, **71**, December 1981, pp. 934–945.

Brock, W.A. (1983). "Contestable Markets and the Theory of Industry Structure: A Review Article," *Journal of Political Economy*, **91**, No. 6, December 1983, pp. 1055–1066.

Brock, W.A., and Scheinkman, J.A. (1983). "Free Entry and the Sustainability of Natural Monopoly: Bertrand Revisited by Cournot," *in* D.S. Evans, ed., *Breaking up Bell: Essays on Industrial Organization and Regulation*, 1983.

Call, G.D., and Keeler, T.E. (1984). "Airline Deregulation, Fares, and Market Behavior: Some Empirical Evidence," in *Analytical Studies in Transport Economics*, 1984.

Cassidy, P.A. (1982). *Australian Overseas Cargo Liner Shipping*, University of Queensland, 1982.

Caves, D.W., Christensen, L.R., and Herriges, J.A. (1984). "Modelling Alternative Residential Peak-Load Electricity Rate Structures," *Journal of Econometrics*, **24**, No. 3, 1984, pp. 249–268.

Caves, D.W., Christensen, L.R., and Swanson, J.A. (1981). "Productivity Growth, Scale Economies, and Capacity Utilization in U.S. Railroads, 1955–74," *American Economic Review*, December 1981, pp. 994–1002.

Caves, D.W., Christensen, L.R., and Tretheway, M.W. (1980). "Flexible Cost Functions for Multiproduct Firms," *Review of Economics and Statistics*, **62**, No. 3, August 1980, pp. 477–481.

Caves, D.W., Christensen, L.R., and Tretheway, M.W. (1983). "Productivity Performance of U.S. Trunk and Local Services Airlines in the Era of Deregulation," *Economic Inquiry*, **21**, July 1983, pp. 312–324.

Caves, R.E., and Porter, M.E. (1977). "From Entry Barriers to Mobility Barriers," *Quarterly Journal of Economics*, **91**, May 1977, pp. 241–261.

Chadwick, E. (1859). "Results of Different Principles of Legislation and Administration in Europe; of Competition for the Field, as Compared with the Competition within the Field of Service," *Journal of the Royal Statistical Society*, 1859, p. 381 and ff.

Chamberlin, E. (1962). *The Theory of Monopolistic Competition*, 8th Ed., Cambridge, Mass.: Harvard University Press, 1962.

Champsaur, P. (1975). "How to Share the Cost of a Public Good," *International Journal of Game Theory*, **4**, 1975, pp. 113–129.

Charnes, A., Cooper, W.W., and Sueyoshi, T. (1985). "A Goal Programing/Constrained Regression Review of the Bell System Breakup," Center for Cybernetic Studies Report 513, University of Texas at Austin, May 1985.

Christensen, L.R., and Green, W.H. (1976). "Economies of Scale in U.S. Electric Power Generation," *Journal of Political Economy*, **84**, No. 4, 1976, pp. 655–676.

Clark, J.M. (1923). *Studies in the Economics of Overhead Costs*, Chicago: Unniversity of Chicago Press, 1923.

Clemens, E. (1950). "Price Discrimination and the Multiproduct Firm," *Review of Economic Studies*, **19**, No. 48, 1950–1951, pp. 1–11; reprinted *in* R.B. Heflebower and G.W. Stocking, eds., *Readings in Industrial Organization and Public Policy*, Homewood, Ill.: American Economic Association, Irwin, 1958, pp. 262–276.

Cournot, A.A. (1960). *Researches into the Mathematical Principles of the Theory of Wealth, 1838,* New York: A.M. Kelly, 1960.

Coursey, D., Isaac, R.M., and Smith, V.L. (1984). "Natural Monopoly and Contested Markets: Some Experimental Results," *Journal of Law and Economics,* **27**, April 1984, pp. 91–113.

Coursey, D., Isaac, R.M., Luke, M., and Smith, V.L. (1984). "Market Contestability in the Presence of Sunk (Entry) Costs," *Rand Journal of Economics,* **15**, Spring 1984, pp. 69–84.

Cowing, T.G., and Holtmann, A.G. (1983). "Multiproduct Short-Run Hospital Cost Functions: Evidence and Policy Implications from Cross-Section Data," *Southern Journal of Economics,* January 1983, pp. 637–653.

Dasgupta, P.S., and Heal, G.M. (1979). *Economic Theory and Exhaustible Resources,* Cambridge, U.K.: Cambridge University Press, 1979.

Dasgupta, P.S., and Stiglitz, J.E. (1986). "Sunk Costs, Competition, and Welfare," mimeo, 1985.

Davies, J.E. (1986). "The Theory of Contestable Markets and its Application to the Liner Shipping Industry," Canadian Transport Commission, Ottawa-Hull, 1986.

Davis, O.A., and Whinston, A.B. (1965). "Welfare Economics and the Theory of the Second Best," *Review of Economic Studies,* **32**, January 1965, pp. 1–14.

Davis, O.A., and Whinston, A.B. (1967). "Piecemeal Policy in the Theory of Second Best," *Review of Economic Studies,* **34**, 1967, pp. 323–331.

Debreu, G. (1959). *Theory of Value,* New York: John Wiley, 1959.

Demsetz, H. (1968). "Why Regulate Utilities?" *Journal of Law and Economics,* **11**, April 1968, pp. 55–65.

Demsetz, H. (1970). "The Private Production of Public Goods," *Journal of Law and Economics,* **13**, October 1970, pp. 293–306.

Demsetz, H. (1973). "Joint Supply and Price Discrimination," *Journal of Law and Economics,* **16**, No. 2, October 1973, pp. 389–405.

Denny, M., and Fuss, M. (1977). "The Use of Approximation Analysis to Test for Separability and the Existence of Consistent Aggregates," *American Economic Review,* **67**, 1977, pp. 404–418.

Denny, M., Fuss, M., Everson, C., and Waverman, L. (1981). "Estimating the Effects of Diffusion of Technological Innovations in Telecommunications: The Production Structure of Bell Canada," *Canadian Journal of Economics,* February 1981, pp. 24–43.

Dewsnup, E.R. (1914). "Railway Rate Making," *American Economic Review,* **4** (Suppl.), 1914, pp. 81–100.

Diamond, P.A. (1973). "Consumption Externalities and Imperfect Corrective Pricing," *Bell Journal of Economics and Management Science,* **4**, Autumn 1973, pp. 526–538.

Diamond, P.A., and Mirrlees, J.A. (1971a). "Optimal Taxation and Public Production I: Production Efficiency," *American Economic Review,* **61**, March 1971, pp. 8–27.

Diamond, P.A., and Mirrlees, J.A. (1971b). "Optimal Taxation and Public Production II: Tax Rules," *American Economic Review,* **61**, June 1971, pp. 261–278.

Dixit, A. (1979). "A Model of Duopoly Suggesting a Theory of Entry Barriers." *Bell Journal of Economics,* **10**, Spring 1979, pp. 20–32.

Dixit, A. (1980). "The Role of Investment in Entry-Deterrence," *The Economic Journal,* **90**, March 1980, pp. 95–106.

Dixit, A., and Stiglitz, J.E. (1977). "Monopolistic Competition and Optimum Product Diversity," *The American Economic Review,* **67**, June 1977, pp. 297–308.

Drèze, J.H. (1964). "Some Post-War Contributions of French Economists to Theory and Public Policy, with Special Emphasis on Problems of Resource Allocation," *American Economic Review,* **54**, No. 4, 1964, pp. 1–64.

Duesenberry, J.S. (1949). *Income, Saving, and the Theory of Consumer Behavior,* Cambridge, Mass.: Harvard University Press, 1949.

Eaton, B.C., and Lipsey, R.G. (1980). "Exit Barriers Are Entry Barriers: The Durability of Capital as a Barrier to Entry," *The Bell Journal of Economics*, **11**, Autumn 1980, pp. 721–729.

Eaton, B.C., and Lipsey, R.G. (1981). " Capital Commitment and Entry Equilibrium," *The Bell Journal of Economics*, **12**, Autumn 1981, pp. 593–604.

Edwards, E.O., and Bell, P.W. (1961). *The Theory and Measurement of Business Income*, Berkeley, Cal.: University of California Press, 1961.

Evans, D.S., and Heckman, J.J. (1984). "A Test for Subadditivity of the Cost Function with an Application to the Bell System," *American Economic Review*, September 1984, pp. 615–623.

Farrell, M.J. (1958). "In Defense of Public Utility Price Theory," *Oxford Economic Papers*, **10**, February 1958, pp. 109–123.

Faulhaber, G. (1971). "Equipment Utilization and Competition," unpublished, Bell Laboratories, 1971.

Faulhaber, G. (1975a). "Increasing Returns to Scale: Optimality and Equilibrium," unpublished Ph.D. dissertation, Princeton University, 1975.

Faulhaber, G. (1975b). "Cross-Subsidization: Pricing in Public Enterprise," *American Economic Review*, **65**, December 1975, pp. 966–977.

Faulhaber, G., and Levinson, S. (1981). "Subsidy Free Prices and Anonymous Equity," *American Economic Review*, **71**, December 1981, pp. 1083–1091.

Federal Communications Commission. (1985). *Notice of Inquiry, in the Matter of Long-Run Regulation at AT & T's Basic Domestic Interstate Sources*, Washington, D.C., 1985.

Federal Trade Commission. (1985). *Final Order: In the Matter of the Echlin Manufacturing Company, and Borg-Warner Corporation*, Docket Number 9157, Washington, D.C., June 28, 1985.

Ferguson, C.E. (1969). *The Neoclassical Theory of Production and Distribution*, Cambridge, U.K.: Cambridge University Press, 1969.

Flaherty, M.T. (1980). "Industry Structure and Cost-Reducing Investment," *Econometrica*, **48**, July 1980, pp. 1187–1209.

Friedlaender, A., Winston, C., and Kung Wang, D. (1983). "Costs, Technology and Productivity in the U.S. Automobile Industry," *Bell Journal of Economics*, Spring 1983, pp. 1–20.

Friedman, J.W. (1977). *Oligopoly and the Theory of Games*, Amsterdam: North-Holland, 1977.

Froeb, L., and Geweke, J. (1984). "Perfect Contestability and the Postwar U.S. Aluminium Industry." Tulane University, December 1984.

Fuss, M.A., and McFadden, D., eds. (1978). *Production Economics: A Dual Approach to Theory and Applications*, Amsterdam and New York: North-Holland, 1978.

Fuss, M.A., and Waverman, L. (1981). "The Regulation of Telecommunications in Canada," Report to the Economic Council of Canada, February 1981.

Gilbert, R.J., and Newbery, D. (1979). "Pre-emptive Patenting and the Persistence of Monopoly," Economic Theory Discussion Paper, University of Cambridge, 1979.

Gilbert, R.J. and Stiglitz, J.E. (1979). "Entry, Equilibrium and Welfare," mimeo, 1979.

Gilligan, T., and Smirlock, M. (1984). "An Empirical Study of Joint Production and Scale Economies in Commercial Banking," *Journal of Banking and Finance*, **8**, 1984, pp. 67–77.

Gilligan, T., Smirlock, M., and Marshall, W. (1984). "Scale and Scope Economies in the Multi-Product Banking Firm," *Journal of Monetary Economics*, **13**, 1984, pp. 393–405.

Ginsberg, W. (1974). "The Multiplant Firm with Increasing Returns to Scale," *Journal of Economic Theory*, **9**, November 1974, pp. 283–292.

Gold, B. (1981). "Changing Perspectives on Size, Scale, and Returns: An Interpretive Survey," *Journal of Economic Literature*, **19**, March 1981, pp. 5–33.

Gorman, I.E. (1985). "Conditions for Economies of Scope in the Presence of Fixed Costs," *Rand Journal of Economics*, **16**, No. 3, Autumn 1985, pp. 431–436.

Graham, D.R., Kaplan, D.P., and Sibley, D.S. (1983). "Efficiency and Competition in the Airline Industry," *Bell Journal of Economics*, Spring 1983, pp. 118–138.

Grossman, S.J. (1981). "Nash Equilibrium and the Industrial Organization of Markets with Large Fixed Costs," *Econometrica*, **49**, September 1981, pp. 1149–1172.

Hadley, A.T. (1886). *Railroad Transportation*, New York and London: G.P. Putnam's Sons, 1886.

Hanoch, G. (1975). "The Elasticity of Scale and the Shape of Average Costs," *American Economic Review*, **65**, June 1975, pp. 492–497.

Harrison, G.W. (1987). "Experimental Evaluation of the Contestable Markets Hypothesis," *in* E.E. Bailey, ed., *Public Regulation: New Perspectives on Institutions and Policies*, Cambridge, Mass.: MIT University Press, 1987, pp. 191–225.

Harrison, G.W., and McKee, M. (1985). "Monopoly Behavior, Decentralized Regulation, and Contestable Markets: An Experimental Evaluation," *Rand Journal of Economics*, **16**, Spring 1985, pp. 51–69.

Harrison, G.W., McKee, M., and Rutström, E.E. (1987). "Experimental Evaluation of Institutions of Monopoly Restraint," *in* L. Green and J.H. Kagel, eds., *Advances in Behavioral Economics*, Norwood, N.J.: Ablex, 1987.

Hay, D.A. (1976). "Sequential Entry and Entry-Deterring Strategies in Spatial Competition," *Oxford Economic Papers*, **28**, July 1976, pp. 240–257.

Helpman, E., and Krugman, P.R. (1985). *Market Structure and Foreign Trade*, 1985.

Hicks, J.R. (1935). "Annual Survey of Economic Theory—Monopoly," *Econometrica*, **3**, 1935, pp. 1–20; reprinted *in* G.J. Stigler and K.E. Boulding, eds., *Readings in Price Theory*, The American Economic Association, Chicago: R.D. Irwin, Inc., 1952, pp. 361–383.

Hirshleifer, J. (1958). "Peak-Loads and Efficient Pricing: Comment," *Quarterly Journal of Economics*, **72**, August 1958, pp. 451–462.

Hotelling, H. (1938). "The General Welfare in Relation to Problems of Taxation and Railway and Utility Rates," *Econometrica*, **6**, 1938, pp. 242–269.

Houthakher, H. (1951). "Electricity Tariffs in Theory and Practice," *Economic Journal*, **61**, March 1951, pp. 1–25.

Interstate Commerce Commission. (1978). Ex Parte No. 347, "Western Coal Investigation—Guidelines for Railroad Rate Structure," **43**, *Federal Register*, 22151, May 22, 1978.

Interstate Commerce Commission. (1979). *Annual Volume Rates on Coal–Wyoming to Flint Creek, Arkansas*, **361**, ICC 539, 1979.

Interstate Commerce Commission. (1980). Ex Parte No. 347 (Sub-No. 1), "Coal Rate Guidelines, Nationwide," **45**, *Federal Register*, 80370, December 4, 1980.

Interstate Commerce Commission. (1981). Ex Parte No. 347 (Sub-No. 1), "Coal Rate Guidelines, Nationwide," December 21, 1981, unpublished.

Interstate Commerce Commission. (1985). Ex Parte No. 347 (Sub-No. 1), "Coal Rate Guidelines, Nationwide," Washington, D.C., decided August 3, 1985.

Jorgenson, D.W. (1963). "Capital Theory and Investment Behavior," *American Economic Review*, **53**, May 1963, pp. 247–259.

Joskow, P.L., and Klevorick, A.K. (1979). "A Framework for Analyzing Predatory Pricing Policy," *Yale Law Journal*, **89**, December 1979, pp. 213–270.

Joskow, P.L., and Noll, R. (1981). "Regulation in Theory and Practice: An Overview," *in* G. Fromm. ed., *Studies in Public Regulation*, Cambridge, Mass.: MIT University Press, 1981, pp. 1–65.

Kahn, A.E. (1970). *The Economics of Regulation: Principles and Institutions*, Vol. 1, New York: Wiley, 1970.

Kahn, A.E. (1971). *The Economics of Regulation: Principles and Institutions*, Vol. 2, New York: Wiley, 1971.

Kamien, M.L., and Schwartz, N.L. (1971). "Limit Pricing and Uncertain Entry," *Econometrica*, **39**, May 1971, pp. 441–454.

Kamien, M.L., and Schwartz, N.L. (1972). "Uncertain Entry and Excess Capacity," *American Economic Review*, **62**, No. 5, December 1972, pp. 918–927.

Keeler, T.E. (1974). "Railroad Costs, Returns to Scale, and Excess Capacity," *Review of Economics and Statistics*, **56**, May 1974, pp. 201–208.

Kellner, S., and Mathewson, G.F. (1983). "Entry, Size Distribution, Scale, and Scope Economies in the Life Insurance Industry," *Journal of Business*, **56**, No. 1, 1983, pp. 25–44.

Kim, H.Y., and Clark, R.M. (1983). "Estimating Multiproduct Scale Economies: An Application to Water Supplies," U.S. Environmental Protection Agency, Municipal Environment Research Laboratory, Cincinnati, Ohio, July 1983.

Knieps, G., and Vogelsang, I. (1982). "The Sustainability Concept under Alternative Behavioral Assumptions," *Bell Journal of Economics*, **13**, No. 1, Spring 1982, pp. 234–241.

Knight, F.H. (1921). *Risk, Uncertainty, and Profit*, Chicago: University of Chicago Press, 1921.

Koenker, R. (1977). "Optimal Scale and American Trucking Firms," *Journal of Transport Economics and Policy*, **11**, No. 1, January 1977, pp. 1–14.

Koenker, R., and Perry, M. (1981). "Product Differentiation, Monopolistic Competition, and Public Policy," *Bell Journal of Economics*, **12**, Spring 1981, pp. 217–231.

Kreps, D., and Scheinkman, J. (1983). "Quantity Precommitment and Bertrand Competition Yield Cournot Outcomes," *Bell Journal of Economics*, **14**, No. 2, Autumn 1983, pp. 326–338.

Kreps, D., and Wilson, R. (1980). "On the Chain-Store Paradox and Predation: Reputation for Toughness," Technical Report No. 317, IMSSS, Stanford University, 1980.

Lancaster, K. (1968). *Mathematical Economics*, New York: Macmillan, 1968.

Levine, M.E. (1987). "Airline Competition in Deregulated Markets: Theory, Firm Strategy, and Public Policy," *Yale Journal on Regulation*, **4**, 1987, 393–494.

Lewis, W.A. (1949). *Overhead Costs; Some Essays in Economic Analysis*, New York: Rinehart, 1949.

Lipsey, R.G., and Lancaster, K. (1956). "The General Theory of Second Best," *Review of Economic Studies*, **24**, 1956–1957, pp. 11–32.

Littlechild, S.C. (1970a). "Peak-Load Pricing of Telephone Calls," *Bell Journal of Economics and Management Science*, **1**, No. 2, Autumn 1970, pp. 191–200.

Littlechild, S.C. (1970b). "Marginal Cost Pricing with Joint Costs," *Economic Journal*, **80**, June 1970, pp. 223–235.

Littlechild, S.C. (1975). "Common Costs, Fixed Charges, Clubs and Games," *Review of Economic Studies*, **42**, January 1975, pp. 117–124.

Loury, G.C. (1979). "Market Structure and Innovation," *Quarterly Journal of Economics*, **93**, No. 3, August 1979, pp. 395–410.

Lowry, E.D. (1973). "Justification for Regulation: The Case for Natural Monopoly," *Public Utilities Fortnightly*, **28**, November 8, 1973, pp. 1–7.

Lutz, F., and Lutz, V. (1951). *The Theory of Investment of the Firm*, Princeton, N.J.: Princeton University Press, 1951.

Manne, A.S. (1952). "Multiple-Purpose Public Enterprises Criteria for Pricing," *Economica*, **19**, August 1952, p . 322–326.

Mansfield, E. (1975). *Micro-Economics*, New York: Norton, 1975.

Marshall, A. (1925). *Principles of Economics*, London and New York: Macmillan, 1925.

Marx, K. (1939). *Grundrisse*, Moscow: Marx-Lenin Institute, 1939.

Mas-Colell, A., ed. (1980). "Non-cooperative Approaches to the Theory of Perfect Competition," Symposium Issue, *Journal of Economic Theory*, **22**, No. 2, April 1980.

Maskin, E., and Tirole, J. (1982). "A Theory of Dynamic Oligopoly," mimeo, 1982.

Mayo, J.W. (1984). "The Technological Determinants of the U.S. Energy Industry Structure," *Review of Economics and Statistics*, January 1984, pp. 51–58.

McFadden, D. (1978). "Cost Revenue and Profit Functions," *in* M.A. Fuss and D. McFadden, eds., *Production Economics: A Dual Approach to Theory and Applications*, Amsterdam: North-Holland, 1978.

McManus, M. (1967). "Private and Social Costs in the Theory of Second Best," *Review of Economic Studies*, **34**, 1967, pp. 317–321.

Menger, K. (1954). "The Logic of the Laws of Return, a Study in Meta-Economics," *in* O. Morgenstern, ed, *Economic Activity Analysis*, New York: John Wiley, 1954.

Mester, L.J. (1985). "A Multiproduct Cost Study of Savings and Loans," Esssay 1, doctoral dissertation, Princeton University, October 1985.

Milgrom, P., and Roberts, J. (1979). "Optimal Limit Price Does Not Deter Entry," unpublished 1979.

Mirman, L.J., Tauman, Y., and Zang, I. (1983). "Monopoly and Sustainable Prices as a Nash Equilibrium in Contestable Markets," mimeo, 1983.

Mirman, L.J., Tauman, Y., and Zang, I. (1985). "Supportability, Sustainability and Subsidy-Free Prices," *Rand Journal of Economics*, **16**, No. 1, Spring 1985, pp. 114–126.

Modigliani, F. (1958), "New Developments on the Oligopoly Front," *Journal of Political Economy*, **66**, June 1958, pp. 215–232.

Morrison, S.A., and Winston, C. (1985). "Empirical Implications and Tests of the Contestability Hypothesis," manuscript, 1985.

Murray, J.D., and White, R.W. (1983). "Economies of Scale and Economies of Scope in Multi-product Financial Institutions: A Study of British Columbian Credit Unions," *Journal of Finance*, June 1983, pp. 887–901.

Negishi, T. (1961). "Monopolistic Competition and General Equilibrium," *Review of Economic Studies*, **28**, June 1960–1961, pp. 196–201.

Negishi, T. (1967). "The Perceived Demand Curve in the Theory of Second Best," *Review of Economic Studies*, **34**, 1967, pp. 315–316.

Novshek, W. (1979). "Essays on Equilibrium Theory with Free Entry," unpublished Ph.D. dissertation, Northwestern University, 1979.

Novshek, W. (1980). "Cournot Equilibrium with Free Entry," *Review of Economic Studies*, **47**, April 1980, pp. 473–486.

Ordover, J.A., and Willig, R.D. (1981). "Economic Definitions of Predatory Product Innovation," *in* S. Salop, ed., *Strategy, Predation, and Antitrust Analysis*, Federal Trade Commission, 1981.

Owen, B.M., and Braeutigam, R. (1978). *The Regulation Game*, Cambridge Mass.: Ballinger, 1978.

Owen, B.M., and Willig, R.D. (1981). "Economics and Postal Pricing Policy," *in* J. Fleishman, ed., *The Future of the Postal Service*, Praeger, 1981.

Panzar, J.C. (1976). "A Neoclassical Approach to Peak Load Pricing," *Bell Journal of Economics*, **7**, Autumn 1976, pp. 521–530.

Panzar, J.C. (1980). "Sustainability, Efficiency and Vertical Integration," *in* P. Kleindorfer and B.M. Mitchell, eds., *Regulated Industries and Public Enterprise*, Lexington, Mass.: D.C. Heath, 1980.

Panzar, J.C. (1981). "Cross Subsidies and Open Entry in Regulated Markets—Comment," *in* G. Fromm, ed., *Studies in Public Regulation*, Cambridge, Mass.: MIT University Press, 1981, pp. 365–369.

Panzar, J.C., and Postlewaite, A. (1985). "The Sustainability of Ramsey Optimal Non-linear Prices," mimeo, 1985.

Panzar, J.C., and Willig, R.D. (1975). "Economies of Scale and Economies of Scope in Multi-Output Production," Bell Laboratories Economic Discussion Paper No. 33, 1975.

Panzar, J.C., and Willig, R.D. (1977a). "Free Entry and the Sustainability of Natural Monopoly," *Bell Journal of Economics*, **8**, Spring 1977, pp. 1–22.

Panzar, J.C., and Willig, R.D. (1977b). "Economies of Scale in Multi-Output Production," *Quarterly Journal of Economics*, **91**, August 1977, pp. 481–494.

Panzar, J.C., and Willig, R.D. (1979). "Economies of Scope, Product-Specific Economies of Scale, and the Multiproduct Competitive Firm," Bell Laboratories Economic Discussion Paper No. 152, August 1979.

Panzar, J.C., and Willig, R.D. (1981). "Economies of Scope," *American Economic Review*, **71**, No. 2, May 1981, pp. 268–272.

Perry, M. (1982). "Sustainable Positive Profit Multiple-Price Strategies in Contestable Markets," mimeo, 1982.

Pigou, A.C. (1920). *The Economics of Welfare*, London: Macmillan, 1920.

Posner, R. (1971). "Taxation by Regulation," *Bell Journal of Economics and Management Science*, **2**, Spring 1971, pp. 22–50.

Posner, R. (1976). *Antitrust Law: An Economic Perspective*, Chicago and London: University of Chicago Press, 1976.

Prescott, E.C., and Visscher, M. (1977). "Sequential Location among Firms with Foresight," *Bell Journal of Economics*, **8**, No. 2, Autumn 1977, pp. 378–393.

Quinzii, M. (1980). "An Existence Theorem for the Core of a Production Economy with Increasing Returns," unpublished.

Quirmbach, H. (1982). "Vertical Integration, Contestable Markets, and the Misfortunes of the Misshaped U," mimeo, 1982.

Ramsey, F.P. (1927). "A Contribution to the Theory of Taxation,"*Economic Journal*, **37**, March 1927, pp. 47–61.

Reay, J.R. (1965). "Generalization of a Theorem of Caratheodory," *Memoirs of the American Mathematical Society*, No. 54, 1965.

Rees, R. (1968). "Second Best Rules for Public Enterprise Pricing," *Economica*, **35**, August 1968, pp. 260–273.

Reynolds, M. (1980). "Predation: The Noisy Price Strategy," unpublished, 1980.

Robinson, J. (1941), "Rising Supply Price," *Economica*, **8**, 1941, pp. 1–8; reprinted *in* G.J. Stigler and K.E. Boulding, eds., *Readings in Price Theory*, The American Economic Association, Chicago: R.D. Irwin, Inc., 1952.

Rosen, S. (1972). "Learning by Experience as Joint Production," *Quarterly Journal of Economics*, **86**, August 1972, pp. 366–382.

Rosenbaum, R.A. (1950). "Sub-Additive Functions," *Duke Mathematical Journal*, **17**, 1950, pp. 227–247.

Sakai, Y. (1974). "Substitution and Expansion Effects in Production Theory: The Case of Joint Production," *Journal of Economic Theory*, **9**, No. 3, November 1974, pp. 255–274.

Salop, S.C. (1979a). "Monopolistic Competition with Outside Goods," *The Bell Journal of Economics*, **10**, No. 1, Spring 1979, pp. 141–156.

Salop, S.C. (1979b). "Strategic Entry Deterrence," *American Economic Review*, **69**, May 1979, pp. 335–338.

Salop, S.C., ed. (1981). *Strategy, Predation, and Antitrust Analysis*, Federal Trade Commission Report, 1981.

Salop, S.C., and Shapiro, C. (1980). "Strategic Predation in Test Markets," unpublished.

Salop, S.C., and Stiglitz, J.E. (1974). "Bargains and Ripoffs: A Model of Monopolistically Competitive Price Dispersion," *Review of Economics Studies*, **44**, October 1974, pp. 493–510.

Samuelson, P.A. (1947). *Foundations of Economic Analysis*, Cambridge, Mass.: Harvard University Press, 1947.

Samuelson, P.A. (1963). *Foundations of Economic Analysis*, 2nd Ed., Cambridge, Mass.: Harvard University Press, 1963.

Samuelson, P.A. (1964). "The Pure Theory of Public Expenditure," *Review of Economic Studies*, **36**, No. 4, November 1964, pp. 387–389.

Samuelson, P.A. (1973). *Economics*, 9th Ed., New York: McGraw-Hill, 1973.

Sandberg, I.W. (1974). "On the Mathematical Theory of Interaction in Social Groups," *IEEE Transactions on Systems, Man and Cybernetics*, Vol. SMC-4, September 1974, pp. 432–445.

Sandberg, I.W. (1975). "Two Theorems on a Justification of the Multiservice Regulated Company," *Bell Journal of Economics*, **6**, Spring 1975, pp. 346–356.

Scarf, H.E., and Hansen, T. (1973). *The Computation of Economic Equilibria*, New Haven: Yale University Press, 1973.

Scherer, F.M. (1976). "Predatory Pricing and the Sherman Act: A Comment," *Harvard Law Review*, **89**, No. 5, March 1976, pp. 869–890.

Scherer, F.M. (1980). *Industrial Market Structure and Economic Performance*, 2nd Ed., Chicago: Rand McNally, 1980.

Scherer, F.M., Beckenstein, A.R., Kaufer, E., and Murphy, R.D. (1975). *The Economics of Multiplant Operation: An International Comparisons Study*, Cambridge, Mass.: Harvard University Press, 1975.

Schmalensee, R. (1978). "Entry Deterrence in the RTE Cereal Industry," *Bell Journal of Economics*, **9**, Autumn 1978, pp. 305–327.

Schwartz, M. (1985). "The Nature and Scope of Contestability Theory," Georgetown University, September 1985, unpublished.

Schwartz, M., and Reynolds, R.J. (1983). "Contestable Markets: An Uprising in the Theory of Industry Structure: Comment," *American Economic Review*, June 1983, **73**, pp. 488–490.

Scott, J.T. (1982). "Multimarket Contract and Economic Performance," *Review of Economics and Statistics*, August 1982, pp. 368–375.

Seade, J. (1980). "On the Effects of Entry," *Econometrica*, **48**, No. 2, March 1980, pp. 479–489.

Shapley, L. (1971). "Cores of Convex Games," *International Review of Game Theory*, **1**, 1971, pp. 11–26.

Sharkey, W.W. (1979). "Existence of a Core when there Are Increasing Returns," *Econometrica*, **47**, July 1979, pp. 869–876.

Sharkey, W.W. (1981). "Existence of Sustainable Prices for Natural Monopoly Outputs," *Bell Journal of Economics*, **12**, Spring 1981, pp. 144–154.

Sharkey, W.W. (1982). *The Theory of Natural Monopoly*, Cambridge: Cambridge University Press, 1982.

Sharkey, W.W., and Telser, L.G. (1978). "Supportable Cost Functions for the Multiproduct Firm," *Journal of Economic Theory*, **18**, No. 1, June 1978, pp. 23–37.

Shephard, R.W. (1970). *Theory of Cost and Production Functions*, Princeton, N.J.: Princeton University Press, 1970.

Shepherd, W.G. (1984). "'Contestability' vs. Competition," *American Economic Review*, **74**, September 1984, pp. 572–587.

Silberberg, E. (1974). "The Theory of the Firm in 'Long-Run' Equilibrium," *American Economic Review*, **64**, September 1974, ppp. 734–741.

Sinden, F. (1979). "Obsolescence, Depreciation and New Technology," *in* H.M. Trebing, ed., *Issues in Public Utilities Regulation*, East Lansing: Michigan State University Public Utilities Papers, 1979, pp. 304–316.

Smith, A. (1776). *An Inquiry into the Nature and Causes of the Wealth of Nations*, London: W. Strahan and T. Cadell, 1776.

Sonnenschein, H. (1980). "Recent Results on the Existence of Cournot Equilibrium when Efficient Scale Is Small Relative to Demand," Address delivered at the 4th World Congress of the Econometric Society, Aix-en-Provence, France, 1980.

Sorenson, J., Tschirhart, J., and Winston, A. (1978). "A Theory of Pricing under Decreasing Costs," *American Economic Review*, **68**, No. 4, September 1978, pp. 614–624.

Spady, R.H., and Friedlaender, A.F. (1978). "Hedonic Cost Functions for the Regulated Trucking Industry," *Bell Journal of Economics*, **9**, No. 1, Spring 1978, pp. 159–179.

Spence, A.M. (1976). "Product Selection, Fixed Costs and Monopolistic Competition," *Review of Economic Studies*, **43**, June 1976, pp. 217–235.

Spence, A.M. (1977). "Entry, Capacity, Investment and Oligopolistic Pricing," *Bell Journal of Economics*, **8**, Autumn 1977, pp. 534–544.

Spence, A.M. (1979). "Investment Strategy and Growth in a New Market," *The Bell Journal of Economics*, **10**, Spring 1979, pp. 1–19.

Spence, A.M. (1983). "Contestable Markets and the Theory of Industry Structure: A Review Article," *Journal of Economic Literature*, **21**, 1983, pp. 981–990.

Spulber, D. (1984). "Scale Economies and Existence of Sustainable Monopoly Prices," *Journal of Economic Theory*, **34**, 1984, pp. 149–163.

Starrett, D.A. (1977). "Measuring Returns to Scale in the Aggregate and the Scale Effect of Public Goods," *Econometrica*, **45**, September 1977, pp. 1439–1456.

Steiner, P.O. (1957). "Peak Loads and Efficient Pricing," *Quarterly Journal of Economics*, **71**, November 1957, pp. 585–610.

Steiner, P.O. (1969). "Peak-Load Pricing Revisited," *in Proceedings of the 1968 Conference of Public Utilities Institute*, Michigan State University, 1969.

Stigler, G.J. (1951). "The Division of Labor Is Limited by the Extent of the Market," *Journal of Political Economy*, **59**, June 1951, pp. 185–193.

Stigler, G.J. (1957). "Perfect Competition, Historically Contemplated," *Journal of Political Economy*, **65**, No. 1, February 1957, pp. 1–17.

Stigler, G.J. (1968). *The Organization of Industry*, Homewood, Ill.: Irwin, 1968.

Stigler, G.J., and Boulding, K.E. eds. (1952). *Readings in Price Theory*, The American Economic Association, Chicago: R.D. Irwin, Inc., 1952.

Stiglitz, J.E. (1981). "Potential Competition May Lower Welfare," *American Economic Review*, **71**, May 1981, pp. 184–189.

Sylos-Labini, P. (1962). *Oligopoly and Technical Progress*, Cambridge, Mass.: Harvard University Press, 1962.

Tauchen, H., Fravel, F.D., and Gilbert, G. (1983). "Cost Structure and the Intercity Bus Industry," *Journal of Transport Economics and Policy*, January 1983, pp. 25–47.

Taussig, F.W. (1913). "Railway Rates and Joint Costs," *Quarterly Journal of Economics*, **27**, No. 4, August 1913, pp. 692–694.

Teece, D. (1980). "Economies of Scope and the Scope of the Enterprise," *Journal of Economic Behavior and Organization*, **1**, September 1980, pp. 223–245.

Teece, D.J. (1982). "Towards an Economic Theory of the Multiproduct Firm," *Journal of Economic Behavior and Organization*, March 1982, pp. 39–63.

Telser, L.G. (1978). *Economic Theory and the Core*, Chicago and London: University of Chicago Press, 1978.

ten Raa, T. (1980). "A Theory of Value and Industry Structure," Ph.D. dissertation, New York University, 1980.

ten Raa, T. (1983). "Supportability and Anonymous Equity," *Journal of Economic Theory*, **31**, 1983, pp. 176–181.

ten Raa, T. (1984). "Resolution of Conjectures on the Sustainability of Natural Monopoly," *Rand Journal of Economics*, **15**, 1, Spring 1984, pp. 135–141.

The Journal of Reprints for Antitrust Law and Economics (1980). **10**, No. 1, 1980.

Tulkens, H. (1968). *Programming Analysis of Postal Service*, Louvain: Libraire Universitaire, 1968.

Turvey, R. (1969). "Marginal Cost," *Economic Journal*, **79**, June 1969, pp. 282–299.

U.S. Department of Justice. (1982). *Merger Guidelines*, Washington, D.C. 1982.

U.S. Supreme Court. (1967). American Commerical Lines, Inc., *et al.* vs. Louisville and Nashville Railroad Co. *et al.* (Ingot Molds case), October term, 1967, pp. 571–597.

Uzawa, H. (1964). "Duality Principles in the Theory of Cost and Production," *International Economic Review*, **5**, May 1964, pp. 216–220.

Varian, H.R. (1978). *Microeconomic Analysis*, New York: W.W. Norton & Company, Inc., 1978.

Varian, H.R. (1980). "A Model of Sales," *American Economic Review*, **70**, No. 4, September 1980, pp. 651–659.

Viner, J. (1931). "Cost Curves and Supply Curves," *Zeitschrift für Nationalökonomie*, **3**, 1931, pp. 23–46; reprinted *in* G.J. Stigler and K.E. Boulding, eds., *Readings in Price Theory*, The American Economic Association, Chicago: R.D. Irwin, Inc., 1952.

von Weizsäcker, C.C. (1980a). "A Welfare Analysis of Barriers to Entry," *The Bell Journal of Economics*, **11**, No. 2, Autumn 1980, pp. 399–420.

von Weizsäcker, C.C. (1980b). *Barriers to Entry: A Theoretical Treatment*, Berlin and New York: Springer-Verlag, 1980.

Walters, A.A. (1963). "Production and Cost Functions: An Econometric Survey," *Econometrica*, **31**, January–April 1963, pp. 1–66.

Weil, R.L., Jr. (1968). "Allocating Joint Costs," *American Economic Review*, **58**, December 1968, pp. 1342–1345.

Weitzman, M.L. (1983). "Contestable Markets: An Uprising in the Theory of Industry Structure: Comment," *American Economic Review*, **73**, June 1983, pp. 486–487.

Williamson, O.E. (1966). "Peak Load Pricing and Optimal Capacity under Indivisibility Constraints," *American Economic Review*, **56**, September 1966, pp. 810–827.

Williamson, O.E. (1976). "Franchise Bidding for Natural Monopolies—In General and with respect to CATV," *Bell Journal of Economics*, **VII**, Spring 1976, pp. 73–104.

Williamson, O.E. (1977). "Predatory Pricing: A Strategic and Welfare Analysis," *Yale Law Journal*, **87**, December 1977, pp. 284–340.

Willig, R.D. (1973). Welfare Analysis of Policies Affecting Prices and Products. Ph.D. dissertation, Stanford University, 1973.

Willig, R.D. (1979a). "Multiproduct Technology and Market Structure," *American Economic Review*, **69**, No. 2, May 1979, pp. 346–351.

Willig, R.D. (1979b). "The Theory of Network Access Pricing," *in* H.M. Trebing, ed., *Issues in Public Utility Regulation*, East Lansing: Michigan State University Public Utilities Papers, 1979, pp. 109–152.

Willig, R.D. (1979c). *Welfare Analysis of Policies Affecting Prices and Products*, New York: Garland, 1979.

Willig, R.D. (1979d). "Consumer Equity and Local Measured Service," *in* J.A. Baude *et al.*, eds., *Perspectives on Local Measured Service*, Kansas City: Telecommunications Industry Workshop, 1979.

Willig, R.D. (1980). "What Can Markets Control?" *in* R. Sherman, ed., *Perspectives on Postal Service Issues*, American Enterprise Institute, 1980.

Willig, R.D., and Bailey, E.E. (1979). "The Economic Gradient Method," *American Economic Review*, **69**, May 1979, pp. 96–101.

Wright, F.K. (1964). "Towards a General Theory of Depreciation," *Journal of Accounting Research*, **2**, Spring 1964, pp. 80–90.

Zajac, E.E. (1978). *Fairness or Efficiency: An Introduction to Public Utility Pricing*, Cambridge, Mass.: Ballinger, 1978.

Copyrights and Acknowledgments

Publications and Writings

Chapters 2–4 and 7 draw extensively from

W.J. Baumol, "Scale Economies, Average Cost, and the Profitability of Marginal-Cost Pricing," *in* R.E. Grieson, ed., *Essays in Urban Economics and Public Finance in Honor of William S. Vickrey*, Lexington, Mass.: D.C. Heath, 1975, pp. 43–57.

J.C. Panzar and R.D. Willig, "Economies of Scale and Economies of Scope in Multi-Output Production," Bell Laboratories Economic Discussion Paper No. 33, 1975.

J.C. Panzar and R.D. Willig, "Economies of Scale in Multi-Output Production," *Quarterly Journal of Economics*, **91**, August 1977, pp. 481–493.

W.J. Baumol, "On the Proper Cost Tests for Natural Monopoly in a Multiproduct Industry," *American Economic Review*, **67**, December 1977, pp. 809–822.

J.C. Panzar and R.D. Willig, "Economies of Scope, Product-Specific Economies of Scale, and the Multiproduct Competitive Firm," Bell Laboratories Economic Discussion Paper No. 152, 1979.

R.D. Willig, "Multiproduct Technology and Market Structure," *American Economic Review*, **69**, May 1979, pp. 346–351.

J.C. Panzar and R.D. Willig, "Economies of Scope," *American Economic Review*, **71**, May 1981, pp. 268–272.

Chapter 5 is, in part, reproduced from

W.J. Baumol and D. Fischer, "Cost-Minimizing Number of Firms and Determination of Industry Structure," *Quarterly Journal of Economics*, **92**, August 1978, pp. 439–467.

Chapter 8 is based upon

J.C. Panzar and R.D. Willig, "Free Entry and the Sustainability of Natural Monopoly," *Bell Journal of Economics*, **8**, Spring 1977, pp. 1–22.

W.J. Baumol, E.E. Bailey, and R.D. Willig, "Weak Invisible Hand Theorems on the Sustainability of Prices in a Multiproduct Monopoly," *American Economic Review*, **67**, June 1977, pp. 350–365.

Chapter 9 draws heavily from

J.C. Panzar and R.D. Willig, "Economies of Scope, Product-Specific Economies of Scale, and the Multiproduct Competitive Firm," Bell Laboratories Economic Discussion Paper No. 152, 1979.

Chapter 10 is to a considerable degree reproduced from
W.J. Baumol and R.D. Willig, "Fixed Costs, Sunk Costs, Entry Barriers, and Sustainability of Monopoly," *Quarterly Journal of Economics*, **96**, August 1981, pp. 405–431.

Chapter 11 draws on
R.D. Willig, "Multiproduct Technology and Market Structure," *American Economic Review*, **69**, May 1979, pp. 346–351.

Chapter 12 draws much material from
R.D. Willig, "What Can Markets Control?" *in* R. Sherman, ed., *Perspectives on Postal Service Issues*, American Enterprise Institute, 1980.

W.J. Baumol, "Quasi Permanence of Price Reductions: A Policy for Prevention of Predatory Pricing," *Yale Law Journal*, **89**, November 1979, pp. 1–26.

Chapter 13 makes use of
W.J. Baumol, "Optimal Depreciation Policy: Pricing the Products of Durable Assets," *Bell Journal of Economics and Management Science*, **2**, August 1971, pp. 365–376.

Chapter 14 has used materials from
W.J. Baumol and R.D. Willig, "Intertemporal Failures of the Invisible Hand: Theory and Implications for International Market Dominance," *Indian Economic Review*, **16**, 1981.

W.J. Baumol and R.D. Willig, "Intertemporal Unsustainability," Princeton University Economic Working Paper, 1980.

Chapter 15 draws upon
W.J. Baumol, D. Fischer, and I.M. Nadiri, "Forms for Empirical Cost Functions to Evaluate Efficiency of Industry Structure" (unpublished).

W.J. Baumol and Y.M. Braunstein, "Empirical Study of Scale Economies and Production Complementarity: The Case of Journal Publication," *Journal of Political Economy*, **85**, October 1977, pp. 1037–1048.

Chapter 16 relies heavily on material in
E.E. Bailey, "Contestability and the Design of Regulatory and Antitrust Policy," *American Economic Review*, **71**, May 1981, pp. 178–183.

Chapter 17 is reproduced from
W.J. Baumol and R.D. Willig, "Contestability: Developments since the Book," *Oxford Economic Papers*, Special Issue, **38**, November 1986, pp. 9–36.

Also for Chapter 17, the authors want to thank the Information Science and Technology Division and the Regulation and Policy Analysis Program of the National Science Foundation and the C.V. Starr Center for Applied Economics for their generous support of this research. We also wish to acknowledge with gratitude the able assistance of Daniel Schult and the useful comments of Marius Schwartz.

Index

A 7
B 8
C 9
D 0
E 1
F 2
G 3
H 4
I 5
J 6

28